N.G. Basov   A.S. Bashkin
V.I. Igoshin   A.N. Oraevsky
V.A. Shcheglov

Chemical Lasers

With 80 Figures

Springer-Verlag
Berlin Heidelberg New York London
Paris Tokyo Hong Kong Barcelona

Professor Dr. Nikolay G. Basov
Dr. Anatoly S. Bashkin
Dr. Valery I. Igoshin
Professor Dr. Anatoly N. Oraevsky
Dr. Vladimir A. Shcheglov

P. N. Lebedev Physics Institute, 53 Lenin Prospect
SU-117924 Moscow, USSR

English by:
Sergei G. Kittell

Institute of Spectroscopy, USSR Academy of Sciences
SU-142092 Troitsk, Moscow Region, USSR

ISBN-13: 978-3-642-70963-0   e-ISBN-13: 978-3-642-70961-6
DOI: 10.1007/978-3-642-70961-6

Library of Congress Cataloging-in-Publication Data. Chemical lasers/N. G. Basov . . . [et al.]. p. cm. Includes bibliographical references (p.) and index. 1. Chemical lasers. I. Basov, N. G. (Nikolaĭ Gennadievich), 1922–.TA1690.C47   1990   621.36'64–dc20   90–38244

© Springer-Verlag Berlin Heidelberg 1990
Softcover reprint of the hardcover 1st edition 1990

2157/3150-543210 – Printed on acid-free paper

Preface

The rapid development of lasers in the past few decades has led to their application in almost every field of science and technology. The idea that it should be possible to convert the energy released in chemical reactions directly into coherent radiation resulted in the advent of chemical lasers in the 1960s. These first chemical lasers, however, consumed much more energy to initiate the reaction than they emitted. The search for more efficient chemical lasing led to the utilization of chain reactions. However, care had to be taken to maintain the appropriate pressure. In 1970, it was demonstrated that the operation of chemical lasers at atmospheric pressure was also feasible, making it easier and cheaper to construct them.

One of the advantages of chemical lasers is the wide range of radiation wavelengths emitted by them: $1.3-26\ \mu m$. The vibrational frequencies of many molecules fall within this range so that they may conveniently be used for the operation of such lasers.

Progress in the development of chemical lasers is intimately connected with advances in related fields such as gas dynamics, chemical reaction kinetics, and research into the energy relaxation and transfer processes in molecular systems.

This book outlines the principles of chemical laser kinetics and relaxation processes in molecular systems, and analyzes recent advances in these areas. It lays emphasis on the specific features of pulsed and continuous-wave chemical lasers. Among the subjects treated are the related physics, experiments involving chemical lasers, important aspects of laser construction, and possible applications of new lasers and their future development.

Moscow, January 1990   *N. G. Basov · A. S. Bashkin · V. I. Igoshin*
A. N. Oraevsky · V. A. Shcheglov

Contents

1. Introduction

The search for ways to develop chemical lasers began in the early 1960s. It was motivated mainly by the need to find a means of converting the energy released in chemical reactions directly into coherent radiation. In other words, a self-contained radiation source requiring no additional power input had to be created.

Chemical lasers passed through several stages in their evolution. In the first paper on chemical lasers, published in 1960, *Polanyi* [1.1] suggested that vibrationally excited molecules could be used to create chemical lasers operating in the infrared region of the spectrum, and introduced the notion of the total and partial population inversion. In 1962, *Basov* and *Oraevsky* [1.2] demonstrated that fast chemical processes could give rise to population inversion as a result of the difference in relaxation rate between the energy sub-levels of molecules. In his paper of 1963 *Oraevsky* [1.3] pointed to some kinetic merits of chain processes in the creation of chemical lasers. A fundamental analysis of the kinetics aspects of this problem was made in 1964 by *Tal'rose* [1.4], who noted, in particular, the advantages of chain and branching-chain chemical processes. Some results of theoretical studies in the field of chemical lasers for the 1960–1965 period were summarized in a special issue of the journal *Applied Optics* [1.5].

A new stage in the development of chemical lasers began in 1965 when *Kasper* and *Pimentel* [1.6] staged a successful experiment for starting a laser on the basis of a mixture of molecular hydrogen and chlorine. A series of investigations followed, and in 1967 lasing was achieved in reactions of atomic fluorine with molecular hydrogen and other molecules (*Deutsch* [1.7]; *Kompa* and *Pimentel* [1.8]). *Anlauf* and *Knutz* et al. [1.9] were the first to achieve continuous-wave amplification due to a chemical reaction in a low-pressure gas flow.

The search for ways to achieve population inversion in chemical lasers as a result of energy transfer from "hot" molecules produced in the course of chemical reactions to "cold" molecules (*Basov* and *Oraevsky* et al. [1.10]) led to the discovery of an effective transfer of energy from hydrogen halides to CO_2 and the development of lasers based on the transfer of energy from DF to CO_2 (*Gross* [1.11]; *Cool* and *Falk* et al. [1.13]), from HCl to CO_2 (*Chen* and *Stephenson* et al. [1.12]), and from N_2 to CO_2 (*Basov* and *Markin* et al. [1.14]).

The first models of cw chemical lasers were constructed in 1969. These lasers depended on the reaction of atomic fluorine with molecular hydrogen (*Spencer* and *Mirels* et al. [1.15]) or with hydrogen chloride (*Airey* and *McKay* [1.16]), which occurred on the mixing of the reagents in a supersonic flow. Atomic fluorine was produced as a result of the thermal dissociation of either SF_6 in an arc discharge [1.15] or F_2 behind a shock wave front [1.16].

Although progress made in that period was quite considerable, it clearly marked only the first step in the development of chemical lasers. True, it really was the energy of a chemical reaction which was converted into radiant energy in the then existing chemical lasers, but it was also true that much more energy went into the initiation of the reaction itself than was emitted as a result. It became evident that in pulsed chemical lasers this difficulty could be overcome only by using chain reactions (*Oraevsky* [1.17]). However, as was demonstrated in that work, not every process that is a chemical chain reaction is also a chain reaction as far as the generation of coherent radiation in concerned. The next essential step, therefore, was the achievement of lasing in a mixture of molecular hydrogen and fluorine. The first successful experiment on obtaining laser action with the aid of these reagents was conducted by *Batovskii* and *Vasil'ev* et al. [1.18]. *Basov* and *Kulakov* et al. [1.19] studied the feasibility of various ways of initiating a reaction in such a mixture and investigated the spectral composition of the ensuing radiation to demonstrate the existence of a laser chain. Using a ruby laser harmonic to excite a hydrogen-fluorine mixture, *Dolgov-Salvel'ev* and *Zharov* et al. [1.20] proved experimentally that the laser chain in this mixture was of great length.

Further work along these lines resulted in the achievement, in the early 1970s, of lasing in a $D_2 + F_2 + CO_2 + He$ mixture at atmospheric pressure (*Basov* and *Zavorotnyi* et al. [1.21]), which provided a high coherent radiant energy output per unit lasing medium volume. The reaction was initiated by UV radiation from an open discharge. Later on, they managed to achieve lasing in $H_2 + F_2$ and $D_2 + F_2$ atmospheric-pressure mixtures. The existence of a long laser chain in the $H_2 + F_2$, $D_2 + F_2$, and $D_2 + F_2 + CO_2$ mixtures made it possible to obtain appreciably more energy in the form of coherent radiation from pulsed hydrogen-fluoride lasers than was deposited in the mixtures in order to initiate the reaction.

It is simpler, in principle, to develop a purely chemical cw laser. As far back as 1967, *Basov* and *Oraevsky* [1.22] suggested using chemical ignition of hydrogen-chlorine and hydrogen-fluorine mixtures by means of easy-to-evaporate alkali metal atoms or free NO radicals under normal conditions. The method was experimentally tested for the first time by *Cool* and *Falk* et al. [1.13, 23], who "ignited" a $D_2 + F_2 + CO_2$ mixture with NO radicals by mixing the reagents in a subsonic jet.

It also proved possible to produce atomic fluorine by purely chemical means. For this purpose the arc discharge or shock tube was replaced by a combustion chamber burning a mixture with excess fluorine molecules (*Meinzer* [1.24]). The heat produced in the process caused molecular fluorine to dissociate into atoms, and the free fluorine atoms thus obtained were then mixed in a supersonic jet with hydrogen or deuterium.

The achievement of pulsed lasing with the use of high-pressure hydrogen-fluorine mixtures is proof of the high efficiency of electron-beam pumping of hydrogen-fluoride lasers (*Zharov* and *Malinovskii* et al. [1.25]).

Progress in the development of purely chemical cw lasers made them rank with the most powerful laser systems. At the same time it brought to light a number of more general problems, among them the problem of energy relaxation and transfer in molecular systems, the necessity of gaining a deeper insight into the chemical reaction kinetics, and some new questions of gas dynamics.

Methods of measuring the cross sections of various chemico-kinetic processes with the aid of chemical lasers were developed. They are based on utilizing both the emission of chemical lasers to excite the object under study, and their unique spectral and temporal characteristics. In this latter case, subject to investigation are the rates of the processes occurring in the chemical laser medium.

One of the advantages of chemical lasers is the wide range of the radiation wavelengths emitted by them: 1.3–26 μm. Some of the wavelengths within the above range cannot be obtained with other types of lasers. The spectral range of the coherent radiation of chemical lasers falls within the vibrational frequency region of many molecules, which is why such lasers can be used for nonthermal stimulation of chemical reactions by way of resonant action upon certain vibrational degrees of freedom of the molecules. Especially interesting in this respect are lasers in which the emitters are the hydrogen halides HF, DF, and HCl whose emission spectra coincide with the absorption region due to various hydrogen bonds in molecules, a subject which is given much space in the present monograph.

The second chapter deals with the fundamentals of chemical laser kinetics. It introduces the reader to problems associated with the realization of the very notion of chemical laser, and gives an idea of the most promising types of chemical processes. The vibrational degrees of freedom of the molecules formed in a chemical reaction constitute a most natural accumulator of the energy released in the process. For this reason, the generation of radiation in chemical lasers is effected mainly on vibrational-rotational transitions. This chapter systematizes the vast amount of experimental material available in the literature on the nonequilibrium vibrational chemical excitation of, and vibrational relaxation processes in, the molecules used in chemical lasers. Ample space is devoted to the discussion of new theoretical results on the mechanisms of vibrational energy conversion (relaxation) upon molecular collisions. And finally, the chapter outlines the general method of numerical calculation of chemico-kinetic laser processes.

Chapter 3 analyzes the most promising chemical laser systems based on chain pumping reactions in $H_2 + F_2$ and $D_2 + F_2 + CO_2$ mixtures. It formulates a kinetic model of the chain-reaction chemical laser, systematizes experimental data on the rate constants of elementary processes, and gives detailed results of numerical calculations of the energetic and temporal emission characteristics. The analysis is restricted to pulsed laser systems. The cw lasers having the same chemico-kinetic basis are considered in Chap. 5.

The fourth chapter is devoted to experimental studies of pulsed chemical lasers. It examines various methods for the initiation and practical realization of chemical lasers. There is a description of the chemical processes taking place when chemical laser mixtures are prepared. These processes govern the choice of a particular laser mixture preparation technique. This chapter also reviews the latest developments and new trends in the studies of pulsed chemical lasers.

Chapter 5 considers the basic schemes and the design and physical characteristics of cw chemical lasers, examines their principles and modes of operation, and discusses their development. Special attention in this chapter is paid to the purely chemical $DF-CO_2$ and the supersonic HF gas-flow lasers, the best-developed chemical lasers today. Information is provided on the various models of such laser systems, with emphasis on their construction features and energy parameters. A detailed analysis is presented of the $DF-CO_2$ laser operating on an initiating reagent and of the HF laser depending on a "cold" reaction. Prospects for the development of the $DF-CO_2$ and HF continuous-wave chemical lasers are discussed with special reference to a purely chemical supersonic model in the former case and a model using a chain excitation mechanism in the latter. Presented in this chapter are also the results of experimental studies of other types of lasers using various initiation techniques and reagents.

One of the most interesting recent achievements in the field is the creation of a laser relying on the transfer of energy from singlet oxygen produced as a result of chlorination of hydrogen peroxide to atomic iodine (*McDermott* and *Pchelkin* et al. [1.26]). The specific features of this laser are considered in Chap. 6.

The book concludes with the description of chemical lasers based on photon-branching reactions (Chap. 7). Photon branching is a new chemical reaction mechanism of great interest. With this mechanism, photons directly take part in the laser-chemical process by initiating the reaction giving rise to the emission of photons. No such laser has as yet been created, but the principal possibilities of its development look promising.

Chemical lasers have now reached a level where it becomes possible to design systems for practical application; and yet fundamental ideas in the field of chemical lasers are far from being exhausted, which is evidenced by the material of Chap. 7.

Over the years since chemical lasers were first developed many ideas have been advanced the realization of which offers new opportunities for progress in the field (*Basov* and *Oraevsky* [1.27]). One of the most interesting problems is the development of a laser operating on the hydrogen oxidation reaction. Such a laser would consume readily available reactants and would be perfectly clean ecological-ly, because the product of the reaction occurring in its active medium would be water. The closest to this laser is the system using the branching-chain reaction in an H_2-O_2-CO mixture yielding excited CO_2 molecules which then lase at a wavelength of 10.6 μm (*Basov* et al. [1.28]). But this system was not developed further. Besides, the use of carbon monoxide in the mixture makes the laser ecologically unhealthy. We therefore consider the problem of the development of the ecologically clean hydrogen-oxygen laser a challenge to the researchers engaged in the field of chemical lasers.

The chemical laser is a complex device based on advances in quantum electronics, chemical kinetics, gas dynamics, spectroscopy, etc. It should also be noted that the question of "relations" between lasers and chemistry is very broad. At present widely used are photodissociation lasers (*Borovich* and *Zuev* et al. [1.29]) and excimer lasers (*Danilychev* and *Kerimov* et al. [1.30]). These lasers do not belong to the class of chemical lasers because their radiant energy originates from the energy of an external source and not from the heat of an exothermic chemical process. But in kinetic terms these lasers are close to chemical lasers.

As far as the dynamics and kinetics of energy exchange processes are concerned, questions relating to chemical lasers are closely connected with those of stimulation of chemical reactions by means of laser radiation (*Moore* [1.31]) and the energy of a nonself-sustained (electric ionization, photoionization) discharge (*Basov* and *Danilychev* et al. [1.32]). Though this monograph is addressed primarily to specialists in the field of chemical lasers, its individual sections are also of interest to a wide circle of scientists, engineers, and postgraduate students.

2. Fundamentals of Chemical Laser Kinetics

Chemical laser studies are at the junction of several avenues of research, namely quantum electronics, chemical and physical kinetics, high-energy chemistry, spectroscopy, and gas dynamics. There are two main aspects to the study of the processes occurring in the active medium of the chemical laser: the dynamics of elementary processes and macroscopic kinetics of the elementary processes in their totality. Both these aspects are important in predicting the operating characteristics of chemical lasers. Three types of elementary process are of principal significance in chemical laser kinetics. They are: absorption and emission of photons (spectroscopy); the chemical transformations proper, including key elementary reactions accompanied by nonequilibrium excitation of the fragments produced (chemical kinetics); and relaxation processes, whereby energy is redistributed without any reconstruction of chemical bonds.

In Sect. 2.1 the main details of chemical laser kinetics are analyzed on the basis of simple mathematical models taking into consideration the more important interactions in the reaction medium. From this analysis the reader can glean the range of emerging problems and fundamental conceptions. Section 2.2 gives a classification of reactions involving chemical activation of the products and systematizes data on vibrational chemical excitation of a number of molecules. The material in the first two sections makes it clear why exchange reactions giving rise to vibrational chemical excitation of the product proved to be most suitable for chemical lasers. It was no accident that the success of experiment was due to the use of this type of reaction, and it was this circumstance that determined the scope of questions treated here. Section 2.3 generalizes some important spectroscopic relationships governing the amplification of radiation in vibrational-rotational transitions. In Sect. 2.4 the results of studies into the vibrational relaxation of the most important molecular systems used in chemical lasers are discussed. Sections 2.5–2.7 are devoted to the analysis of questions related to the macroscopic kinetics of chemical lasers, Sect. 2.5 formulating the equations that are at the root of current chemical laser models and Sects. 2.6 and 2.7 discussing important general conclusions from these equations about vibrational kinetics and induced emission in multilevel lasers (which rely on vibrational-rotational transitions in molecules) and formulating efficient methods for calculating energy characteristics of pulsed and continuous-wave chemical lasers.

2.1 Qualitative Analysis of Chemical Laser Operation

Some important conclusions as to the specific features of coherent radiation generation in chemical reactions can be drawn without considering the details of a concrete reaction mechanism. In this section a number of formal kinetic questions of the chemical laser theory are treated on the basis of a simple kinetic reaction model. In particular, the principal relaxation effects in chemical pumping are revealed and the general requirements of the chemical process are formulated, the satisfaction of which is essential for the development of lasing and for efficient laser operation.

Although the nature of excitation (rotational, vibrational, or electronic) is not specified here, it should be noted that the emergence of the very idea of the chemical laser [2.1], the first successful experiment [2.2], and the further development of chemical lasers are intimately related to the vibrational chemical excitation of molecules.

2.1.1 Specific Power of Coherent Radiation

From the standpoint of kinetics, one can single out two groups of laser systems relying on chemical pumping:

a) systems with direct production of population inversion in an elementary "lasing" chemical reaction event; and

b) systems with the necessary population inversion produced through energy transfer from the "hot" molecules excited by chemical reaction to the "cold" molecules constituting the laser medium.

In the case of direct excitation of laser levels, knowledge of the character of molecular distribution over the excited levels of the reaction product is essential. To achieve the threshold inversion density, it is necessary that the rate of pumping to the upper level be higher than the rate of population of the lower level.

As the excited molecules are accumulated, the rate of energy relaxation processes increases. For this reason, no matter what the relationship between the specific rates of the pumping reaction channels, a molecular distribution is established that excludes the possibility of radiation amplification, although the energy store of the excited states may be considerable.

In the second case, the fact of inversion distribution of the reaction products among energy levels is of secondary significance, the only essential thing being that the energy store is in hyperequilibrium. The molecules excited during reaction serve as an energy "reservoir". With the lasing molecules being suitably selected (a high energy transfer rate is convenient for creating a level inversion system), an effective conversion of the reservoir energy into coherent radiation is possible. Specifically, use can be made of reactions accompanied by excitation of metastable states from which radiative transitions are forbidden.

By their mode of operation chemical lasers are classified as pulsed and continuous-wave (cw) types, but their chemico-kinetic basis is one and the same.

Let us consider the kinetics of processes occurring in a chemically reacting gas placed in an optical cavity. In pulsed lasers, the gas is stationary, and, with the reaction being initiated in a uniform manner, all the processes depend on time and not on the spatial position, whereas in cw lasers the gas is moving so that the coherent radiation is directed across the gas flow (Fig. 2.1). The main quantity determining the energy characteristics of a laser is the stimulated emission power density (specific power) P_l [W/cm^3]. For pulsed lasers, P_l is a function of time, $P_l = f(t)$, while for cw lasers it is a function of the coordinate along the gas flow: $P_l = f(x)$. The output radiation intensity of a laser, no matter what its mode of operation (pulsed or cw), is related to P_l by the relation (one of the cavity mirrors is fully reflecting)

$$I_l^{out} = P_l l \; , \tag{2.1}$$

where l is the active medium length in the direction of radiation. The total power Q_l emitted by a pulsed laser through two mirrors is

$$Q_l = P_l V \; ,$$

where V is the active medium volume, and the total laser energy is determined by time integrating Q_l:

$$E_l = \int_{t_0}^{t_1} Q_l(t)dt = V\varepsilon_l \; . \tag{2.2a}$$

Here

$$\varepsilon_l = \int_{t_0}^{t_1} P_l(t)dt \tag{2.2b}$$

is the laser energy per unit active medium volume and t_0 and t_1 are the lasing (amplification) onset and quenching moments, respectively. The output power of a cw laser can be determined by considering radiation emitted from a stream tube with a cross-sectional area of $A(x) = ly(x)$, where l is the tube width (taken to be constant) and $y(x)$ its height, which in the general case varies as a result of the gas

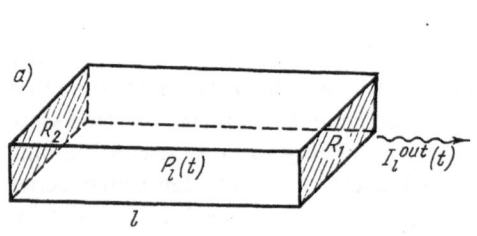

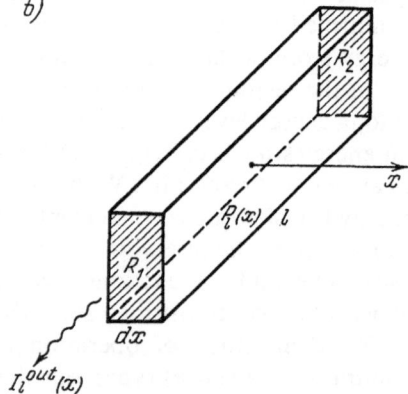

Fig. 2.1. Emission geometry in (a) a pulsed laser and (b) a stream tube in a cw laser

expansion. The output power emitted through two mirrors from the layer shown in Fig. 2.1b is

$$dQ_l = P_l(x)A(x)dx \; . \tag{2.3}$$

Integrating (2.3) between x_0 (the point marking the onset of lasing) and x_1 (that marking its quenching), we get the total laser power

$$Q_l = \int_{x_0}^{x_1} P_l(x)A(x)dx \; . \tag{2.4}$$

Calculation of the function $P_l(t)$ [or $P_l(x)$] and studies of the effects of the main parameters (chemical composition of the reagents, gas mixture pressure and temperature, etc.) on the energy characteristics of the laser constitute the central task of theoretical analysis. In this section use will be made of a simple flow-and-physico-chemical-kinetics model making it possible to reveal the most important details of chemical laser kinetics due to competition between excitation and relaxation processes.

Systems with Direct Inversion Production. The model includes the following fundamental processes: chemical pumping of lasing levels, relaxation, and stimulated emission. For simplicity, we consider a two-level system, which does not alter the essence of the conclusions drawn. Let us write down balance equations in the form

$$\frac{dn^*}{dt} = \alpha^* W - \sigma_{21}n^* - \sigma c \varrho_{ph}(n^* - n) \; ,$$

$$\frac{dn}{dt} = \alpha W + \sigma_{21}n^* + \sigma c \varrho_{ph}(n^* - n) \; , \tag{2.5}$$

$$\frac{d\varrho_{ph}}{dt} = \sigma c \varrho_{ph}(n^* - n) - \frac{\varrho_{ph}}{\tau_{ph}} + \frac{\sigma c}{V} n^* \; ,$$

where n^* and n are the population densities of the upper and lower laser levels, respectively, α^* and α the probabilities of the reaction product being formed at the upper and lower laser levels, respectively, ϱ_{ph} is the photon density in the cavity, W the chemical reaction rate, σ_{21} the relaxation probability, σ the stimulated transition cross section, c the velocity of light, τ_{ph} the photon lifetime in the cavity, and V the cavity volume.

If d/dt in (2.5) means a Lagrangian derivative, the set of equations (2.5) describes the kinetics of the processes occurring in lasers, both pulsed and continuous-wave. Indeed, for a steady-state flow, $dt = dx/u$ (u being the flow velocity). Therefore, the time dependence of any physical quantity, found from (2.5), is transformed into the coordinate dependence upon the change of variable $t = x/u$, provided that the flow density, temperature, and velocity remain the same. A more general flow model requires introduction of hydrodynamic conservation equations, and is treated in Sect. 2.7.

It should also be noted that the simple chemical laser model under consideration assumes that the laser mixture components are intermixed beforehand. But in cw laser systems using highly reactive reagents, some of them in the atomic state, the components, as a rule, cannot be preliminarily mixed, and so one has to mix them directly in the flow. The laser model being considered here describes only systems for which the reagent mixing time is shorter than the reaction time. If the characteristic mixing time is commensurable with, or is longer than, the characteristic reaction time, the mathematical model of the processes occurring in such lasers must allow for transfer phenomena. Chemical lasers of this type (diffusion chemical lasers) are considered in Chap. 5.

Let us analyze the set of equations (2.5). The rate of photon production by stimulated emission is described by the first term in the equation for $d\varrho_{ph}/dt$. The contribution of spontaneous emission to the photon production rate, equal to $\sigma c n^*/V$, is substantial at the initial stage of the emission process. The term ϱ_{ph}/τ_{ph} corresponds to the specific rate of photon emission through the cavity mirrors. The power density of the radiation coupled out of the cavity is therefore given by

$$P_l = \hbar\omega_l \frac{\varrho_{ph}}{\tau_{ph}} , \qquad (2.6)$$

where $\hbar\omega_l$ is the laser photon energy. It follows from (2.1) and (2.6) that the radiation intensity I_l inside the cavity, equal to $\hbar\omega_l\varrho_{ph}c$, is related to the output radiation intensity by the relation

$$I_l = I_l^{out}(c\tau_{ph}/l) . \qquad (2.7)$$

The set of equations (2.5) is nonlinear, and to find $P_l(t)$, it must be solved numerically. The quantity $P_l(t)$ can be found much more easily if a physically reasonable assumption is made that the rate of production of photons in the course of lasing is equal to that of their extraction from the cavity. This assumption is at the root of what is known as the "quasistationary laser" model. Spontaneous emission disregarded, it follows from (2.5) that in the quasistationary approximation

$$n^* - n = \Delta = 1/\sigma c\tau_{ph} ,$$

i.e., the difference between the populations of the lasing levels remains constant and equal to a threshold value. The constancy of the inversion density is due to the fact that its decrease as a result of the action of the emission field is compensated for by pumping. Since Δ is constant, it is legitimate to put

$$\frac{dn^*}{dt} = \frac{dn}{dt} ,$$

and then it is not very difficult to find from (2.5) and (2.6) that in the quasistationary approximation the power density of laser radiation is given by the expression

$$P_l(t) = \tfrac{1}{2}\hbar\omega_l \left[(\alpha^* - \alpha)W - \sigma_{21}\left(\int_0^t W(t')dt' + \Delta_{thr} \right) \right] , \qquad (2.8)$$

where $\Delta_{thr} = 1/\sigma c \tau_{ph}$ is the threshold inversion density and

$$n^* = \left(\int_0^t W \, dt' + \Delta_{thr} \right) \Big/ 2$$

the population density of the upper level in the course of lasing, account being taken of the fact that

$$n^* + n = \int_0^t W \, dt' \, , \qquad n^* = n + \Delta_{thr} \, .$$

Formula (2.8) means that the lasing follows in a quasistationary fashion the course of the chemico-kinetic process, and the problem of calculating the radiation power thus boils down to the calculation of the chemical kinetics involved.

Relation (2.8) allows a number of important inferences to be drawn about specific features of lasing under chemical pumping. In particular, it can be seen from (2.8) that the presence of relaxation processes limits the chemical reaction rate to some threshold value of W_{thr} below which no lasing can take place:

$$(\alpha^* - \alpha) W_{thr} = \frac{\sigma_{21}}{\sigma c \tau_{ph}} \, . \tag{2.9}$$

The time τ_{ph} may be expressed as

$$\tau_{ph} = \frac{2l}{c} \ln^{-1} \left(\frac{1}{R_1 R_2} \right) , \tag{2.10}$$

where l is the cavity length and R_1 and R_2 are the mirror reflectivities.

Referring to (2.8), one can infer an important feature of laser systems with direct chemical pumping of the lasing molecules that makes them different from non-chemical lasers. This feature is due to the fact that in this type of chemical laser there occurs a constant increase in the number of the molecules constituting the lasing medium, and hence a constant growth of the rate of relaxation processes, equal to $\sigma_{21} n^*$. The question of the rate of chemical reaction therefore becomes important. For the laser to operate effectively, it is necessary that the rate of the reaction should either rise in line with its progress or have a high initial value. A detailed analysis of this question requires consideration of a definite reaction mechanism and is made in Chap. 3.

Systems with Energy Transfer. The model allows for the chemical pumping of the reaction product, energy transfer, relaxation, and stimulated emission of radiation. We write down the balance equations in the form

$$\frac{dN^*}{dt} = W - (\sigma_1 + \sigma_3) N^* \, ,$$

$$\frac{dn^*}{dt} = \sigma_3 N^* - \sigma_2 n^* - \sigma c \varrho_{ph}(n^* - n) \, , \tag{2.11}$$

$$\frac{d\varrho_{ph}}{dt} = \sigma c \varrho_{ph}(n^* - n) - \frac{\varrho_{ph}}{\tau_{ph}} \, .$$

Here N^* is the density of the excited molecules produced in the reaction proceeding at a rate of W, σ_1 the relaxation probability of the excited product, σ_2 the relaxation probability of the upper laser level, and σ_3 the probability that an excitation quantum will be transferred from the excited product to the laser molecules. No equation is specified for the population rate of the lower laser level.

In the same quasistationary approximation, for the laser power density we get

$$P_l(t) = \hbar\omega_l \left[\frac{\sigma_3}{\sigma_1 + \sigma_3} W - \sigma_2 (n + \Delta_{thr}) - \frac{dn}{dt} \right] , \qquad (2.12)$$

where account is taken of the fact that during lasing $n^* = n + \Delta_{thr}$. As follows from (2.12), for lasing to occur, the following condition must be satisfied:

$$\frac{\sigma_3}{\sigma_1 + \sigma_3} W > \sigma_2 (n + \Delta_{thr}) , \qquad (2.13)$$

where n is the equilibrium population of the lower laser level. This condition determines the threshold reaction rate W_{thr}. In deriving (2.13), we neglected the term dn/dt in (2.12). The quantity dn/dt is governed by the relaxation processes at the lower level. The population of this level can, in principle, rise as a result of the heating of the mixture. The threshold condition (2.13), therefore, is in the general case necessary, but not sufficient.

It follows from (2.12) that for a laser with energy transfer to operate effectively, the condition $\sigma_3 \gg \sigma_1$ must be fulfilled. This condition imposes a limitation on the composition of the laser mixture in so far as the probabilities σ_i depend on the mixture component concentrations. It is evident from comparison between (2.8) and (2.12) that there is an essential difference in kinetics between chemical lasers with and without energy transfer due to the fact that in systems with energy transfer the total laser molecule concentration remains constant, whereas in lasers with direct inversion production it increases with time. If condition (2.13) is satisfied, lasing in a chemical laser relying on energy transfer can be sustained for a period of time longer than the characteristic relaxation time, even if the chemical process does not accelerate with time. This means that chemical lasers with energy transfer must have a higher chemical efficiency compared to their no-energy-transfer counterparts in the region of slow reaction rates. However, if the thermal population of the lower level is substantial and the laser medium gets highly heated in the course of the reaction, to maintain lasing then requires that the reaction rate W should be progressively increased.

2.1.2 Laws Governing Population Inversion Development in the Case of Chemical Pumping

With chemical pumping, the excitation rate of the laser levels is limited by the chemical reaction rate. An inevitable stage in the process is the relaxation of the lasing transition. The relationship between the chemical pumping and relaxation rates determines the course of the reaction (with or without inversion production) and the energy characteristics of the system. These questions were analyzed by

Oraevsky [2.3, 6, 15, 18], *Tal'rose* [2.4, 16], *Igoshin* and *Oraevsky* [2.5], *Polanyi* [2.6], *Basov* and *Igoshin* et al. [2.8, 9], *Dzhidzhoev* and *Platonenko* et al. [2.10, 17], *Pimentel* [2.11], *Shuler* and *Carrington* et al. [2.12], *My* [2.13], *Young* [2.14], and *Bashkin* and *Igoshin* et al. [2.19]. Laying aside some details, one can obtain a number of important conditions which are imposed on the rates of elementary reactions (or acts) and provide for a progressive growth of inversion in the course of an explosive reaction. Disregarding the stimulated emission field, the temporal evolution of the laser level populations is described by the equations

$$\frac{dn^*}{dt} = W - \sigma_{21} n^* , \tag{2.14}$$

$$\frac{dn}{dt} = \sigma_{21} n^* , \tag{2.15}$$

where n^* and n are the populations of the excited and unexcited levels, respectively, W is the reaction rate, and σ_{21} the relaxation probability. Let us consider process kinetics for various types of temporal behavior of the reaction rate W.

In developing chemical lasers, of prime importance are self-sustained chemical processes involving nonequilibrium excitation of the products: chain reactions [2.3], branching chain reactions [2.4], and thermal explosion [2.5]. In the works cited above and also in the work reported in [2.6], a rather general analysis of the kinetics of the processes of this type was for the first time made from the standpoint of the possibility of producing inverted media, and the promises of such processes were pointed out.

Nonbranching Chain Reaction. Let us consider one of the simplest possible chain reactions following the scheme

0) $A \xrightarrow{k_0} 2N_1$ (chain initiation),

1) $N_1 + C \xrightarrow{k_1} P + N_2$

2) $N_2 + A \xrightarrow{k_2} P + N_1$ } (chain propagation),

3) $N_1 + R + M \xrightarrow{k_3} N_1 R + M$ (chain termination),

where A and C are the reactants, P is the reaction product molecule, N_1 and N_2 are two active intermediate particles (active centers), R may be either N_1 or N_2 or else some other mixture component causing the decay of the active centers, and M is any mixture component, Stages 1) and 2) make up a chain link. The reactions of halogens with hydrogen can exemplify the reactions going in accordance with the above scheme. Let us assume that at the initial moment the initiating pulse produces some concentration of the active centers. It is known from the chain reaction kinetics [2.20] that

$$W = 2k_{ch}[A][N]\exp(-\sigma_{dec}t) , \tag{2.16}$$

where k_{ch} is the chain propagation rate constant, equal to the rate constant for the slowest stage of the chain link; $[A]$ is the reactant concentration;

$$[N] = \int_0^\infty \gamma(t)\,dt$$

is the initial active center concentration, $\gamma(t)$ describing the time dependence of the initiating pulse; and σ_{dec} is the decay rate coefficient of the active centers of the chain. Letting ξ_i represent the relative concentration of the ith component, we may write

$$[A] = \xi_A[M], \quad [N] = \xi_N[M], \quad [R] = \xi_R[M] ,$$

$$\sigma_{21} = \left(\sum_i k_{21}^i \xi_i\right)[M], \quad \sigma_{dec} = k_{rec}\xi_R[M]^2 ,$$

(2.17)

where $[M]$ is the total mixture concentration, k_{21}^i is the rate constant for the deactivation of the laser molecules upon collision with the ith component, and k_{rec} is the rate constant for the active center recombination.

For simplicity, let us consider the following two extreme cases:

Case 1. $\sigma_{dec} \ll \sigma_{21}$, which is satisfied at not very high mixture pressures. After integrating (2.14) in the linear approximation, i.e., without considering the burn-out of the substance, we have

$$n^*(t) = \frac{2k_{ch}[A][N]}{\sigma_{21}}[1 - \exp(-\sigma_{21}t)] ,$$

(2.18)

$$n(t) = \sigma_{21}\int_0^t n^*(t')\,dt' .$$

(2.19)

During the time $t_m = \ln(2/\sigma_{21})$ the inverted population reaches its maximum:

$$(n^* - n)_m = (1 - \ln 2)\frac{2k_{ch}[A]}{\sigma_{21}}\int_0^\infty \gamma(t)\,dt .$$

(2.20)

It can be seen from (2.20) that the inversion maximum depends on the chain propagation rate. If the latter exceeds the vibrational relaxation rate, a substantial inversion can be reached. As $[A] \sim [M]$, $[N] \sim [M]$, $\sigma_{21} \sim [M]$, the inversion maximum is proportional to the mixture pressure: $(n^* - n)_m \sim [M]$.

Case 2. $\sigma_{dec} \gg \sigma_{21}$. As seen from (2.17), this case is realized at sufficiently high magnitudes of $[M]$. The transition region of pressures $[M]^*$ that separates the cases under consideration is determined from the condition: $\sigma_{dec} = \sigma_{21}$. We have from this condition that

$$[M]^* = \frac{\sum_i k_{21}^i \xi_i}{k_{rec}\xi_R} .$$

(2.21)

Integrating (2.14) in the domain $[M] > [M]^*$ yields

$$n^*(t) = \frac{2k_{\text{ch}}[A][N]}{\sigma_{\text{dec}}}[1 - \exp(-\sigma_{\text{dec}}t)] , \tag{2.22}$$

$$(n^* - n) = \frac{2k_{\text{ch}}[A]}{\sigma_{\text{dec}}} \int_0^\infty \gamma(t)\,dt . \tag{2.23}$$

Since $\sigma_{\text{dec}} \sim [M]$, the inversion maximum in this pressure region is independent of the mixture pressure. Thus, the recombination of the active centers in triple collisions is the very mechanism which is responsible for the principal limitation on the maximum possible specific energy output in chemical lasers based on chain reactions.

Branching Chain Reaction. The chemical pumping rate in the case of branching chains may be written as [2.20]

$$W = W_0 e^{st} ,$$

where s is the branching ratio. In the same linear approximation, we get the following solution of the rate equations:

$$n^*(t) = n_0^* e^{st}, \quad n(t) = \frac{\sigma_{21}}{s} n_0^* e^{st} .$$

If account should be taken of the lower level relaxation, then

$$\frac{dn}{dt} = \sigma_{21} n^* - \sigma_{10} n ,$$

where σ_{10} is the relaxation probability of the lower level. In this case

$$n(t) = n_0^* \frac{\sigma_{21}}{\sigma_{10} + s} e^{st} .$$

It can be seen from this solution that population inversion grows exponentially with time, provided that $s > \sigma_{21} - \sigma_{10}$. At the same time, the inequality $s > 0$ expresses the self-ignition condition. For this reason, if $\sigma_{21} > \sigma_{10}$, the reaction occurs, although inversion is not produced everywhere in the ignition region. This is illustrated schematically in Fig. 2.2a. If $\sigma_{21} < \sigma_{10}$, the inversion region coincides with the entire ignition region. But this obtains if the inequality $\sigma_{21} < \sigma_{10}$ is fulfilled at any temperature. If, for example, $\sigma_{21} < \sigma_{10}$ for $T < T_1$, the inversion region may look as shown in Fig. 2.2b. Finally, if a temperature T_2 exists such that $s > \sigma_{21} - \sigma_{10}$ at $T_1 < T < T_2$ and $s < \sigma_{21} - \sigma_{10}$ at $T > T_2$, there arises an inversion island (Fig. 2.2c). When the inversion region does not coincide with the ignition region, a finite energy must be spent to enter the inversion region. It cannot be ruled out that the mixture may enter the inversion region by self-heating. If, on the other hand, the

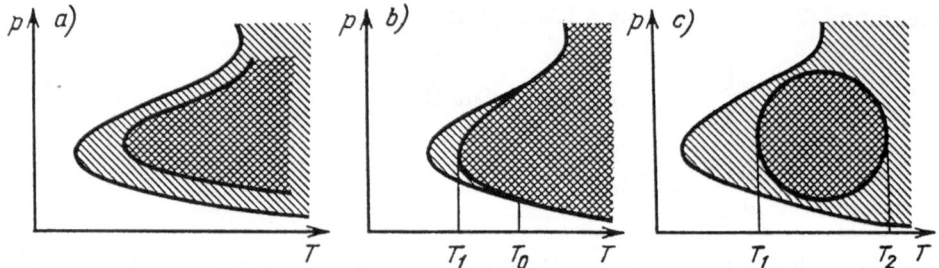

Fig. 2.2. Mixture self-combustion region in the pressure-temperature coordinates ($p-T$ diagram) for branching chain reactions. The inversion region lies inside the self-combustion region and is cross-hatched

inversion region is identical with the ignition region, a purely chemical laser can be created with the inversion region being entered by changing the mixture pressure.

The kinetics of laser systems depending on branching chain reactions was considered by *Tal'rose* [2.4, 16], *Oraevsky* [2.6, 18], *Basov* and *Igoshin* et al. [2.8, 9], *Dzhidzhoev* and *Pimenov* et al. [2.17], and *Bashkin* and *Igoshin* et al. [2.19]. In particular, *Oraevsky* [2.6] analyzed the inversion region equation in the $p-T$ coordinates for a number of concrete reaction mechanisms and demonstrated that the inversion maximum in branching chain systems depends but little on the initiation energy. This circumstance means that the use of branching chain reactions makes it possible to develop an oscillator close to the ideal chemical laser consuming practically no energy from the outside. Later in the text we will discuss how fully the capabilities of branching chain reactions are being utilized in actually existing schemes.

Thermal Self-Combustion. Thermal explosion is the most common self-combustion mechanism. If the molecules produced in a reaction are subject to chemical activation, the thermal acceleration of the reaction can be ensured at the expense of the translational and rotational energies released in the elementary reaction, and the energy of excited states can be used to generate coherent radiation. The possibility of combining chemical pumping with thermal explosion in order to produce an inverted medium was examined by *Igoshin* and *Oraevsky* [2.5]. The temperature dependence of the upper level population rate obeys the Arrhenius law:

$$W = k[A]^n \exp\left(\frac{-\varepsilon}{RT}\right) ,$$

where n is the kinetic order of the reaction and ε the activation energy. The set of equations (2.14), (2.15) must be supplemented with an equation for the mixture temperature. Using the thermal explosion theory routine of expanding the Arrhenius function as

$$\exp\left(\frac{-\varepsilon}{RT}\right) = \exp\left(\frac{-\varepsilon}{RT_0}\right)\exp(\theta) ,$$

where $\theta = (\varepsilon/RT_0^2)(T - T_0)$, T_0 being the reactor wall temperature, and assuming that the temperature dependence of the relaxation rate is weak compared to the Arrhenius law, we write down the heat-balance equation in the form

$$\frac{d\theta}{dt} = \sigma_3 e^{\theta} - \sigma_4 \theta + \sigma_5 \ . \tag{2.24}$$

Here σ_3 is the reduced initial heat liberation rate given by

$$\sigma_3 = (1 - \eta)\frac{Q}{C}\frac{\varepsilon}{RT_0^2}\frac{1}{[A]}\sigma_1 \ ,$$

$(1 - \eta)$ being the molecular energy fraction distributed among the rotational-translational degrees of freedom, Q the heat of reaction, C the specific heat, and $\sigma_1 = k[A]^n \exp(-\varepsilon/RT_0)$ the reaction rate at the initial temperature; the quantity $\sigma_4 = hS/[A]CV$, h being the heat-transfer coefficient, S the reactor surface area, and V the reactor volume, characterizes heat emission; and σ_5 is the rate constant for heat liberation in the course of relaxation, expressed as

$$\sigma_5 = \eta\frac{Q}{C}\frac{\varepsilon}{RT^2}\frac{1}{[A]}\sigma_{21} \ .$$

It follows from the analysis of the set of equations (2.14), (2.24) that the progressive growth of inversion is possible if the following two inequalities are satisfied [2.5]:

$$\frac{\sigma_3}{\sigma_4} > e^{-1} \ , \tag{2.25}$$

$$\frac{\sigma_3}{\sigma_{21}} > v_{cr} \simeq 0.45 \ . \tag{2.26}$$

Formula (2.25) expresses the thermal instability condition. The physical meaning of inequality (2.26) may be explained as follows: The lasing transition relaxation time is proportional to $1/\sigma_{21}$. The reaction induction period (the time it takes to reach pre-explosion reaction rates) is proportional to $1/\sigma_3$. Condition (2.26) means that the ratio of these quantities must exceed some critical value, otherwise the decay of inversion will occur at the initial, isothermal stage of the reaction. Relations (2.25), (2.26) may be reduced to the form including directly the pressure, temperature, kinetic constants, and heat of reaction:

$$(1 - \eta)Qk \exp(-\varepsilon/RT_0)[A]^n \frac{\varepsilon}{RT_0^2}\frac{r^2}{\lambda} > \delta_{cr} \ , \tag{2.27}$$

$$(1 - \eta)\frac{Q}{C}\frac{\varepsilon}{RT_0^2}\frac{k\exp(-\varepsilon/RT_0)}{k_R}[A]^{n-2} > v_{cr} \ , \tag{2.28}$$

where λ is the heat conductivity of the medium, r the reactor radius, $\delta_{cr} = 2$ for a cylinder, and $k_R[A] = \sigma_{21}$ is the relaxation probability.

The conditions formulated above yield an important conclusion about the shape of the inversion region in the $p - T$ coordinates. Where the kinetic reaction order $n < 2$, this region looks like a peninsula (Fig. 2.3a); the upper limit, which rises with temperature, is described by inequality (2.28); the position of the lower limit is set by the thermal self-ignition condition and depends on the geometry of the reactor. If $n \geqslant 2$, the inversion region is restricted only by the lower pressure limit (Figs. 2.3b and c).

The physical cause of these phenomena is that as the reaction rate increases in proportion to the factor $[A]^n$ and, at the same time, proportionally with the rise of pressure, the heat capacity of the starting mixture grows higher (which in turn leads to a decrease in the heating rate); besides, the relaxation rate increases linearly with pressure. Competition between these factors gives rise to the inversion region in the $p - T$ coordinates. The fact that the inversion region comes close to the slow, steady-state reaction region makes it possible to prepare the mixture in the vicinity of the lower limit of the inversion peninsula and thus minimize the expenditure of energy in initiating the reaction. This circumstance makes promising the use of thermal explosion to create a chemical laser.

2.1.3 Energy Characteristics and Their Dependence on the Physico-Chemical Reaction Mechanism

Let us now consider, in addition to the specific laser energy ε_l defined by formula (2.2b), the chemical and engineering laser efficiencies. The chemical efficiency is equal to the ratio of the specific laser energy to the specific chemical energy store:

$$\eta_{\text{chem}} = \frac{\varepsilon_l}{(-\varDelta H)[A]_0} , \tag{2.29}$$

where $(-\varDelta H)$ is the heat of reaction and $[A]_0$ the initial concentration of the reagent A in the mixture (it is assumed that $[A]_0 < [C]_0$; if $[A]_0 > [C]_0$, (2.29) will include $[C]_0$ instead of $[A]_0$). The engineering laser efficiency is defined as the ratio between the laser energy and the expenditure of energy to produce atoms:

$$\eta_{\text{eng}} = \frac{\varepsilon_l \eta_{\text{in}}}{\mathscr{E}_a [N]} , \tag{2.30}$$

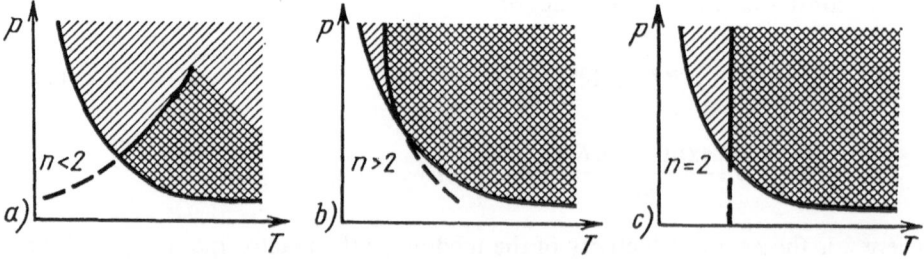

Fig. 2.3. The same as Fig. 2.2, but for reactions proceeding by the thermal explosion mechanism

where η_{in} is the efficiency of the initiation source and $\mathscr{E}_a$ the energy required to produce a reaction center. Apart from ε_l, η_{chem}, and η_{eng}, of much interest is also the effective chain length, defined as the number of molecules per active center that contribute to coherent radiation:

$$v_{eff} = \frac{\varepsilon_l}{\hbar\omega_l}\frac{1}{[N]} ,\qquad(2.31)$$

where $[N]$ is the concentration of the active centers produced by the initiating pulse.

Let us consider the expression for the specific laser energy in systems with inversion produced directly in the course of the reaction. Using (2.2a) and (2.7), we get

$$\varepsilon_l = \tfrac{1}{2}\hbar\omega_l \int_{t_0}^{t_1} [(\alpha^* - \alpha)W(t) - \sigma_{21}(\int_{t_0}^{t} W(\tau)d\tau + \varDelta_{thr})]dt .\qquad(2.32)$$

Two important conclusions follow from (2.8) and (2.32) about the dynamics of the systems of the above-considered type. First, in the absence or relaxation ($\sigma_{21} = 0$, $t_1 \to \infty$), the specific laser energy and chemical laser efficiency are independent of the chemical reaction rate and reach their ultimate values, given by

$$\varepsilon_l^{lim} = \tfrac{1}{2}\hbar\omega_l(\alpha^* - \alpha)\int_0^\infty W(t)dt ,\qquad(2.33)$$

$$\eta_{chem}^{lim} = \frac{1}{2}\hbar\omega_l\frac{(a^* - \alpha)}{(-\varDelta H)} .\qquad(2.34)$$

Secondly, taking account of relaxation makes the specific laser energy dependent on the chemical reaction time. The quantity

$$\sigma_{21}\int_{t_0}^{t_1}\left(\int_0^t W(\tau)d\tau + \varDelta_{thr}\right)dt$$

in (2.32) characterizes the loss of energy due to relaxation processes. The finiteness of the quantity

$$\int_0^t W(\tau)d\tau$$

makes this loss vanishingly small as $t_1 \to 0$. Thus, as the reaction time is reduced (and the reaction rate increased accordingly), both the specific laser energy and chemical laser efficiency grow higher, approaching their ultimate values defined by (2.33) and (2.34). Analysis of (2.32) shows that $\varepsilon_l = \varepsilon_l^{lim}, \eta_{chem} = \eta_{chem}^{lim}$ at $t_0 < t_1 \ll (\alpha^* - \alpha)/\sigma_{21}$.

Thus, the energy parameters of a laser depend materially on the reaction mechanism and relaxation processes. Let us calculate the specific laser energy as a function of the system parameters for the most typical cases. It should be noted that the relaxation probability σ_{21} in general depends on time through the agency of the mixture component concentration. Consider the following two typical relaxation mechanisms: $\sigma_{21} = \text{const}$, which is possible where the different mixture components are equally effective in quenching the excited product or where the decisive contribution to the relaxation probability is made by a component whose concentration varies but little in the course of the reaction, and $\sigma_{21} = k_{21}^{(P)}[P]$, i.e., the main relaxation mechanism is the self-relaxation of the product. Let us calculate ε_l for each of these relaxation mechanisms for the cases where the chain reaction rate remains constant and where the chemical process is self-accelerating, i.e., where the temporal dependence of the reaction rate W can be represented in the form $W = W_0 e^{st}$. The mechanism responsible for the growth of the reaction rate with time can be of two types: chain explosion and thermal explosion. In the case of a branching chain reaction, the quantity s, known as the branching ratio, depends on the branching mechanism. Thus, if branching is of an energetic nature,

$$P^* + A \xrightarrow{k_{br}} P + 2N_1 \ ,$$

k_{br} being the branching constant, one may get the following expression [2.16] for the factor s deep inside the self-ignition region:

$$s_{ch} = (2k_{ch}k_{br})^{1/2}[A] \ .$$

In the case of thermal explosion, as follows from (2.24), $s_{ther} \simeq \sigma_3$, i.e., for a chain reaction, we have

$$s_{ther} = (1-\eta)\frac{Q}{C}\frac{\varepsilon}{RT_0^2}\frac{1}{[M]}W_0 \ ,$$

where $W_0 = 2k_{ch}[N][A] = 4k_{ch}f[A]^2$, f being the fraction of the molecules A dissociated into atoms under the effect of the initiating pulse.

And so, we have the following four typical cases, differing in the reaction and relaxation mechanisms:

1) $\sigma_{21} = \text{const}$, $W = W_0$. Disregarding Δ_{thr} in (2.8) and putting $t_0 = 0$, we get from (2.8), (2.31), and (2.32) the following relations for the lasing time, specific laser energy, and the effective chain length, respectively:

$$t_1 = \frac{\alpha^* - \alpha}{\sigma_{21}} = \frac{\alpha^* - \alpha}{k_{21}}\frac{1}{[M]} \ , \tag{2.35a}$$

$$\varepsilon_l = \frac{1}{4}\hbar\omega_l(\alpha^* - \alpha)^2\frac{W_0}{\sigma_{21}} = \hbar\omega_l(\alpha^* - \alpha)^2\frac{k_{ch}f\xi_A^2}{k_{21}}[M] \ , \tag{2.35b}$$

$$v_{eff} = \frac{1}{2}\frac{k_{ch}}{k_{21}}(\alpha^* - \alpha)^2\xi_A \ . \tag{2.35c}$$

These relations show that in the case of a simple chain reaction where the relaxation probability is independent of time, the lasing time does not depend on the reaction rate, the specific laser energy is proportional to the mixture pressure and the initiation level f and inversely proportional to the relaxation rate constant, and the effective chain length is independent of pressure. It is easy to see that in this case the efficiencies η_{chem} and η_{eng} do not depend on pressure, because in the conditions considered both the chemical and the initiating energy are proportional to pressure.

2) $\sigma_{21} = \text{const}$, $W = W_0 e^{st}$. It follows from (2.8) that if $s > \sigma_{21}/(\alpha^* - \alpha)$, the reaction rate "outstrips" the relaxation rate and the lasing time is in this case determined by the fuel store. The quantity t_1 can be estimated proceeding from the particle number conservation condition

$$\int_0^{t_1} W(t)dt = 2[A]_0 \tag{2.36}$$

(it is assumed that $[A]_0 \leqslant [C]_0$; otherwise $[A]_0$ in (2.36) should be replaced by $[C]_0$), whence

$$t_1 = \frac{1}{s} \ln\left(1 + \frac{2s[A]_0}{W_0}\right) = \frac{1}{s} \ln\left(1 + \frac{s}{2k_{\text{ch}}f\xi_A} \frac{1}{[M]}\right) . \tag{2.37a}$$

Taking the integral in (2.32), we get

$$\varepsilon_l = \frac{1}{2} \hbar\omega_l W_0 \left\{ \frac{\alpha^* - \alpha}{s}\left[1 - \frac{\sigma_{21}}{(\alpha^* - \alpha)s}\right](e^{st_1} - 1) + \frac{\sigma_{21}}{s} t_1 \right\} . \tag{2.37b}$$

In the limit $st_1 \to 0$, (2.37b) changes to (2.35b). The main contribution to the quantity ε_l is made by the first term in the braces. Taking into consideration (2.37a), ε_l may be written down in the form

$$\varepsilon_l = \hbar\omega_l(\alpha^* - \alpha)\left[1 - \frac{\sigma_{21}}{(\alpha^* - \alpha)s}\right][A]_0 . \tag{2.37c}$$

Comparing (2.37c) and (2.33), we find that

$$\frac{\varepsilon_l}{\varepsilon_l^{\text{lim}}} = \left[1 - \frac{\sigma_{21}}{(\alpha^* - \alpha)s}\right] . \tag{2.38}$$

Relation (2.38) indicates that the use of self-accelerating processes is especially promising for creating chemical lasers, for such processes allow the ultimate chemical efficiencies to be approached most closely.

3) $\sigma_{21} = k^{(P)}[P]$, $W = W_0$. For ε_l, we have

$$\varepsilon_l = \frac{1}{2} \hbar\omega_l\left[(\alpha^* - \alpha)W_0 t_1 - k_{21}^{(P)} W_0^2 t_1^3/3\right] . \tag{2.39}$$

The lasing time in the case of a simple chain reaction with self-relaxation of the reaction product is

$$t_1 = \frac{1}{2}\left[\frac{\alpha^* - \alpha}{k_{21}^{(P)} k_{\text{ch}}f}\right]^{1/2} \frac{1}{\xi_A[M]} , \tag{2.40a}$$

i.e., depends not only on the relaxation rate, as in Case 1, but also on the chain reaction rate. Taking into consideration (2.40a), we get

$$\varepsilon_l = \frac{1}{3}\hbar\omega_l\left[\frac{(\alpha^* - \alpha)W_0}{k_{21}^{(P)}}\right]^{1/2} \tag{2.40b}$$

and hence

$$\nu_{\text{eff}} = \frac{1}{3}\left[\frac{(\alpha^* - \alpha)k_{\text{ch}}}{k_{21}f}\right]^{1/2}. \tag{2.40c}$$

Thus, with the reaction mechanism considered here, the effective chain length, and hence the engineering laser efficiency, are inversely proportional to $f^{1/2}$, whereas ε_l and the chemical laser efficiency are directly proportional to $f^{1/2}$. Consequently, the degree of dissociation of A_2 should be chosen as a trade-off of ε_l against η_{eng}. If we take the condition $\eta_{\text{eng}} > 1$ to be the criterion of the veritably chemical laser, it can be satisfied with

$$\nu_{\text{eff}} > \frac{\hbar\omega_l}{\mathscr{E}_a}\eta_{\text{in}}.$$

Where relaxation is intense, it may turn out that $\nu_{\text{eff}} < 1$. In that case, no use at all is made of the process self-sustention effect and to achieve a sufficiently high inversion density requires a high initiation energy deposition.

4) $\sigma_{21} = k_{21}^{(P)}[P]$, $W = W_0 e^{st}$. Taking the integral in (2.32) yields

$$\varepsilon_l = \frac{1}{2}(\alpha^* - \alpha)\hbar\omega_l W_0\left\{\frac{1}{s}(e^{st_1} - 1) - \left[\frac{k_{21}^{(P)} W_0}{\alpha^* - \alpha s^2}\right]\right.$$

$$\left. \times\left[\frac{1}{2s}(e^{2st_1} - 1) - \frac{2}{s}(e^{st_1} - 1) + t_1\right]\right\}. \tag{2.41}$$

In the limit $st \to 0$, (2.41) changes to (2.39).

Let us consider first of all the case of moderate values of the ratio s:

$$\left[k_{21}^{(P)}\frac{W_0}{\alpha^* - \alpha}\right]^{1/2} < s < 2[A]_0\frac{k_{21}^{(P)}}{\alpha^* - \alpha}. \tag{2.42}$$

If $s < [k_{21}^{(P)} W_0/(\alpha^* - \alpha)]^{1/2}$, the self-acceleration of the process is unessential and this case is then considered similarly to the preceding one. With the magnitudes of the ratio s satisfying inequality (2.42), the relaxation rate at some moment "outstrips" the pumping rate, so that lasing is quenched before the substance is consumed completely. It is not very difficult to find that

$$t_1 = \frac{1}{s}\ln\left[1 + \frac{(\alpha^* - \alpha)s^2}{k_{21}^{(P)} W_0}\right]. \tag{2.43a}$$

In that case,

$$\varepsilon_l = \frac{1}{4}\hbar\omega_l(\alpha^* - \alpha)\frac{s}{k_{21}^{(P)}} \, , \tag{2.43b}$$

and so the specific laser energy is proportional to the factor s.

Finally, if

$$s > 2[A]_0\frac{k_{21}^{(P)}}{\alpha^* - \alpha} \, , \tag{2.44}$$

t_1 is determined by the time it takes for the substance to burn out, as given by (2.37a), and

$$\varepsilon_l = \hbar\omega_l(\alpha^* - \alpha)[A]_0\left\{1 - \frac{k_{21}^{(P)}[A]_0}{(\alpha^* - \alpha)s}\right\} \, . \tag{2.45}$$

It follows from (2.45) and (2.33) that

$$\frac{\varepsilon_l}{\varepsilon_l^{\lim}} = \left\{1 - \frac{k_{21}^{(P)}[A]_0}{(\alpha^* - \alpha)s}\right\} \, . \tag{2.46}$$

Note that expression (2.46) is a particular case of formula (2.38), where only one component contributes to relaxation.

Condition (2.44) is rather stringent. No branching chain reactions have yet been found that would satisfy this requirement. But the thermal acceleration of the chemical process is essential, for example, for the hydrogen halide lasers considered in Chaps. 3–5.

So, in this section we have analyzed the relationships between the energy characteristics and kinetic parameters of chemical lasers and established some formal criteria that a given reaction system must meet for it to be considered potentially suitable for the development of a chemical laser. In conclusion, let us formulate qualitatively the main requirements.

First, reactions are required with highly exothermic acts (i.e., elementary processes or reactions) and having a considerable proportion of their energy manifest in the form of nonequilibrium excitation of the fragments formed. Secondly, the ratio between the reaction rate and the rate of relaxation of the excited states must be high, otherwise the free-energy of the reactants will mainly be expended in the thermal heating of the mixture. Thirdly, to surpass the self-excitation threshold and ensure an acceptable specific power of the induced radiation, a sufficiently high absolute reaction rate is necessary. And finally, it is necessary to have a self-sustaining reaction mechanism providing gain in the ratio of the useful radiation energy to the energy consumed on initiating the reaction, and in the ideal case allowing a self-contained laser system to be realized, depending on chemical energy alone.

2.2 Nonequilibrium Excitation in Chemical Reactions

The main specific feature of the chemical laser is the production of an inverted population of the lasing levels on account of nonequilibrium excitation of the internal degrees of freedom of the molecules formed in the course of a reaction. A great variety of reactions are known today that lead to the formation of molecules with a large internal energy store. Detailed reviews of investigations in this field can be found in [2.21, 22]. Given below is a classification of the types of reactions accompanied by rotational, vibrational, and electronic excitation [2.22].

Rotational Excitation. Nonequilibrium rotational excitation is fairly common and, as a rule, occurs concurrently with vibrational and electronic excitation. But the high rates of the relaxation processes involved make rotational transitions energetically unpromising for creating chemical lasers[1].

Vibrational Excitation. 1) Exchange reactions taking place with the participation of atoms and polyatomic groups:

$$A + BC \rightarrow AB^v + C \ .$$

2) Exchange reactions taking place with the formation of four-center complexes:

$$AB + CD \rightarrow AC^v + BD \ .$$

3) Recombination reactions involving vibrational excitation:

$$A + B \rightarrow AB^v \ ,$$

$$AB^v + M \rightarrow AB + M \ .$$

4) Decomposition reactions involving vibrational excitation:

$$A \rightarrow B^v + C$$

Electronic Excitation. 1) Radiative recombination of atoms:

$$A + B \rightarrow AB^e \rightarrow AB + h\nu \ .$$

2) Excitation of third particles upon recombination of atoms:

$$A + B + C \rightarrow AB + C^e \ .$$

---

[1] This does not mean that lasing on rotational transitions excited in the course of chemical reactions cannot be realized in principle. What is more, such lasing has been observed, in [2.36–38]. We only wish to emphasize here that one can hardly expect high energy performance from lasers using chemically pumped rotational transitions. Nevertheless, in view of the extensive possibilities of using laser light in the submillimeter and far infrared regions, great interest is being shown in lasers relying on rotational transitions.

3) Radiative recombination of atoms with binary molecules:

$$A + BC \rightarrow ABC^e \rightarrow ABC + hv \ .$$

4) Exchange reactions taking place with the participation of atoms:

$$A + BC \rightarrow AB^e + C \ ,$$

$$A + BC \rightarrow AB + C^e \ ,$$

$$A + BCD \rightarrow AB^e + CD \ .$$

The symbols v and e in the reaction equations denote vibrational and electronic excitation, respectively.

Consequently, pumping in chemical lasers can, in principle, excite electronic, vibrational-rotational, and purely rotational lasing transitions. One can expect that electronic population inversion will be observed fairly seldom, for electronic transitions are characterized by rather high energies as compared with the heat of reaction. Rotational inversion is energetically possible in many reactions, but even if it occurs, it is difficult to realize because of the high rate at which rotational-translational equilibrium is established. The use of vibrational excitation allows both these difficulties to be avoided to a certain extent. In the case of vibrational excitation, only moderate energies are necessary, and it frequently takes many thousands of molecular collisions for vibrational equilibrium to be established. Moreover, an inverted population of vibrational levels can, in a number of instances, be produced in the process of vibrational relaxation. Thus, there are rich possibilities of developing chemical lasers with vibrational population inversion.

Table 2.1 lists energy distributions obtained experimentally for a number of important elementary chemical reactions accompanied by nonequilibrium vibrational excitation of molecules (hydrogen halides, carbon monoxide). It is seen from these data that the fraction of the heat of reaction accounted for by vibrational excitation of newly formed bonds in exchange reactions can be fairly considerable and that inversed excitation is realized for some pairs of vibrational levels. As we have already seen, both these facts, as well as high rates of elementary lasing processes are essential to lasing under chemical pumping. Another important circumstance drawing attention to elementary exchange reactions is the fact that many chain processes include elementary lasing reactions as chain links. Among these are, for example, reactions of halogens with hydrogen. It is important that the rate of chain reactions is proportional to both the concentration of free atoms (active centers) and that of the starting reactants. Therefore, if the reactant concentration is high enough, high chemical pumping rates can be attained with moderate concentrations of free atoms.

To achieve the necessary pumping rate in the case of atomic recombination reactions, much higher concentrations of free atoms or radicals must be produced, which in itself is a difficult problem. What is more, atomic recombination, like decomposition reactions, is not a chain process. It is clear from the foregoing that exchange reactions accompanied by nonequilibrium vibrational excitation meet most fully the requirements for the "lasing" reactions in chemical lasers. It is no accident that the first successful experiments and the most important advances in the field of chemical lasers are associated with the use of reactions of this type (see Chap. 3). Further in this chapter we will restrict ourselves to chemical lasers depending for their operation on vibrational molecular transitions. The data in Table 2.1 on vibrational energy distribution have been obtained by two experimental techniques.

The first – the method of studying infrared chemiluminescence in a flow of reacting gases, developed by *Polanyi* and co-workers – is at present the main and as yet the most reliable technique, as it allows the excited molecule and its energetic state to be monitored directly.

To study the distribution of vibrational energy in the reaction $A + BC \rightarrow AB(v)$ $+ C$, the atoms A are mixed in a flow with the molecules BC and then the relative radiation intensities are measured in the different absorption bands of AB. The populations of the vibrational levels of AB are then calculated, using the Einstein coefficients, from the intensity data. The measurements and extraction of information on the primary distribution are carried out by two methods: the method of "frozen relaxation", in which the spectral line intensities are measured before the effect of vibrational relaxation becomes manifest, and the method of "measured relaxation", where population data are extracted at several points along the flow and then extrapolated to a point at the beginning of the flow in order to take into account the effect of relaxation processes. Note that such experiments provide only information about the relative populations $AB(v)$ and cannot furnish any data on the proportion q of molecules formed at the zero level. It follows from additional information on absorption that the number of unexcited molecules $AB(0)$ produced in the case of hydrogen halides is, as a rule, negligibly small.

The main difficulty with this technique is to eliminate, or take into consideration (if possible), the vibrational relaxation process. To do so, one has to experiment at low gas pressures, which places heavy demands on the optical system and instrumentation used. Figures 2.4 and 2.5 show distributions of the excitation rates for various vibrational-rotational levels of the HF molecules formed in the reactions $F + H_2$ and $H + F_2$, obtained by the method of studying infrared chemiluminescence.

Another technique, used for the first time in [2.24, 25, 133], consists in examining the relative intensities of chemical laser lines. The technique makes it possible under certain conditions to obtain the ratio between the partial rate constants for the formation of excited molecules in two or more vibrational states participating in lasing, and so it can only help in the investigation of inverted energy distributions. Studies into vibrational energy distribution in crossing molecular beams investigating the angular distribution of reaction products are discussed in [2.26].

The current status of the theory does not, as a rule, allow the distribution of energy among the various energy states of reaction products to be obtained by way of an *a priori* calculation. Nevertheless, there are a number of useful models on the basis of which an analysis has been made of reactive collisions, and some general inferences have been drawn about the relationships between the distribution of

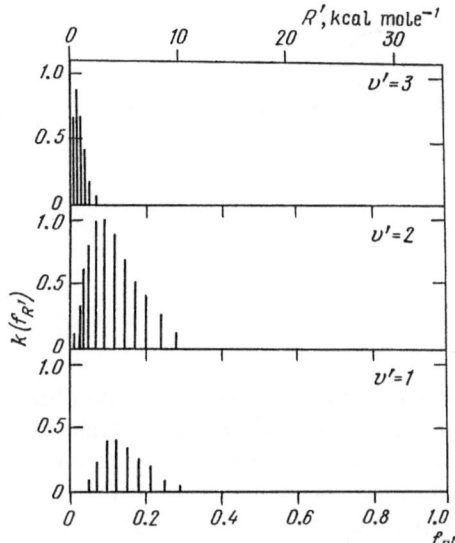

Fig. 2.4. Rate constants $k(f_{R'})$ [or $k(R')$ on the top scale] for the reaction $F + H_2 \rightarrow HF(v') + H$, [2.134]

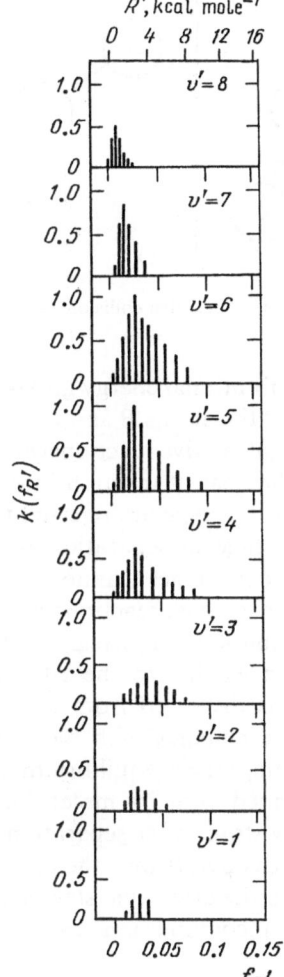

Fig. 2.5. Rate constants $k(f_{R'})$ [or $k(R')$ on the top scale] for the reaction $H + F_2 \rightarrow HF(v') + F$, [2.152]

energy in reaction products and the kinematic and energy characteristics of the reactants. These questions are treated in [2.22, 27, 240].

Figure 2.6 shows a schematic picture of the reaction between the F atom and H_2 molecule. The fluorine atom approaches the hydrogen molecule. Sometimes the chemical bond between the hydrogen atoms in the molecule is shifted upon collision from a hydrogen atom to the fluorine atom; after such a collision the fluorine atom links up with the hydrogen atom and the other hydrogen atom recedes alone.

The reactive collision process $A + BC$ described above can be schematically divided into three stages: (1) the approach of the atom A to the complex AB, (2) the intermediate stage wherein the approach continues but the bond B–C starts stretching, and (3) the separation of the reaction products AB and C. Let $\varepsilon_1, \varepsilon_2$, and ε_3 denote the energies released in these stages, respectively. Calculations show that the energy ε_1 manifests itself in the form of vibrations of AB, and the energy ε_3, in the form of the translation of the products. The fraction of the energy ε_2 that is

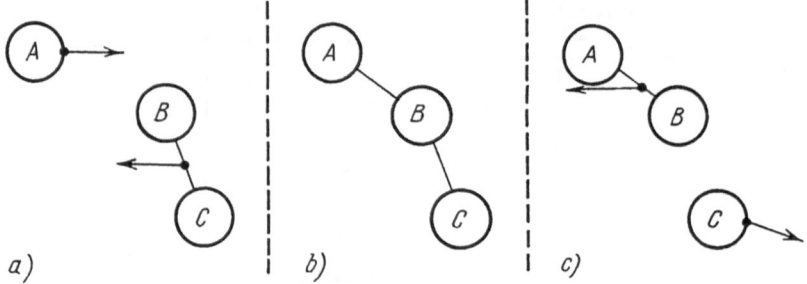

Fig. 2.6a–c. Illustrating a reactive collision: (**a**) prior to collision; (**b**) upon collision; (**c**) after collision

converted into vibrational energy grows with the increasing ratio of the mass of the atom A to that of the atoms B and C. Reactions in which energy is released mainly in the first stage (attractive reactions) are characterized by the conversion of a large proportion of the heat of reaction into the vibrations of the newly formed bond. Reactions of another type are referred to as repulsive reactions: energy in them is released in the stage of separation of the reaction products. Finally, reactions wherein the above stages cannot be separated are called mixed type. This classification of exchange reactions was suggested by *Polanyi* and *Rosner* [2.28].

From the standpoint of physics, we may use the following intuitive picture of a reactive collision leading to the vibrational excitation of AB [2.29]. When the chemical bond is displaced from BC to AB, there occurs a stepwise electronic change (charge exchange), with the atoms A and B being at that moment at a distance exceeding their equilibrium separation in the molecule AB. So, the molecule is formed with its molecular bond being stretched, and hence has a potential energy store. After separation of the products AB and C, this stretching potential energy converts into the energy of vibrations along the new bond.

According to the electronic step change model, the reaction $A + BC$ described above is an ion recombination reaction:

$$A + B - C \rightarrow A^+ \ldots B^- \ldots C \rightarrow A^+ B^- + C \ .$$

A graphic term – "harpooning" – has been proposed to describe this process [2.27, 29]: the attacking atom A throws out its valence electron, gets hold of the atom B, and attracts it, on account of the Coulomb interaction.

The development of research on chemical lasers has aroused considerable interest in the simple exchange reactions giving rise to hydrogen fluoride, namely, the reaction of atomic fluorine with molecular hydrogen and that of atomic hydrogen with molecular fluorine. These reactions are at the root of a very efficient chemical laser [2.16]. The harpooning model is inapplicable to reactions of the type $F + H_2$, and they are calculated in the adiabatic approximation (motion over a single potential energy surface). We would like to draw the reader's attention to the works by *Wilkins* [2.30, 31], who carried out, in order to determine the dynamics of the reactions $F + H_2 \rightarrow HF + H$ [2.30] and $H + F_2 \rightarrow HF + F$ [2.31], a classical calculation of three-dimensional trajectories, making use in the description of the interatomic interaction by force of a semiempirical potential-energy surface. The initial data for each trajectory were selected using the Monte-Carlo technique. Based on the analysis of a large number of these trajectories, rate constants were calculated for the reactions yielding the hydrogen fluoride molecules in various vibrational-rotational states. The results of these semiempirical calculations fit the available kinetic information, both as regards the form of the nonequilibrium distribution of the HF molecules among their vibrational levels and the proportion of the energy of the reactions localized in the vibrations of the molecules being formed. The investigations carried out to date can serve as a basis for launching systematic studies into the factors influencing the distribution of energy in elementary exchange reactions and stimulate the development of more rigorous quantum-mechanical approaches [2.32–35].

Levine and *Bernstein* [2.39] have developed a theoretical-informational approach to the description of energy distribution in chemical reactions, which makes it possible to simplify the presentation of nonequilibrium distributions among the states of the product. The key notion in this approach is the concept of "surprisal", deviation of the actually observed distribution from the one expected on the basis of microcanonical equilibrium. The quantitative measure of the surprisal of some or other event has been suggested by *Shannon* and *Weaver* [2.40] in their fundamental paper on the theory of information.

Let $P^0(A)$ be the *a priori* probability of the event A, which reflects any *a priori* information we can have about the distribution of events. The surprisal $I(A)$ of the event A is defined as

$$I(A) = -\ln\left[\frac{P(A)}{P^0(A)}\right] .$$

If the event A is observed to occur with the probability expected [i.e., if $P(A) = P^0(A)$], then $I(A) = 0$. The surprisal $I(A)$ is, therefore, the measure of deviation of the observed probability $P(A)$ from the probability $P^0(A)$ expected *a priori*, so that

$$P(A) = P^0(A)\exp[-I(A)] .$$

Instead of providing information about the probability of an event, one can furnish data on its surprisal – deviation from the expected. Let us apply this idea to the distribution of the products of a reaction among vibrational states. With the total energy known and no additional *a priori* information available, all the possible quantum states of a product can be considered equally probable, because there is little probability that some strict selection rules exist in the process of chemical reaction. The total energy E of the colliding particles is a conserved quantity. If the reaction products are an atom and a binary molecule, then $E = T + V + R$, where T is the translational energy of the products and V and R are, respectively, the vibrational and rotational energies of the molecule being formed. Let us introduce the ratios

$$f_V = \frac{V}{E} , \quad f_T = \frac{T}{E} , \quad \text{and} \quad f_R = \frac{R}{E} = 1 - f_V - f_T ,$$

i.e., the total energy fractions distributed among the vibrational, translational, and rotational degrees of freedom of the products. The number of the quantum states of the products, which correspond to the vibrational level v, the rotational level J, and the translational energy in the interval from T to $T+dT$ is equal to $A(2J+1)\,T^{1/2}\,dT$, where A is the normalization constant. The rotational energy $R = BJ(J+1)$ can be regarded as a continuous variable. The number of the quantum states corresponding to the total energy E and the vibrational level v can be found by integration with respect to all R and T such that $R+T=E-V$:

$$P^0(V)=\int dR \int dT A' T^{1/2}\,\delta(E-V-R-T)$$

$$= A' \int_0^{E-V} dR\,(E-V-R)^{1/2} = A''(E-V)^{3/2} \, ,$$

where A' and A'' are combinations of constants. Since

$$\int_0^1 P(f_V)\,df_V = 1 \, ,$$

we get

$$P^0(f_V) = \tfrac{5}{2}(1-f_V)^{3/2} \, . \tag{2.47}$$

The main quantitative conclusion from (2.47) is that the *a priori* $P^0(V)$ is a decreasing function of V, which is typical of the statistical reaction models wherein the final quantum states are always assumed to be equally probable. As this conclusion is not necessarily confirmed by observations, vibrational states are determined not only by statistics, but, to some extent, also by the dynamics of collisions. Based on the *a priori* distribution $P^0(V)$, one may find the surprisal of the actually observed populations by the formula

$$I(f_V) = -\ln\left[\frac{P(f_V)}{P^0(f_V)}\right] . \tag{2.48}$$

To illustrate (2.48), Figs. 2.7 and 2.8 present "vibrational surprisal" curves for the reactions $F+HBr$ and $O+CS$, along with the curves of the observed and expected populations $P(f_V)$ and $P^0(f_V)$.

It can be seen that the vibrational surprisal is a linear function of f_V. The slope of the curve $I(f_V)$ is determined by the parameter $\lambda_v = dI(f_V)/df_V$. The linearity of $I(f_V)$ is characteristic of many exchange reactions, but is not a universal rule. Exceptions are, for example, reactions of the type $H+X_2 \rightarrow HX+X$, where X is a halogen atom. But in this case, too, the variation of $I(f_V)$ is much more simple and regular than the variations of the populations themselves [2.41]. Thus, for reactions of the form $X+LY \rightarrow LX+Y$ (X and Y denoting some halogen atoms and L, an H or D atom), as well as for the reaction $O+CS$, we have

$$P(V) = P^0(V)\frac{\exp(-\lambda_v f_V)}{\exp(\lambda_0)} . \tag{2.49}$$

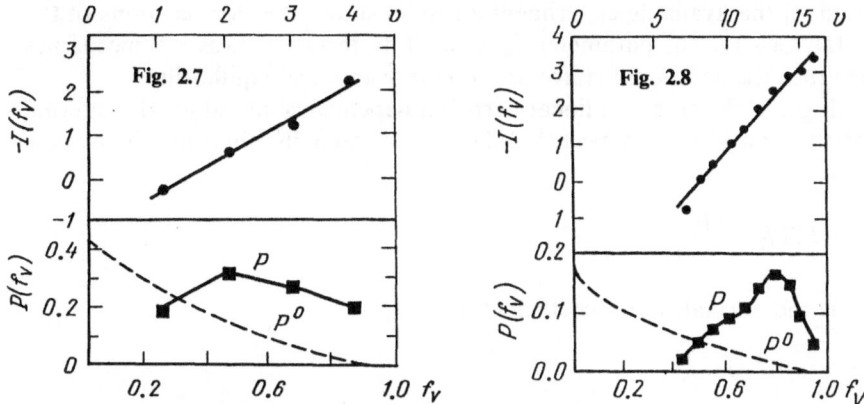

Fig. 2.7. Analysis of the surprisal of the product distribution among vibrational states for the reaction $F + HBr \rightarrow HF(v) + Br$; $\lambda_v = -4$ [2.39]. *Top*: vibrational surprisal plot; *bottom*: comparison between the observed population probabilities (P) and the expected ones (P^0)

Fig. 2.8. Surprisal analysis for the reaction $O + CS \rightarrow CO(v) + S$; $\lambda_v = -7.7$ [2.39]

Since $P^0(V)$ is a decreasing function of V, then λ_v is negative. Setting λ_v is sufficient for the description of the function $P(V)$. In this sense, λ_v (or, to be more exact, λ_v^{-1}) plays the part of a temperature-like parameter. Note that $\exp(\lambda_0)$ performs the role of a statistical sum:

$$Q_V = \exp(\lambda_v) = \sum_V P^0(V) \exp(-\lambda_v f_V) \; .$$

To analyze the rotational distribution at a given vibrational level, it is necessary to investigate the (conditional) surprisal

$$I\left(\frac{R}{V}\right) = -\ln\left[\frac{P(R/V)}{P^0(R/V)}\right] \; ,$$

where $P(R/V)$ and $P^0(R/V)$ are the rotational energy distributions at a given vibrational energy. In the simple linear "rotational surprisal" case

$$I\left(\frac{R}{V}\right) = \text{const} + \theta_R g_R \; ,$$

where $g_R = f_R/(1 - f_R)$ and θ_R is a rotational temperature-like parameter, the joint vibrational-rotational population distribution takes the form

$$P(R, V) = P\left(\frac{R}{V}\right) P(V) = P^0\left(\frac{R}{V}\right) P^0(V) \frac{\exp(-\lambda_v f_V - \theta_R g_R)}{Q_{VR}} \; , \tag{2.50a}$$

Q_{VR} being the normalization coefficient,

$$P^0\left(\frac{R}{V}\right) \sim f_T^{1/2} = (1 - f_V - f_R)^{1/2} = (1 - f_V)^{1/2}(1 - g_R)^{1/2} \; . \tag{2.50b}$$

The analysis of the available experimental results shows that for reactions of the type $Cl + HI$, $Cl + DI$, the parameter θ_R is small. In terms of statistical mechanics this means that the rotational states are in microcanonical equilibrium.

By analogy, in the case of a linear surprisal dependence found for the internal energy distribution, one can expect that the translational distribution will have the form

$$P(f_T) = P^0(f_T) \frac{\exp(-\lambda_T f_T)}{Q_T} ,$$

where Q_T is the normalization coefficient, or

$$I(f_T) = \lambda_0 + \lambda_T f_T ,$$

$P^0(f_T)$ being given by

$$P^0(f_T) = \tfrac{15}{4} f_T^{1/2}(1 - f_T) .$$

It can be demonstrated that in the limit $\theta_R \to 0$ and at not very large λ_v (for the reactions considered, λ_v ranges between -5 and -10), $\lambda_T \simeq -\lambda_v/2$. So, those reactions which are accompanied by population inversion and are characterized by negative λ_v will have positive λ_T.

It is of interest to find the average value of surprisal of a distribution, i.e., the average deviation of the observed distribution from the anticipated one. It turns out that the average surprisal coincides with the entropy of the distribution. In the case of vibrational levels, the distribution entropy is

$$S^{(\text{vib})} = - R \sum_V P(V) \ln \left[\frac{P(V)}{P^0(V)} \right] ,$$

where R is the gas constant. It is clear that $S^{(\text{vib})} \leqslant 0$, the equality holding at $P(V) = P^0(V)$. The deficit of entropy may be defined as the non-negative quantity

$$\Delta S^{(\text{vib})} = S^{0(\text{vib})} - S^{(\text{vib})} ,$$

where $S^{0(\text{vib})}$ is obtained from $S^{(\text{vib})}$ at $P = P^0$. The deficit of entropy is a convenient measure of the deviation of an observed distribution from the expected one. This measure is independent of whether the surprisal dependence is linear or not.

Table 2.2 lists the results of the analysis of the reactions $Cl + HI$ and $Cl + DI$. The considerable vibrational population inversion in these reactions is reflected by both high negative values of λ_v and large deficits of entropy. The introduction of the notion of the deficit of entropy provides us with an important conceptual tool. Any experiment on molecular dynamics involves studies of numerous independent, individual collisions. The observed distribution of the product populations will be that which conforms to the minimum deficit of entropy (at a given total energy released in the reactive collision).

The theoretical-informational approach can be used not only in the analysis of the details of energy release in the course of reactions, but also when examining the selectivity of energy absorption processes and studying the kinetics of energy transfer processes [2.42–44].

2.3 Population Inversion and Amplification of Radiation in Vibrational-Rotational Transitions

2.3.1 Amplification Conditions. Total and Partial Population Inversion

The following simplified vibrational-level model is adequate for the analysis of a number of questions relating to the operation of the chemical laser. The molecular energy is determined by the formula

$$E_{v,J} = E_v + F(J)hc \; , \tag{2.51}$$

where v and J are the vibrational and rotational quantum numbers, respectively, h is Planck's constant, c the velocity of light, and E_v the vibrational energy of the vth level. For the diatomic molecules of primary interest to us, we have (account being taken of anharmonicity of vibrations)

$$E_v = \omega_e v - \omega_e x_e v(v+1) \; , \tag{2.52a}$$

$$F(J) = B_v J(J+1) \; ,$$

$$B_v = B_e - \alpha_e(v + \tfrac{1}{2}) \; , \tag{2.52b}$$

where ω_e, $\omega_e x_e$, B_e, and α_e are the well-known spectroscopic constants [2.45]. It is assumed that the rotational-level system is in equilibrium, so that the population $n_{v,J}$ of the level v, J is described by the Boltzmann formula

$$n_{v,J} = g_J \frac{N_v}{Q} \exp\left[-\frac{\theta_r}{T} J(J+1) \right] \; , \tag{2.53}$$

where g_J is the statistical weight of the state being considered, $\theta_r = B_v hc/k$ is the characteristic rotational temperture, k the Boltzmann constant, and Q the rotational statistical sum. For linear molecules at sufficiently high temperatures

$$Q = T/\sigma\theta_r \; , \tag{2.54}$$

where σ is the symmetry factor equal to unity for molecules having no inversion center and two for centrally symmetric molecules. Expression (2.54) holds true at $Q \gg 1$.

Amplification of radiation in the transition v', $J' \to v$, J is possible if the inequality

$$\Delta_{v,J}^{v',J'} = n_{v',J'} - \frac{g_{J'}}{g_J} n_{v,J} > 0 \tag{2.55}$$

is satisfied. Relation (2.55) imposes restrictions on the possible spectral composition of the radiation being amplified in vibrational-rotational transitions.

Let us apply relations (2.51)–(2.55) to analyze the transitions from the level $v' = v + 1$, $J' = J + m$ to the level v, J, where $m = -2, -1, 0, 1$, and 2 for the O-, P-, Q-, R-, and S-branch, respectively. In the case of dipole emission, $\Delta J = 0, \pm 1$, and in the general case, there are observed three branches (P, Q, and R). But for most two-atom molecules in the electronic ground state (namely, for molecules having zero electronic momentum, i.e., being in the state $^1\Sigma$), the transitions with $\Delta J = 0$ are additionally forbidden, and so there are observed only two branches $-R$ and P [2.45]. The statistical weights of the states v, J and $v + 1, J + m$ are

$$g_J = 2J + 1 \quad \text{and} \quad g_{J+m} = 2J + 1 + 2m \, . \tag{2.56}$$

Then

$$\Delta_{v,J}^{v+1,J+m} = \frac{2}{Q}\left[\frac{1}{2} + J + m\right]\left\{N_{v+1}\exp\left[\frac{-F(J+m)\,hc}{kT}\right]\right.$$

$$\left. - N_v\exp\left[\frac{-F(J)\,hc}{kT}\right]\right\} \, . \tag{2.57}$$

We have from formula (2.55) the following amplification condition:

$$\frac{N_{v+1}}{N_v} > \exp\left\{[F(J+m) - F(J)]\,hc\,\frac{1}{kT}\right\} = \exp\left[m(2J+1+m)\frac{\theta_r}{T}\right] \, . \tag{2.58}$$

At $N_{v+1}/N_v > 1$ (total inversion), amplification is possible for the P- and Q-branch transitions and for the three R-branch transitions for which $J < J_{\max}$, where $J_{\max} = (1/2)(T/\theta_r)\ln(N_{v+1}/N_v) - 1$.

At $N_{v+1}/N_v < 1$ (partial inversion) amplification is possible for the P-branch transitions with $J > J_{\min}$, where $J_{\min} = (1/2)(T/\theta_r)\ln(N_v/N_{v+1})$, but not for the Q- and R-branch transitions. If we introduce the vibrational temperature T_v defined by the formula

$$T_v = \theta_v \ln^{-1}\left(\frac{N_v}{N_{v+1}}\right) \, , \tag{2.59}$$

where $\theta_v = (E_{v+1} - E_v)/k$ is the characteristic vibrational temperature, the amplification condition for the P-branch transitions may be written as

$$J > \frac{1}{2}\frac{\theta_v}{\theta_r}\frac{T}{T_v} \, . \tag{2.60}$$

Condition (2.60) bounds J only below. The upper limit to J, to which amplification of radiation is possible, is defined by the population of rotational sublevels that must be large enough for the electromagnetic wave to interact effectively with the medium. Another condition limiting the magnitude of J in the P-branch is the energetic condition $E_{v+1,J-1} > E_{v,J}$, equivalent to the inequality $J < (1/2)(\theta_v/\theta_r)$. The latter inequality becomes invalid only at J in excess of 10^2. As a rule, the region of sufficiently populated rotational sublevels corresponds to much lower values of J.

When deriving condition (2.60), it has been assumed that rotational levels are in equilibrium, and the ratio between the populations of the upper and lower vibrational levels has been defined in terms of the vibrational temperature T_v. Generally speaking, the vibrational temperature in the laser medium differs from one pair of levels to another, and so a large number of parameters need to be introduced for multilevel systems.

It is interesting to derive the amplification condition for the extreme case where the vibrational-rotational distribution is formed by the chemical reaction itself and is not distorted by relaxation processes. As shown in Sect. 2.2, such distributions can be characterized by one or two temperature-like parameters. If the inversion density is expressed in terms of these parameters, condition (2.55) will enable one to predict the possible emission spectrum of a given chemical laser. Based on formulas (2.49) and (2.50), the distribution of the product among vibrational and rotational levels at $\theta_R = 0$ may be described by the relation

$$n_{v,J} = P(f_v, J)n = nQ_{VR}^{-1}(2J+1)[1-f_v-f_J(v)]^{1/2}\exp(-\lambda_r f_v) ,$$

where $f_J(v)$ is the fraction of energy in rotation, n the product concentration, and $f_v = E_v/E$. The pre-exponential in this formula is proportional to the density of states of an atom-binary molecule system with a total energy of E, the molecule being in the state v, J. Disregarding the rotational relaxation, we may obtain the following expression for the inversion density [2.236]:

$$\Delta_{v,J}^{v+1,J-1} = c(2J-1)\exp(-\lambda_v f_v)\{q_v^{-1/2}[S_{v+1}-J(J-1)]^{1/2}$$
$$-[S_v-J(J+1)]^{1/2}\} ,$$

where c is the proportionality constant and q_v and S_v are dimensionless parameters defined as

$$q_v = \left(\frac{b_v}{b_{v+1}}\right)^{1/2}\exp(2\lambda_v \Delta f_v) , \qquad S_v = \frac{1-f_v}{b_v} ,$$

where $\Delta f_v = f_{v+1}-f_v$ and $b_v = B_v/E$.

The number J corresponding to the maximum of the distribution function $P(f_v, J)$ is equal to $(S_v/2)^{1/2}$. The maximum value of J in the P-branch for which condition (2.55) is fulfilled, i.e., for which inversion remains positive, is defined by the formula

$$J_m = (2r_v)^{-1}\left[1+\left(1+4r_v^2\frac{S_{v+1}-q_v S_v}{1-q_v}\right)^{1/2}\right]\simeq(2r_v)^{-1}+\left[\frac{S_{v+1}-q_v S_v}{1-q_v}\right]^{1/2} ,$$

where $r_v = (1-q_v)/(1+q_v)$.

For the reaction $F+H_2$, $E = -\Delta H + \varepsilon + (5/2)RT = 34.7$ kcal/mole $= 12\,150$ cm^{-1}, $\lambda = -6.5$, and so in the bands 1–0, 2–1, and 3–2, we have $J_m = 20$, 15, and 7, respectively. In the emission spectrum of the HF-laser pumped by the same reaction, there are observed transitions with the values of the number J ranging between 1 and 17 in the 1–0 band, between 1 and 16 in the 2–1 band, and between 2 and 8 in the 3–2 band, which agrees almost completely with the predicted transition range.

Finally, consider the situation where the distribution among vibrational levels is described by formula (2.49) and the rotational distribution corresponds to the equilibrium one (fast rotational relaxation). In that case, for $n_{v,J}$, we may write

$$n_{v,J} \sim (1-f_v)^{3/2}(2J+1)\exp[-\lambda_v f_v - \beta b_v J(J+1)] .$$

The maximum of $n_{v,J}$ corresponds to $J \simeq (2\beta b_v)^{-1/2} - (1/2)$. Amplification condition (2.55) in this case is equivalent to the inequality (P-branch)

$$- \lambda_v \Delta f_v + \beta J \left\{ 2b_v + \alpha(J-1) - \frac{3}{2} \ln \left[1 - \frac{\Delta f_v}{1-f_v} \right] \right\} > 0 \ ,$$

where $\alpha = \alpha_e/E$ and $\beta = E/kT$.

For the R-branch ($J+1 \to J$), the second term in this formula is replaced by $-\beta(J+1)[2b_v - \alpha(J+2)]$, which is negative except for very large J (the respective levels are practically unpopulated). This limits the possible values of J. In the case of the reaction $F + H_2$, the amplification condition is satisfied for all the P-branch transitions, even in the 3–2 band wherein the upper vibrational level is less populated than the lower one.

2.3.2 Gain

The small-signal gain in the transition $v', J' \to v, J$ is defined by the formula

$$\alpha_{v,J}^{v',J'} (v) = \sigma_{v,J}^{v',J'} \Delta_{v,J}^{v',J'} \ , \quad \text{where} \tag{2.61}$$

$$\sigma_{v,J}^{v',J'} = \frac{c^2}{8\pi} A_{v,J}^{v',J'} S(v) v^{-2} \tag{2.62}$$

is the induced transition cross section at the frequency v. The normalized form factor at the emission line center of the vibrational-rotational transition ($v = v_{v,J}^{v',J'} = v_0$) is given by

$$S(v_0) = \left(\ln \frac{2}{\pi} \right)^{1/2} \frac{1}{\Delta v_D} \exp(a^2) \operatorname{erfc}(a) \ , \quad \text{where} \tag{2.63}$$

$$a = \frac{\Delta v_L}{\Delta v_D} (\ln 2)^{1/2} = 0.833 \frac{\Delta v_L}{\Delta v_D} \ . \tag{2.64}$$

Formula (2.63) allows for the joint action of the broadening mechanisms responsible for both the Lorentzian and the Doppler emission line shapes. At $a \gg 1$, when the line width is determined by collisions (in practice at $a > 4$), asymptotic expansion of (2.64) yields

$$S(v_0) = 1/\pi \Delta v_L \ .$$

Conversely, at $a \ll 1$

$$S(v_0) = \left(\ln \frac{2}{\pi} \right)^{1/2} \frac{1}{\Delta v_D} \ .$$

When making practical calculations, it is convenient to express the error integral as [2.46]

$$\operatorname{erfc}(a) = f(a) \exp(-a^2) \ , \tag{2.65}$$

where for $a > 1.5$, expansion into the following rapidly converging continued fraction is valid:

$$f(a) = \frac{1}{\sqrt{\pi}} \cfrac{1}{a + \cfrac{1/2}{a + \cfrac{1}{a + \cfrac{3/2}{a + \cfrac{2}{a + \ldots}}}}} \;,$$

and for $a < 1.5$,

$$f(a) = \frac{1}{\pi^{1/2}} (b_1 t + b_2^2 t^2 + b_3^3 t^3) \;,$$

where $t = 1/(1 + b_0 a)$, $b_0 = 0.47047$, $b_1 = 0.61686$, $b_2 = -0.16994$, and $b_3 = 1.32554$. Considering (2.65), the quantity $S(v_0)$ takes the form

$$S(v_0) = \frac{a}{\pi^{1/2} \Delta v_L} f(a) \;. \tag{2.66}$$

In the general case, for the quantum transition $u \to l$, the collisional half-width Δv_L of the molecular species A colliding with the molecular species B is defined as [2.47]

$$\Delta v_L = \frac{1}{4\pi} (2Z_u^{A-B} + 2Z_l^{A-B}) \;, \quad \text{where} \tag{2.67}$$

$$Z_j^{A-B} = 3.37 \times 10^{10} \sigma_A^j \sigma_B \left(\frac{1}{M_A} + \frac{1}{M_B} \right)^{1/2} \frac{p_B}{T^{1/2}} \;, \quad j = u, l$$

in compliance with the gas-kinetic theory. In formula (2.67), $Z_u^{A-B}(Z_l^{A-B})$ is the collision frequency $[\text{s}^{-1}]$, σ_A and σ_B are the optical diameters of the colliding molecules $[\text{Å}]$, M_A and M_B the respective molecular masses $[\text{g/mole}]$, and p_B is the total pressure $[\text{atm.}]$ of the broadening component B, which is related to the particle density N_B $[\text{cm}^{-3}]$ and the absolute temperature T $[\text{K}]$ by the relation $N_B = 7.34 \times 10^{21} p_B/T$. Assuming that $Z_u^{A-B} = Z_l^{A-B} = Z^{A-B}$, we have

$$\Delta v_L = Z^{A-B}/\pi \;. \tag{2.68a}$$

Relation (2.68a) can naturally be generalized to the case of multicomponent mixtures:

$$\Delta v_L = \frac{1}{\pi} \sum_B Z^{A-B} \;. \tag{2.68b}$$

The Doppler half-width Δv_D is defined by the formula

$$\Delta v_D = \frac{v_0}{c} \left(2 \ln 2 k N_A \frac{T}{M} \right)^{1/2} = 3.581 \times 10^{-7} v_0 \left(\frac{T}{M} \right)^{1/2} \;, \tag{2.69}$$

where M is the molecular weight and N_A the Avogadro number. With the particle density being specified, $\Delta v_L \sim T^{1/2}$, and so in that case the parameter a is independent of temperature.

The Einstein coefficient $A_{v,J}^{v',J'}$ (s^{-1}) for spontaneous emission in transition from the upper level (v', J') to the lower level (v, J) is

$$A_{v,J}^{v',J'} = \frac{64\pi^4 v^3}{3\hbar c^3} \frac{1}{(2J'+1)} |\mathcal{M}|^2 \, , \tag{2.70}$$

where $\mathcal{M}$ is the dipole-moment matrix element for the transition between the initial and final states. The quantity $|\mathcal{M}|^2$ may be expressed as

$$|\mathcal{M}|^2 = S_J |R_v^{v'}|^2 F_{v,J}^{v',J'} \, , \tag{2.71}$$

where S_J is the "line strength", equal to J' for the R-branch transitions and $J'+1$ for the P-branch ones, $R_v^{v'}$ is the vibrational matrix element, and $F_{v,J}^{v',J'}$ is the factor allowing for the interaction of vibrations and rotations. In the harmonic oscillator approximation,

$$|R_v^{v+1}|^2 = \mu_1^2 \frac{v+1}{2\alpha} \, , \qquad 2\alpha = \frac{\omega_e}{B_e r_{eq}^2} \, , \tag{2.72}$$

where r_{eq} is the equilibrium internuclear separation. Within the framework of the rigid rotator model, $F_{v,J}^{v',J'} = 1$. Allowing for vibrational-rotational interaction leads to the following approximate expression for the factor F:

$$F_{v,J}^{v',J'} = 1 - 4\gamma\theta m \, ,$$

where $\gamma = 2B_e/\omega_e$, $m = J'$ for the R-branch transitions and $m = -(J'+1)$ for the P-branch ones, and $\theta = \mu_0/\mu_1 r_{eq}$. In (2.71) and (2.72), μ_0 and μ_1 are the first two coefficients in the dipole moment expansion

$$\mu = \sum_{i=0}^{\infty} \mu_i (r - r_{eq})^i \, ,$$

where $(r - r_{eq})$ is the deviation of the internuclear separation from the equilibrium one. More exact calculations of the quantity $\mathcal{M}$ in the 0–1, 1–2, and 0–2 transitions for binary molecules with zero electronic momentum in the ground state have been presented in the monograph by *Persky* [2.47] [formulas (7.76)–(7.78)].

Combining (2.54), (2.56), (2.57), (2.61)–(2.63), (2.69), and (2.70), we get the following expressions for the gain [cm^{-1}] and induced transition cross section:

$$\alpha_{v,J}^{v+1,J+m} = 0.546 \times 10^{24} \sigma \theta_r \frac{M^{1/2}}{T^{3/2}} |\mathcal{M}|^2 \exp(a)^2 \mathrm{erfc}(a)$$

$$\times \left\{ N_{v+1} \exp\left[-\frac{F(J+m)hc}{kT}\right] - N_v \exp\left[-\frac{F(J)hc}{kT}\right] \right\} \, , \tag{2.73}$$

$$\sigma_{v,J}^{v+1,J+m} = 0.546 \times 10^{24} \frac{M^{1/2}}{T^{1/2}} \frac{1}{2J+1+2m} |\mathcal{M}|^2 \exp(a^2) \mathrm{erfc}(a) \, , \tag{2.74}$$

where the numerical multiplier is $8\pi^{5/2}/[3h(2kN_A)^{1/2}]$. The total gain G over the active medium length l is

$$G = \exp(\alpha_{v,J}^{v',J'} l) .$$

Thus, the active molecule is characterized by the following main parameters: σ, θ_r, M, $|\mathcal{M}|$, and a. Out of these parameters, σ, $M^{1/2}$, $|\mathcal{M}|^2$, and a appear, together with l, only into the quantity

$$g = 0.546 \times 10^{24} \sigma \theta_r \frac{M^{1/2}}{T^{3/2}} |\mathcal{M}|^2 \exp(a^2) \mathrm{erfc}(a) l.$$

For this reason, g is an important similarity parameter. Variations of σ, M, $|\mathcal{M}|$, and a, can be compensated for by changing l appropriately. Besides, the parameter g serves as a relative measure of the ability of the molecule to lase. Another important parameter is θ_r/T. The lasing frequency v_0 enters into the parameter a. In the case of purely Doppler broadening, $a = 0$, and g is independent of the lasing frequency.

Using (2.73), it is not very difficult to demonstrate that at a specified ratio N_{v+1}/N_v the P-branch transitions provide for the maximum gain.

It follows from (2.73) that for a given molecule the function

$$\tau = F_{v,J}^{v',J'} J \left\{ \exp\left[-\frac{\theta_r}{T} J(J-1) \right] - \frac{N_v}{N_{v+1}} \exp\left[-\frac{\theta_r}{T} J(J+1) \right] \right\} \left(\frac{300}{T} \right)^{3/2}$$

is almost independent of the transition being considered and is determined only by N_v/N_{v+1}, T, and J. The quantity $\tau(N_v/N_{v+1}, T, J)$ includes the entire dependence of the gain in the P-branch on T and J, and determines which transition will provide for the maximum gain at a given gas temperature T. In this connection, when considering the threshold conditions, it is advisable to introduce the function τ_{max} [2.48] determined in an operational manner: we select from the one-dimensional array

$$\tau\left(\frac{N_v}{N_{v+1}}, T, 1 \right) , \quad \tau\left(\frac{N_v}{N_{v+1}}, T, 2 \right), \dots$$

the element with the maximum value and denote it by $\tau(N_v/N_{v+1}, T, J_{max})$; then we assign this value to the quantity τ_{max}:

$$\tau_{max} = \tau\left(\frac{N_v}{N_{v+1}}, T, J_{max} \right) .$$

For any molecular laser, τ_{max} depends only on θ_r, T, and the ratio N_v/N_{v+1}. For example, for a laser in which the lasing emitters are the vibrationally excited HF molecules, τ_{max} takes on the values shown in Fig. 2.9.

The curves with $N_v/N_{v+1} < 1$ correspond to total inversion, whereas those with $N_v/N_{v+1} > 1$ relate to a system with a partial inversion. Indicated on each curve is the value of J providing for the maximum gain. It is interesting to note that although gain on a transition with a fixed J, given by the function τ, decreases rapidly with the rising T, the temperature dependence of the quantity τ_{max} is much

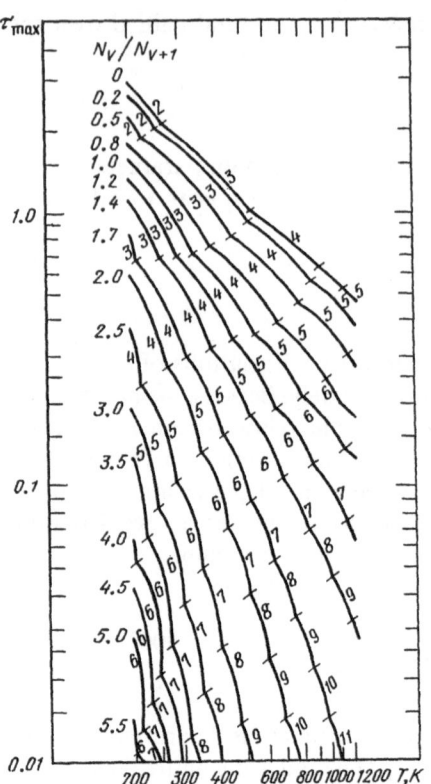

Fig. 2.9. τ_{max} as a function of the temperature T for the HF molecule, with the quantity N_v/N_{v+1} taken as a parameter [2.48]. The value of J for the transition $P(J)$ ensuring the maximum gain in the given temperature range is indicated on each curve

weaker than that of τ. The increase of both the ratio N_v/N_{v+1} and the gas temperature T causes J_{max} to shift toward higher values. This circumstance is of importance in the interpretation of the emission spectrum dynamics of chemical lasers.

2.4 Elementary Vibrational Relaxation Processes

2.4.1 Experimental Data

The vibrational-translational $(V \rightarrow T)$ and vibrational-vibrational $(V \rightarrow V)$ energy transfer processes govern the behavior of a great number of molecular systems of interest from both scientific and practical points of view. In particular, knowledge of energy transfer rates is important in the prognosis of the operation of molecular and chemical lasers where the emitters are mainly hydrogen halide molecules, CO, and CO_2 and the lasing media components may be inert monatomic gases such as He, Ar, Ne, and Xe, chemically active atoms like F, H, Cl, and O, diatomic gases such as H_2, O_2, and N_2, and polyatomic gases like SF_6, C_2F_6, ClF_5, and CH_4. For this reason, questions relating to energy transfer have recently been the subject of much investigation in chemical physics.

For a series of molecules such as N_2, O_2, CO, and Cl_2, there is a vast wealth of experimental material available, and theoretical calculations of their relaxation times fit the measured data over a wide temperature range to an accuracy better than an order of magnitude. But the vibrational relaxation of the molecules used as emitters in chemical lasers was until recently hardly studied. Only in the past few years has substantial information about the vibrational relaxation of hydrogen halides appeared. The data on the $V \to T$ ($V \to R$, T) and $V \to V$ energy transfer processes with the participation of hydrogen halide molecules are listed in Table 2.3.

In accordance with the method of producing vibrationally excited molecules, the following two main groups of experiments can be singled out: shock-tube experiments and laser-induced fluorescence experiments.

It has been found experimentally that the vibrational relaxation rate of HF has a peculiar temperature dependence with a minimum observed near 1000 K. Perhaps the most specific feature of the relaxation of HF (DF) is its fast rate, irrespective of whether it is a matter of self-deactivation or of energy transfer to other molecules. *Airey* and *Fried* [2.49] were the first to measure the relaxation rate of HF at 350 K by the laser-induced fluorescence technique. The measured relaxation rate is by several orders of magnitude greater than predicted by the vibrational-translational energy transfer theory allowing only for short-range repulsive forces [2.50]. At the present time, there is an abundance of data available on the vibrational relaxation of hydrogen halides, including information about their relaxation upon collisions with various partners at various temperatures. Deactivation (relaxation) processes studied experimentally may be classified as follows [2.51]:

1) self-relaxation;
2) relaxation on an isotope;
3) relaxation on atoms: (a) hydrogen H, (b) halogens X, (c) inert gases He, Ne, Ar, and Xe, and (d) other atoms;
4) relaxation on molecules: (a) halides, (b) deuterides, (c) homopolar diatomic molecules, (d) heteropolar diatomic molecules, and (e) polyatomic molecules.

2.4.2 Theoretical Studies

Conversion of Vibrational Energy into Translational and Rotational Energy. The rapid development of chemical lasers has posed a number of interesting problems for theoreticians. Information about the vibrational relaxation rates of hydrogen halides can serve as a sensitive criterion for testing most vibrational energy transfer theories, because these molecules, compared to their homonuclear or almost symmetrical heteronuclear counterparts, possess particular physical and chemical properties (high rotation rate, distinct tendency to hydrogen bonding, large vibrational amplitude comparable with the radius of short-range repulsive forces, high values of electric dipole and quadrupole moments).

The widespread vibrational energy relaxation theory expounded by *Schwartz*, *Slawsky*, and *Herzfeld* (SSH theory) [2.50], and generalized by *Tranczos* [2.52] to the case of polyatomic molecules, demonstrates how relaxation time depends on individual molecular constants, such as the masses of collision partners, their vibrational frequencies, and the shapes of their potential energy curves. In this theory based on the Landau-Teller one-dimensional model, the interaction of colliding partners is described by the exponential repulsive potential

$$\Phi = W \exp(-\alpha d) + \Phi_0 \ ,$$

where α^{-1} is the characteristic action radius of the interaction potential and d the distance between colliding atoms in molecules, for example, the distance d_{BC} in the collision $AB + C$ (the interaction of the other atoms is disregarded).

The oscillator undergoing a vibrational transition under the effect of a perturbation due to a molecular collision is modeled by a radially oscillating sphere. The relaxation mechanism here is the $V \rightarrow T$ energy transfer process, the conversion of the vibrational energy into rotational energy being neglected. According to the SSH theory, the probability that a pair of molecules, a and b, which initially were in the i_a and i_b states of the ground vibrational modes with frequencies v_a and v_b, will undergo a transition as a result of collision to the states f_a and f_b is

$$
P(a, b) = P_o(a) P_o(b) P_c V^2(a) V^2(b) 8 \left(\frac{\pi}{3}\right)^{1/2} \left[\frac{8\pi^3 \mu \Delta E^2}{(\alpha^*)^2 h^2}\right]
$$

$$
\times \chi^{1/2} \exp\left[-\left(3\chi - \frac{\Delta E}{2kT} + \frac{\Phi_0}{kT}\right)\right] , \tag{2.75}
$$

where

$$
\chi = \frac{\mu(v_0^*)^2}{2kT} = \left[\frac{2\pi^4 \mu(\Delta E)^2}{(\alpha^*)^2 h^2 kT}\right]^{1/3} ,
$$

$$
\Delta E = h v_a(i_a - f_a) + h v_b(i_b - f_b) .
$$

Equation (2.75) is written in the general form applicable to both the $V \rightarrow T$ and $V \rightarrow V$ processes discussed below. In this equation, μ is the reduced mass of the collision partners, ΔE is the change of the translational energy in collision, Φ_0 is the minimum value of the potential function used, equal to $(-\varepsilon)$, ε is the parameter either of the Lennard-Jones potential

$$
\Phi(r) = 4\varepsilon[(r_0/r)^{12} - (r_0/r)^6]
$$

describing the interaction of nonpolar molecules or of the Krieger potential describing the interaction of polar molecules:

$$
\Phi(r) = 4\varepsilon[(r_0/r)^{12} - (r_0/r)^6 - \delta^*(r_0/r)^3] ,
$$

where the last term allows for the dipole-dipole attraction, and α^* is the force constant for the repulsive potential replacing the Lennard-Jones or Krieger potential and calculated by the technique indicated below at v_0 equal to v_0^* — the "most favorable" relative velocity of the partners. The emergence of the most favorable collision velocity v_0^* at a given ΔE is due to the fact that, as v_0 is reduced, the transition probability decreases exponentially, while, as v_0 is increased, the number of molecules possessing such a velocity drops rapidly; at $v_0 = v_0^*$ the product of the transition probability and the molecular velocity distribution function reaches its maximum.

The factors in (2.75) are interpreted as follows: P_o is the orientational (steric) factor given by

$$P_o(i_a \rightarrow i_a) = 1, \qquad P_o(i_a \rightarrow i_a \pm 1, i_a \pm 2) = N_s/6 \; ,$$

where N_s is the number of atoms on the surface of the molecule. P_c is the factor allowing for the change in the cross-sectional area of spherical molecule for those collision partners which draw together at the most favorable velocity:

$$P_c = 1.364 \left(1 + \frac{C}{T} \right)^{-1} (r_c/r_0)^2 \; ,$$

where C is the Sutherland constant and r_c and r_0 are the distances between the molecular centers at the point of the closest approach and at the point where the interaction potential passes through zero, respectively. V^2 is the vibrational matrix element defined for the different quantum transitions as

$$V^2(i_a \rightarrow i_a) = 1 \; ,$$

$$V^2(i_a \rightarrow i_a \pm 1) = \left[h \left(i_a + \frac{1}{2} \pm \frac{1}{2} \right) \middle/ 8\pi^2 v_a \right] \frac{\alpha^2}{N_s} \sum_s \frac{A_{sa}^2}{m_s} \; ,$$

$$V^2(i_a \rightarrow i_a \pm 2) = \left[h(i_a + 1 \pm 1) \frac{i_a \pm 1}{16\pi^2 v_a} \right]^2 \frac{\alpha^2}{N_s} \sum_s \left(\frac{A_{sa}^2}{m_s} \right)^2 \; ,$$

where the subscripts s denote surface atoms and m_s is the mass of the sth atom on the surface. The expression

$$\frac{1}{N_s} \sum_s \frac{A_{sa}^2}{m_s}$$

is referred to as the amplitude factor. The coefficients A_{sa}^2 are given by the relation

$$A_{sa}^2 = \sum_l (a_{la}^{(s)})^2 \; ,$$

where the subscript l has three values corresponding to the Cartesian coordinates of the surface atom s. The matrix a, with the elements a_{la}, is the orthonormal transformation matrix relating the $3N$ reduced Cartesian coordinates $\| q \|$ of the atomic displacement from the equilibrium position to the $3N$ normal coordinates $\| Q \|$:

$$\| q \| = a \| Q \| \; .$$

The displacements $(r_t - r_t^0)$ and the reduced displacements q_t of the atom t from its equilibrium position are related by the relation

$$q_t = m_t^{1/2} (r_t - r_t^0) \; .$$

In the case of the two-atom molecule BC, there is only one vibrational mode, and the reduced displacements of the atoms B and C are related to the normal coordinate Q ($Q = m_{BC}^{1/2} S$, where S is the departure of the interatomic separation from its equilibrium value and m_{BC} the reduced molecular mass) by the relations

$$q_B = -Q\left(\frac{m_C}{m_B + m_C}\right)^{1/2}, \quad q_C = Q\left(\frac{m_B}{m_B + m_C}\right)^{1/2}$$

(the other coordinates q are zero). Then

$$A_B^2 = \frac{m_C}{m_B + m_C}, \quad A_C^2 = \frac{m_B}{m_B + m_C}, \quad \frac{1}{N_s}\sum_s \frac{A_s^2}{m_s} = \frac{1}{2}\left(\frac{A_B^2}{m_B} + \frac{A_C^2}{m_C}\right).$$

In the general case, the elements of the matrix a are calculated by the equation

$$\tilde{a} = L^{-1}B ,$$

where $\tilde{a}$ is the transposed matrix a, L is the transformation matrix relating the $3N$ natural coordinates $\|R\|$, of which $3N-6$ are internal ones (bond lengths and valence angles), to the normal coordinates:

$$\|R\| = L\|Q\| ,$$

and B is the transformation matrix between the coordinates $\|q\|$ and $\|R\|$:

$$\|R\| = B\|q\| .$$

The procedure for calculating the matrices B and L is presented in the molecular vibration theory [2.53]. The rest of expression (2.75) is called the translational factor. In the SSH theory the effect of attractive forces is taken into consideration by the factor $\exp(-\Phi_0/kT)$. The force constant $\alpha^* = \alpha(v_0^*)$ is found by equating at the point r_c the exponential repulsive potential and the Lennard-Jones (or Krieger) potential and their first-order derivatives (Method A) or the potentials at the points r_c and r_0 (Method B). Then $\alpha(v_0)$ for the Lennard-Jones potential is

$$\alpha = \frac{12}{r_0}\left\{\frac{1}{2}\left[1 + \left(\frac{\mu v_0^2}{2\varepsilon} + 1\right)^{1/2}\right]\right\}^{1/3}\left[1 + \left(\frac{\mu v_0^2}{\varepsilon} + 1\right)^{-1/2}\right]$$

(Method A) and

$$\alpha = \frac{1}{r_0}\left[\ln\left(\frac{\mu v_0^2}{2\varepsilon} + 1\right)\right]\left[1 - \left\{\frac{1}{2}\left[1 + \left(\frac{\mu v_0^2}{2\varepsilon} + 1\right)^{1/2}\right]\right\}^{-1/2}\right]$$

(Method B), and for the Krieger potential, we have

$$\alpha = \frac{12}{r_0}\frac{r_0}{Y r_c}\left\{1 + Y + 3\delta^*\left(\frac{r_0}{r_c}\right)^3 + \left[1 + Y + \delta^*\left(\frac{r_0}{r_c}\right)^3\right]^{1/2}\right\}$$

(Method A), where

$$\frac{r_0}{r_c} = \left\{ \frac{1}{2} \left(1 + \left[1 + Y + 4\delta^* \left(\frac{r_0}{r_c} \right)^3 \right]^{1/2} \right) \right\}^{1/6}, \quad Y = \left(\frac{\mu v_0^2}{2} - \Phi_0 \right) \bigg/ \varepsilon.$$

The effective interaction radius usually lies close to $\alpha^{-1} = 0.18$ Å.

The SSH theory allows one correctly to calculate the relaxation rates for a broad class of molecules. The success of the theory is mainly due to the fact that it predicts the order of magnitude of the probability of vibrational-translational ($V \to T$) energy transfer and its temperature dependence in the region of high temperatures. The specific features of the $V \to T$ energy transfer process predicted by the SSH theory are a decrease in relaxation rate with the increasing vibrational frequency of the relaxing molecules and mass of the oncoming partner and an increase in relaxation rate with temperature.

Comparison between experimental data on the vibrational relaxation of hydrogen halides and predictions of the SSH theory has shown that these data cannot be explained within the framework of the $V \to T$ energy transfer mechanism. The SSH theory is incapable of predicting not only the absolute value of the relaxation rate of hydrogen halides and its temperature dependence, but also the relative relaxation rates of isotope-substituted pairs. The calculated relaxation rate is too low compared with the measured rate, the rate of relaxation itself being not a monotonically increasing function of temperature. Besides, according to this theory, deuterium halides must relax faster than their hydrogen counterparts, whereas the reverse situation is observed in experiment over a wide range of temperatures, except for the highest ones.

The above discrepancies stimulated, in the 1970s, extensive studies into vibrational relaxation mechanisms in gases, which led to critical revision of the fundamental theories and the development of new, more consistent methods for calculating vibrational relaxation times [2.51].

The review given below of the literature material on these matters requires some preliminary comments. First of all, it should be borne in mind that there is as yet no satisfactory explanation of many a relaxation phenomenon observed. Only some molecular systems have been examined using the new, more advanced theoretical models. What leaps to the eye when considering the possible approaches to the problems of the molecular collision theory is the almost boundless variety of the analytical methods used. *Ormonde* [2.51] has noted that this variety is explained by the complexity of the problems being tackled and their rich physical substance, allowing different approaches to be used in different situations, and not by the whim of theoreticians, every one of whom develops a standpoint of his own. But the choice of a particular approach is, unfortunately, often governed by a theoretical school, hence largely by the individual predilection for some or other theoretical scheme, and not by impartial assessment of the problem in hand. This, naturally, makes it difficult both constructively to discuss the results obtained and to use them.

To treat the relaxation of hydrogen halides realistically, one needs a correct expression (angular relationship included) for the intermolecular potential, with account being taken of dipole-dipole attraction and hydrogen bonding. What is more, the collisional dynamics of the system must allow for the possibility of energy transfer to both the translation and rotation of both partners. The notion that the vibrational-rotational $(V \rightarrow R)$ energy transfer mechanism may be an important vibrational relaxation mechanism in rapidly rotating molecules has been advanced comparatively recently [2.54, 55]. For example, the rotational velocity of hydrogen-containing molecules can be higher than their translational velocity. In the simplest $V \rightarrow R$ theory version [2.56], the dynamic parameters of the $V \rightarrow T$ theories (reduced mass, velocity) that characterize the translational motion of the partners are replaced by the corresponding parameters characterizing their rotational motion. In this approach, vibrational energy is first transferred to rotation (intermolecular $V \rightarrow R$ energy transfer process), the role of translational motion consisting simply in the drawing of the molecules together.

The $V \rightarrow R$ theories hold more promise for explaining the data on vibrational relaxation of hydrogen halides, but discrepancies between the simplest $V \rightarrow R$ theory [2.56] and experimental data are still too great. In particular, this theory is incapable of explaining the temperature dependence of the relaxation rate of HF and DF. The more detailed theories [2.57–59] take into consideration the specific features of the intermolecular potential and the possibility of simultaneous conversion of vibrational energy into both rotational and translational energy.

The most successful explanation of the vibrational relaxation mechanism of hydrogen halides has been offered by *Shin* [2.57, 60–63] who developed the theory of $V \rightarrow R, T$ relaxation of rapidly rotating molecules. To derive the probability of vibrational self-relaxation of such molecules, he [2.57, 63] has considered, within the framework of classical mechanics, a collisional system consisting of a rigid rotator AB and an oscillator $A'B'$ (Fig. 2.10). It is assumed that the relative velocity of the molecules AB and $A'B'$ is directed along the line connecting their centers of mass. The intermolecular potential is expressed as the sum of the Morse potentials between the pairs AA', AB', BA', and BB':

$$U(r, x, \theta, \theta') = \frac{1}{4} \sum_i D \left[\exp\left(1 - \frac{r_i}{a}\right) - 2\exp\left(\frac{l}{2} - \frac{r_i}{2a}\right) \right], \qquad (2.76)$$

where r_i are interatomic separations, r is the distance between the centers of mass of

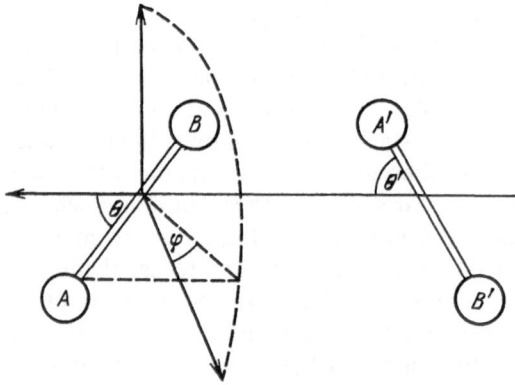

Fig. 2.10. $V \rightarrow R$ energy transfer model

AB and $A'B'$, x the vibrational coordinate of the oscillator, θ and θ' are the angles of rotation of the rotator and oscillator, a is the spatial parameter determining the rate of change of the potential at short distances, and $l = r_{eq}/a$ (r_{eq} being the equilibrium distance between the colliding particles).

Potential (2.76) is averaged over the orientations θ' of the rotating oscillator on the grounds that the rotational velocity is high compared with translational velocity, so that $U = U(r, \theta, x)$. In other words, the collisional system consists of the rigid rotator AB interacting with the rapidly rotating oscillator $A'B'$ treated as a rough sphere by means of the potential $U(r, \theta, x)$. Further, the linearization of the potential with respect to the vibrational coordinate or, to be more exact, with respect to the parameter x/d, where d is the equilibrium distance between the oscillator atoms, allows the interaction potential to be expressed in the form

$$U(r, \theta, x) = U(r, \theta) - F(r, \theta)x \ ,$$

where $F(r, \theta)$ is the angle-dependent disturbing force. The potential energy $U(r, \theta)$ determines the collision trajectory and time. This energy must also include the dipole-dipole attraction forces, whose potential (after averaging over the angle θ') is written as

$$U_{dd}(r, \theta) = -\frac{\pi \tilde{\mu}^2}{4r^3} \sin \theta$$

($\tilde{\mu}$ being the dipole moment of the molecule) and is then reduced to a form convenient for integrating the classical equations of motion of a system with the energy $V(r, \theta) = U(r, \theta) + U_{dd}(r, \theta)$. The energy transferred to the initially unexcited oscillator in the course of collision is calculated by the formula

$$\Delta E = \frac{1}{2\mu} \left| \int_{-\infty}^{+\infty} e^{i\omega t} F(t)\, dt \right|^2 \ ,$$

where $F(t) = F(r(t), \theta(t))$, and, as usual, the excitation probability is given by

$$P_{01}(E_r, E_0) = \frac{\Delta E}{\hbar \omega} \ ,$$

where E_r and E_0 denote the initial rotational energy of the rotator and the initial relative translational energy. The excitation probability as a function of temperature is found by averaging $P_{01}(E_r, E_0)$ over the Boltzmann distribution:

$$P_{01}(T) = \left[\frac{1}{(kT)^2} \right] \iint P_{01}(E_r, E_0) \exp\left(-\frac{E_r + E_0}{kT} \right) dE_r dE_0 \ .$$

The deactivation probability $P_{10}(T)$ is related to $P_{01}(T)$ by the detailed balancing principle: $P_{10}(T) = P_{01}(T)\exp(\hbar\omega/kT)$ and is given by

$$P_{10}(T) = \frac{8\omega}{\hbar M} \frac{\pi I a}{d^2} \left(\frac{a}{qd} \right)^2 (1+\alpha)^4 \left[1 - \frac{qd}{a} \mathrm{cth} \left(\frac{qd}{a} \right) \right]^2 f_{VT} \left(\frac{4\pi}{3} \frac{\chi}{kT} \right)^{1/2}$$

$$\times \exp \left\{ -\frac{3\chi}{kT} + \frac{\pi \tilde{\mu}^2}{r_{eq}^3} \left[1 + \frac{121}{72} \left(\frac{a}{qd} \right)^2 \right] \left[\frac{c_1 - c_2 + c_3}{(1+\alpha)kT} \right] + \frac{8}{\pi} \left(\frac{a}{qd} \right)^{1/2} \right.$$

$$\left. \times \frac{\mathrm{sh}(qd/2a)}{[\mathrm{sh}(qd/a)]^{1/2}} \frac{(D\chi)^{1/2}}{kT} + \frac{64}{3\pi^2} \frac{a}{qd} \frac{\mathrm{sh}^2(qd/2a)}{\mathrm{sh}(qd/a)} \frac{D}{kT} + \frac{\hbar\omega}{2kT} \right\} , \qquad (2.77)$$

where

$$\chi = \left\{ \left(\frac{I}{2q} \right)^{1/2} \left[\pi(1+\alpha) \frac{a\omega kT}{d} \right] \right\}^{2/3} ,$$

$$\alpha = \frac{9}{2} \left(\frac{a}{qd} \right)^2 \left[1 + \frac{4}{3} \left(\frac{a}{qd} \right)^2 \right] ,$$

$$f_{VT} = \frac{1}{A} \left[1 - 2 \frac{B}{A} + 2 \left(\frac{B}{A} \right)^2 \right] ,$$

$$A = 1 - \frac{\pi^2 M}{12\mu} (1+\alpha) ,$$

$$B = \frac{t_r^* kT}{\omega \mu a^2} , \quad q = \frac{m_A}{(m_A + m_B)} ,$$

$$c_1 = \frac{9}{2} c + 24c^2 + 3c^3 , \quad c_2 = 9c + 30c^2 + 30c^3 ,$$

$$c_3 = 1 + \frac{11}{2} c + 12c^2 + 10c^3 , \quad c = \frac{2\alpha}{r_{eq}} ,$$

$$t_r^* = \pi \left(\frac{I}{2E_r^*} \right)^{1/2} \alpha_0 \left[1 - \frac{2}{\pi} \left(\frac{D}{E_r^*} \right)^{1/2} g \right] ,$$

$$E_r^* = \chi - \frac{8g}{3\pi} (D\chi)^{1/2} , \quad \alpha_0 = (1+\alpha) \frac{a}{qd} ,$$

$$g = \frac{a}{qd} \frac{\mathrm{sh}^2(qd/2a)}{\mathrm{sh}(qd/a)} .$$

m_A and m_B are the masses of the atoms A and B ($m_B \ll m_A$, $m_{B'} = m_B$, $m_{A'} = m_A$), M is the reduced mass of the molecule, μ the reduced mass of the collisional system, t_r^* the rotational collision time calculated for the most favorable energy, ω the angular vibrational frequency, and I the moment of inertia of the rotator.

To perform the necessary calculations, it is required that the intermolecular interaction potential parameters D, r_{eq}, and a should be set. The parameter D is taken equal to the quantity ε in the Lennard-Jones potential. The parameter r_{eq} is found as the equilibrium value r of the potential $4\varepsilon[(r_0/r)^{12} - (r_0/r)^6] - 2\tilde{\mu}^2/r^3$. The parameter a is expressed in terms of r_0 and ε by equating the greatest term (as regards its absolute value) in the exponential of formula (2.77), i.e., $3\chi/kT$, to the analogous term obtained for the Lennard-Jones potential [2.63–65]:

$$a = \frac{r_0}{\pi^{1/2}} \left(\Gamma\frac{7}{12} \middle/ \Gamma\frac{1}{12} \right) \left(\frac{4D}{\chi_{LJ}} \right)^{1/12} ,$$

where

$$\chi_{LJ} = \left[\left(\Gamma\frac{19}{12} \middle/ \Gamma\frac{1}{12} \right) (4D)^{1/12} (2\pi M)^{1/2} r_0 \frac{\Delta E}{\hbar} kT \right]^{12/19} ,$$

$$\Delta E/\hbar = \omega .$$

The quantity $f_{VT} \leqslant 1$ reflects the effect of translational motion on the transition probability. For a "frozen" translation, $f_{VT} = 1$. The exponential in (2.77) represents the effect of the forces of molecular repulsion (the first term) and attraction (the second and the third term, the latter corresponding to the dipole-dipole attraction). The hydrogen bonding forces, not taken into account in deriving (2.77), play an important part in the vibrational relaxation of hydrogen halides. The angular dependence of the potential energy V_{HB} of this interaction may be approximately represented in the form

$$V_{HB} = -\tfrac{1}{4} V_0 (1 + \cos \theta')(1 + \cos \theta) , \tag{2.78}$$

where $(-V_0)$ is the maximum hydrogen bonding energy. According to (2.78), the energy V_{HB} is a maximum for the configuration $X - H \ldots X - H$ ($\theta = \theta' = 0$), whereas for the configuration $H - X \ldots X - H$, $V_{HB} = 0$, where H is the hydrogen atom and X a halogen atom. The orientation-averaged hydrogen bonding energy $\bar{V}_{HB} = -\tfrac{1}{4} V_0$. The effect of the hydrogen bonding forces on the transition probability can then be taken into account by means of the factor $f_{HB} = \exp(V_0/4kT)$ by analogy with the SSH theory, which considers the forces of attraction by means of the factor $\exp(\varepsilon/kT)$. Obviously in a more strict approach the potential V_{HB} should be included along with the dipole-dipole attraction potential in the total interaction energy. This more strict approach leads to the following expression for the exponential entering into equation (2.77) [2.66]:

$$f_{HB} = \exp\left[\left(1 + \frac{121a^2}{72q^2d^2} \right) \left(1 + \frac{a}{qd} \right) \frac{V_0}{4(1+\alpha)kT} \right] .$$

For hydrogen halides, this factor is numerically close to $\exp(V_0/4kT)$.

The vibrational relaxation time is expressed in terms of $P_{10}(T)$ as

$$p\tau = \left\{ ZP_{10}(T) \left[1 - \exp\left(-\frac{\hbar\omega}{kT} \right) \right] \right\}^{-1} ,$$

where Z is the number of collisions per second per molecule at $p = 1$ atm. Figure 2.11 presents the results of calculation of the vibrational relaxation time of the HF molecules, together with experimental data. It can be seen that calculation fits experiment satisfactorily. The negative temperature dependence of the relaxation rate of HF, i.e., the increase of $p\tau$ with rising T, in the range 300–1000 K is due to the influence of the dipole-dipole attraction and hydrogen bonding forces. At 300 K the second exponential factor in (2.77) is equal to 29, whereas $\exp(V_0/4kT)$ $= 4.3$, so that if these interactions had not been taken into account, the relaxation time would have been 128 times that in Fig. 2.11. At 3000 K they reduce the relaxation time by a factor of 1.6 only. At low temperatures the influence of translational motion is insignificant. Thus, at 300 K $f_{VT} = 0.61$. So, the vibrational relaxation of HF in the temperature interval 300–4000 K is satisfactorily explainable within the framework of the above-considered intermolecular $V \rightarrow R$ energy transfer model, including the action of the dipole-dipole attraction and hydrogen bonding forces. In that case, the strongest interaction leading to energy transfer occurs upon collisions between peripheral hydrogen atoms.

Based on a similar approach, *Shin* [2.67] has calculated the rate of vibrational relaxation of $CO_2(00^01)$ on HF and DF and found that the observed temperature dependence of the relaxation rate of CO_2 is well explainable within the framework of the $V \rightarrow R$ energy transfer model allowing for the effect of the hydrogen bonding $O=C=O \ldots H–F$.

The above theory ignores the fact that the energy levels of the rotator and oscillator are quantized. Calculating the vibrational-rotational energy transfer probability of rapidly rotating molecules in a quasiclassical approximation allowing for the quantization of rotational and vibrational levels yields results close to purely classical calculations [2.68].

At low temperatures the HF molecules, because of their strong hydrogen bonding ($\simeq 6$ kcal/mole), form dimers with a sufficiently long lifetime. The molecules in such dimers cannot execute complete rotation, but oscillate about the equilibrium orientation HF $\ldots$ HF^v. In that case, as shown by calculations of the dynamics of such a system, the vibrational energy of HF^v can be transferred to oscillatory motion in an intramolecular fashion, i.e., to the oscillations of the initially excited HF^v molecule about its equilibrium position in the dimer

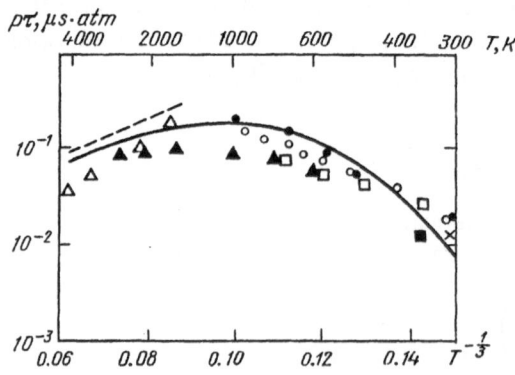

Fig. 2.11. Temperature dependence of the vibrational relaxation time of HF. Calculation data [2.63]: (———) Experimental data: (---) [2.175]; $\bigcirc$, $\square$, $\blacktriangle$ [2.63]; $\times$ [2.195]; $\blacksquare$ [2.49]; $\triangle$ [2.177]

HF ... HF^v [2.69, 70]. Thus, at low temperatures ($T = 200$–500 K) the vibrational relaxation of HF probably proceeds by an intramolecular vibrational-oscillatory energy transfer mechanism ($V \rightarrow O$ energy transfer). The important role of the collisional system configurations of the form HF ... HF^v, which come into being owing to hydrogen bonding and give rise to the intramolecular $V \rightarrow R$ relaxation of HF, is also confirmed by the numerical calculations of the collisional dynamics of the HF molecules carried out within the framework of a two-dimensional model by *Berend* and *Thommarson* [2.71]. These calculations, however, take no account of the dipole-dipole interaction, and so make a worse fit to experiment than those performed by *Shin* [2.63].

Experimental studies of the vibrational relaxation of DF and HF at low temperatures (200–300 K) show that the process under such conditions cannot be explained by the monomer and dimer relaxation models alone. The effect of relaxation on the polymers (DF)$_n$, where $n \leqslant 6$, is important for the collisional quenching of HF or DF on DF, the polymer relaxation being more pronounced in DF than in HF [2.72].

Shin and *Young* [2.73] and *Berend* and *Thommarson* [2.74] have investigated the relaxation of HF on Ar. The main mechanism by which HF loses its vibrational energy upon collision with Ar at temperatures up to 3000 K is the intramolecular $V \rightarrow R$ energy transfer.

One of the important processes of deactivation of vibrationally excited hydrogen halide molecules HX (DX) in chemical lasers is quenching on the chemically active atoms H (D), X. It has been suggested that the relaxation of HX (DX) on X and H is largely due to reactive collisions of the type X' + HX$^v \rightarrow$ X'H + X. The most comprehensive studies [2.75–82] of the mechanism of relaxation of hydrogen halides on the atoms H and X are based on computation of the classical three-dimensional trajectories using a semiempirical potential energy surface and the Monte-Carlo technique under specified initial conditions for each trajectory. The results obtained show that the relaxation processes under consideration go, at least to a noticeable degree, via the reactive channel, the latter being sometimes predominant, as, for example, in the case of relaxation of HF on F and H. These results, however, are sensitive to the choice of the potential energy surface parameters. The recent measurements of the rate of relaxation of HF on H [2.83–85] fail to agree with the results of the earlier trajectory calculations, and so the mechanism of relaxation of hydrogen halides on chemically active atoms still remains debatable [2.76–82, 86–89].

Vibrational-Vibrational Energy Exchange. The foregoing discussion has been concerned with the conversion of vibrational energy to translational or rotational energy. The vibrational-vibrational ($V \rightarrow V$) energy exchange processes play an important part in the operation of chemical and molecular lasers. The $V \rightarrow V$ exchange processes are essential in chemical kinetics, especially in reactions involving vibrationally excited molecules. Consequently, the numerical methods for calculating the probability of $V \rightarrow V$ processes are of great interest. Many experimental data obtained in recent years on $V \rightarrow V$ processes have been collected by *Ormonde* [2.51] and *Moore* [2.90].

Because of the adiabatic character of the long-range intermolecular forces, it has initially been suggested that of most effect in $V \rightarrow V$ processes are short-range repulsive forces. The above-discussed quantum-mechanical theory of the $V \rightarrow T$ and $V \rightarrow V$ energy exchange processes (the SSH theory) constructed within the framework of perturbation theory (the first-order Born approximation allowing for the distortion of the wave being scattered) is based precisely on this suggestion. The probability of a nonresonant $V \rightarrow V$ process is described in the SSH theory by relation (2.75). In the case of resonant collision where $\Delta E = 0$, equation (2.75) should be replaced by the following one:

$$P(a, b) = P_0(a)P_0(b) V^2(a) V^2(b) \frac{64\pi^2 \mu k T}{\alpha^2 h^2} \exp\left[-(\Phi_0/kT) \right] . \tag{2.79}$$

For a resonant collision, the quantity $\alpha = \alpha(v_0)$ is determined at the average thermal relative velocity of the partners. *Rapp* and *Golden* [2.91] have developed a semiclassical $V \rightarrow V$ exchange theory based on a collinear collision model, and falling outside the scope of the perturbation theory.

These theories disregard the effect of long-range forces and rotational motion on energy transfer. As can be seen from relations (2.75) and (2.79), the probability of $V \rightarrow V$ energy exchange rapidly (exponentially) decreases with the increasing amount of energy transferred to translational motion (energy defect). This probability also sharply decreases where great changes in the vibrational quantum number are required. Besides, the theories based on the description of the intermolecular interaction by the short-range potential predict a positive temperature dependence of the $V \rightarrow V$ exchange probability. But it is seen from relation (2.79) that the transition probability under resonant exchange conditions only slightly depends on the interaction radius α^{-1}, and therefore the suggestion that it is mainly short-range forces that are responsible for energy exchange seems groundless.

The probability that vibrational energy transfer will be observed in the course of a collision described by the classical trajectory $R(t)$ may be written in the first-order perturbation theory approximation as

$$P_{if} = h^{-2} \left| \int_{-\infty}^{+\infty} e^{i\omega t} \langle f | V(R(t)) | i \rangle \, dt \right|^2 , \tag{2.80}$$

where $\hbar\omega$ is the amount of energy transferred to translational motion (the change in the vibrational-rotational energy of the system during collision), $|i\rangle$ and $|f\rangle$ are the initial and final vibrational-rotational states, respectively, and $V(R(t))$ is the intermolecular interaction potential, including the relationship between the interaction and the vibrational and rotational coordinates. The long-range part of the intermolecular potential can contribute substantially to the transition probability for resonant collisions, but this contribution rapidly decreases down to zero as ω is increased. *Mahan* [2.92] has been the first to draw attention to the fact that the long-range dipole-dipole interaction depending on the distance between the molecules as R^{-3} can give rise to high $V \rightarrow V$ energy exchange probabilities. In the general case, the long-range intermolecular forces are expressed in terms of the molecular multipole moments. The qualitative difference between the short- and long-range interactions is that the short-range contribution to the intermolecular potential brings about both transitions and bending of the trajectory of the relative molecular motion, whereas its long-range counterpart causes only transitions, having no material effect on the motion trajectory.

Sharma and *Brau* [2.93] have developed the theory of quasiresonant $V \to V$ relaxation caused by long-range multipole interactions. In this theory, referred to as the SB theory, the molecules are assumed to follow a classical rectilinear trajectory, while their multipole interaction causing vibrational-rotational transitions is treated quantum-mechanically. To calculate the transition probability, use is made of the first-order Born approximation, which leads to the same rotational selection rules as in the case of optical transitions: $\Delta J = \pm 1$ for dipole transitions and $\Delta J = \pm 1, \pm 2$ for quadrupole transitions. Since the vibrational energy defect, measured from the midband, must be transferred to translational motion or to rotation (or taken off them), the rotational selection rules impose certain restrictions on how much of the vibrational energy transferred to translational motion can be compensated at the expense of rotation.

Let us present the final results of the SB theory [2.93–95]. The velocity-averaged $V \to V$ transition probability for the dipole-dipole interaction has the form

$$P^{dd}_{n,m;n'm'}(T) = \frac{16}{3}(h^2 d^4)^{-1}\frac{\tilde{m}}{kT}|\langle n'|Q_1|n\rangle|^2|\langle m'|Q_1|m\rangle|^2$$

$$\times \sum_{\substack{J_1,J_2; \\ J_1',J_2'}} C^2(J_1 1 J_1';00)C^2(J_2 1 J_2';00)p_{J_1}p_{J_2}I_2(\omega, d, T) \ , \tag{2.81}$$

where d is the diameter of the molecule treated as a solid sphere, $\tilde{m}$ the reduced mass of the system, $\langle n'|Q_1|n\rangle$ the dipole vibrational matrix element for the $n \to n'$ transition, C^2 the Clebsch-Gordan coefficient, p_J the population probability of the Jth rotational state, and $I_2(\omega, d, T)$ the Fourier transform of the dipole interaction potential discussed in detail in [2.93]; the quantity ω is calculated by the formula

$$\omega = 2\pi c|E(n) + F(n, J_1) + E(m) + F(m, J_2)$$

$$- E(n') - F(n', J_1') - E(m') - F(m', J_1')| \ ,$$

where E and F denote the vibrational and rotational energies of the molecules (in cm^{-1}), respectively; $\Delta J_1 = J_1' - J_1$ and $\Delta J_2 = J_2' - J_2$ take on the values ± 1. For the dipole-quadrupole interaction, the transition probability is

$$P^{dq}_{n,m;n',m'}(T) = 4\frac{1}{h^2 d^2}(\tilde{m}/kT)|\langle n'|Q_1|n\rangle|^2|\langle m'|Q_2|m\rangle|^2$$

$$\times \sum_{\substack{J_1,J_2 \\ J_1',J_2'}} C^2(J_1 1 J_1';00)C^2(J_2 2 J_2';00)p_{J_1}p_{J_2}I_3(\omega, d, T) \ . \tag{2.82}$$

Equation (2.82) is written for the case where molecule 1 makes a dipole transition and molecule 2 a quadrupole transition, with $\Delta J_1 = \pm 1$ and $\Delta J_2 = \pm 1, \pm 2$.

In contrast to the SSH theory, the SB theory predicts that the probability of the resonant and quasiresonant $V \rightarrow V$ exchange processes will decrease with rising temperature. This latter theory has succeeded in explaining the magnitude and temperature dependence of the rate of vibrational energy transfer from N_2 to CO_2, the first process for which a decrease in the $V \rightarrow V$ exchange cross section with rising temperature was observed in the range 300–1000 K.

In their subsequent works [2.96, 97], *Sharma* and co-workers have developed a procedure for calculating the probabilities of the $V \rightarrow V$ energy transfer resulting from multipole interactions, based on the first-order Born approximation allowing for the distortion of the incident wave on the scattering potential. This procedure requires numerical calculations. Comparison between the results of calculation allowing for the wave distortion and those based on the rectilinear trajectory has shown them to agree well for quasiresonant collisions in the region $\omega < 40$ cm^{-1} in the case of dipole-dipole interaction and in the region $\omega < 80$ cm^{-1} in the case of dipole-quadrupole interaction.

The experimental data amassed in recent years on the $V \rightarrow V$ processes [2.51, 90] demonstrate that the measured $V \rightarrow V$ exchange probabilities decrease with increasing energy defect much more slowly than those calculated by the SB theory taking into consideration the concurrent rotational transitions with $\Delta J = \pm 1$. At first glance this bears out the previous idea that the long-range forces are responsible for the $V \rightarrow V$ exchange only in resonance conditions, the most important part in the $V \rightarrow V$ exchange under off-resonance conditions being played by the short-range repulsive forces [2.98]. But thorough comparison between calculation and experiment has shown that this idea is wrong, or at least cannot aspire to universality.

Since the time when experimental data on the probability of the $V \rightarrow V$ exchange in CO,

$$CO(v = n-1) + CO(v = 1) \leftrightarrow CO(v = n) + CO(v = 0)$$

$$+ (n-1)\,[26.58 - 0.033(n+1)]\ \text{cm}^{-1}\ ,$$

became available for a wide range of n values and various temperatures, several investigators have compared the predictions of the existing $V \rightarrow V$ exchange theories with the results of measurements. *Jeffers* and *Kelley* [2.95] have explained the slow lowering of the $V \rightarrow V$ exchange probability in CO with increasing energy defect on the basis of two mechanisms: the long-range interaction predominant at small energy defects and the short-range repulsive interaction dominating at large defects ($n > 7$). Calculation fits well the experimental data on the $V \rightarrow V$ exchange in CO at 300 K, but the predicted temperature dependence of the exchange probability has proved much stronger than measured in the later experiments by *Witting* and *Smith* [2.99]. The calculated transition probabilities for high vibrational quantum numbers have turned out to be lower by an order of magnitude than measured in [2.99] at 100 K. Allowing for the higher-order multipole interactions over the entire range of energy defects ΔE, as well as for the distortion of the scattered wave [2.100], has also failed to make calculation agree with experiment.

Studies of the vibrational energy exchange among HF, DF, and CO_2 have yielded high rates, a fact which also cannot be understood on the basis of the $V \to V$ energy transfer theories developed earlier. The predicted $V \to V$ exchange rates are lower by several orders of magnitude than those observed in collisions involving large energy defects.

The data obtained may point to the participation of multiple-quantum rotational transitions in nonresonant $V \to V$ energy transfer. Such transitions can raise the energy exchange probability and reduce the dependence on the energy defect of vibrational quanta by providing for an additional energy release in the $V \to V$ exchange, and thus decreasing the amount of energy transferred directly to translational motion.

Significant in this respect are the calculations by *Lev-On, Palke*, and *Millikan* [2.101] of the $V \to V$ exchange rates in CO, based on a phenomenological description of multiple quantum rotational transitions. These investigators have assumed that the transfer of energy at all the energy defects under consideration is determined by the dipole-dipole interaction and treated it in the quasiclassical approximation for the rectilinear trajectory by formula (2.80) with the quantity ω replaced by the difference in energy between the initial and final states:

$$\omega = \omega_0 + 2\pi c \{ B_e^{(1)} \Delta J^{(1)} [2J_i^{(1)} + 1 + \Delta J^{(1)}] - B_e^{(2)} \Delta J^{(2)} [2J_i^{(2)} + 1 + \Delta J^{(2)}] \} \ ,$$

where $\hbar\omega_0$ is the purely vibrational energy defect and the terms in braces represent the contributions to $\hbar\omega$ from rotational levels as functions of the parameters J_i, ΔJ, and B_e of the molecules being considered. It is believed (and this fact allows the present approach to be characterized as a phenomenological one) that the exchange probability is nonzero for transitions with $\Delta J \neq \pm 1$, but drops with varying ΔJ by a certain (Gaussian) law:

$$\xi(\Delta J) = \frac{\exp\{-\sigma^2[(\Delta J)^2 - 1]\}}{\sum\limits_{\Delta J} \exp\{-\sigma^2[(\Delta J)^2 - 1]\}} \ .$$

Considering the thermal population of rotational sublevels, the total exchange probability has the form

$$P(\omega_0) = \sum\limits_{J_i^{(1)}, J_i^{(2)}} P_{J_1} P_{J_2} \sum\limits_{\Delta J^{(1)}, \Delta J^{(2)}} \xi(\Delta J^{(1)}) \xi(\Delta J^{(2)}) P(\omega) \ .$$

Using this approach, the above investigators [2.101] have demonstrated that at a certain σ for a given molecule ($\sigma = 0.5$ for CO) it proves possible not only correctly to describe the lowering of the energy exchange probability with the increasing vibrational energy defect $\hbar\omega_0$, but also to reproduce exactly the temperature dependence of the probability observed at various energy defects. These results show that the data on the $V \to V$ exchange probability, both as a function of energy defect and as a function of temperature, can probably be explained within the framework of a single interaction mechanism, namely, the long-range force interaction, but with the inclusion of concurrent multiple-quantum rotational transitions. Consistent calculation of the probabilities of processes of this type must be based on a theory lying outside the scope of the first-order perturbation theory. Two approaches differing in the character of approximations have been suggested to calculate the probability of nonresonant collisions.

Sharma and co-workers [2.102] have obtained an expression for nonresonant energy transfer in the second-order Born approximation allowing for the distortion of the wave being scattered on a spherically symmetric Lennard-Jones potential. It has been suggested that the transition is caused by multipole interactions. The use of the second-order approximation has made it possible to take account of multiple-quantum rotational transitions realized via virtual states. The scattering dynamics has been analyzed numerically in order to eliminate further simplifications in describing relative motion, the scattering cross section being a functional of the scattered wave. Using this method, they have calculated the cross section for energy exchange in the process

$$CO(v = 9) + CO(v = 1) \rightarrow CO(v = 10) + CO(v = 0) \ ,$$

with a vibrational energy defect of 236 cm^{-1}. In view of the fact that computation takes much computer time, it has been performed only for the most probable relative velocity at room temperature. It has also been assumed that both molecules are initially in the rotational state with $J = 7$, which is the most probable state at this temperature. Since most of the energy defect is converted to rotational energy when the number J of both molecules is increased by three, the cross section of the process has been calculated only for the case where the final state of both molecules is characterized by $J' = 10$. The calculated energy exchange cross section is 3×10^{-18} cm^2, which fits well the measured cross section equal to 2.6×10^{-18} cm^2.

Dillon and *Stephenson* [2.103] have developed a $V \rightarrow V$ energy exchange theory based on an exponential (unitary) expression for the scattering operator S and making it possible to estimate the elements of the S matrix in any order of perturbation and hence calculate multiple-quantum rotational and vibrational transitions. Based on this approach, they have calculated the cross sections for excitation transfer from HF, DF, and HCl to CO_2, which agree well with experimental data. Let us present the main physical assumptions made by them, which contributed much to the development of the vibrational relaxation theory.

The rate constant for the molecular transition from the initial state i to the final state f is proportional to the scattering operator matrix element

$$k_{if} \sim \langle |\langle i| S |f \rangle|^2 \rangle_{av} \ ,$$

where $\langle \rangle_{av}$ denotes averaging over the thermal distribution of the translational degrees of freedom. The scattering operator may be represented in the form of a chronologically ordered exponent defined by the expansion

$$S = \theta \exp\left[(-i/\hbar) \int\limits_{-\infty}^{\infty} \tilde{V}_{12}(t) \, dt \right] = \sum_{n=0}^{\infty} S_n \ , \tag{2.83}$$

where θ is the chronological operator, $\tilde{V}_{12} = \exp[(i/\hbar)Ht] V_{12} \exp[-(i/\hbar)Ht]$ is the intermolecular potential in the interaction representation, V_{12} is the interaction potential of molecules 1 and 2, and S_n is the n-fold multiple time-ordered integral

$$S_n = (-i/\hbar)^n \int\limits_{-\infty}^{\infty} dt_1 \int\limits_{-\infty}^{\infty} dt_2 \ldots \int\limits_{-\infty}^{t_{n-1}} dt_n \tilde{V}_{12}(t_1)\tilde{V}_{12}(t_2) \ldots \tilde{V}_{12}(t_n) \, ,$$

where $-\infty < t_n \leqslant t_{n-1} \leqslant \ldots \leqslant t_2 \leqslant t_1 < \infty$. The theories in frequent use are based on the termination of series (2.83) at $n = 1$ or 2, which holds true subject to the condition

$$\left| \int\limits_{-\infty}^{+\infty} \exp(i\omega_{ab}t) \langle a| V_{12}(t)|b\rangle \, dt \right| \ll \hbar$$

for all the intermediate states a and b. But the fact that this inequality is satisfied for the initial and the final state is not enough for the series to converge rapidly. Where the interaction and the total energy defect are great, the main contribution may be from the subsequent terms of the series. The perturbation theory retaining only S_1 or S_2 will in such cases yield wrong results, because the terms including the matrix elements S_n with $n > 1$ or 2 may give rise to multiple-quantum rotational transitions absorbing a large energy defect. Thus, in the case of $V \to V$ exchange between HF and CO_2, the maximum absolute value of $\langle i|S_n|f\rangle$ is reached for $n = 10$–15 [2.103].

Exact calculation of the matrix elements S_n includes summation over the complete set of intermediate states and comes up against computational difficulties even for small n. The theory under consideration uses an approximation retaining the unitary character of the scattering operator. This approximation consists in the replacement of the energy defects ω_{ab} for the intermediate states by the average value $\bar{\omega}_{if}$ depending only on the initial and the final state. The average value $\bar{\omega}_{if}$ is selected to make the approximation precise for the most probable transitions. For other transitions, this approximation proves exact only in a certain domain of impact parameter and impact velocity values. The use of the constant average value $\bar{\omega}_{if}$ allows the scattering operator to be written as the exponential operator

$$\langle i|S|f\rangle = \left\langle i \left| \exp\left[(-i/\hbar) \int\limits_{-\infty}^{\infty} \exp(i\bar{\omega}_{if}t)V_{12}(t)\, dt \right] \right| f \right\rangle$$

having no restrictions due to time ordering. Since the potential $V_{12}(t)$ is expressed as a sum of the products of vibrational operators for the production and decay of a^+ and a, S_n can then be written in the form of an ordered sequence of such operators, which simplifies the computation of the matrix elements. The further formalization of the theory consists in changing over to some diagonalized presentation in which the action of the vibrational operators is formally reduced to rotational operators and the scattering operator is expanded in a series in spherical tensor operators by a method ensuring unitarity and facilitating the computation of the matrix elements. The transition probability is calculated using the classical motion trajectory defined by the Lennard-Jones potential. It is assumed that the $V \to V$ exchange is due to the dipole-dipole interaction, but when considering the $V \to V$ exchange between the HX and CO_2 molecules, the purely rotational energy exchange due to the dipole-quadrupole interaction is also included.

Calculations by this method have shown that the large cross sections for the $V \to V$ exchange between HX and CO_2 result from a change in the rotational state, in which most of the energy defect ΔE goes to rotational excitation. For example, for the HF molecules the initial state of which is characterized by $J_i = 2$, the most probable final state is $J_f = 9$. The main contribution to the cross section is made by collisions in which less than 50 cm^{-1} of energy is converted to translational motion, while the total energy defect is 1612 cm^{-1}.

In a certain domain of the impact energy and impact parameter values, scattering angles may exceed 2π (spiral trajectory). *Dillon* and *Stephenson* [2.104] and *Cait* [2.105] have demonstrated that collisions causing one to three turns of spiral motion contribute much to the rate of the $V \to V$ exchange between HX and CO_2 and are responsible for the negative temperature dependence of the quantum transfer cross section. *Dillon* and *Stephenson* [2.103] have also calculated the cross sections for the transfer of two quanta from HF, HCl, and DF to CO_2. In the case of HF and HCl, these cross sections are small, but for DF, the two-quantum exchange cross section is only a factor of 2–6 smaller than single-quantum transition cross sections.

This formalism has been used by *Ormonde* [2.51] to calculate the $V \to V$ relaxation of CO, and to show that the quantum transfer cross section decreases with increasing energy defect much slower than calculated on the basis of the first-order Born approximation. These calculations [2.51] have also yielded large cross sections for the two-quantum exchange in CO.

Thus, calculations of the nonresonant $V \to V$ energy exchange rate, which fall outside the scope of the first-order perturbation theory, have borne out the decisive role of the long-range multipole interactions and multiple-quantum rotational transitions allowing the energy defect to be compensated for.

In conclusion, let us present the results of theoretical studies of the $V \to V$ exchange processes in hydrogen halides. *Sentman* [2.106] has proposed a classical theory of the $V \to R$ and $V \to V$ energy exchange processes that allows for contribution from rotational motion. The relaxation times calculated for hydrogen halides make a reasonable fit to the measured values in the region of temperatures exceeding 1000 K. At lower temperatures account should be taken of attractive forces. A more adequate model of the $V \to V$ relaxation of hydrogen halides has been developed by *Shin* [2.66]. In this model, the translational and rotational degrees of freedom are treated classically, and vibrational motion, quantum-mechanically. The collisional dynamics is similar to the one considered above for the $V \to R, T$ relaxation of the rapidly rotating molecules AB and $A'B'$, but account now is taken of the vibrational motion of the rotator AB. The interaction potential $U(r, x, x', \theta, \theta')$ including the repulsive, attractive, and dipole-dipole energy terms, as well as the hydrogen bonding energy, is averaged over θ' and transformed to $U(r, x, x', \theta) = U(r, \theta) + V(r, \theta, x, x')$. The transition probability is calculated by the formula

$$P_{v,v';v-1,v'+1}(E_r, E_0) = \hbar^{-2} \left| \int_{-\infty}^{\infty} \langle v, v' | V(r, x, x', \theta) | v-1, v'+1 \rangle \exp(i\Delta Et/\hbar) \, dt \right|^2$$

and averaged over thermal distribution. The potential $U(r, \theta)$ defines the classical trajectory. Based on this approach, *Shin* [2.66] has calculated the probability of the quasiresonant process

$$HCl(v = 2) + HCl(v = 0) \leftrightarrow HCl(v = 1) + HCl(v = 1), \quad \Delta E = 102 \text{ cm}^{-1}$$

in the temperature range 300–1000 K. He has demonstrated that the probability of this process decreases almost linearly with increasing temperature in the range 300–600 K, which agrees well with experiment. After reaching its minimum at $T = 700$ K, the probability starts increasing with temperature. *Sharma* and *Chen* [2.107] have calculated the probability of the same process by the SB theory discussed above. The calculations approximately fit experiment, provided the $V \rightarrow V$ exchange in HCl is caused by the dipole-quadrupole interaction and not by the dipole-dipole one. It should be noted that the $V \rightarrow V$ exchange rate in HCl calculated by the SSH theory fits experimental data fairly well at 300 K, but the theory yields a wrong temperature dependence of the $V \rightarrow V$ exchange rate.

The interpretation of the $V \rightarrow V$ exchange in HF, analyzed by *Shin* [2.108, 109] and *Wilkins* [2.110], is fairly complicated. As mentioned earlier, the hydrogen fluoride molecules are subject to strong attraction owing to hydrogen bonding, and effective energy transfer between the molecules, especially at low temperatures, can take place as a result of formation of dimers. With the quasiequilibrium configuration HF ... HF^v, the molecules in the dimers execute limited rotational and reciprocal motions, so that the energy defect of the $V \rightarrow V$ process can be transferred to a limited translational motion. This idea has been used by *Shin* [2.108] to calculate the $V \rightarrow V$ relaxation rate at a temperature of 300 K. The calculations have shown that such a $V \rightarrow V$ exchange mechanism leads to high [$\simeq 10^6$ s^{-1} (mm Hg)$^{-1}$] exchange rates, fitting the results of measurements taken at the same temperature. At higher temperatures the HF molecules execute complete rotation. The model describing the $V \rightarrow V$ energy transfer in these conditions has been briefly discussed above. Based on these two $V \rightarrow V$ energy transfer mechanisms, *Shin* [2.109] has investigated the rate constant of the $V \rightarrow V$ exchange process in HF,

$$HF(v) + HF(0) \rightarrow HF(v-1) + HF(1) ,$$

as a function of temperature in the range 200–2000 K and the vibrational quantum number v in the range 2–5. The rate constant of this process at $v = 2$ or 3 has a strong negative temperature dependence in the region of 200–400 K where the dimer energy exchange mechanism is most important. After reaching its minimum at around 400 K, the rate constant starts increasing with the rising T. For $v = 5$, the rate constant increases with T throughout the temperature range studied. At high temperatures, it is the rotational motion of HF that is responsible for the absorption of the energy defect ΔE. Another interesting result is that the rate constant decreases with the increasing v at low temperatures and grows at high temperatures. These two temperature regions separate at approximately $T = 300$ K.

Thus, the above review of the literature data shows that the rapid progress in the field of chemical and molecular lasers has recently aroused great interest in the development of theoretical models of molecular relaxation processes. Attention here has been concentrated on some of the most important molecular systems, the experimental and theoretical studies of which have demonstrated that many of the actually observed characteristic features of vibrational relaxation in gases require a more adequate explanation of collisional dynamics, compared to the earlier approaches. This has led in recent years to the advent of a fairly broad spectrum of vibrational relaxation models, some of which are reflected in this review. It is as yet rather difficult to establish a logical connection between the various approaches, since models are frequently built on a particular basis in order to portray some or other effect. But a more general picture of molecular relaxation processes has already started coming into view.

Current literature is discussing the following main ways in which vibrational energy can be transferred:

1) $V \to T$ and $V \to R, T$ processes;
2) resonant (quasiresonant) and nonresonant $V \to V$ processes;
3) intermolecular $V \to R$ and/or $V \to V$ processes;
4) intramolecular $V \to R$ and/or $V \to V$ processes;
5) multiple-quantum $V \to V$ process;
6) multiple-quantum $V \to R$ process.

These energy transfer routes are being interpreted within the scope of the following types of force interactions (isotropic and anisotropic) and mechanisms:

1) short-range repulsion;
2) multipole interactions; (a) dipole-dipole, (b) dipole-quadrupole, (c) dipole-octupole, and (d) others;
3) detachment of an H atom;
4) detachment of an X (hydrogen halide) atom;
5) hydrogen bonding;
6) polymerization;
7) spiral motion;
8) complexing;
9) collisional interference.

It is likely that new theoretical approaches, and a great many new experiments as well, will be required before all the essential questions concerning the relaxation processes in gases can be answered. The differences between intermolecular forces, vibrational modes and frequencies, molecular structures, masses, and moments of inertia are too great for all the details of energy transfer to be explained within the framework of one of the existing theories. Nevertheless, the above theoretical and experimental works on the relaxation of hydrogen halides, CO, and CO_2 have materially contributed to our knowledge of vibrational relaxation mechanisms and produced a number of quantitative results facilitating the analysis of the kinetics of chemical and molecular lasers.

2.5 General Equations Describing Physico-Chemical Processes Occurring in the Laser Medium

The temporal evolution of the lasing process in a chemical laser is determined by the kinetics of the chemical reaction involved, the distribution of the reaction energy among the molecular degrees of freedom, relaxation processes, and changes in the temperature of the mixture in the course of the reaction.

This section formulates the equations forming the basis of the mathematical models of chemical lasers and describing the main physico-chemical processes taking place in the laser medium: emission of radiation, chemical reactions, vibrational-rotational relaxation, and gas-dynamic motion. In the case of pulsed systems, the medium is at rest and its components are premixed, so that the effects of gas-dynamic motion and transfer phenomena in them are unimportant, but when calculating continuous-wave chemical lasers, the motion of the medium, transfer phenomena (diffusion), and physico-chemical kinetics need to be considered simultaneously.

Let us first consider the kinetics of processes in the resting medium where all of them are time-dependent (pulsed operation) and then generalize our treatment to the case of emission of radiation in gas flows (continuous-wave operation).

2.5.1 Radiative Processes

A sufficiently accurate calculation of the energy characteristics of a laser is possible within the scope of rate equations in the case of a laser oscillator or within the scope of radiation transfer equations in the case of a laser amplifier.

Laser oscillator rate equations have the form

$$\frac{d\varrho_{r-s,J}^{r,J+m}}{dt} = c\sigma_{r-s,J}^{r,J+m} \Delta_{r-s,J}^{r,J+m} \varrho_{r-s,J}^{r,J+m} - \frac{\varrho_{r-s,J}^{r,J+m}}{\tau_{ph}} + \frac{c\sigma_{r-s,J}^{r,J+m}}{V} n_{v,J} , \qquad (2.84)$$

where $n_{r,J}$ is the population density of the vibrational-rotational level with the quantum numbers r and J, $\Delta_{r-s,J}^{r,J+m}$ and $\varrho_{r-s,J}^{r,J+m}$ are respectively, the inversion density and the density of the photons emitted in the cavity from the transition $r, J+m \rightarrow r - s, J$. $\sigma_{r-s,J}^{r,J+m}$ is the cross section for the stimulated emission of radiation in the above transition; c is the velocity of light; τ_{ph} the photon lifetime in the cavity; and V the cavity volume.

The total specific stimulated emission power is

$$P_1 = \frac{1}{\tau_{ph}} \sum_{r,s,J,m} \hbar \omega_{r-s,J}^{r,J+m} \varrho_{r-s,J}^{r,J+m} . \qquad (2.85)$$

The main contribution to the emission of radiation is usually made only by the transitions with $s = 1$ (the fundamental harmonic of molecular vibrations) and $m = -1$ (P-branch).

The traveling wave amplification is described by the following radiation transfer equations

$$\frac{1}{c}\frac{\partial I_{r-s,J}^{r,J+m}}{\partial t}+\frac{\partial I_{r-s,J}^{r,J+m}}{\partial z}=\sigma_{r-s,J}^{r,J+m}\,\varDelta_{r-s,J}^{r,J+m}(I_{r-s,J}^{r,J+m}+S_{r-s,J}^{r,J+m})\ ,\tag{2.86}$$

where $I_{r-s,J}^{r,J+m}$ is the intensity of the light wave propagating along the z-axis, and $S_{r-s,J}^{r,J+m}$, the function describing the effect of spontaneous emission of radiation, given by

$$S_{r-s,J}^{r,J+m}=\left[\frac{2\hbar}{c^2}(\omega_{r-s,J}^{r,J+m})^3\right]\frac{n_{r,J+m}}{n_{r,J+m}-(g_{J+m}/g_J)n_{r-s,J}}\ .$$

When solving equations (2.86), account is taken of the initial and boundary conditions of the problem in hand.

In the general case, the inversion density entering into (2.84) is defined by the relation

$$\varDelta_{r-s,J}^{r,J+m}=n_{r,J+m}-\frac{g_{J+m}}{g_J}\,n_{r-s,J}\ .$$

Under rotational equilibrium conditions, the rotational sublevel population is expressed in terms of the total vibrational state population n_v by formula (2.53), and so it is only the finiteness of the vibrational level relaxation rate that is to be taken into consideration when calculating the laser kinetics. In a more general case, to determine $n_{v,J}$, it is necessary to consider both vibrational and rotational kinetics.

The effect of spontaneous and stimulated emission of radiation on the population of the vibrational and rotational levels is described by the equations

$$\frac{dn_r}{ds}=\sum_{s,J,m}n_{r+s,J+m}\,A_{r,J}^{r+s,J+m}-\sum_{s,J,m}n_{r,J}\,A_{r-s,J-m}^{r,J}$$

$$+\sum_{s,J,m}c\sigma_{r,J}^{r+s,J+m}\,\varDelta_{r,J}^{r+s,J+m}\,\varrho_{r,J}^{r+s,J+m}-\sum_{s,J,m}c\sigma_{r-s,J}^{r,J+m}\,\varDelta_{r-s,J}^{r,J+m}\,\varrho_{r-s,J}^{r,J+m}\ ,\tag{2.87}$$

$$\frac{dn_{r,J}}{dt}=\sum_{s,m}n_{r+s,J+m}\,A_{r,J}^{r+s,J+m}-\sum_{s,m}n_{r,J}\,A_{r-s,J-m}^{r,J}$$

$$+\sum_{s,m}c\sigma_{r,J}^{r+s,J+m}\,\varDelta_{r,J}^{r+s,J+m}\,\varrho_{r,J}^{r+s,J+m}-\sum_{s,m}c\sigma_{r-s,J-m}^{r,J}\,\varDelta_{r-s,J-m}^{r,J}\,\varrho_{r-s,J-m}^{r,J}\ ,$$

$$\tag{2.88}$$

where $A_{r,J}^{r+s,J+m}$ is the Einstein coefficient for spontaneous emission.

To be able to predict not only the energy characteristics of a laser, but also its beam divergence, one needs a more adequate, wave-equation description of the electromagnetic field. *Rensch* [2.111] has developed an effective method of calculating near- and far-field distributions, which allows for the effect of the active medium in the optical cavity and is based on the solution of a parabolic wave equation. The method is applicable to both stable and unstable optical resonators.

2.5.2 Chemical Kinetics

The mathematical description of reactions using classical chemical kinetics is based on a phenomenological law which establishes the relationship between the time derivatives of substance concentrations (reaction rate) and the concentrations themselves. This law (the law of mass action) states that the rate of chemical reaction is proportional to the product of the concentrations of the reagents. It is assumed that the mass law describes an elementary process in which the reactant molecules actually participate. Most reactions involve a number of elementary stages. The rate of formation of a substance is therefore equal to the sum of the rates of formation of this substance in the individual stages. Let us formulate the main chemical kinetics equations for homogeneous reactions in closed systems.

A complex chemical reaction consisting of L elementary stages may be described by a set of L stoichiometric equations:

$$\sum_{i=1}^{N} v'_{ij} M_i \rightarrow \sum_{i=1}^{N} v''_{ij} M_i , \qquad j = 1, \ldots, L , \tag{2.89}$$

where the arrow indicates the direction of the reaction, v'_{ij} and v''_{ij} are the stoichiometric coefficients of the ith substance taking part in the jth reaction as the reagent and the product, respectively, and M_i is the chemical symbol of the ith substance. Every reaction is reversible. In (2.89) the forward and back reactions are treated as separate stages. The set of equations (2.89) is frequently replaced by an equivalent set where the forward and back reactions are grouped together:

$$\sum_{i=1}^{N} v'_{ij} M_i \leftrightarrow \sum_{i=1}^{N} v''_{ij} M_i , \qquad j = 1, \ldots, M , \tag{2.90}$$

where $M = L/2$. According to the phenomenological law of chemical kinetics, for reactions of general form (2.90), one may write

$$\frac{dn_i}{dt} = \sum_{j=1}^{M} (v''_{ij} - v'_{ij}) \left(k_{fj} \prod_{i=1}^{N} n_i^{v'_{ij}} - k_{bj} \prod_{i=1}^{N} n_i^{v''_{ij}} \right) , \tag{2.91}$$

where n_i is the concentration of the ith substance and k_{fj} and k_{bj} are the rate constants of the forward and back reactions, respectively. Equation (2.91) holds true subject to the condition that the change in concentration is only caused by chemical reactions and not by gas-dynamic motion or diffusion.

2.5.3 Vibrational-Rotational Kinetics

In rotational equilibrium conditions, it is only necessary to monitor the total population of the vibrational levels. A molecule can change its vibrational energy as a result of collision with some partner whose vibrational state can either change or remain unchanged upon this collision. In the former case, we speak of the $V \rightarrow V$ transfer, whereby at least some excitation energy is transferred to the vibrational degrees of freedom of the partner. The latter is the case of the $V \rightarrow R, T$ transfer or conversion of the vibrational energy into the translational and/or rotational energy of the collision partners. If molecules in different vibrational states are treated as different mixture components, and the elementary $V \to V$ and $V \to R, T$ relaxation processes as "reactions", the kinetics equations can formally be written in the form of (2.91). Since all the processes under consideration are bimolecular, the contribution of the relaxation processes to the rate of deactivation of any vibrational level in the general case may then be represented, for greater clarity, in the form

$$\frac{dn_r^A}{dt} = \sum_{s,p,q,M} k_{s,p;r,q}^{A-M} n_s^A n_p^M - \sum_{s,p,q,M} k_{r,q;s,p}^{A-M} n_r^A n_q^M , \tag{2.92}$$

where n_r^A is the concentration of the molecular species A at the level r, n_p^M that of the molecular species M at the level p, and $k_{s,p;r,q}^{A-M}$ the rate constant of the process in which the molecule A moves from the level s to the level r and its collision partner M moves from the level p to the level q. In the case of the $V \to R, T$ process, the subscripts p and q coincide. For a polyatomic molecule, each of the subscripts p, q, r, and s is described by a set of vibrational quantum numbers $(v_1, v_2, \ldots, v_n)$, where v_k is the vibrational quantum number of the kth vibrational mode. For any diatomic and, the more so, polyatomic molecule, the number of levels and kinetic constants in the set of equations (2.92) is very great. This circumstance makes the direct analysis of set (2.92) rather difficult. It is common practice to use the following main assumptions and simplifications:

– consideration is given only to transitions between the closest levels;
– the effect of excited electronic states is disregarded;
– use is made of the simplest molecular potential models, namely, the harmonic oscillator or Morse oscillator model;
– the detailed-balance principle is applied to all $V \to R, T$ and $V \to V$ processes because the translational and rotational degrees of freedom are considered to be in equilibrium.

The grounds for these assumptions and their field of application have been discussed in the review literature [2.27, 112–115].

For single-quantum transitions,

$$A(r) + M(s) \underset{k_{r+1,s-1;r,s}^{A-M}}{\overset{k_{r,s;r+1,s-1}^{A-M}}{\rightleftarrows}} A(r+1) + M(s-1) ,$$

$$A(r) + M(s) \underset{k_{r+1,r}^{A-M}}{\overset{k_{r,r+1}^{A-M}}{\rightleftarrows}} A(r+1) + M(s) .$$

Considering the detailed balance principle, equations (2.92) take the form

$$\frac{dn_r^A}{dt} = \sum_M \left\{ k_{r+1,r}^{A-M} \left[n_{r+1}^A n^M - n_r^A n^M \exp\left(-\frac{E_{r+1}^A - E_r^A}{kT} \right) \right] \right.$$

$$\left. - k_{r,r-1}^{A-M} \left[n_r^A n^M - n_{r-1}^A n^M \exp\left(-\frac{E_r^A - E_{r-1}^A}{kT} \right) \right] \right]$$

$$+ \sum_s k_{r+1,s-1;r,s}^{A-M} \left[n_{r+1}^A n_{s-1}^M - n_r^A n_s^M \exp\left(-\frac{E_{r+1}^A + E_{s-1}^M - E_r^A - E_s^M}{kT} \right) \right]$$

$$- \sum_s k_{r,s;r-1,s+1}^{A-M} \left[n_r^A n_s^M - n_{r-1}^A n_{s+1}^M \exp\left(-\frac{E_r^A + E_s^M - E_{r-1}^A - E_{s+1}^M}{kT} \right) \right] \right\} .$$

(2.93)

Let us express a collision leading to changes in the vibrational, rotational, and translation energies as

$$A(r, l) + M(q, j) \leftrightarrow A(s, k) + M(p, i) ,$$

where r, l and q, j are the vibrational and rotational quantum numbers of the molecules A and M, respectively. As applied to the vibrational-rotational kinetics processes, equations (2.91) may be written in the form

$$\frac{dn_{r,l}^A}{dt} = \sum_{\substack{s,k,p, \\ i,q,j,M}} k_{s,k;p,i \to l;q,j}^{A-M} n_{s,k}^A n_{p,i}^M - \sum_{\substack{q,j,s,k \\ p,i,M}} k_{r,l;q,j \to s,k;p,i}^{A-M} n_{r,l}^A n_{q,j}^M ,$$

(2.94)

where $k_{s,k;p,i \to r,l;q,j}^{A-M}$ is the rate constant of the process in which the molecule A moves from the level $v = s, J = k$ to the level $v = r, J = l$; and its collision partner, from the level $v = p, J = i$ to the level $v = q, J = j$.

These equations take into consideration in the general form all the main types of energy exchange processes:

$$V \to T, \quad V \to R, T, \quad V \to V, \quad V \to V, R, \quad R \to R, \quad R \to T .$$

At the current stage of investigations, the detailed rate constants of the elementary processes entering into (2.94) are, as a rule, unknown. For this reason, a number of assumptions are introduced that simplify the analysis while making it possible to take account of the principal kinetics features resulting from the finiteness of the rotational relaxation rate. These assumptions are as follows:

- the (relatively) slower $V \to T$ and $V \to V$ energy exchange processes are considered to be independent of J;
- the $R \to R$ (rotational-rotational) energy exchange process is ignored, i.e., the collision partner M is treated as a structureless particle;
- the multiple-quantum $R \to T$ exchange is disregarded, i.e., consideration is only given to collisional transitions between neighboring rotational levels.

Bearing these assumptions in mind and allowing for the detailed balance principle, which connects the rate constants of the forward and back reactions, let us write equations (2.94) in the more graphic form

$$\frac{dn_{r,J}^A}{dt} = \sum_M \left\{ k_{J+1,J}^{A-M} \left[n_{r,J+1}^A n^M - \frac{g_{J+1}}{g_J} \exp\left(-\frac{E_{r,J+1} - E_{r,J}}{kT} \right) n_{r,J}^A n^M \right] \right.$$

$$- k_{J,J-1}^{A-M} \left[n_{r,J}^A n^M - \frac{g_J}{g_{J-1}} \exp\left(-\frac{E_{r,J} - E_{r,J-1}}{kT} \right) n_{r,J-1}^A n^M \right]$$

$$+ k_{r+1,r}^{A-M} \left[n_{r+1,J}^A n^M - \exp\left(-\frac{E_{r+1}^A - E_r^A}{kT} \right) n_{r,J}^A n^M \right]$$

$$- k_{r,r-1}^{A-M} \left[n_{r,J}^A n^M - \exp\left(-\frac{E_r^A - E_{r-1}^A}{kT} \right) n_{r-1,J}^A n^M \right]$$

$$+ \sum_s k_{r+1,s-1;r,s}^{A-M} \left[n_{r+1,J}^A n_{s-1}^M \right.$$

$$- \exp\left(-\frac{E_{r+1}^A + E_{s-1}^M - E_r^A - E_s^M}{kT} \right) n_{r-1,J}^A n_s^M \right]$$

$$- \sum_s k_{r,s;r-1,s+1}^{A-M} \left[n_{r,J}^A n_s^M \right.$$

$$\left. - \exp\left(-\frac{E_r^A + E_s^M - E_{r-1}^A - E_{s+1}^M}{kT} \right) n_{r-1,J}^A n_s^M \right] \right\} , \qquad (2.95)$$

where the first two lines describe the single-quantum $R \to T$ exchange (transitions between neighboring rotational levels), the next two lines, the $V \to T$ exchange, and the last four lines, the $V \to V$ exchange.

The next step allowing the description of the $R \to T$ energy exchange process to be improved is the inclusion of the multiple-quantum $R \to T$ processes

$$A(v, J) + M \underset{k_{J'J}^{A-M}}{\overset{k_{JJ'}^{A-M}}{\rightleftharpoons}} A(v, J') + M$$

the formal consideration of which in equations of form (2.95) presents no difficulties, but complicates calculations.

The elementary process rate constants entering into equations (2.95) are taken from experimental data and extrapolated, if necessary, to a wider temperature interval on the basis of the basic theories. The available information on the $V \to T$ and $V \to V$ processes in the most important molecular systems has been discussed in Sect. 2.4. Much less is known about the rate constants of the $R \to T$ processes. When making kinetics calculations, it is advisable to use the empirical $R \to T$ energy transfer model suggested by *Polanyi* and *Woodall* [2.116], which describes the vibrational relaxation kinetics of hydrogen halides very well. The model is based on the exponential energy defect dependence of the $R \to T$ process rate constants:

$$k_{JJ'} = ZN \exp[-C(E_J - E_{J'})] ,$$

where $J < J'$. Z is the unit molecular collision frequency, and N and C are temperature-independent parameters. The studies of the rotational kinetics of the HCl and HF hydrogen halides in the $HCl + H_2$ and $HF + Ar$ systems [2.116, 117] by the laser-induced IR luminescence technique have yielded $C \simeq 4.5 \times 10^{-3}$ cm^{-1} for HCl and $C \simeq 4.4 \times 10^{-3}$ cm^{-1} (1.55 ± 0.25 kcal^{-1} mole) for HF. The parameter N for HCl is equal to 1.42. The most important transitions are those with $\Delta J = 1, \ldots, 5$ for HCl and $\Delta J = 1, \ldots, 3$ for HF. In some studies, the $R \to T$ process rate constants are written in the form

$$k_{JJ'} = ZN(2J'+1)\exp\left(-C\frac{E_{J'}-E_J}{kT}\right) \quad \text{or}$$

$$k_{JJ'} = ZN(2J'+1)\exp[-C(E_{J'}-E_J)] \ ,$$

the parameters N and C being recalculated accordingly.

The $R \to R$ processes can be taken into consideration on the basis of the empirical model suggested by *Hinchen* and *Hobbs* [2.118], who studied rotational relaxation in $HF + HF$ collisions. The $R \to R$ exchange rate constants in (2.94) are those whose subscripts corresponding to the vibrational quantum number remain unchanged: $s = r$, $p = q$. Omitting for simplicity these subscripts, let us write the $R \to R$ exchange rate constant in the general case as $k_{k,i;l,j}$. The following relation satisfying the detailed balance principle has been suggested for these constants:

$$k_{k,i;l,j} = K_0 \bar{n}_l \bar{n}_j \exp\left(-\alpha\frac{E_j - E_i + E_l - E_k}{kT}\right) \ ,$$

where K_0 and α are empirical parameters, E_j and E_i denote the purely rotational energy of the final and initial states, respectively, and

$$\bar{n}_j = (2j+1)N_v(T)\exp(-E_j/kT) \ ,$$

$$N_v(T) = \left[\sum_j (2j+1)\exp(-E_j/kT)\right]^{-1} \ .$$

In the case $l = k$, the model also describes the $R \to T$ energy exchange process. The two empirical parameters for HF are: $K_0 = (2 \pm 0.4) \times 10^8$ s^{-1} (mm Hg)$^{-1}$, $\alpha = 0.96$ [2.118].

With account being taken of all types of interaction, the kinetics of the vibrational and vibrational-rotational states of the active molecules are described by the equations

$$\frac{dn_r}{dt} = \chi_r^{rad} + \chi_r^{chem} + \chi_r^{rel} \ , \tag{2.96a}$$

$$\frac{dn_{r,J}}{dt} = \chi_{r,J}^{rad} + \chi_{r,J}^{chem} + \chi_{r,J}^{rel} \ , \tag{2.96b}$$

where the quantities χ_r^{rad}, χ_r^{chem}, and χ_r^{rel} describing the contribution of radiative processes, chemical reactions, and relaxation processes to the kinetics are defined by the right-hand sides of equations (2.87), (2.91), and (2.93) and the analogous quantities $\chi_{r,J}^{rad}$, $\chi_{r,J}^{chem}$, and $\chi_{r,J}^{rel}$, by the right-hand sides of equations (2.88), (2.91), and (2.95), respectively.

2.5.4 Energy Conservation Equation

The liberation of energy in reaction and relaxation processes cause the laser medium temperature to change. Some energy is converted into radiation. Inasmuch as the rate of elementary processes depends materially on temperature, the temperature change in the course of reaction must be taken into account. This change is determined by the energy conservation equation and the thermodynamic properties of the laser mixture components.

As applied to a gas mixture of constant density, the energy conservation equation, which is an analytical expression of the first law of thermodynamics, may be written in the form

$$\sum_i n_i c_{pi} \frac{dT}{dt} - \frac{dp}{dt} = -P_l - \sum_i h_i \frac{dn_i}{dt} , \tag{2.97}$$

where p is the gas mixture pressure, c_{pi} the specific heat at constant pressure of the ith mixture component, and h_i the specific enthalpy of the ith component (including molecules in different quantum states).

Equation (2.97) must be supplemented with the perfect gas law

$$p = R^0 T \sum_i n_i \tag{2.98}$$

and the caloric equation of state

$$h_i = h_i^0 + \int_{T_0}^{T} c_{pi} dt , \tag{2.99}$$

where R^0 is the universal gas constant and h_i^0 the standard specific heat of formation of the ith component at a temperature of T_0.

Equations (2.84)–(2.99) formulated in this section allow one to calculate the energy characteristics of a pulsed chemical laser and analyze in detail the kinetics of the processes occurring in the active medium, provided that the reaction mechanism, the elementary process rate constants, and the spectroscopic and thermodynamic properties of the laser mixture components are known.

2.5.5 Generation of Radiation in Gas Flows

Population inversion in continuous-wave (cw) chemical lasers arises in the mixing zone of initially unmixed reagent flows. The mixing and subsequent chemical reactions produce vibrationally excited molecules which emit energy.

Three generations of models can be singled out in the history of the development of the theory of chemical lasers of this type. The first-generation models either disregard the finite character of the mixing rate (instantaneous mixing) [2.119, 120] or consider mixing within the scope of the "flame front" concept discussed in detail in Chap. 5, which simplifies the analysis. The flow in that case is treated as being quasiunidimensional. The second-generation models make it possible to carry out more detailed gas-dynamic calculations on the basis of solution of boundary layer equations [2.121, 122]. To date these two types of models have contributed most to the interpretation of the cw chemical laser mechanism. The third generation of theoretical models, now at the initial stage of development, is based on the solution of the complete set of Navier–Stokes equations, which in principle allows one to perform a most consistent analysis of cw chemical lasers. Computations in that case are naturally much more laborious, and so far only limited results have been obtained with the use of the Navier–Stokes equations [2.123, 124].

Let us present the basic equations of the quasiunidimensional model of the gas-dynamic flow in a chemical laser with the instantaneous mixing of the reagents. The merit of this model is that it enables one most simply to calculate the upper attainable limit of the energy performance of a given laser system and thus assess its prospects.

Let us introduce the following notation:

x – coordinate along the flow;
ϱ – gas density;
u – mean mass velocity;
p – hydrostatic pressure;
T, T_0 – temperature of the outer degrees of freedom, standard temperature;
y^i, y_v^L – mass fraction of the ith component, of the active laser component L in the vth vibrational state;
A – cross-sectional area of flow;
m – mass flow of mixture;
μ, μ_i – molecular weight of mixture, of the ith component;
h^i, h_v^L – specific enthalpy of the ith component, of the active laser component L in the vth vibrational state;
M, N – total number of reactions, total number of mixture components.

The conservation equations describing the steady-state quasiunidimensional flow of an energy-emitting multicomponent reacting gas mixture have the following form:

General continuity equation

$$\frac{1}{\varrho}\frac{d\varrho}{dx}+\frac{1}{u}\frac{du}{dx}+\frac{1}{A}\frac{dA}{dx}=0 \ , \tag{2.100a}$$

or, in integral form,

$$\varrho u A = \dot{m} \ ; \tag{2.100b}$$

momentum conservation equation

$$u\frac{du}{dx}=-\frac{1}{\varrho}\frac{d\varrho}{dx} \ ; \tag{2.101}$$

energy conservation equation

$$\varrho u \frac{dh}{dx} - u \frac{dp}{dx} = -P_1 , \quad \text{where} \tag{2.102}$$

$$h = \sum_{i=1}^{N} h^i \frac{y^i}{\mu_i} + \sum_{v=0}^{R} h_v^L \frac{y_v^L}{\mu_L} ;$$

equations of continuity of chemical components (here, for simplicity, we will restrict ourselves to the case of rotational equilibrium; the generalization to the case where rotational equilibrium is absent presents no difficulties and is attained by increasing the number of variables, as for the resting medium above)

$$\frac{dy^i}{dx} = \frac{\mu_i}{\varrho u} W_i , \quad i = 1, 2, \ldots, N , \tag{2.103}$$

$$\frac{dy_v^L}{dx} = \frac{\mu_L}{\varrho u} (\chi_v^{\text{rad}} + \chi_v^{\text{chem}} + \chi_v^{\text{rel}}) , \quad v = 0, 1, \ldots, R . \tag{2.104}$$

According to the phenomenological law of chemical kinetics, the quantities W_i for reactions of general form (2.90) are defined by the equations

$$W_i = \sum_{j=1}^{M} (v_{i,j}'' - v_{i,j}') \left[k_{f,j} \prod_{i=1}^{N} \left(\frac{\varrho}{\mu_i} y^i \right)^{v_{i,j}'} - k_{b,j} \prod_{i=1}^{N} \left(\frac{\varrho}{\mu_i} y^i \right)^{v_{i,j}''} \right] ,$$

where account is taken of the fact that the concentrations n^i are related to the mass fractions y^i by the relation

$$n^i = \frac{\varrho}{\mu_i} y^i .$$

Similarly, the following substitution should be made in the expressions for χ_v^{rad}, χ_v^{chem}, and χ_v^{rel} entering into (2.104):

$$n_v = \frac{\varrho}{\mu_i} y_v^L .$$

Equations (2.103) and (2.104) hold true both for resting media and for moving media where density is time-dependent. Here resides the advantage of using the variables y^i and y_v^L instead of the variables n^i and n_v.

The photon density in the cavity is defined by equations (2.84), where for steady-state flow, $d/dt = u(d/dx)$, and the specific radiation power is given by formula (2.85). The set of $4 + N + R$ equations (2.100a), (2.101)–(2.104) contains $6 + N + R$ unknowns p, ϱ, u, A, T, y^i ($i = 1, 2, \ldots, N$), and y_v^L ($v = 0, 1, \ldots, R$). Use is also made of the following two additional equations: the equation of state

$$\frac{1}{p}\frac{dp}{dx} - \frac{1}{\varrho}\frac{d\varrho}{dx} - \frac{1}{T}\frac{dT}{dx} + \frac{1}{\mu}\frac{d\mu}{dx} = 0 , \tag{2.105a}$$

or, in integral form,

$$p = (\varrho R^0 T)/\mu \ , \quad \text{where} \tag{2.105b}$$

$$\mu^{-1} = \sum_{i=1}^{N} y^i/\mu_i \ ;$$

and the flow model equation

$$\frac{a_1}{T}\frac{dT}{A} + \frac{a_2}{A}\frac{dA}{dx} + \frac{a_3}{\varrho}\frac{d\varrho}{dx} + \frac{a_4}{p}\frac{dp}{dx} = b(x) \ . \tag{2.106}$$

The meaning of (2.106) can be most simply explained by considering specific examples. For restricted flows (e.g., in nozzles), one should put in (2.106) $a_1 = a_3 = a_4 = 0, a_2 = 1$. In that case, the area $A(x)$ is actually a known function of x. In the case of free flows, the assumption formally reflected by the conditions $a_1 = a_2 = a_3 = 0$, $b = 0$, $a_4 = 1$ frequently proves adequate. The area $A(x)$, like the other variables, is then found from the solution of the set of equations.

The specific enthalpies of the mixture components are defined by the caloric equation of state (2.99). Since the active molecules in different vibrational states are treated as individual mixture components, it is convenient to represent (2.99) in the form

$$h^i = h^i_0 + \int_{T_0}^{T} c^i_p dT \ , \quad i = 1, 2, \ldots, N \ , \quad i \neq L \ ,$$

$$h^L_v = h^L_0 + \varepsilon_v + \int_{T_0}^{T} c^L_p dT \ , \quad v = 0, 1, \ldots, R \ ,$$

where the specific heat at constant pressure c^L_p of the active laser component allows only for the rotational and translational motion of the molecules and ε_v is the energy of the vth vibrational level.

The radiation power Q_l as a function of the coordinate along the flow is given by

$$Q_l(x) = \int_{x_0}^{x} P_l(x')A(x')dx' \ ,$$

where x_0 is defined by the condition $P_l(x) > 0$ at $x > x_0$.

With numerical calculations, the set of equations is solved for the derivatives of the unknowns [2.120], which makes it possible to carry out numerical integration by standard techniques.

2.6 Calculation of the Generation and Amplification of Radiation in Multi-Level Chemical Lasers in Quasistationary Approximation

To perform the quantitative analysis of the kinetics of the processes taking place in the active medium of a pulsed chemical laser and calculate the laser performance characteristics, it is necessary to solve a set of differential equations including the energy conservation and balance equations which describe the reaction mechanism and the main interactions in the system of active molecules: excitation of the laser levels, vibrational relaxation, and radiative transitions. This approach was formulated as far back as 1969 by *Igoshin* and *Oraevsky* [2.125] using by way of example the chain reaction between hydrogen and chlorine. When calculating cw chemical laser properties, the motion of the medium and the physico-chemical kinetics involved must be considered simultaneously.

Computations of the characteristics of particular systems by way of direct numerical integration of a closed set of the equations of chemical kinetics, vibrational relaxation equations, balance equations for the density of photons in the cavity, and hydrodynamic conservation equations presented in Sect. 2.5 involve a number of difficulties. In particular, these difficulties are associated with the use of a great number of differential equations to describe the radiative processes occurring in such a multi-level system as the chemical laser. To solve these equations takes much computer time, even where modern high-speed computers are used. What is more, the stability of numerical solution cannot always be guaranteed because of the discrete nature of computations. All these difficulties have stimulated the development of a more effective numerical procedure for computing the kinetics of chemical lasers, based on a quasistationary approximation in calculating the radiation intensity [2.119, 126]. An important step in the development of this technique has been taken by *Airey* [2.126], who analyzed the simple $Cl_2 + HI$ system with the HCl molecules lasing only in the 1–0 band. Generalizing the approach used in [2.126] to the case of multi-level systems, *Turner*, *Adams*, and *Emanuel* [2.119] have formulated a method for calculating conditions existing in a chemically reacting gas placed in an optical cavity. The results obtained by these investigators are applicable to molecular lasers using nonchemical excitation. Calculation by this method can be made both for resting gas mixtures (where all processes depend only on time and not on spatial position) and for moving media. In the latter case, the motion is considered to be steady and unidimensional, while the coherent radiation is taken to be at right angles to the flow. The physical assumptions at the root of the analysis are as follows.

To calculate the radiation intensity, use is made of what is known as the constant gain technique [2.127]. With this technique, gain is kept constant throughout the cavity and is determined by the condition of equality between gain and loss. The spatial dependence of interaction between the radiation field and the active medium, which exists in actual cavities, is not allowed for. Moreover, the technique essentially uses the assumption that the stimulated emission of radiation is quasistationary, i.e., no account is taken of transients in the emission process.

The populations of the rotational sublevels of a given vibrational level are described by the Boltzmann distribution at the gas temperature, with only one transition within a given vibrational band, namely, that which provides for the maximum gain, oscillating at any given moment of time. This transition is not fixed and can shift to other rotational states in the course of lasing.

The consistent use of these assumptions [2.119] leads to the formulation of an algorithm circumventing the difficulty of direct numerical integration of the differential balance equations for the photon density in the cavity by replacing them with algebraic equations. On the whole, the problem consists in the simultaneous solution of a set of differential and nonlinear algebraic equations. A complicating factor is the large number of discontinuities that the unknowns suffer when the inversion density reaches its threshold value, when the radiation spectrum changes (J-shifting), and when oscillation is quenched. The occurrence of discontinuities is due to the fact that the technique fails to describe the evolution and quenching of oscillations in time. The radiation field saturating the lasing transition is taken to arise or decay instantaneously, the moments this happens being determined by means of special criteria. Thus, the criterion for the onset of oscillation is the gain reaching its threshold value. The ensuing calculation algorithm is much more complicated in its logical scheme than the algorithm for the direct numerical integration of the basic equations, which can be performed by routine programs. This complication is probably justified, for stability of numerical solution is attained without losing much in accuracy, while the computation speed is increased.

The above formalism makes it possible to analyze the basic stages in the total process taking place in the active medium of a chemical laser, starting with the onset of reaction and ending with the establishment of complete equilibrium in the system, particularly the production and decay of the inverted population of vibrational-rotational states, the onset and temporal behavior of lasing, and redistribution of the molecules among vibrational levels. Of course, calculation is only possible if the reaction mechanism is established, the elementary process rate constants are known, and the necessary thermodynamic and spectroscopic data are available. But in that case, it is necessary to develop cumbersome kinetics computer programs.

A simpler technique for calculating stimulated emission of radiation in multilevel molecular systems has been developed by *Igoshin* and *Masterov* [2.120, 128]. Using the quasistationary approximation, they have obtained an analytical expression for the oscillation power as a function of the partial rates of excitation of vibrational states, spectroscopic constants of the active molecule, and kinetic constants of the relaxation processes. This has made it possible to reduce the task of calculating the oscillation power to the calculation of the chemical kinetics involved (as in the case of the two-level medium considered in Sect. 2.1), which is a much more simple matter. This technique disregards some details typical of the behavior of a real system, such as the finite character of the threshold inversion accumulation time and nonsynchronous onset and quenching of oscillation in different vibrational bands. However, typical calculations show that over a wide range of conditions of practical interest these effects are of little importance to the total energy characteristics of the laser. At the same time, the analysis technique, despite its simple calculation algorithm, enables one to consider the main factors governing the behavior of a laser with vibrational-rotational population inversion (under rotational equilibrium conditions).

To calculate the stimulated emission power and energy, it is necessary to consider the vibrational kinetics of the active molecules, while allowing for the effect of the stimulated emission field. For the laser oscillator mode, the equations of the population density of vibrational states and the density of photons in the cavity may be written as

$$\frac{d\varrho_{v-1,J}^{v,J+m}}{dt} = c\sigma_{v-1,J}^{v,J+m} \Delta_{v-1,J}^{v,J+m} \varrho_{v-1,J}^{v,J+m} - \frac{1}{\tau_{ph}} \varrho_{v-1,J}^{v,J+m} , \tag{2.107a}$$

$$\frac{dn_v}{dt} = W_v + c\sigma_{v,J}^{v+1,J+m} \Delta_{v,J}^{v+1,J+m} \varrho_{v,J}^{v+1,J+m} - c\sigma_{v-1,J}^{v,J+m} \Delta_{v-1,J}^{v,J+m} \varrho_{v-1,J}^{v,J+m}$$
$$+ n_{v+1} g_{v+1,v} - n_v(g_{v,v-1} + g_{v,v+1}) + n_{v-1} g_{v-1,v} , \tag{2.107b}$$

where W_v is the rate of excitation of the vth level by the pumping source; n_v $= \Sigma_J n_{v,J}$ is the population density of the vth vibrational level; $\Delta_{v-1,J}^{v,J+m}$, $\varrho_{v-1,J}^{v,J+m}$, and $\sigma_{v-1,J}^{v,J+m}$ are the inverted population density, the photon density in the cavity, and the stimulated emission cross section, respectively, corresponding to the transition $(v, J +m) \to (v-1, J)$;

$$g_{v\pm 1,v} = \sum_M k_{v\pm 1,v}^M n_M + \sum_s n_{s\pm 1} k_{s\pm 1,v\pm 1;s,v}$$

is the coefficient describing the effect of the $V \to T$ and $V \to V$ processes on the population of the vth level; $k_{v\pm 1,v}^M$ is the rate constant for the $V \to T$ relaxation of the active molecules by collision with the component M present in a concentration of n_M; and $k_{s\pm 1,v\pm 1;s,v}$ is the rate constant for the $V \to V$ process occurring upon collisions of the active molecules between themselves.

When writing (2.107a) and (2.107b) and performing the subsequent analysis, the following assumptions are essential:

1) Oscillation occurs only on transitions between adjacent vibrational levels and simultaneously in all the bands $V \to V-1$ under consideration.

2) Within each vibrational band, only one rotational line is generated, its rotational quantum number J being independent of the vibrational quantum number v. This assumption is made first of all to simplify calculations, although the analysis can be carried out in a more general form. The study of the radiation spectrum dynamics of a chemical HF laser [2.129, 130] has also shown that oscillation within different vibrational bands at any given moment of time occurs in transitions with close J values. It is unlikely that analyzing the oscillation for a more complex radiation spectrum will yield qualitatively new results. Indeed, from the qualitative point of view, the presence of the rotational degree of freedom in molecules only manifests itself in one effect typical of molecular lasers, namely, in the possibility of lasing in the absence of full population inversion of vibrational levels (partial inversion phenomenon). This feature of molecular lasers is allowed for even within the framework of the simple radiation model adopted here. From the quantitative standpoint, the adopted assumption is quite justified, provided that the difference ΔJ between the laser lines in various bands satisfies the condition $\Delta J \theta_r / T \ll 1$, where θ_r and T are the characteristic rotational and gas temperatures, respectively. A generalization to the case of a more complex radiation spectrum where the quantity J depends on v in a random way has been carried out by *Igoshin* and *Masterov* [2.120].

3) Consideration is only given to single-quantum collisional transitions between vibrational levels.

4) The analysis covers the interaction of a finite number of vibrational levels, numbered $0, 1, 2, \ldots, R$. The position of the uppermost level under consideration is determined by the problem in hand. In the case of chemical pumping, only those levels which become excited in the course of the reaction are taken into consideration. However, higher levels can also undergo excitation as a result of the $V \rightarrow V$ energy exchange. According to current knowledge, such an inversion mechanism plays an important role in electrically-pumped CO lasers. In principle, no special restrictions other than the common quantum-mechanical constraints are imposed on the possible value of R in the model considered here.

5) The effect of spontaneous emission of radiation on the population of vibrational levels in the laser oscillator mode is neglected.

6) The system is considered to be in rotational-translational equilibrium.

Let us define the transition matrix $W_{v'v}$ to ensure that the rate W_v of excitation (or depletion if $W_v < 0$) of the level numbered v is expressed in terms of the elements of this matrix as follows:

$$W_v = \sum_{v' \neq v} (W_{v'v} - W_{vv'}) + W_{vv} . \tag{2.108}$$

The diagonal elements of the transition matrix correspond to the excitation of the vibrational levels of the active molecule in the nascent state (chemical pumping). The nondiagonal elements correspond to transitions between various vibrational levels occurring under other types of excitation (optical pumping, electron collision). Obviously, in the case of chemical pumping, only the diagonal transition matrix elements can be other than zero, whereas with nonchemical excitation, it is only the nondiagonal matrix elements that can be nonzero. The actual form of the matrix elements $W_{v'v}$ is determined by the excitation mechanism. Thus, in the case of electrical pumping,

$$W_{v'v} = n_{v'} n_e \left(\frac{2}{m_e} \right)^{1/2} \int \sigma_{v'v}(\varepsilon) f(\varepsilon) \varepsilon^{1/2} d\varepsilon ,$$

where $\sigma_{v'v}(\varepsilon)$ is the cross section of the process $A(v') + e \rightarrow A(v) + e$, which depends on the electronic energy ε, A is the chemical formula of the active molecule, n_e is the electron density, $f(\varepsilon)$ is the electron energy distribution function found from the solution of the Boltzmann kinetic equation, and m_e is the electron mass.

To characterize pumping in a more graphic way, let us represent the rate of excitation of the vth level in the equivalent form

$$W_v = \alpha_v W , \tag{2.109}$$

where the quantity

$$W = \sum_{v=0}^{R} \sum_{v'=0}^{R} W_{v'v} \tag{2.110}$$

is equal to the total number of elementary pumping events per unit volume per unit time and

$$\alpha_v = \frac{1}{W}\left[\sum_{v'\neq v}(W_{v'v} - W_{vv'}) + W_{vv}\right] . \tag{2.111}$$

The coefficients α_v describe the excitation rate distribution among vibrational levels, the quantity

$$\varepsilon_1 = \sum_{v=0}^{R} v\alpha_v \tag{2.112}$$

being equal to the average number of vibrational quanta acquired by the molecule during one elementary pumping event. In the case of chemical pumping, the coefficients α_v are non-negative, satisfy the normalization condition

$$\sum_{v=0}^{R} \alpha_v = 1 , \tag{2.113}$$

and can be regarded as the probabilities of the molecule being formed at the vth level in an elementary chemical laser process event. With nonchemical pumping, the condition

$$\sum_{v=0}^{R} \alpha_v = 0 \tag{2.114}$$

is satisfied, which is the condition for conservation of the number of molecules.

Let us analyze the set of equations (2.107), which contains $2R + 1$ equations in as many variables n_v and ϱ_{v-1}^v. Here we treat the rotational quantum number J as a known parameter of the problem in hand and omit this subscript on the quantities Δ_{v-1}^v, ϱ_{v-1}^v, and σ_{v-1}^v. The mathematical analysis is aimed mainly at deriving an expression for the coherent radiation power density P_l as a function of the quantities α_v, J, m, $g_{v\pm1,v}$ and W. The quantity P_l is defined as

$$P_l = \frac{1}{\tau_{ph}}\sum_{v=1}^{R} \hbar\omega_{v,v-1}(J)\varrho_{v-1}^v , \tag{2.115}$$

where $\hbar\omega_{v,v-1}(J)$ is the energy of the quantum emitted on the transition $(v, J + m) \rightarrow (v - 1, J)$. To calculate the energy of vibrational-rotational levels, we use for the sake of simplicity the harmonic oscillator, rigid rotator approximation. In this approximation,

$$\hbar\omega_{v,v-1}(J) = \hbar\omega + Bhcm(2J + 1 + m) , \tag{2.116}$$

where the frequency ω is independent of J, $m = -1, 0, 1$ for the P-, Q-, and R-branch, respectively, and B is the rotational constant. The results of the basic analysis allow generalization to the case of anharmonic oscillator, which has been carried out by $Igoshin$ and $Masterov$ [2.128]. In the presence of equilibrium in the rotational level system,

$$\Delta_{v-1}^v = A_J(n_v - B_J n_{v-1}) , \tag{2.117}$$

where

$$A_J = \frac{2}{Q}\left(\frac{1}{2} + J + m\right)\exp\left[-Bhc\frac{(J+m)(J+m+1)}{kT}\right] \tag{2.118}$$

and

$$B_J = \exp\left(Bhcm\frac{2J+1+m}{kT}\right) , \tag{2.119}$$

$Q = kT/(Bhc)$ being the rotational statistical sum.

In the laser oscillator mode, the distribution of molecules among vibrational levels is determined mainly by the interaction between the molecules and the radiation field. In the course of lasing, the difference in population between any pair of levels involved in stimulated transitions is maintained at a practically constant, threshold level. To be more exact, it is the gain, equal to $\sigma_{v-1}^v \Delta_{v-1}^v$, that remains constant. But as a rule, the stimulated transition cross section changes only little during oscillation, and with this effect disregarded, one may speak of the inversion density constancy. What is more, the change of the pumping rate during the characteristic stimulated photon emission time is usually negligibly small, i.e.,

$$\frac{1}{W}\frac{dW}{dt}(\sigma_{v-1}^v \varrho_{v-1}^v c)^{-1} \ll 1 . \tag{2.120}$$

These two circumstances allow the quasistationary approximation to be used to calculate the emission power P_l.

The physical meaning of the quasistationary approximation is that the coherent radiation power in this approximation "follows" the pumping rate (see Sect. 2.1). If condition (2.120) is satisfied, the quasistationary approximation will suffice for calculating the energy parameters of a laser.

In the quasistationary approximation, we may put

$$\frac{d\varrho_{v-1}^v}{dt} = 0 , \tag{2.121}$$

$$\frac{d\Delta_{v-1}^v}{dt} = 0 , \tag{2.122}$$

and so the set of differential equations for the inverted population density and the density of photons in the cavity is reduced to a set of algebraic equations. Differentiating the quantity Δ_{v-1}^v, we assume for the time being that the gas temperature T is constant and the changes in the populations of vibrational levels are due only to elementary physico-chemical processes and not to the expansion of the medium or diffusion. It is only if the latter condition is satisfied that equation (2.107b) holds true.

Taking into consideration (2.107a), we find from (2.121) that in the course of lasing

$$\Delta_{v-1}^v = 1/c\sigma_{v-1}^v \tau_{ph} . \tag{2.123}$$

Using recurrence relation (2.123), let us express the population of the vth level in terms of that of the zero level:

$$n_v = \frac{1}{A_J} \sum_{k=1}^{v} B_J^{v-k} \Delta_{k-1}^k + B_J^v n_0 \ . \tag{2.124a}$$

The more general expression for the distribution of molecules among vibrational levels in the laser oscillator mode in the case where the quantum number $J(v)$ of the laser line differs between the different bands $v \to v-1$ involved in the radiation process has the form

$$n_v = \sum_{k=1}^{v} \frac{\Delta_{k-1}^k}{A_{J(k)} B_{J(k)}} \prod_{s=k}^{v} B_{J(s)} + \prod_{s=1}^{v} B_{J(s)} n_0 \ . \tag{2.124b}$$

The population of the zero level during the process of oscillation may be obtained from the distribution law (2.124) found above and the normalization condition

$$\sum_{v=0}^{R} n_v = n \ . \tag{2.125}$$

Carrying out the summation on the left-hand side of (2.125), we have

$$n_0 = n \frac{1-B_J}{1-B_J^{R+1}} - \frac{1}{A_J} \sum_{v=1}^{R} \Delta_{v-1}^v \frac{1-B_J^{R+1-v}}{1-B_J^{R+1}} \ . \tag{2.126}$$

In the case of chemical pumping, the total number of active molecules produced by reaction in unit volume by the moment t is

$$n = \int_0^t W(\tau) d\tau \ .$$

Further, taking into consideration the right-hand sides of equations (2.107b), we may find from the R equations (2.122) the density of photons emitted in the transition $v \to v-1$:

$$\varrho_{v-1}^v = \frac{1}{1-B_J} \left[(1-B_J^v)\varrho_0^1 - \tau_{ph} \sum_{k=1}^{v-1} (1-B_J^{v-k})a_k \right] \ , \tag{2.127}$$

where

$$\varrho_0^1 = \frac{\tau_{ph}}{1-B_J^{R+1}} \sum_{k=1}^{R} (1-B_J^{R-k+1})a_k \tag{2.128}$$

and

$$a_k = W(\alpha_k - B_J \alpha_{k-1}) + n_{k+1}g_{k+1,k} - n_k(g_{k,k+1}+g_{k,k-1}) + n_{k-1}g_{k-1,k}$$
$$- B_J[n_k g_{k,k-1} - n_{k-1}(g_{k-1,k}+g_{k-1,k-2}) + n_{k-2}g_{k-2,k-1}] \ . \tag{2.129}$$

Based on the relations (2.127)–(2.129) found above, the expression for the emission power density is reduced, after algebraic manipulations, to the form allowing for a graphic physical interpretation:

$$P_l = \hbar\omega(J)\left[W\left(\varepsilon_1 - \varepsilon_2 \sum_{v=0}^{R} \alpha_v\right) - G\right] ;$$
(2.130)

here

$$\varepsilon_2 = \frac{B_J + RB_J^{R+2} - (R+1)B_J^{R+1}}{(1-B_J)(1-B_J^{R+1})}$$
(2.131)

and

$$G = \sum_{v=1}^{R} (n_v g_{v,\,v-1} - n_{v-1} g_{v-1,\,v}) .$$
(2.132)

The quantity ε_1 is defined by formula (2.112) and, as mentioned earlier, represents the average number of vibrational quanta acquired by the active molecule during an elementary pumping event. The quantity ε_2 is the average number of vibrational quanta emitted per molecule in the process of stimulated emission of radiation in the limiting case where the threshold population is equal to zero. The latter is only possible in a loss-free cavity where the condition $1/c\tau_{ph}\sigma_{v-1}^{v} = 0$ is satisfied, or under radiation amplification conditions.

The interpretation of the quantity ε_2 can be substantiated. Indeed, in the limiting case where the threshold population Δ_{v-1}^{v} is equal to zero, it is not very difficult to find from distribution (2.124a) that the expression $\Sigma_{v=0}^{R} v n_v/n$ coincides with the right-hand side of formula (2.131). The difference $(\varepsilon_1 - \varepsilon_2 \Sigma_{v=0}^{R} \alpha_v)$ is equal to the average number of photons that can be emitted by a single active molecule in the absence of relaxation processes. It is obvious from expressions (2.130) and (2.114) that the case of chemical pumping differs essentially from that of non-chemical pumping. In the latter case, the quantity ε_2 has no effect on the energy characteristics of a laser. This difference can be interpreted as follows. The quantity $\hbar\omega\varepsilon_2 n$ is the nonequilibrium energy store in the medium necessary to sustain oscillation. With nonchemical excitation, the number of molecules constituting the active medium is constant. For this reason, the pumping power is consumed to effect nonequilibrium excitation of the active medium only at the initial stage till the onset of oscillation. Subsequently [and it is precisely this stage that is described by formula (2.130)] pumping must only maintain the nonequilibrium distribution that the relaxation processes try to destroy. On the contrary, in the case of chemical pumping, the number n of particles increases with time. Therefore, in the course of oscillation, the pumping power is expended not only to compete with relaxation, but also continuously to pump up the medium to the necessary nonequilibrium excitation level. Finally, the quantity G describes the effect of the $V \to T$ and $V \to V$ processes on the stimulated emission power.

Let us analyze the function G for molecules modeled by harmonic oscillators. Within the scope of the harmonic oscillator model, the following relations hold true:

$$g_{v,\,v+1} = (v+1)g_{0,\,1} ,$$
(2.133a)

$$g_{v+1,v} = (v+1)g_{1,0} , \quad \text{where} \tag{2.133b}$$

$$g_{0,1} = \sum_M k_{0,1}^M n_M + \sum_{r=1}^R n_r r k_{0,1;1,0} , \tag{2.134a}$$

$$g_{1,0} = \sum_M k_{1,0}^M n_M + \sum_{r=1}^R n_{r-1} r k_{0,1;1,0} , \tag{2.134b}$$

$$k_{0,1}^M = k_{1,0}^M \exp\left(-\frac{\hbar\omega}{kT}\right) . \tag{2.135}$$

Taking into consideration (2.133)–(2.135), the quantity G may be written in the form

$$G = \left(\sum k_{1,0}^M n_M\right)\left\{\left[1 - \exp\left(-\frac{\hbar\omega}{kT}\right)\right]\left(\varepsilon_2 n + \frac{1}{A_J}\sum_{v=1}^R \Delta_{v-1}^v f_v\right)\right.$$
$$\left. - \exp\left(-\frac{\hbar\omega}{kT}\right)[n - (R+1)n_R]\right\}, \tag{2.136}$$

where

$$f_v = \frac{1 - B_J^{R-v+1}}{1 - B_J}(v - \varepsilon_2) + \frac{B_J + (R-v)B_J^{R-v+2} - (R-v+1)B_J^{R-v+1}}{(1 - B_J)^2} .$$

An essential feature of the expression obtained for the quantity G is that it does not include the probabilities of the $V \to V$ energy exchange. Thus, the above analysis leads to an important conclusion: within the framework of the harmonic oscillator model, the $V \to V$ energy exchange under quasistationary oscillation conditions has no effect on the total stimulated emission power. The main function of the $V \to V$ exchange is the redistribution of the total emission power among the vibrational bands of the harmonic oscillator. It is important to note that this conclusion is true for a random distribution of molecules among vibrational levels and is only based on relations (2.133)–(2.135) for the rate constants of collisional processes and on the assumption that the photons emitted in the different $v \to v-1$ bands are of equal energies.

If the rotational quantum number J differs between different emitting states, then, with the number J varying over a wide range, this circumstance has the greatest effect on the vibrational energy distribution function and not on the energy of the emitted quanta. It can, therefore, be expected that in the case of a more complex oscillation spectrum, the total emission power in the quasistationary mode under consideration will also depend only weakly on the rate of the $V \to V$ processes.

As applied to chemical pumping, formula (2.136) enables one to calculate the energy parameters of a chemical laser when one reaction only contributes to the emission of radiation. The generalization of (2.130) to the case where the lasing molecules are excited in M reactions simultaneously has the form

$$P_l = \hbar\omega(J)\left[\sum_{l=1}^M W^{(l)}(\varepsilon_1^{(l)} - \varepsilon_2) - G\right], \tag{2.137}$$

where $W^{(l)}$ is the rate of the lth pumping reaction, $\varepsilon_1^{(l)} = \Sigma_{v=0}^R v\alpha_v^{(l)}$ is the average number of vibrational quanta excited in the active molecule in the lth reaction, $\alpha_v^{(l)}$ is the probability that the active molecule will be formed at the vth level in the lth reaction, and

$$\sum_{v=0}^R \alpha_v^{(l)} = 1 \ .$$

By similar manipulations, we easily find that in the case of varying macroscopic parameters – temperature and gas density – the specific stimulated emission power is [2.120]

$$P_l = \hbar\omega(J)\left\{ \sum_{l=1}^M W^{(l)}(\varepsilon_1^{(l)} - \varepsilon_2) - G + \frac{1}{c\tau_{ph}}\frac{1}{A_J} \right.$$

$$\times \left[\left(J(J-1)\frac{Bhc}{kT} - 1 \right)\frac{1}{T}\frac{dT}{dt} + \frac{1}{S(v_0)}\frac{dS(v_0)}{dt} + \frac{1}{\varrho}\frac{d\varrho}{dt} \right]$$

$$\times \frac{B_J}{1-B_J}\sum_{v=1}^R \frac{1}{\sigma_{v-1,J}^{v,J-1}}\left[v - \frac{(R+1)B_J^{R-v+1}(1-B_J^v)}{1-B_J^{R+1}} \right]$$

$$\left. - \frac{2JBhc}{kT^2}\frac{dT}{dt}\frac{B_J}{1-B_J}\sum_{v=0}^R n_v\left[v+1 - \frac{(R+1)B_J^{R-v}(1-B_J^{v+1})}{1-B_J^{R+1}} \right] \right\} \ , \tag{2.138}$$

where

$$\frac{1}{S(v_0)}\frac{dS(v_0)}{dt} = \left(\frac{1}{\varrho}\frac{d\varrho}{dt} + \frac{\varrho R^0 T^{1/2}}{\Delta v_L}\sum\frac{\gamma_i}{\mu_i}\frac{dy^i}{dt} \right)$$

$$\times \left[2a^2 - a\exp(-a^2) \middle/ \int_a^\infty \exp(-\xi)d\xi \right] - \frac{1}{2T}\frac{dT}{dt} \ ,$$

$S(v_0)$ is the normalization line form factor at the emission line center defined by formula (2.63),

$$a = \frac{\Delta v_L}{\Delta v_D}(\ln 2)^{1/2} \ ,$$

$$\Delta v_L = \frac{p}{T^{1/2}}\mu\sum_i\frac{\gamma_i}{\mu_i}y^i \ ,$$

$$\Delta v_D = \frac{v_0}{c}\left[2\ln 2kN_A T\left(\frac{1}{\mu_L}\right) \right]^{1/2} \ ,$$

and γ_i is the constant of the line broadening due to the component i.

It should be noted that, as a rule, the main contribution to the emission power is made by the terms in the first line of formula (2.138). When calculating the characteristics of a pulsed laser in the quasistationary approximation considered above, they perform simultaneous numerical integration of the chemical kinetics equations (2.91) and the energy conservation equation (2.97). In the case of cw mode, it is the hydrodynamic conservation equations (2.100)–(2.104), and (2.105) and (2.106) that undergo simultaneous numerical integration. The distribution of the active molecules among vibrational levels is in both cases specified by relations (2.124a) and (2.126), and the total specific emission power, by relation (2.137) or, more accurately, (2.138), where for steady-state unidimensional flow, $dt = (1/u)dx$.

In making numerical calculations, one can take into consideration the non-synchronous onset and quenching of oscillation in different vibrational bands by taking numerical summation over the positive values of ϱ^v_{v-1} only, where ϱ^v_{v-1} are defined by relations (2.127) and (2.128). In that case, information about the rates of the $V \to V$ processes must be read in the computer program. Note that if some ϱ^v_{v-1} becomes negative, then, strictly speaking, the vibrational energy distribution will not obey one and the same law (2.124a) at all vs. In that case, the distribution of form (2.124a) is a first approximation to the actual distribution.

In the above calculation technique, the rotational quantum number J [or a set of numbers $J(v)$, where $v = 1, 2, \ldots, R$] of the emitting states is considered to be a known parameter of the problem in hand. The computation method in which the number J [or a set of numbers $J(v)$] remains unchanged in the course of computation is applicable to an amplifier operating under saturation conditions or an oscillator with a mode selective cavity. In the case of the amplifier, the stimulated emission spectrum is determined by the spectrum of the input signal. After calculating the specific emission power, the radiation intensity I_1 at the exit from the amplifier is determined by the formula $I_1 = lP_1$, where l is the amplifying medium length.

In a laser oscillator, the coherent radiation spectrum is usually formed by the active medium itself, so that oscillation in each $v + 1 \to v$ band occurs on the transition providing for the maximum gain (nonselective cavity). The computation method for this mode of operation is as follows. On each integration step, we compute the gain $\alpha^{v+1, J-1}_{v, J}$ as a function of J [see (2.73)]. In (2.73), the distribution among the levels is defined by formulas (2.124a) and (2.126), the quantity $J(v)$ entering into these formulas being already determined on the preceding integration step. Should it turn out that the maximum gain corresponds to some $J(v)$ value other than that of the preceding step, the value of $J(v)$ in the next step is then replaced by the new one ensuring maximum gain. Initially the number $J(v)$ is taken to have the minimum possible value, i.e., unity. Thus, the number $J(v)$ itself becomes a result of calculation and can vary in the course of a reaction owing to the heating of the mixture.

Such a computation scheme can be implemented most easily for the case where J is the same for all the bands $v \to v - 1$. In that case, it is necessary to search for the rotational transition providing the maximum gain in only one band, e.g., $1 \to 0$. This approximate calculation technique gives a sufficiently good idea of the emission spectrum and does not change in any appreciable way predictions about the energy characteristics of a given laser, compared with the more stringent approach which takes into consideration the dependence of J on v.

Table 2.1. Vibrational distributions in some elementary exchange reactions $A + BC \xrightarrow{k(v)} AB(v) + C$

Reaction and vibrational distribution	f_v	Reference
$AB = $ HF, DF		
F + H$_2$		
$k(0):k(1):k(2):k(3) = 0.06:0.29:1:0.63$	0.67	[2.131]
$\quad k(1):k(2):k(3) = 0.28:1:0.55$	0.664	[2.132]
$\qquad\qquad 0.18:1:1.33$		[2.133]
$\qquad\qquad 0.31:1:0.47$	0.66	[2.134–2.136]
$\qquad\qquad 0.31:1:0.57$		[2.136]
$\qquad\qquad 0.3:1:0.5$		[2.137–2.138]
$\qquad\qquad 0.36:1::0.47$		[2.139]
$\qquad\qquad 0.31:1:0.51$	0.69	[2.146]
F + p-H$_2$		
$T = 77$ K $\quad k(1):k(2):k(3) = 0.25:1:0.56$		[2.140]
$T = 290$ K $\;k(1):k(2):k(3) = 0.26:1:0.59$		[2.140]
F + o-H$_2$		
$T = 77$ K $\quad k(1):k(2):k(3) = 0.28:1:0.45$		[2.140]
$T = 290$ K $\;k(1):k(2):k(3) = 0.3:1:0.47$		[2.140]
F + n-H$_2$		
$T = 236–364$ K $\quad k(3):k(2) = 0.39 \exp(117/RT)$		[2.141, 2.142]
$T = 172–432$ K $\quad k(2):k(1) = 2.14 \exp(254/RT)$		[2.141, 2.142]
F + p-H$_2$		
$T = 263–400$ K $\quad k(3):k(2) = 0.42 \exp(139/RT)$		[2.141]
$T = 120–449$ K $\quad k(2):k(1) = 2.83 \exp(262/RT)$		[2.141]
F + H$_2(J)$		
$T = 100–200$ K, $\quad J = 0$		
$\quad k(1):k(2):k(3) = 0.25:1:0.56$		[2.140]
$T = 100–200$ K, $\quad J = 1$		
$\quad k(1):k(2):k(3) = 0.28:1:0.45$		[2.140]
$T = 290$ K, $\quad J = 0$		
$\quad k(1):k(2):k(3) = 0.25:1:0.56$		[2.140]
$T = 290$ K, $\quad J = 1$		
$\quad k(1):k(2):k(3) = 0.3:1:0.47$		[2.140]
$T = 290$ K, $\quad J = 2$		
$\quad k(1):k(2):k(3) = 0.27:1:0.63$		[2.140]
F + H$_2$		
$T = 298$ K $\quad k(1):k(0) > 10$		[2.143]
$T = 172$ K $\quad k(2):k(1) = 5$		[2.143]
$T = 298$ K $\quad k(2):k(1) = 3.8$		[2.143]
$T = 432$ K $\quad k(2):k(1) = 3$		[2.143]
$T = 298$ K $\quad k(3):k(2) = 0.45–0.62$		[2.143]
F + D$_2$		
$T = 300$ K		
$\quad k(1):k(2):k(3):k(4) = 0.15:0.52:1:0.59$	0.665	[2.132]
$\quad k(0):k(1):k(2):k(3):k(4) = 0.1:0.24:0.56:1:0.4$	0.6	[2.131]
$\quad k(1):k(2):k(3):k(4) = 0.28:0.65:1:0.71$	0.66	[2.134]
$\quad k(2):k(3):k(4) = 0.63:1:1.2$		[2.133]

Table 2.1 (continued)

Reaction and vibrational distribution	f_v	Reference
F + D$_2$		
$T = 301$ K		
$\quad k(3):k(2) = 1.51$		[2.142]
$T = 567$ K		
$\quad k(3):k(2) = 2.63$		[2.142]
F + HD → HF(v) + D		
$k(1):k(2):k(3) = 0.3:1:0.14$	0.588	[2.132]
$k(0):k(1):k(2):k(3) = 0.06:0.32:1:0.15$	0.59	[2.131]
F + HD → DF(v) + H		
$k(0):k(1):k(2):k(3):k(4) = 0.13:0.28:0.63:1:0.3$	0.55	[2.131]
$k(1):k(2):k(3):k(4) = 0.18:0.54:1:0.61$	0.626	[2.132]
F + CH$_4$		
$k(1):k(2):k(3) = 0.36:1:0.14$		[2.139]
$\qquad 0.33:1:0.23$		[2.137, 2.138]
$\qquad 0.24:1:0.11$		[2.136]
$\qquad 0.42:1:0.28$		[2.136]
$\qquad 0.38:1:0.26$		[2.144]
$\qquad 0.34:1:0.07$	0.55–0.58	[2.145]
$\qquad 0.34:1:0.2$	0.6	[2.146]
F + C$_2$H$_6$		
$k(1):k(2):k(3) = 0.34:1:0.73$		[2.139]
$\qquad 0.26:0.9:1$		[2.147]
$\qquad 0.28:1:0.72$	0.62	[2.146]
$\qquad 0.3:1:0.72$	0.6–0.63	[2.145]
F + C(CH$_3$)$_4$		
$k(1):k(2):k(3) = 0.29:1:0.43$		[2.136]
$\qquad 0.39:1:0.25$	0.56	[2.146]
F + CH$_3$Cl		
$k(1):k(2):k(3) = 0.32:1:0.95$	0.68	[2.146]
F + CH$_2$Cl$_2$		
$k(1):k(2):k(3) = 0.54:0.54:1$		[2.136]
$\qquad 0.87:0.76:1$	0.43–0.59	[2.146]
F + CH$_3$Br		
$k(1):k(2):k(3) = 0.44:0.78:1$	0.65–0.69	[2.146]
F + CH$_3$CF$_3$		
$k(1):k(2):k(3) = 0.32:1:0.27$	0.67	[2.146]
F + Si(CH$_3$)$_4$		
$k(1):k(2):k(3) = 0.51:1:0.19$	0.49–0.5	[2.146]
F + c-C$_6$H$_{12}$		
$k(1):k(2):k(3) = 0.22:1:0.34$	0.53	[2.146]
F + HCCl$_3$		
$k(1):k(2):k(3) = 1:0.8:0.37$	0.29–0.45	[2.146]
F + (CH$_3$)$_2$O		
$k(1):k(2):k(3) = 0.95:1:0.43$	0.36–0.45	[2.145]

Table 2.1 (continued)

Reaction and vibrational distribution	f_V	Reference
F + (CH₃)₂S		
$k(1):k(2):k(3) = 1:0.88:0.48$	0.4–0.49	[2.145]
F + (CH₃)₃N		
$k(1):k(2):k(3) = 0.68:1:0.39$	0.49–0.57	[2.145]
F + H₂O		
$k(1):k(2) = 1:0$		[2.144, 2.145]
$\quad\quad\quad\quad 1:0$	0.32–0.64	[2.145]
F + H₂O₂		
$k(1):k(2):k(3):k(4) = 0.97:1:0.81:0.48$	0.44–0.52	[2.145]
$\quad\quad\quad\quad 1:1:0.86:0.65$		[2.144]
$\quad\quad\quad\quad 1:0.44:0.27:0.06$		[2.144]
F + H₂S		
$k(1):k(2):k(3):k(4) = 0.9:1:0.9:0.6$	0.48–0.57	[2.145]
$\quad\quad\quad\quad 0.91:1:0.91:0.68$		[2.144]
$\quad\quad\quad\quad 1:0.73:0.92:0.43$		[2.144]
F + NH₃		
$k(1):k(2) = 1:0.89$	0.36–0.5	[2.145]
$\quad\quad\quad\quad 1:0.7$		[2.139]
F + N₂H₄		
$k(1):k(2):k(3):k(4):k(5) = 1:0.8:0.2:0.06:0.02$	0.2–0.3	[2.145]
F + PH₃		
$k(1):k(2):k(3):k(4):k(5) = 0.64:0.89:1:0.86:0.21$	0.45–0.49	[2.145]
F + SiH₄		
$k(1):k(2):k(3):k(4) = 0.45:0.67:1:0.26$	0.44–0.49	[2.149]
F + HCl		
$k(0):k(1):k(2):k(3) = 0.1:0.5:1:0.17$		[2.138]
$k(1):k(2):k(3) = 0.47:1:0.09$		[2.148]
$\quad\quad\quad\quad 0.5:1:0.17$	0.59	[2.149]
F + HBr		
$k(1):k(2):k(3):k(4) = 0.2:0.8:1:0.95$	0.55	[2.149]
F + HI		
$k(1):k(2):k(3):k(4):k(5):k(6) = 0.1:0.2:0.3:0.6:1:0.8$	0.62	[2.149]
F + GeH		
$k(1):k(2):k(3):k(4):k(5) = 0.24:0.44:0.71:1:0.05$		[2.150]
H + F₂		
$k(0):\ldots:k(10) = 0.04:0.09:0.11:0.13:0.45:0.89:1:$		
$\quad\quad\quad\quad 0.45:0.02:<0.04:<0.04$	0.58	[2.151]
$k(1):\ldots:k(9) = 0.12:0.13:0.25:0.35:0.78:1:0.4:$		
$\quad\quad\quad\quad 0.26:0.16$	0.53	[2.152]
$k(0):\ldots:k(9) = 0.025:0.052:0.072:0.075:0.334:$		
$\quad\quad\quad\quad 0.76:1:0.114:0.052:0.075$	0.53	[2.153]

Table 2.1 (continued)

Reaction and vibrational distribution	f_V	Reference
AB = HCl, DCl		
H + Cl$_2$		
$k(1): \ldots : k(6) = 0.2:0.8:1:0.2:0.03:0.003$	0.39	[2.154]
H + SCl$_2$		
$k(0): \ldots : k(5) = 0.3:0.53:0.72:1:0.83:0.25$	0.43	[2.155]
H + ICl		
$k(0): \ldots : k(7) = 0.16:0.3:0.63:0.79:0.81:0.96:1:0.75$	0.59	[2.156]
H + BrCl		
$k(1): \ldots : k(7) = 0.13:0.36:0.83-0.89:1:$		
$\qquad 0.78-0.85:0.28-0.3:0.015$	0.55	[2.156]
Cl + HI		
$k(1):k(2):k(3):k(4) = 0.2:0.4:1:0.8$	0.7	[2.157]
Cl + DI		
$k(1): \ldots : k(6) = 0:0.056:0.22:0.46:1:0.11$	0.7	[2.157]
AB = HBr		
H + Br$_2$		
$k(2):k(3):k(4):k(5) = 0.15:1:1:0.15$	0.55	[2.152, 2.154]
H + BrCl		
$k(0): \ldots : k(7) = 0.05:0.13:0.36:0.85:1:0.79:0.3:0.015$	0.58	[2.156]
AB = HI		
H + I$_2$		
$k(0): \ldots : k(6) = 0:0:0.04:0.42:1:0.19:0$		
(calculation, $T = 600\ K$)		
AB = CO		
O + CS		
$k(7): \ldots : k(18) = 0.06:0.27:0.61:0.66:0.8:0.87:$		
$\qquad 1:0.64:0.2:0:0:0$		[2.159]
$k(5): \ldots : k(18) = 0:0.05:0.17:0.32:0.41:0.55:0.65:$		
$\qquad 0.85:1:0.9:0.58:0.32:0.18:0$		[2.159]
$k(11): \ldots : k(18) = 0.6:0.87:1:0.72:0.28:0:0:0$		[2.159]

Table 2.2. Measures of energy release specificity in chemical reactions [2.39]

Reaction	$Cl + HI \rightarrow HCl + I$	$Cl + DI \rightarrow DCl + I$
E [kcal/mole]	34	34
f_V	0.71	0.71
λ_v	-8	-8
$\Delta S^{(vib)}$ [cgs units]	3.66	3.66

Table 2.3. Experimental data on vibrational relaxation rates of hydrogen halides

M	Process and kinetic data	T [K]	Reference
Relaxation of HF			

$$HF\ (v=1) + M \xrightarrow{\ k_{V \rightarrow RT}\ } HF\ (v=0) + M$$

M	Process and kinetic data	T [K]	Reference
HF	$p\tau = 0.014\ \mu s$ atm.	350	[2.49]
	$[9.5 \times 10^4\ s^{-1}\ (mm\ Hg)^{-1}]$		
	$(7 \pm 1) \times 10^4\ s^{-1}\ (mm\ Hg)^{-1}$	300	[2.160]
	$(4.95 \pm 0.4) \times 10^4\ s^{-1}\ (mm\ Hg)^{-1}$	350	[2.160]
	$p\tau = 13.9 \pm 1.6\ \mu s$ mm Hg	294	[2.161]
	$[7.2 \times 10^4\ s^{-1}\ (mm\ Hg)^{-1}]$		
	$8.7 \times 10^4\ s^{-1}\ (mm\ Hg)^{-1}$	300	[2.162]
	$p\tau = 0.06\ \mu s$ atm.	600	[2.162]
	$p\tau = 0.09\ \mu s$ atm.	900	[2.162]
	$p\tau = 0.1\ \mu s$ atm.	1900	[2.162]
	$p\tau = 0.09\ \mu s$ atm.	2400	[2.162]
	$p\tau = 0.024\ \mu s$ atm.	295	[2.163]
	$(5.6 \pm 0.5) \times 10^4\ s^{-1}\ (mm\ Hg)^{-1}$		
	$p\tau = 0.0572\ \mu s$ atm.	460	[2.163]
	$p\tau = 0.0876\ \mu s$ atm.	555	[2.163]
	$p\tau = 0.118\ \mu s$ atm.	720	[2.163]

$$DF\ (v=1) + M \xrightarrow{\ k_{V \rightarrow RT}\ } DF\ (v=0) + M$$

M	Process and kinetic data	T [K]	Reference
He	$< 30\ s^{-1}\ (mm\ Hg)^{-1}$	295	[2.167]
	$p\tau = 1.9 \times 10^{-3} \exp(101.5/T^{1/3})\ \mu s$ atm.	1500–3500	[2.188]
	$p\tau = 1.82 \times 10^{-4} \exp(135/T^{1/3})\ \mu s$ atm.	900–2600	[2.205]
Ar	$< 30\ s^{-1}\ (mm\ Hg)^{-1}$	295	[2.167]
	$p\tau = (7.1 \pm 1) \times 10^{-3} \exp[(128.6 \pm 6)/T^{1/3}]\ \mu s$ atm.	800–4000	[2.201]
	$p\tau = 1.46 \times 10^{-2} \exp(78.04/T^{1/3})\ \mu s$ atm.	1500–5000	[2.176]
	$p\tau = 1.27 \times 10^{-3} \exp(119/T^{1/3})\ \mu s$ atm.	910–1150	[2.205]
Relaxation of DF*			

$$DF\ (v=1) + M\ (v=0) \xrightarrow{\ k_{V \rightarrow V} + k_{V \rightarrow RT}\ } \begin{cases} DF(v=0) + M(v=1) \\ DF(v=0) + M(v=0) \end{cases}$$

$$DF\ (v=0) + M\ (v=1) \xrightarrow{\ k_{V \rightarrow RT}^{M-DF}\ } DF\ (v=0) + M\ (v=0)$$

*For additional data on relaxation of DF see pp. 97–101

Table 2.3 (continued)

M	Process and kinetic data	T [K]	Reference
HCl	$k_{V \to V} = (0.4 \pm 0.04) \times 10^6 \text{ s}^{-1} \text{ (mm Hg)}^{-1}$	295	[2.206]
	$k_{V \to RT}^{M-DF} + 0.86 k_{V \to RT}$		
	$= (1.7 \pm 0.6) \times 10^4 \text{ s}^{-1} \text{ (mm Hg)}^{-1}$	295	[2.206]
	$k_{V \to V} = (2.8 \pm 0.2) \times 10^{12} \text{ cm}^3/\text{mole s}$	745	[2.206]
HCl	$k_{V \to RT}^{M-DF} + (0.95 \pm 0.5) k_{V \to RT}$		
	$= (1.8 \pm 0.6) \times 10^{11} \text{ cm}^3/\text{mole s}$	745	[2.206]
HF	$p\tau = 0.178 \ \mu\text{s atm.}$	1030	[2.163]
	$5.6 \times 10^4 \text{ s}^{-1} \text{ (mm Hg)}^{-1}$	295	[2.164]
	$10^5 \text{ s}^{-1} \text{ (mm Hg)}^{-1}$	295	[2.165]
	$7.8 \times 10^4 \text{ s}^{-1} \text{ (mm Hg)}^{-1}$	350	[2.165]
	$p\tau = 0.07 \ \mu\text{s atm.}$	730	[2.165]
	$(8.74 \pm 0.1) \times 10^4 \text{ s}^{-1} \text{ (mm Hg)}^{-1}$ (dilution in Ar)	294	[2.166]
	$(4.4 \pm 0.3) \times 10^4 \text{ s}^{-1} \text{ (mm Hg)}^{-1}$ (pure HF)	294	[2.166]
	$(6.1 \pm 0.4) \times 10^4 \text{ s}^{-1} \text{ (mm Hg)}^{-1}$	295	[2.167]
	$(6.4 \pm 0.6) \times 10^4 \text{ s}^{-1} \text{ (mm Hg)}^{-1}$ (dilution in Ar)	295	[2.167]
	$(5.8 \pm 0.7) \times 10^4 \text{ s}^{-1} \text{ (mm Hg)}^{-1}$ (dilution in He)	295	[2.167]
	$(8.4 \pm 1) \times 10^4 \text{ s}^{-1} \text{ (mm Hg)}^{-1}$	297	[2.168]
	$(7.1 \pm 0.5) \times 10^4 \text{ s}^{-1} \text{ (mm Hg)}^{-1}$	321	[2.168]
	$(4.3 \pm 0.5) \times 10^4 \text{ s}^{-1} \text{ (mm Hg)}^{-1}$	395	[2.168]
	$(2.7 \pm 0.3) \times 10^4 \text{ s}^{-1} \text{ (mm Hg)}^{-1}$	475	[2.168]
	$(2.1 \pm 0.3) \times 10^4 \text{ s}^{-1} \text{ (mm Hg)}^{-1}$	570	[2.168]
	$(2 \pm 0.3) \times 10^4 \text{ s}^{-1} \text{ (mm Hg)}^{-1}$	678	[2.168]
	$(8.4 \pm 0.7) \times 10^4 \text{ s}^{-1} \text{ (mm Hg)}^{-1}$	295	[2.169]
	$(6.7 \pm 0.4) \times 10^4 \text{ s}^{-1} \text{ (mm Hg)}^{-1}$	324	[2.169]
	$(5.2 \pm 0.3) \times 10^4 \text{ s}^{-1} \text{ (mm Hg)}^{-1}$	350	[2.169]
	$(3.9 \pm 0.4) \times 10^4 \text{ s}^{-1} \text{ (mm Hg)}^{-1}$	420	[2.169]
	$(2.8 \pm 0.4) \times 10^4 \text{ s}^{-1} \text{ (mm Hg)}^{-1}$	500	[2.169]
	$(2.4 \pm 0.4) \times 10^4 \text{ s}^{-1} \text{ (mm Hg)}^{-1}$	580	[2.169]
	$(2 \pm 0.4) \times 10^4 \text{ s}^{-1} \text{ (mm Hg)}^{-1}$	670	[2.169]
	$p\tau = 0.022 \pm 0.002 \ \mu\text{s atm.}$	300	[2.170]
	$p\tau = 0.125 \pm 0.015 \ \mu\text{s atm.}$	940	[2.170]
	$(6.75 \pm 0.5) \times 10^4 \text{ s}^{-1} \text{ (mm Hg)}^{-1}$	295	[2.171]
	(average of data [2.160, 2.166, 2.169] with allowance made for gas-dynamic corrections)		
	$(8.05 \pm 0.8) \times 10^4 \text{ s}^{-1} \text{ (mm Hg)}^{-1}$	295	[2.171]
	(average of data [2.160, 2.166, 2.169] without allowance made for gas-dynamic corrections)		
	$(5.8 \pm 0.5) \times 10^4 \text{ s}^{-1} \text{ (mm Hg)}^{-1}$	295	[2.171]
	(average of data [2.163, 2.167, 2.172] where gas-dynamic effects are insignificant)		
	$5.6 \times 10^4 \text{ s}^{-1} \text{ (mm Hg)}^{-1}$	r.t.*	[2.172]
	$5.25 \times 10^4 \text{ s}^{-1} \text{ (mm Hg)}^{-1}$	350	[2.173]
	$(0.98 \pm 0.4) \times 10^{12} \text{ cm}^3/\text{mole s}$	298	[2.174]
	$5.3 \times 10^4 \text{ s}^{-1} \text{ (mm Hg)}^{-1}$		
	$1.38 \times 10^5 \text{ s}^{-1} \text{ (mm Hg)}^{-1}$	330	[2.182]
	$p\tau = 1.02 \times 10^{-2} \exp(34.39/T^{1/3}) \ \mu\text{s atm.}$	1350–4000	[2.175]
	$p\tau = 5.74 \times 10^{-3} \exp(42/T^{1/3}) \ \mu\text{s atm.}$	1500–5000	[2.176]
	$p\tau = 6.3 \times 10^{-4} \exp(64/T^{1/3}) \ \mu\text{s atm.}$	1400–4000	[2.177]

*r.t. – room temperature

Table 2.3 (continued)

M	Process and kinetic data	T [K]	Reference
HF	HF $(v=2)+M \xrightarrow{k_{V \to RT}}$ HF $(v=1)+M$ $k_{2,1} \simeq 6k_{1,0}=7.5 \times 10^{12}$ cm^3/mole s		[2.178]
HF	HF $(v=1)+M$ $(v=1)$ $\xrightarrow{k_{V \to V}}$ HF $(v=0)+M$ $(v=2)$ 2.2×10^{13} cm^3/mole s	295	[2.163]
HF	HF $(v=2)+M$ $(v=0) \to$ HF $(v<2)+M(v \geqslant 0)$ (joint effect of $V \to V$ and $V \to RT$ relaxation processes) $(9.9 \pm 3) \times 10^{12}$ cm^3/mole s 1.1×10^{13} cm^3/mole s 1.5×10^{13} cm^3/mole s $p\tau = 6.6 \pm 1.7$ μs mm Hg 7.8×10^{12} cm^3/mole s	298 r.t.* r.t.* 294 295	[2.174] [2.179] [2.172] [2.161] [2.180]
HF	HF $(v=3)+M$ $(v=0) \to$ HF $(v<3)+M$ $(v \geqslant 0)$ $(15.7 \pm 6) \times 10^{12}$ cm^3/mole s 1.2×10^{13} cm^3/mole s 2.9×10^{13} cm^3/mole s 1.1×10^{13} cm^3/mole s $(1.7 \pm 0.2) \times 10^{13}$ cm^3/mole s	298 298 298 298 293	[2.174] [2.179] [2.172] [2.180] [2.181]
HF	HF $(v=4)+M$ $(v=0) \to$ HF $(v<4)+M$ $(v \geqslant 0)$ $(16.3 \pm 6) \times 10^{12}$ cm^3/mole s $>3.2 \times 10^{13}$ cm^3/mole s 2.6×10^{13} cm^3/mole s $(4.3 \pm 0.3) \times 10^{13}$ cm^3/mole s 1.9×10^{13} cm^3/mole s	298 298 r.t.* 293 298	[2.174] [2.179] [2.172] [2.181] [2.180]
HF	HF $(v=5)+M$ $(v=0) \to$ HF $(v<5)+M$ $(v \geqslant 0)$ $(5.2 \pm 3) \times 10^{12}$ cm^3/mole s $>4.8 \times 10^{13}$ cm^3/mole s 2.8×10^{13} cm^3/mole s	298 300 298	[2.174] [2.179] [2.180]
HF	HF $(v=3, 4, 5)+M \xrightarrow{k_{V \to RT}}$ HF $(v-1)+M$ $k_{V \to RT} \sim v^{2.7}$	r.t.*	[2.241]
HF	HF $(v=3, 4, 5)+M \xrightarrow{k_{V \to RT}}$ HF $(v-1)+M$ $k_{V \to RT} \sim v^{2.7}$ to v^3	300–700	[2.242]
HF	HF $(v=1$ to $6)+M \xrightarrow{k_{V \to RT}}$ HF $(v-1)+M$ $k_{V \to RT} \sim v^{2.3}$	300–1000	[2.243]

*r.t. – room temperature

Table 2.3 (continued)

M	Process and kinetic data	T [K]	Reference
	$\text{HF}(v=1)+M \xrightarrow{k_{V \to RT}} \text{HF}(v=0)+M$		
DF	$<(4.5\pm0.7)\times10^4 \text{ s}^{-1} \text{ (mm Hg)}^{-1}$	300	[2.160]
	$<(3.3\pm0.5)\times10^4 \text{ s}^{-1} \text{ (mm Hg)}^{-1}$	350	[2.160]
	$\text{HF}(v=1)+M(v=0) \xrightarrow{k_{V \to V}+k_{V \to RT}} \begin{cases} \text{HF}(v=0)+M(v=1) \\ \text{HF}(v=0)+M(v=0) \end{cases}$		
DF	$(13.3\pm1)\times10^4 \text{ s}^{-1} \text{ (mm Hg)}^{-1}$	300	[2.160]
	$(9.8\pm1.5)\times10^4 \text{ s}^{-1} \text{ (mm Hg)}^{-1}$	350	[2.160]
	$(7.7\pm0.4)\times10^4 \text{ s}^{-1} \text{ (mm Hg)}^{-1}$	295	[2.164]
	$(6.7\pm0.5)\times10^4 \text{ s}^{-1} \text{ (mm Hg)}^{-1}$	295	[2.167]
	$(13.5\pm1)\times10^4 \text{ s}^{-1} \text{ (mm Hg)}^{-1}$	297	[2.168]
	$(11\pm0.8)\times10^4 \text{ s}^{-1} \text{ (mm Hg)}^{-1}$	321	[2.168]
	$(7.3\pm0.5)\times10^4 \text{ s}^{-1} \text{ (mm Hg)}^{-1}$	395	[2.168]
	$(5.1\pm0.3)\times10^4 \text{ s}^{-1} \text{ (mm Hg)}^{-1}$	475	[2.168]
	$(3.9\pm0.4)\times10^4 \text{ s}^{-1} \text{ (mm Hg)}^{-1}$	570	[2.168]
	$(3.2\pm0.3)\times10^4 \text{ s}^{-1} \text{ (mm Hg)}^{-1}$	678	[2.168]
	$\text{HF}(v=1)+M(v=0) \xrightarrow{k_{V \to V}} \text{HF}(v=0)+M(v=1)$		
DF	$7.1\times10^4 < k_{V \to V} < 14.3\times10^4 \text{ s}^{-1} \text{ (mm Hg)}^{-1}$	300	[2.160]
	$4.5\pm10^4 < k_{V \to V} < 11.3\times10^4 \text{ s}^{-1} \text{ (mm Hg)}^{-1}$	350	[2.160]
	$(2.4\pm1.8)\times10^4 \text{ s}^{-1} \text{ (mm Hg)}^{-1}$	295	[2.167]
	$\text{HF}(v=1)+M \to \begin{cases} \text{HF}(v=0)+M \\ MF(v=0)+H \end{cases}$		
H	$<0.5\times10^3 \text{ s}^{-1} \text{ (mm Hg)}^{-1}$	300	[2.183]
	$<9\times10^9 \text{ cm}^3/\text{mole s}$		
	$(7\pm4)\times10^{11} \text{ cm}^3/\text{mole s}$	295	[2.184]
	$(1.4\pm0.4)\times10^{11} \text{ cm}^3/\text{mole s}$	295	[2.83, 2.185]
D	$(1.8\pm1.6)\times10^{10} \text{ cm}^3/\text{mole s}$	295	[2.83]
	$\text{HF}(v=2)+M \to \begin{cases} \text{HF}(v<2)+M \\ MF(v<2)+H \end{cases}$		
H	$(5\pm2)\times10^{11} \text{ cm}^3/\text{mole s}$	295	[2.185]
	$(6.6\pm3)\times10^{11} \text{ cm}^3/\text{mole s}$	295	[2.85]
	$0.9\times10^{13} \text{ cm}^3/\text{mole s}$	295	[2.184]
	$\text{HF}(v=3)+M \to \begin{cases} \text{HF}(v<3)+M \\ MF(v<3)+H \\ HM+F \end{cases}$		
H	$1.4\times10^{13} \text{ cm}^3/\text{mole s}$	295	[2.184]
	$(6.3\pm1.5)\times10^{13} \text{ cm}^3/\text{mole s}$	295	[2.185, 2.85]
	$10^{14} \text{ cm}^3/\text{mole s}$	200	[2.186]
D	$5.9\times10^{13} \text{ cm}^3/\text{mole s}$	295	[2.186]
	$7.9\times10^{13} \text{ cm}^3/\text{mole s}$	200	[2.186]
	$\text{HF}(v=1)+M \to \begin{cases} \text{HF}(v=0)+M \\ MH(v=0)+F \end{cases}$		

Table 2.3 (continued)

M	Process and kinetic data	T [K]	Reference
F	$(0.9 \pm 0.2) \times 10^4 \text{ s}^{-1} \text{ (mm Hg)}^{-1}$	300	[2.183]
	$(1.7 \pm 0.4) \times 10^{11} \text{ cm}^3/\text{mole s}$	300	[2.187]
	$p\tau = 0.03 \ \mu\text{s atm.}$	2000	[2.162]
Cl	$(4.5 \pm 1) \times 10^{11} \text{ cm}^3 (\text{mole s})^{-1}$	300	[2.187]
O	$(1 \pm 0.2) \times 10^5 \text{ s}^{-1} \text{ (mm Hg)}^{-1}$	300	[2.183]
	$(1.9 \pm 0.4) \times 10^{12} \text{ cm}^3/\text{mole s}$	300	[2.187]

$$\text{HF } (v=1) + M \xrightarrow{k_{V \to RT}} \text{HF } (v=0) + M$$

M	Process and kinetic data	T [K]	Reference
He	$p\tau = 1.52 \times 10^{-4} \exp(133.3/T^{1/3}) \ \mu\text{s atm.}$	1350–4000	[2.175]
	$p\tau = 7.7 \times 10^{-4} \exp(117.2/T^{1/3}) \ \mu\text{s atm.}$	1500–3500	[2.188]
	$< 60 \text{ s}^{-1} \text{ (mm Hg)}^{-1}$	295	[2.167]
Ar	$p\tau = 1.62 \times 10^{-3} \exp(111.97/T^{1/3}) \ \mu\text{s atm.}$	1350–4000	[2.175]
	$p\tau = 8.49 \times 10^{-3} \exp(89.47/T^{1/3}) \ \mu\text{s atm.}$	1500–5000	[2.176]
	$40 > p\tau > 5 \ \mu\text{s atm.}$	800–2400	[2.162]
	$< 60 \text{ s}^{-1} \text{ (mm Hg)}^{-1}$	295	[2.166, 2.167]
	$< 10^2 \text{ s}^{-1} \text{ (mm Hg)}^{-1}$	350	[2.165]

$$\text{HF } (v=1) + M \xrightarrow{k_{V \to RT} + k_{V \to V}} \begin{cases} \text{HF } (v=0) + M \\ \text{HF } (v=0) + M^v \end{cases}$$

M	Process and kinetic data	T [K]	Reference
HCl	$(2.2 \pm 0.4) \times 10^4 \text{ s}^{-1} \text{ (mm Hg)}^{-1}$	350	[2.160]
	$(2.9 \pm 0.3) \times 10^4 \text{ s}^{-1} \text{ (mm Hg)}^{-1}$	300	[2.160]
	$(1.7 \pm 0.1) \times 10^4 \text{ s}^{-1} \text{ (mm Hg)}^{-1}$	295	[2.164]
	$(p\tau)^{-1} = 10^{5.7 \pm 0.2} \exp[(-156 \pm 20)/T^{1/3}]$ $(\mu\text{s atm.})^{-1}$	1600–3000	[2.189]
	$k_{V \to RT} < 1.2 \times 10^4 \text{ s}^{-1} \text{ (mm Hg)}^{-1}$	350	[2.160]
	$k_{V \to RT} < 1.6 \times 10^4 \text{ s}^{-1} \text{ (mm Hg)}^{-1}$	300	[2.160]

$$\text{HF } (v=3) + M \to \text{HF}(v<3) + M, \ M^v$$

M	Process and kinetic data	T [K]	Reference
HCl	$(0.4 \pm 0.04) \times 10^6 \text{ s}^{-1} \text{ (mm Hg)}^{-1}$	295	[2.191]

$$\text{HF}(v=1) + M \xrightarrow{k_{V \to V}} \text{HF } (v=0) + M^v$$

M	Process and kinetic data	T [K]	Reference
HCl	$(1.2 \text{ to } 2.5) \times 10^4 \text{ s}^{-1} \text{ (mm Hg)}^{-1}$	350	[2.160]
	$(1.6 \text{ to } 3.2) \times 10^4 \text{ s}^{-1} \text{ (mm Hg)}^{-1}$	300	[2.160]

$$\text{HF}(v=1) + M \xrightarrow{k_{V \to RT} + k_{V \to V}} \begin{cases} \text{HF } (v=0) + M \\ \text{HF}(v=0) + M^v \end{cases}$$

M	Process and kinetic data	T [K]	Reference
HBr	$(0.35 \pm 0.2) \times 10^4 \text{ s}^{-1} \text{ (mm Hg)}^{-1}$	350	[2.160]
	$(0.75 \pm 0.2) \times 10^4 \text{ s}^{-1} \text{ (mm Hg)}^{-1}$	300	[2.160]
	$(0.75 \pm 0.1) \times 10^4 \text{ s}^{-1} \text{ (mm Hg)}^{-1}$	295	[2.164]
	$k_{V \to RT} < 0.28 \times 10^4 \text{ s}^{-1} \text{ (mm Hg)}^{-1}$	350	[2.160]
	$k_{V \to RT} < 0.47 \times 10^4 \text{ s}^{-1} \text{ (mm Hg)}^{-1}$	300	[2.160]
	$k_{V \to V} = (0.28 \text{ to } 0.55) \times 10^4 \text{ s}^{-1} \text{ (mm Hg)}^{-1}$	350	[2.160]
	$k_{V \to V} = (0.47 \text{ to } 0.95) \times 10^4 \text{ s}^{-1} \text{ (mm Hg)}^{-1}$	300	[2.160]
HI	$(0.2 \pm 0.15) \times 10^4 \text{ s}^{-1} \text{ (mm Hg)}^{-1}$	350	[2.160]
	$(0.35 \pm 0.2) \times 10^4 \text{ s}^{-1} \text{ (mm Hg)}^{-1}$	300	[2.160]
DBr	$(0.44 \pm 0.15) \times 10^4 \text{ s}^{-1} \text{ (mm Hg)}^{-1}$	295	[2.190]

Table 2.3 (continued)

M	Process and kinetic data	T [K]	Reference
N_2	$(1.52 \pm 0.150) \times 10^2$ s^{-1} (mm Hg)$^{-1}$	295	[2.164]
	$(1.25 \pm 0.6) \times 10^2$ s^{-1} (mm Hg)$^{-1}$	294	[2.166, 2.195]
	$(2 \pm 1) \times 10^2$ s^{-1} (mm Hg)$^{-1}$	295	[2.165]
	$(2 \pm 1) \times 10^2$ s^{-1} (mm Hg)$^{-1}$	350	[2.165]
	$(1.45 \pm 0.15) \times 10^2$ s^{-1} (mm Hg)$^{-1}$	295	[2.191]
	$p\tau = 1.9$ μs atm.	2000	[2.193]
	$p\tau_{HF-D_2} < p\tau_{HF-N_2} < p\tau_{HF-He}$	1350–4000	[2.175]
	$HF(v=2) + M \rightarrow HF(v=1) + M, M^v$		
N_2	$(8.1 \pm 1) \times 10^2$ s^{-1} (mm Hg)$^{-1}$	295	[2.191]
	$HF(v=3) + M \rightarrow HF(v=2) + M, M^v$		
N_2	$(2.92 \pm 0.3) \times 10^3$ s^{-1} (mm Hg)$^{-1}$	295	[2.191]
	$HF(v=1) + M \xrightarrow{k_{V \rightarrow V} + k_{V \rightarrow RT}} \begin{cases} HF(v=0) + M \\ HF(v=0) + M^v \end{cases}$		
H_2	$(1.7 \pm 0.1) \times 10^4$ s^{-1} (mm Hg)$^{-1}$	295	[2.164]
	$(2.4 \pm 0.3) \times 10^4$ s^{-1} (mm Hg)$^{-1}$	294	[2.166]
	$(2.39 \pm 0.6) \times 10^4$ s^{-1} (mm Hg)$^{-1}$	295	[2.167]
	$(1.43 \pm 0.15) \times 10^4$ s^{-1} (mm Hg)$^{-1}$	295	[2.191]
	$k_{V \rightarrow RT} = (3.8 \pm 1.1) \times 10^2$ s^{-1} (mm Hg)$^{-1}$	295	[2.192]
	$HF(v=2) + M \rightarrow HF(v=1) + M, M^v$		
H_2	$(1.23 \pm 0.1) \times 10^4$ s^{-1} (mm Hg)$^{-1}$	295	[2.191]
	1.26×10^{11} cm^3/mole s	298	[2.180]
	$HF(v=3) + M \rightarrow HF(v=2) + M, M^v$		
H_2	$(1.13 \pm 0.1) \times 10^4$ s^{-1} (mm Hg)$^{-1}$	295	[2.191]
	0.9×10^{11} cm^3/mole s	298	[2.180]
	$(1.87 \pm 0.36) \times 10^{11}$ cm^3/mole s	293	[2.237]
	$HF(v=4) + M \rightarrow HF(v=3) + M, M^v$		
H_2	1.26×10^{11} cm^3/mole s	298	[2.180]
	$(2.83 \pm 0.72) \times 10^{11}$ cm^3/mole s	293	[2.237]
	$HF(v=5) + M \rightarrow HF(v=4) + M, M^v$		
H_2	2.94×10^{11} cm^3/mole s	298	[2.180]
	$HF(v=6) + M \rightarrow HF(v=5) + M, M^v$		
H_2	5.96×10^{11} cm^3/mole s	298	[2.180]
	$HF(v=7) + M \rightarrow HF(v=6) + M, M^v$		
H_2	9.63×10^{11} cm^3/mole s	298	[2.180]
	$HF(v=1) + M \rightarrow HF(v=0) + M, M^v$		
D_2	$(0.31 \pm 0.06) \times 10^4$ s^{-1} (mm Hg)$^{-1}$	295	[2.164]
	$(0.37 \pm 0.04) \times 10^4$ s^{-1} (mm Hg)$^{-1}$	294	[2.166, 2.195]
	$p\tau = 5.1 \times 10^{-4} \exp(96.6/T^{1/3})$ μs atm.	1350–4000	[2.175]

Table 2.3 (continued)

M	Process and kinetic data	T [K]	Reference
D$_2$	$HF(v=2)+M \rightarrow HF(v=1)+M, M^v$ 1.56×10^{11} cm^3/mole s	298	[2.180]
D$_2$	$HF(v=3)+M \rightarrow HF(v=2)+M, M^v$ 3.54×10^{11} cm^3/mole s $(8.06 \pm 1.56) \times 10^{11}$ cm^3/mole s	298 293	[2.180] [2.237]
D$_2$	$HF(v=4)+M \rightarrow HF(v=3)+M, M^v$ 7.2×10^{11} cm^3/mole s $(2 \pm 0.36) \times 10^{12}$ cm^3/mole s	298 293	[2.180] [2.237]
D$_2$	$HF(v=5)+M \rightarrow HF(v=4)+M, M^v$ 1.86×10^{12} cm^3/mole s	298	[2.180]
D$_2$	$HF(v=6)+M \rightarrow HF(v=5)+M, M^v$ 4.94×10^{12} cm^3/mole s	298	[2.180]
D$_2$	$HF(v=7)+M \rightarrow HF(v=6)+M, M^v$ 3.67×10^{12} cm^2/mole s	298	[2.180]
O$_2$	$HF(v=1)+M \rightarrow HF(v=0)+M, M^v$ $(4.5 \pm 0.6) \times 10^2$ s^{-1} (mm Hg)$^{-1}$ $p\tau = 9.7 \times 10^{-2} \exp(46.7/T^{1/3})$ μs atm. $p\tau = 3.3$ μs atm. $(3.5 \pm 0.25) \times 10^2$ s^{-1} (mm Hg)$^{-1}$	295 1200–3000 2000 294	[2.164] [2.188] [2.193] [2.194]
O$_2$	$HF(v=3)+M \rightarrow HF(v=2)+M, M^v$ $(7.5 \pm 1) \times 10^2$ s^{-1} (mm Hg)$^{-1}$	295	[2.191]
F$_2$	$HF(v=1)+M \rightarrow HF(v=0)+M, M^v$ $<10^2$ s^{-1} (mm Hg)$^{-1}$	350	[2.165]
NO	$(6 \pm 1) \times 10^3$ s^{-1} (mm Hg)$^{-1}$ $(6.2 \pm 0.3) \times 10^3$ s^{-1} (mm Hg)$^{-1}$ $p\tau = 10^{-1.1 \pm 0.2} \exp[(16 \pm 1)/T^{1/3}]$ μs atm.	295 294 1000–4100	[2.164] [2.194] [2.189]
CO	$(1.8 \pm 0.2) \times 10^3$ s^{-1} (mm Hg)$^{-1}$	295	[2.164]
ClF	$(8 \pm 2) \times 10^2$ s^{-1} (mm Hg)$^{-1}$	295	[2.190]
CO$_2$	6.6×10^{11} cm^3/mole s 4×10^4 s^{-1} (mm Hg)$^{-1}$ 3.7×10^4 s^{-1} (mm Hg)$^{-1}$ $(5.9 \pm 0.2) \times 10^4$ s^{-1} (mm Hg)$^{-1}$ $(3.6 \pm 0.2) \times 10^4$ s^{-1} (mm Hg)$^{-1}$ 4.5×10^{11} cm^3/mole s 6×10^{11} cm^3/mole s $(7 \pm 0.5) \times 10^4$ s^{-1} (mm Hg)$^{-1}$ $(5 \pm 0.4) \times 10^4$ s^{-1} (mm Hg)$^{-1}$ $(3.6 \pm 0.3) \times 10^4$ s^{-1} (mm Hg)$^{-1}$ $(3.1 \pm 0.3) \times 10^4$ s^{-1} (mm Hg)$^{-1}$ $(2.2 \pm 0.3) \times 10^4$ s^{-1} (mm Hg)$^{-1}$ $(2 \pm 0.3) \times 10^4$ s^{-1} (mm Hg)$^{-1}$	300 300 300 294 295 700 1000 295 324 350 420 500 580	[2.196] [2.179] [2.173] [2.166] [2.164] [2.164] [2.164] [2.169] [2.169] [2.169] [2.169] [2.169] [2.169]

Table 2.3 (continued)

M	Process and kinetic data	T [K]	Reference
CO_2	$(1.8 \pm 0.3) \times 10^4$ s^{-1} (mm Hg)$^{-1}$	670	[2.169]
	$(3.9 \pm 0.4) \times 10^4$ s^{-1} (mm Hg)$^{-1}$	295	[2.191]
	7.7×10^{11} cm^3/mole s	298	[2.197]
	4×10^4 s^{-1} (mm Hg)$^{-1}$	298	[2.200]
	$HF(v=2) + M \rightarrow HF(v<2) + M, M^v$		
CO_2	2.9×10^{12} cm^3/mole s	298	[2.197]
	1.6×10^5 s^{-1} (mm Hg)$^{-1}$	300	[2.179]
	$(19 \pm 2) \times 10^4$ s^{-1} (mm Hg)$^{-1}$	295	[2.191]
	2.05×10^{12} cm^3/mole s	298	[2.180]
	$HF(v=3) + M \rightarrow HF(v<3) + M, M^v$		
CO_2	3.1×10^{12} cm^3/mole s	298	[2.197]
	2.4×10^5 cm^3/mole s	300	[2.179]
	$(38 \pm 4) \times 10^4$ s^{-1} (mm Hg)$^{-1}$	295	[2.191]
	4.22×10^{12} cm^3/mole s	298	[2.180]
	$(6.1 \pm 0.96) \times 10^{12}$ cm^3/mole s	293	[2.237]
	$HF(v=4) + M \rightarrow HF(v<4) + M, M^v$		
CO_2	4.8×10^5 s^{-1} (mm Hg)$^{-1}$	300	[2.179]
	7.52×10^{12} cm^3/mole s	298	[2.180]
	$(13.5 \pm 2.6) \times 10^{12}$ cm^3/mole s	293	[2.237]
	$HF(v=5) + M \rightarrow HF(v<5) + M, M^v$		
CO_2	$>8 \times 10^5$ s^{-1} (mm Hg)$^{-1}$	300	[2.179]
	1.62×10^{13} cm^3/mole s	298	[2.180]
	$HF(v=6) + M \rightarrow HF(v<6) + M, M^v$		
CO_2	2.1×10^{13} cm^3/mole s	298	[2.180]
	$HF(v=7) + M \rightarrow HF(v<7) + M, M^v$		
CO_2	1.92×10^{13} cm^3/mole s	298	[2.180]
	$HF(v=1) + M \rightarrow HF(v=0) + M, M^v$		
H_2O	$(4.1 \pm 0.5) \times 10^6$ s^{-1} (mm Hg)$^{-1}$	294	[2.166]
	$(p\tau)^{-1} = 10^{2.8 \pm 0.1} \exp[-(0.57 \pm 0.03)$		
	$\times 10^{-3} T](\mu s \, atm.)^{-1}$	1600–3000	[2.189]
D_2O	$(4.1 \pm 0.5) \times 10^6$ s^{-1} (mm Hg)$^{-1}$	294	[2.166]
	$(p\tau)^{-1} = 10^{2.4 \pm 0.1} \exp[-(0.76 \pm 0.02)$		
	$\times 10^{-3} T](\mu s \, atm.)^{-1}$	1000–3500	[2.189]
N_2O	$(3.4 \pm 0.3) \times 10^4$ s^{-1} (mm Hg)$^{-1}$	295	[2.190]
	2.3×10^{11} cm^3/mole s	298	[2.197]
	$HF(v=2) + M \rightarrow HF(v<2) + M, M^v$		
N_2O	4.1×10^{11} cm^3/mole s	298	[2.197]
	$HF(v=1) + M \rightarrow HF(v=0) + M, M^v$		
SO_2	$(2.4 \pm 0.3) \times 10^4$ s^{-1} (mm Hg)$^{-1}$	295	[2.190]
H_2S	$(6.1 \pm 0.6) \times 10^4$ s^{-1} (mm Hg)$^{-1}$	295	[2.190]
	1.4×10^{12} cm^3/mole s	298	[2.197]

Table 2.3 (continued)

M	Process and kinetic data	T [K]	Reference
	$HF(v=2)+M \to HF(v<2)+M, M^v$		
H_2S	5.2×10^{12} cm^3/mole s	298	[2.197]
	$HF(v=3)+M \to HF(v<3)+M, M^v$		
H_2S	6.1×10^{12} cm^3/mole s	298	[2.197]
	$HF(v=1)+M \to HF(v=0)+M, M^v$		
COF_2	$(5.74\pm0.5) \times 10^4$ s^{-1} (mm Hg)$^{-1}$	295	[2.190]
BF_3	$(1.53\pm0.15) \times 10^3$ s^{-1} (mm Hg)$^{-1}$	295	[2.190]
NF_3	3×10^2 s^{-1} (mm Hg)$^{-1}$	295	[2.190]
ClF_3	$(1.13\pm0.17) \times 10^5$ s^{-1} (mm Hg)$^{-1}$	295	[2.198]
C_2H_2	5.9×10^4 s^{-1} (mm Hg)$^{-1}$	300	[2.199]
	$HF(v=2)+M \to HF(v<2)+M, M^v$		
C_2H_2	2×10^5 s^{-1} (mm Hg)$^{-1}$	300	[2.199]
	$HF(v=1)+M \to HF(v=0)+M, M^v$		
$CBrF_3$	$(3.9\pm0.6) \times 10^2$ s^{-1} (mm Hg)$^{-1}$	295	[2.190]
SiF_4	$(3.4\pm0.5) \times 10^2$ s^{-1} (mm Hg)$^{-1}$	295	[2.190]
CF_4	$(4.2\pm0.6) \times 10^2$ s^{-1} (mm Hg)$^{-1}$	295	[2.190]
	$<7 \times 10^9$ cm^3/mole s	298	[2.197]
	$HF(v=2)+M \to HF(v<2)+M, M^v$		
CF_4	1.3×10^{10} cm^3/mole s	298	[2.197]
	$HF(v=1)+M \to HF(v=0)+M, M^v$		
SO_2F_2	$(1.41\pm0.15) \times 10^4$ s^{-1} (mm Hg)$^{-1}$	295	[2.190]
CH_4	$(5.3\pm0.8) \times 10^4$ s^{-1} (mm Hg)$^{-1}$	295	[2.198]
	6.4×10^4 s^{-1} (mm Hg)$^{-1}$	r.t.*	[2.199]
	3.9×10^{11} s^{-1} (mm Hg)$^{-1}$	298	[2.197]
	3.5×10^{11} s^{-1} (mm Hg)$^{-1}$	298	[2.200]
	$(2.6\pm0.3) \times 10^4$ s^{-1} (mm Hg)$^{-1}$	r.t.*	[2.238]
	$HF(v=2)+M \to HF(v<2)+M, M^v$		
CH_4	2.3×10^5 s^{-1} (mm Hg)$^{-1}$	r.t.*	[2.199]
	1.3×10^{12} cm^3/mole s	298	[2.197]
	$HF(v=3)+M \to HF(v<3)+M, M^v$		
CH_4	1.6×10^{12} cm^3/mole s	298	[2.197]
	$HF(v=1)+M \to HF(v=0)+M, M^v$		
C_2H_4	5×10^4 s^{-1} (mm Hg)$^{-1}$	r.t.*	[2.199]
	4.9×10^4 s^{-1} (mm Hg)$^{-1}$	298	[2.200]
C_2F_4	3.5×10^3 s^{-1} (mm Hg)$^{-1}$	298	[2.200]
C_2HF_3	1.9×10^4 s^{-1} (mm Hg)$^{-1}$	298	[2.200]
C_2H_3F	3.4×10^4 s^{-1} (mm Hg)$^{-1}$	298	[2.200]
$1,1-C_2H_2F_2$	2.7×10^4 s^{-1} (mm Hg)$^{-1}$	298	[2.200]

*r.t. – room temperature

Table 2.3 (continued)

M	Process and kinetic data	T [K]	Reference
cis-1,2-$C_2H_2F_2$	3.1×10^4 s^{-1} (mm Hg)$^{-1}$	298	[2.200]
trans-1,2-$C_2H_2F_2$	2.9×10^4 s^{-1} (mm Hg)$^{-1}$	298	[2.200]
	$HF(v=2)+M \rightarrow HF(v<2)+M, M^v$		
C_2H_4	2.2×10^5 s^{-1} (mm Hg)$^{-1}$	r.t.*	[2.199]
	1.3×10^5 s^{-1} (mm Hg)$^{-1}$	298	[2.200]
C_2F_4	8.8×10^3 s^{-1} (mm Hg)$^{-1}$	298	[2.200]
C_2HF_3	4.2×10^4 s^{-1} (mm Hg)$^{-1}$	298	[2.200]
C_2H_3F	8.7×10^4 s^{-1} (mm Hg)$^{-1}$	298	[2.200]
1,1-$C_2H_2F_2$	6.5×10^4 s^{-1} (mm Hg)$^{-1}$	298	[2.200]
cis-1,2-$C_2H_2F_2$	5.9×10^4 s^{-1} (mm Hg)$^{-1}$	298	[2.200]
trans-1,2-$C_2H_2F_2$	6.6×10^4 s^{-1} (mm Hg)$^{-1}$	298	[2.200]
	$HF(v=1)+M \rightarrow HF(v=0)+M, M^v$		
PF_5	$(7.3 \pm 1) \times 10^3$ s^{-1} (mm Hg)$^{-1}$	295	[2.190]
SF_6	$(0.9 \pm 0.75) \times 10^2$ s^{-1} (mm Hg)$^{-1}$	295	[2.165]
	< 50 s^{-1} (mm Hg)$^{-1}$	295	[2.190]
	$HF (v=2)+M \rightarrow HF (v<2)+M, M^v$		
SF_6	$< 3 \times 10^9$ cm^3/mole s	298	[2.197]
	$HF (v=3)+M \rightarrow HF (v<3)+M, M^v$		
SF_6	1.7×10^{10} cm^3/mole s	298	[2.197]
	$HF (v=1)+M \rightarrow HF (v=0)+M, M^v$		
C_2F_6	$(1.6 \pm 0.5) \times 10^2$ s^{-1} (mm Hg)$^{-1}$	295	[2.190]
C_2H_6	5.6×10^4 s^{-1} (mm Hg)$^{-1}$	r.t.*	[2.199]
	$(1.1 \pm 0.16) \times 10^5$ s^{-1} (mm Hg)$^{-1}$	295	[2.198]
	9.3×10^4 s^{-1} (mm Hg)$^{-1}$	298	[2.200]
	$(5.9 \pm 0.6) \times 10^4$ s^{-1} (mm Hg)$^{-1}$	300	[2.238]
	$HF (v=2)+M \rightarrow HF (v<2)+M, M^v$		
C_2H_6	2.8×10^5 s^{-1} (mm Hg)$^{-1}$	300	[2.199]
	$HF (v=1)+M \rightarrow HF (v=0)+M, M^v$		
C_3H_5	$(3.2 \pm 0.5) \times 10^5$ s^{-1} (mm Hg)$^{-1}$	295	[2.198]
C_3H_8	$(1.35 \pm 0.2) \times 10^5$ s^{-1} (mm Hg)$^{-1}$	295	[2.198])
	8.3×10^4 s^{-1} (mm Hg)$^{-1}$	r.t.*	[2.199]
	$(8.4 \pm 0.9) \times 10^4$ s^{-1} (mm Hg)$^{-1}$	300	[2.238]
	$HF (v=2)+M \rightarrow HF (v<2)+M, M^v$		
C_3H_8	3×10^5 s^{-1} (mm Hg)$^{-1}$	300	[2.199]
	$HF (v=1)+M \rightarrow HF (v=0)+M, M^v$		
C_4F_8	$(6.3 \pm 1) \times 10^2$ s^{-1} (mm Hg)$^{-1}$	295	[2.190]
C_4H_{10}	$(1.7 \pm 0.25) \times 10^5$ s^{-1} (mm Hg)$^{-1}$	295	[2.198]
	$(1.28 \pm 0.13) \times 10^5$ s^{-1} (mm Hg)$^{-1}$	300	[2.238]

*r.t. – room temperature

Table 2.3 (continued)

M	Process and kinetic data	T [K]	Reference

Relaxation of DF

$$DF\ (v=1)+M\ \xrightarrow{k_{v\to RT}}\ DF\ (v=0)+M$$

M	Process and kinetic data	T [K]	Reference
DF	$(1.8\pm0.3)\times10^4\ \text{s}^{-1}\ (\text{mm Hg})^{-1}$	350	[2.160]
	$(2.6\pm0.4)\times10^4\ \text{s}^{-1}\ (\text{mm Hg})^{-1}$	300	[2.160]
	$(2.4\pm0.3)\times10^4\ \text{s}^{-1}\ (\text{mm Hg})^{-1}$ (pure DF)	295	[2.167]
	$(2.7\pm0.5)\times10^4\ \text{s}^{-1}\ (\text{mm Hg})^{-1}$ (dilution in Ar)	295	[2.167]
	$(2.1\pm0.6)\times10^4\ \text{s}^{-1}\ (\text{mm Hg})^{-1}$ (dilution in He)	295	[2.167]
	$(2.5\pm0.4)\times10^4\ \text{s}^{-1}\ (\text{mm Hg})^{-1}$	297	[2.168]
	$(1.9\pm0.4)\times10^4\ \text{s}^{-1}\ (\text{mm Hg})^{-1}$	321	[2.168]
	$(1\pm0.3)\times10^4\ \text{s}^{-1}\ (\text{mm Hg})^{-1}$	395	[2.168]
	$(0.8\pm0.1)\times10^4\ \text{s}^{-1}\ (\text{mm Hg})^{-1}$	475	[2.168]
	$(0.5\pm0.2)\times10^4\ \text{s}^{-1}\ (\text{mm Hg})^{-1}$	570	[2.168]
	$(0.5\pm0.2)\times10^4\ \text{s}^{-1}\ (\text{mm Hg})^{-1}$	678	[2.168]
	$(2.7\pm0.3)\times10^4\ \text{s}^{-1}\ (\text{mm Hg})^{-1}$	299	[2.169]
	$(2.1\pm0.3)\times10^4\ \text{s}^{-1}\ (\text{mm Hg})^{-1}$	325	[2.169]
	$(2\pm0.2)\times10^4\ \text{s}^{-1}\ (\text{mm Hg})^{-1}$	350	[2.169]
	$(1.2\pm0.3)\times10^4\ \text{s}^{-1}\ (\text{mm Hg})^{-1}$	399	[2.169]
	$(0.87\pm0.2)\times10^4\ \text{s}^{-1}\ (\text{mm Hg})^{-1}$	473	[2.169]
	$(0.77\pm0.15)\times10^4\ \text{s}^{-1}\ (\text{mm Hg})^{-1}$	566	[2.169]
	$(0.64\pm0.14)\times10^4\ \text{s}^{-1}\ (\text{mm Hg})^{-1}$	670	[2.169]
	$(2.65\pm0.3)\times10^4\ \text{s}^{-1}\ (\text{mm Hg})^{-1}$ (average of data [2.160, 2.169] without allowance made for gas-dynamic corrections)	300	[2.171]
	$(2.23\pm0.3)\times10^4\ \text{s}^{-1}\ (\text{mm Hg})^{-1}$ (average of data [2.160, 2.169] with allowance made for gas-dynamic corrections)	300	[2.171]
	$p\tau=(1.4\pm0.15)\times10^{-3}$ $\times\exp[(63.7\pm3)/T^{1/3}]\ \mu\text{s atm.}$	> 2000	[2.201]
	$p\tau=0.37\ \mu\text{s atm.}$	800–1500	[2.201]
	$p\tau=10^{-2.6\pm0.2}\exp[(56\pm3)/T^{1/3}]\ \mu\text{s atm.}$	1500–4000	[2.103]
	$p\tau=2.26\times10^{-3}\exp(57.25/T^{1/3})\ \mu\text{s atm.}$	1500–5000	[2.176]
	$p\tau=0.055\pm0.005\ \mu\text{s atm.}$	300	[2.170]
DF	$p\tau=0.32\pm0.05\ \mu\text{s atm.}$	700	[2.170]
	$p\tau=0.063\pm0.005\ \mu\text{s atm.}$	295	[2.204]
	$p\tau=0.36\pm0.03\ \mu\text{s atm.}$	900	[2.204]
	$4.4\times10^{14}\ T^{-1.24}+6.1\times10^3\ T^{2.46}\ \text{cm}^3/\text{mole s}$	295–5000	[2.204]

$$DF\ (v=1)+M(v=1)\ \underset{k_{v\to v}^-}{\overset{k_{v\to v}}{\rightleftarrows}}\ DF\ (v=2)+M(v=0)$$

M	Process and kinetic data	T [K]	Reference
DF	$k_{v\to v}^-=6.2\times10^5\ \text{s}^{-1}\ (\text{mm Hg})^{-1}$	r.t.*	[2.202]
	$k_{v\to v}^-=1.9\times10^{13}\ \text{cm}^3/\text{mole s}$	295	[2.203]
	$k_{v\to v}^-=1.01\times10^{13}\ \text{cm}^3/\text{mole s}$	444	[2.203]
	$k_{v\to v}=1.35\times10^{13}\ \text{cm}^3/\text{mole s}$	444	[2.203]
	$k_{v\to v}^-=0.95\times10^{13}\ \text{cm}^3/\text{mole s}$	450	[2.203]
	$k_{v\to v}=1.27\times10^{13}\ \text{cm}^3/\text{mole s}$	450	[2.203]
	$k_{v\to v}^-=0.94\times10^{13}\ \text{cm}^3/\text{mole s}$	479	[2.203]
	$k_{v\to v}=1.23\times10^{13}\ \text{cm}^3/\text{mole s}$	479	[2.203]

*r.t. – room temperature

Table 2.3 (continued)

M	Process and kinetic data	T [K]	Reference
DF	$k_{\bar{V} \to V} = (0.73 \pm 0.02) \times 10^{13}$ cm^3/mole s	739	[2.203]
	$k_{V \to V} = (0.87 \pm 0.02) \times 10^{13}$ cm^3/mole s	739	[2.203]
	$k_{V \to V} = (1.96 \pm 0.2) \times 10^3$ cm^3/mole s	295	[2.204]

$$\text{DF } (v=1) + M \xrightarrow{k_{V \to RT}} \text{DF } (v=0) + M$$

M	Process and kinetic data	T [K]	Reference
HF	7×10^4 s^{-1} (mm Hg)$^{-1}$	298	[2.179]
	$(4.5 \pm 0.7) \times 10^4$ s^{-1} (mm Hg)$^{-1}$	300	[2.160]
	$(3.3 \pm 0.5) \times 10^4$ s^{-1} (mm Hg)$^{-1}$	350	[2.160]
	$(4.1 \pm 0.4) \times 10^4$ s^{-1} (mm Hg)$^{-1}$	297	[2.168]
	$(3.4 \pm 0.4) \times 10^4$ s^{-1} (mm Hg)$^{-1}$	321	[2.168]
	$(2.3 \pm 0.3) \times 10^4$ s^{-1} (mm Hg)$^{-1}$	395	[2.168]
	$(1.5 \pm 0.2) \times 10^4$ s^{-1} (mm Hg)$^{-1}$	475	[2.168]
	$(1.2 \pm 0.2) \times 10^4$ s^{-1} (mm Hg)$^{-1}$	570	[2.168]
	$(0.9 \pm 0.2) \times 10^4$ s^{-1} (mm Hg)$^{-1}$	678	[2.168]
	$(3.4 \pm 0.3) \times 10^4$ s^{-1} (mm Hg)$^{-1}$	295	[2.167]
	$(0.96 \pm 0.4) \times 10^{12}$ cm^3/mole s	298	[2.174]
	$p\tau = 0.04 \pm 0.005$ μs atm.	300	[2.170]
	$p\tau = 0.14 \pm 0.03$ μs atm.	600	[2.170]

$$\text{DF } (v=2) + M \xrightarrow{k_{V \to RT}} \text{DF } (v<2) + M$$

M	Process and kinetic data	T [K]	Reference
HF	9.5×10^4 s^{-1} (mm Hg)$^{-1}$	298	[2.179]
	$(2.8 \pm 1) \times 10^{12}$ cm^3/mole s	298	[2.174]

$$\text{DF } (v=3) + M \xrightarrow{k_{V \to RT}} \text{DF } (v<3) + M$$

M	Process and kinetic data	T [K]	Reference
HF	$(5 \pm 2) \times 10^{12}$ cm^3/mole s	298	[2.174]

$$\text{DF } (v=4) + M \xrightarrow{k_{V \to RT}} \text{DF } (v<4) + M$$

M	Process and kinetic data	T [K]	Reference
HF	$(1.6 \pm 0.6) \times 10^{12}$ cm^3/mole s	298	[2.174]

M	Process and kinetic data	T [K]	Reference
	DF $(v=1) + M \to$ DF $(v=0) + M$		
H	$(6.7 \pm 1.8) \times 10^{10}$ cm^3/mole s	295	[2.83]
D	$\simeq 10^{10}$ cm^3/mole s	295	[2.83]

$$\text{DF} + M \to \begin{cases} \text{D}M + \text{F} \\ \text{M}\text{F} + \text{D} \end{cases}$$

M	Process and kinetic data	T [K]	Reference
H	$2.9 \times 10^{14} \exp(-33700/RT)$ cm^3/mole s	2100–3900	[2.84)

$$\text{DF } (v=1) + M \to \begin{cases} \text{DF } (v=0) + M \\ \text{M}\text{D } (v=0) + \text{F} \end{cases}$$

M	Process and kinetic data	T [K]	Reference
F	$(3.9 \pm 0.7) \times 10^{11}$ cm^3/mole s	300	[2.187]
	$(1.55 \pm 0.3) \times 10^{13}$ cm^3/mole s	2000–3000	[2.201]

Table 2.3 (continued)

M	Process and kinetic data	T [K]	Reference
Cl	$(1.2 \pm 0.2) \times 10^{12}$ cm^3/mole s	300	[2.187]
O	$(4.7 \pm 1.3) \times 10^{12}$ cm^3/mole s	300	[2.187]
HCl*	$k_{V \to V} = 2.62 \times 10^{12}$ cm^3/mole s	558	[2.206]
	$k_{V \to RT}^{M - DF} + (0.95 \pm 0.5) k_{V \to RT}$		
	$= (2.2 \pm 0.5) \times 10^{11}$ cm^3/mole s	558	[2.206]
	$k_{V \to V} = (3 \pm 0.1) \times 10^{12}$ cm^3/mole s	475	[2.206]
	$k_{V \to RT}^{M - DF} = (0.95 \pm 0.5) k_{V \to RT}$		
	$= (2.1 \pm 0.7) \times 10^{11}$ cm^3/mole s	475	[2.206]
HBr	$k_{V \to V} + k_{V \to RT} = (7.1 \pm 0.7) \times 10^4$ s^{-1} (mm Hg)$^{-1}$	295	[2.206]
	$k_{V \to V} + B k_{V \to RT} = 6.2 \times 10^{11}$ cm^3/mole s,		
	$0.64 < B < 0.79$	740	[2.206]
	$k_{V \to V} + B k_{V \to RT} = 6.86 \times 10^{11}$ cm^3/mole s,		
	$0.64 < B < 0.79$	558	[2.206]
	$k_{V \to V} + B k_{V \to RT} = 7 \times 10^{11}$ cm^3/mole s,		
	$0.64 < B < 0.79$	471	[2.206]
DBr	$k_{V \to V} + k_{V \to RT} = (5.4 \pm 1) \times 10^3$ s^{-1} (mm Hg)$^{-1}$	295	[2.206]
	$k_{V \to V} + k_{V \to RT} = 7.7 \times 10^{10}$ cm^3/mole s	466	[2.206]
	$k_{V \to V} + k_{V \to RT} = 5.5 \times 10^{10}$ cm^3/mole s	565	[2.206]
	$k_{V \to V} + k_{V \to RT} = 9.1 \times 10^{10}$ cm^3/mole s	740	[2.206]
	DF $(v = 1) + M \to$ DF $(v = 0) + M, M^v$		
N$_2$	$(0.2 \pm 0.05) \times 10^4$ s^{-1} (mm Hg)$^{-1}$	295	[2.167]
	$p\tau = 10^{-3.9 \pm 0.5} \exp[(127 \pm 12)/T^{1/3}]$ μs atm.	1500–4000	[2.193]
	$p\tau = 4$ μs atm.	1300	[2.201]
	$p\tau = 1$ μs atm.	4000	[2.201]
	$(0.91 \pm 0.09) \times 10^3$ s^{-1} (mm Hg)$^{-1}$	295	[2.206]
	1.58×10^{10} cm^3/mole s	472	[2.206]
	$(1.93$ to $2.06) \times 10^{10}$ cm^3/mole s	630	[2.206]
	$(2.44$ to $2.63) \times 10^{10}$ cm^3/mole s	849	[2.206]
	$(3.66$ to $3.84) \times 10^{10}$ cm^3/mole s	1114	[2.206]
H$_2$	$(0.44 \pm 0.08) \times 10^4$ s^{-1} (mm Hg)$^{-1}$	295	[2.167]
	$p\tau = 1.9 \times 10^{-2} \exp(35/T^{1/3})$ μs atm.	800–4000	[2.201]
	$(0.66 \pm 0.07) \times 10^3$ s^{-1} (mm Hg)$^{-1}$	295	[2.192]
	$p\tau = 1.82$ to 1.85 μs atm.	445	[2.192]
	$p\tau = 1.3$ to 1.42 μs atm.	560–564	[2.192]

$$\text{DF } (v = 1) + M(v = 0) \underset{k_{\bar{V} \to V}}{\overset{k_{V \to V}}{\rightleftharpoons}} \text{DF } (v = 0) + M(v = 1)$$

$$\text{DF } (v = 1) + M (v = 0) \xrightarrow{k_{V \to RT}} \text{DF } (v = 0) + M (v = 0)$$

$$\text{DF } (v = 0) + M (v = 1) \xrightarrow{k_{V \to RT}^{M - DF}} \text{DF } (v = 0) + M (v = 0)$$

M	Process and kinetic data	T [K]	Reference
D$_2$	$k_{V \to V} + 0.74 k_{V \to RT} = (0.0188 \pm 0.002)$		
	$\times 10^6$ s^{-1} (mm Hg)$^{-1}$	295	[2.206]
	$k_{\bar{V} \to V} = 5.5 \times 10^{11}$ cm^3/mole s	300–730	[2.206]

*For other data on the relaxation of DF on HCl see p. 88

Table 2.3 (continued)

M	Process and kinetic data	T [K]	Reference
D_2	$k_{V \to RT} + 0.667 k_{V \to RT}^{M-DF}$		
	$= (100 \pm 40) \text{ s}^{-1} \text{ (mm Hg)}^{-1}$	295	[2.192]
	$k_{V \to V} = (2.02 \pm 0.33) \times 10^4 \text{ s}^{-1} \text{ (mm Hg)}^{-1}$	295	[2.167]
	$DF\,(v = 2) + M \to DF\,(v < 2) + M,\ M^v$		
D_2	$(3.5 \pm 1.2) \times 10^{11} \text{ cm}^3/\text{mole s}$	298	[2.174]
	$DF\,(v = 3) + M \to DF\,(v < 3) + M,\ M^v$		
D_2	$(4 \pm 1.5) \times 10^{11} \text{ cm}^3/\text{mole s}$	298	[2.174]
	$DF\,(v = 4) + M \to DF\,(v < 4) + M,\ M^v$		
D_2	$(4.5 \pm 1.6) \times 10^{11} \text{ cm}^3/\text{mole s}$	298	[2.174]
	$DF\,(v = 1) + M \to DF\,(v = 0) + M,\ M^v$		
O_2	$(0.75 \pm 0.08) \times 10^2 \text{ s}^{-1} \text{ (mm Hg)}^{-1}$	295	[2.206]
	(2.1) to $2.15) \times 10^9 \text{ cm}^3/\text{mole s}$	550	[2.206]
	$(2.9$ to $3.28) \times 10^9 \text{ cm}^3/\text{mole s}$	702	[2.206]
	$(5.13$ to $6.77) \times 10^9 \text{ cm}^3/\text{mole s}$	870–885	[2.206]
	$p\tau = 4.4 \times 10^{-2} \exp(60.8/T^{1/3}) \ \mu\text{s atm.}$	1200–3000	[2.188]
NO	$(8.2 \pm 0.8) \times 10^3 \text{ s}^{-1} \text{ (mm Hg)}^{-1}$	295	[2.206]
	$(1.16$ to $1.28) \times 10^{11} \text{ cm}^3/\text{mole s}$	709–725	[2.206]
	$(0.95$ to $0.98) \times 10^{11} \text{ cm}^3/\text{mole s}$	544–560	[2.206]
	$(1.07$ to $1.08) \times 10^{11} \text{ cm}^3/\text{mole s}$	470–476	[2.206]
	$p\tau = 10^{-4.3 \pm 0.2} \exp[(103 \pm 5)/T^{1/3}] \ \mu\text{s atm.}$	1000–3000	[2.189]
CO	$(3.9 \pm 0.4) \times 10^3 \text{ s}^{-1} \text{ (mm Hg)}^{-1}$	295	[2.206]
	$(7$ to $7.1) \times 10^{10} \text{ cm}^3/\text{mole s}$	710–725	[2.206]
	$(5.4$ to $5.9) \times 10^{10} \text{ cm}^3/\text{mole s}$	555–567	[2.206]
	$5.2 \times 10^{10} \text{ cm}^3/\text{mole s}$	473	[2.206]
CO_2	$(0.9 \pm 0.3) \times 10^{12} \text{ cm}^3/\text{mole s}$	300	[2.207]
	$(1.75 \pm 0.25) \times 10^5 \text{ s}^{-1} \text{ (mm Hg)}^{-1}$	350	[2.173]
	$1.3 \times 10^5 \text{ s}^{-1} \text{ (mm Hg)}^{-1}$	300	[2.179]
	$1.33 \times 10^{14} \ T^{-0.68} \text{ cm}^3/\text{mole s}$	295–570	[2.204]
	$50(2 \times 10^{11} - k_{V \to V}^{D_2 - CO_2}) \text{ cm}^3/\text{mole s}$	300	[2.196]
	$(24 \pm 5) \times 10^4 \text{ s}^{-1} \text{ (mm Hg)}^{-1}$	299	[2.169]
	$(19 \pm 3) \times 10^4 \text{ s}^{-1} \text{ (mm Hg)}^{-1}$	325	[2.169]
	$(14.6 \pm 2.5) \times 10^4 \text{ s}^{-1} \text{ (mm Hg)}^{-1}$	399	[2.169]
	$(11.7 \pm 2.5) \times 10^4 \text{ s}^{-1} \text{ (mm Hg)}^{-1}$	473	[2.169]
	$(9.7 \pm 2) \times 10^4 \text{ s}^{-1} \text{ (mm Hg)}^{-1}$	566	[2.169]
	$(8.6 \pm 1.5) \times 10^4 \text{ s}^{-1} \text{ (mm Hg)}^{-1}$	670	[2.169]
	$2 \times 10^5 \text{ s}^{-1} \text{ (mm Hg)}^{-1}$	295	[2.208]
	$DF\,(V = 2) + M \to DF\,(v < 2) + M,\ M^v$		
CO_2	$2.2 \times 10^5 \text{ s}^{-1} \text{ (mm Hg)}^{-1}$	300	[2.179]
	$5.1 \times 10^5 \text{ s}^{-1} \text{ (mm Hg)}^{-1}$	295	[2.208]
	$DF\,(v = 3) + M \to DF\,(v < 3) + M,\ M^v$		
CO_2	$5.2 \times 10^5 \text{ s}^{-1} \text{ (mm Hg)}^{-1}$	300	[2.179]
	$DF\,(v = 1) + M \to DF\,(v = 0) + M,\ M^v$		
SO_2	$(1.27 \pm 0.15) \times 10^4 \text{ s}^{-1} \text{ (mm Hg)}^{-1}$	295	[2.238]
NF_3	$(5.1 \pm 0.6) \times 10^2 \text{ s}^{-1} \text{ (mm Hg)}^{-1}$	295	[2.238]

Table 2.3 (continued)

M	Process and kinetic data	T [K]	Reference
BF$_3$	$(7.1 \pm 0.9) \times 10^3$ s^{-1} (mm Hg)$^{-1}$	295	[2.238]
C$_2$H$_2$	$(4 \pm 0.6) \times 10^4$ s^{-1} (mm Hg)$^{-1}$	295	[2.238]
CH$_4$	$(0.22 \pm 0.03) \times 10^6$ s^{-1} (mm Hg)$^{-1}$	295	[2.238]
CF$_4$	$(1 \pm 0.1) \times 10^3$ s^{-1} (mm Hg)$^{-1}$	295	[2.238]
CF$_3$H	$(1.95 \pm 0.25) \times 10^3$ s^{-1} (mm Hg)$^{-1}$	295	[2.238]
CBrF$_3$	$(5.6 \pm 0.5) \times 10^2$ s^{-1} (mm Hg)$^{-1}$	295	[2.238]
CH$_3$F	$(0.36 \pm 0.04) \times 10^6$ s^{-1} (mm Hg)$^{-1}$	295	[2.238]
C$_2$H$_4$	$(0.175 \pm 0.02) \times 10^6$ s^{-1} (mm Hg)$^{-1}$	295	[2.238]
C$_2$H$_2$F$_2$	$(1.86 \pm 0.2) \times 10^4$ s^{-1} (mm Hg)$^{-1}$	295	[2.238]
C$_2$H$_6$	$(0.61 \pm 0.1) \times 10^6$ s^{-1} (mm Hg)$^{-1}$	295	[2.238]
C$_4$H$_{10}$	$(1.26 \pm 0.15) \times 10^6$ s^{-1} (mm Hg)$^{-1}$	295	[2.238]

Relaxation of HCl

$$\text{HCl} \, (v = 1) + M \xrightarrow{k_{v \to RT}} \text{HCl} \, (v = 0) + M$$

HCl	$(0.9 \pm 0.2) \times 10^3$ s^{-1} (mm Hg)$^{-1}$	350	[2.160]
	$(0.12 \pm 0.03) \times 10^4$ s^{-1} (mm Hg)$^{-1}$	300	[2.160]
	$(0.85 \pm 0.1) \times 10^4$ s^{-1} (mm Hg)$^{-1}$	295	[2.164]
	$(0.83 \pm 0.08) \times 10^3$ s^{-1} (mm Hg)$^{-1}$	296	[2.209]
	5.9×10^3 s^{-1} (mm Hg)$^{-1}$	144	[2.210]
	3.1×10^3 s^{-1} (mm Hg)$^{-1}$	169	[2.210]
	1.96×10^3 s^{-1} (mm Hg)$^{-1}$	190	[2.210]
	0.9×10^3 s^{-1} (mm Hg)$^{-1}$	273	[2.210]
	0.78×10^3 s^{-1} (mm Hg)$^{-1}$	298	[2.210]
	0.69×10^3 s^{-1} (mm Hg)$^{-1}$	355	[2.210]
	0.65×10^3 s^{-1} (mm Hg)$^{-1}$	407	[2.210]
	0.65×10^3 s^{-1} (mm Hg)$^{-1}$	472	[2.210]
	0.82×10^3 s^{-1} (mm Hg)$^{-1}$	584	[2.210]
	$p\tau = 1 \, \mu$s atm.	1000	[2.211, 2.212]
	$p\tau = 0.2 \, \mu$s atm.	2000	[2.211, 2.212]

$$\text{HCl}(v = 1) + M(v = 1) \underset{k_{\bar{v} \to v}}{\overset{k_{v \to v}}{\rightleftharpoons}} \text{HCl}(v = 0) + M(v = 2)$$

HCl	$(1.5 \text{ to } 3) \times 10^{12}$ cm^3/mole s	300	[2.213]
	$k_{\bar{v} \to v} = (0.91 \text{ to } 1.82) \times 10^{12}$ cm^3/mole s	300	[2.213]
	$k_{\bar{v} \to v} = (1 \pm 0.1) \times 10^5$ s^{-1} (mm Hg)$^{-1}$	296	[2.214]
	0.025 (collisions)$^{-1}$	300	[2.215]
	0.02 (collisions)$^{-1}$	400	[2.215]
	0.008 (collisions)$^{-1}$	650	[2.215]
	$k_{\bar{v} \to v} = (0.9 \pm 0.2) \times 10^5$ s^{-1} (mm Hg)$^{-1}$	300	[2.216]

$$\text{HCl}^{35}(v = 1) + \text{HCl}^{37}(v = 0) \xrightarrow{k_{v \to v}} \text{HCl}^{35}(v = 0) + \text{HCl}^{37}(v = 1)$$

HCl	$(6.2 \pm 1.7) \times 10^5$ s^{-1} (mm Hg)$^{-1}$	296	[2.214]

$$\text{HCl}(v = 1) + M(v = 0) \xrightarrow{k_{v \to v}} \text{HCl}(v = 0) + M(v = 1)$$

DCl	$(0.325 \pm 0.02) \times 10^4$ s^{-1} (mm Hg)$^{-1}$	296	[2.209]

Table 2.3 (continued)

M	Process and kinetic data	T [K]	Reference
	$HCl(v=1)+M\rightarrow\begin{cases}HCl(v=0)+M\\HM+Cl\end{cases}$		
H	$(4.1\pm1)\times10^{12}$ cm^3/mole s	295	[2.83]
	$(3.9\pm1.3)\times10^{12}$ cm^3/mole s	300	[2.217]
Cl	$(3.5\pm2)\times10^5$ s^{-1} (mm Hg)$^{-1}$	300	[2.218]
	$(3.2\pm0.3)\times10^4$ s^{-1} (mm Hg)$^{-1}$	300	[2.219]
	$(2.8\pm0.9)\times10^5$ s^{-1} (mm Hg)$^{-1}$	294	[2.239]
	$(4.8\pm0.18)\times10^{12}$ cm^3/mole s	294	[2.239]
O	$(2.2\pm0.7)\times10^{12}$ cm^3/mole s	300	[2.217]

	$HCl(v=1)+M\xrightarrow{k_{V\rightarrow RT}}HCl(v=0)+M$		
^{3}He	$4^{+0.7}_{-1.5}$ s^{-1} (mm Hg)$^{-1}$	295	[2.220]
^{4}He	1.8 ± 0.2 s^{-1} (mm Hg)$^{-1}$	295	[2.220]
He	$\log_{10}[p\tau(\mu s\,atm.)]=-3.07+36.7\,T^{-1/3}$ (mixture 10% HCl + 90% He)	1000–2000	[2.221]
Ne	0.9 ± 0.25 s^{-1} (mm Hg)$^{-1}$	295	[2.220]
	$\log_{10}[p\tau(\mu s\,atm.)]=-2.48+32.4\,T^{-1/3}$ (mixture 10% HCl + 90% Ne)	1000–2000	[2.221]
Ar	0.11 ± 0.02 s^{-1} (mm Hg)$^{-1}$	295	[2.220]
	$\log_{10}[p\tau(\mu s\,atm.)]=-2.41+33.5\,T^{-1/3}$ (mixture 10% HCl + 90% Ar)	1000–2000	[2.221]
Kr	$\log_{10}[p\tau(\mu s\,atm.)]=-2.53+40.8\,T^{-1/3}$ (mixture 10% HCl + 90% Kr)	1000–2000	[2.221]
HF	$(2\pm0.3)\times10^4$ s^{-1} (mm Hg)$^{-1}$	300	[2.160]
	$(1.6\pm0.4)\times10^4$ s^{-1} (mm Hg)$^{-1}$	350	[2.160]
	$(1.5\pm0.2)\times10^4$ s^{-1} (mm Hg)$^{-1}$	295	[2.164]
	$(1.29\pm0.2)\times10^4$ s^{-1} (mm Hg)$^{-1}$	295	[2.222]
	6×10^3 s^{-1} (mm Hg)$^{-1}$	640	[2.222]
DF	$(1.6\pm0.3)\times10^4$ s^{-1} (mm Hg)$^{-1}$	295	[2.206]

	$HCl(v=1)+M(v=0)\xrightarrow{k_{V\rightarrow V}}HCl(v=0)+M(v=1)$		
HBr	$(3.1\pm0.3)\times10^4$ s^{-1} (mm Hg)$^{-1}$	295	[2.222]
	4×10^{11} cm^3/mole s	550	[2.222]
	3.6×10^{12} cm^3/mole s	750	[2.222]
	$(3.4\pm0.3)\times10^4$ s^{-1} (mm Hg)$^{-1}$	296	[2.224]
HI	$(5.3\pm0.8)\times10^3$ s^{-1} (mm Hg)$^{-1}$	296	[2.224]
N$_2$	$(8.7\pm0.8)\times10^2$ s^{-1} (mm Hg)$^{-1}$	296	[2.224]
	$(8.7\pm0.8)\times10^2$ s^{-1} (mm Hg)$^{-1}$	295	[2.222]
	6.67×10^2 s^{-1} (mm Hg)$^{-1}$	632	[2.222]

	$HCl(v=1)+M\xrightarrow{k_{V\rightarrow RT}}HCl(v=0)+M$		
H$_2$	$(1.7\pm0.3)\times10^2$ s^{-1} (mm Hg)$^{-1}$	296	[2.209]
	$(1.79\pm0.2)\times10^3$ s^{-1} (mm Hg)$^{-1}$	295	[2.222]
D$_2$	17 ± 2 s^{-1} (mm Hg)$^{-1}$	296	[2.225]
HD	81 ± 11 s^{-1} (mm Hg)$^{-1}$	296	[2.225]
H$_2$	$(11.5$ to $12.6)\times10^2$ s^{-1} (mm Hg)$^{-1}$	770	[2.222]

Table 2.3 (continued)

M	Process and kinetic data	T [K]	Reference
	$$HCl(v = 1) + M(v = 0) \underset{k_{V \to V}^{-}}{\overset{k_{V \to V}}{\rightleftharpoons}} HCl(v = 0) + M(v = 1)$$		
D_2	$k_{V \to V}^{-} = (9.1 \pm 1) \times 10^3 \, s^{-1} \, (mm \, Hg)^{-1}$	296	[2.224, 2.226]
	$k_{V \to V}^{-} = (10.7 \pm 1) \times 10^3 \, s^{-1} \, (nm \, Hg)^{-1}$	295	[2.222]
	$3.57 \times 10^3 \, s^{-1} \, (mm \, Hg)^{-1}$	742	[2.222]
	$k_{V \to V} = 4.38 \times 10^3 \, s^{-1} \, (mm \, Hg)^{-1}$	742	[2.222]
	$k_{V \to V}^{-} = 2 \times 10^4 \, s^{-1} \, (mm \, Hg)^{-1}$	196	[2.226]
	$(2 \pm 0.2) \times 10^{11} \, cm^3/mole \, s$	295–740	[2.222]
HD	$97 \pm 10 \, s^{-1} \, (mm \, Hg)^{-1}$	296	[2.225]
	$$HCl(v = 1) + M \xrightarrow{k_{V \to RT}} HCl(v = 0) + M$$		
Cl_2	$1.8 \times 10^2 \, s^{-1} \, (mm \, Hg)^{-1}$	295	[2.222]
	$(1.8 \pm 0.3) \times 10^2 \, s^{-1} \, (mm \, Hg)^{-1}$	r.t.*	[2.218]
	$2.2 \times 10^3 \, s^{-1} \, (mm \, Hg)^{-1}$	r.t.*	[2.219]
	$5.1 \times 10^2 \, s^{-1} \, (mm \, Hg)^{-1}$	679	[2.222]
	$HCl(v = 1) + M \to HCl(v = 0) + M, \, M^v$		
CO	$(2.7 \pm 0.3) \times 10^3 \, s^{-1} \, (mm \, Hg)^{-1}$	296	[2.224]
NO	$4 \times 10^{-4} \, (collisions)^{-1}$	295	[2.90]
CO_2	$8.7 \times 10^4 \, s^{-1} \, (mm \, Hg)^{-1}$	298	[2.227]
H_2O	$(5 \pm 3) \times 10^5 \, s^{-1} \, (mm \, Hg)^{-1}$	296	[2.209]
H_2S	$(7.4 \pm 0.6) \times 10^4 \, s^{-1} \, (mm \, Hg)^{-1}$	r.t.*	[2.228]
H_2Se	$(2.6 \pm 0.2) \times 10^4 \, s^{-1} \, (mm \, Hg)^{-1}$	r.t.*	[2.228]
$SOCl_2$	$(1.1 \pm 0.1) \times 10^4 \, s^{-1} \, (mm \, Hg)^{-1}$	r.t.*	[2.228]
SiH_2Cl_2	$(1.4 \pm 0.1) \times 10^4 \, s^{-1} \, (mm \, Hg)^{-1}$	r.t.*	[2.228]
$SiHCl_3$	$(0.42 \pm 0.02) \times 10^4 \, s^{-1} \, (mm \, Hg)^{-1}$	r.t.*	[2.228]
NO_3Cl	$(2.6 \pm 0.4) \times 10^4 \, s^{-1} \, (mm \, Hg)^{-1}$	r.t.*	[2.228]
HNO_3	$10^5 \, s^{-1} \, (mm \, Hg)^{-1}$	r.t.*	[2.228]
CH_4	$(8.4 \pm 0.6) \times 10^4 \, s^{-1} \, (mm \, Hg)^{-1}$	296	[2.224]
PF_3H_2	$(3 \pm 1) \times 10^4 \, s^{-1} \, (mm \, Hg)^{-1}$	r.t.*	[2.228]
CF_3OCl	$(0.22 \pm 0.02) \times 10^4 \, s^{-1} \, (mm \, Hg)^{-1}$	r.t.*	[2.228]
$1,1\text{-}C_2H_2Cl_2$	$(1 \pm 0.2) \times 10^4 \, s^{-1} \, (mm \, Hg)^{-1}$	r.t.*	[2.228]
$trans\text{-}C_2H_2Cl_2$	$(1 \pm 0.2) \times 10^4 \, s^{-1} \, (mm \, Hg)^{-1}$	r.t.*	[2.228]

Relaxation of DCl

M	Process and kinetic data	T [K]	Reference
	$$DCl(v = 1) + M \xrightarrow{k_{V \to RT}} DCl(v = 0) + M$$		
DCl	$(0.25^{+0.03}_{-0.09}) \times 10^3 \, s^{-1} \, (mm \, Hg)^{-1}$	296	[2.209]
	$(0.22 \pm 0.03) \times 10^3 \, s^{-1} \, (mm \, Hg)^{-1}$	296	[2.230]
HCl	$(0.575 \pm 0.06) \times 10^3 \, s^{-1} \, (mm \, Hg)^{-1}$	296	[2.209]
^3He	$7^{+1}_{-1.5} \, s^{-1} \, (mm \, Hg)^{-1}$	295	[2.220]
^4He	$1.9 \pm 0.2 \, s^{-1} \, (mm \, Hg)^{-1}$	295	[2.220]
Ne	$0.32 \pm 0.04 \, s^{-1} \, (mm \, Hg)^{-1}$	295	[2.220]
Ar	$0.06 \pm 0.02 \, s^{-1} \, (mm \, Hg)^{-1}$	295	[2.220]
n-H_2	$(0.69 \pm 0.07) \times 10^3 \, s^{-1} \, (mm \, Hg)^{-1}$	295	[2.230]

*r.t. – room temperature

Table 2.3 (continued)

M	Process and kinetic data	T [K]	Reference
p-H$_2$	$(0.64 \pm 0.06) \times 10^3\,\text{s}^{-1}\,(\text{mm Hg})^{-1}$	295	[2.230]
HD	$(0.27 \pm 0.03) \times 10^3\,\text{s}^{-1}\,(\text{mm Hg})^{-1}$	295	[2.230]
D$_2$	$(0.59 \pm 0.05) \times 10^2\,\text{s}^{-1}\,(\text{mm Hg})^{-1}$	295	[2.230]
	$\text{DCl}(v=1) + M \rightarrow \text{DCl}(v=0) + M,\ M^v$		
DBr	$2.8 \times 10^{-3}\,(\text{collisions})^{-1}$	295	[2.90, 2.230]
DI	$2.7 \times 10^{-4}\,(\text{collisions})^{-1}$	295	[2.90, 2.230]
N$_2$	$2.2 \times 10^{-4}\,(\text{collisions})^{-1}$	295	[2.90, 2.230]
O$_2$	$8.6 \times 10^{-5}\,(\text{collisions})^{-1}$	295	[2.90, 2.230]
^{13}CO	$8.8 \times 10^{-3}\,(\text{collisions})^{-1}$	295	[2.90, 2.230]
CO	$6.4 \times 10^{-3}\,(\text{collisions})^{-1}$	295	[2.90, 2.230]
NO	$5.1 \times 10^{-3}\,(\text{collisions})^{-1}$	295	[2.90, 2.230]
CO$_2$	$2.9 \times 10^4\,\text{s}^{-1}\,(\text{mm Hg})^{-1}$	295	[2.227]
CH$_4$	$2.8 \times 10^{-3}\,(\text{collisions})^{-1}$	295	[2.90, 2.229]
CD$_4$	$5.5 \times 10^{-3}\,(\text{collisions})^{-1}$	295	[2.90, 2.229]

Relaxation of HBr

	Process and kinetic data	T [K]	Reference
	$\text{HBr}(v=1) + M \xrightarrow{k_{V \to RT}} \text{HBr}(v=0) + M$		
HBr	$(0.6 \pm 0.2) \times 10^3\,\text{s}^{-1}\,(\text{mm Hg})^{-1}$	300	[2.160]
	$(0.6 \pm 0.2) \times 10^3\,\text{s}^{-1}\,(\text{mm Hg})^{-1}$	350	[2.160]
	$(0.8 \pm 0.1) \times 10^3\,\text{s}^{-1}\,(\text{mm Hg})^{-1}$	295	[2.164]
	$(0.571 \pm 0.05) \times 10^3\,\text{s}^{-1}\,(\text{mm Hg})^{-1}$	296	[2.223]
	$1.71 \times 10^3\,\text{s}^{-1}\,(\text{mm Hg})^{-1}$	169	[2.210]
	$1.2 \times 10^3\,\text{s}^{-1}\,(\text{mm Hg})^{-1}$	189	[2.210]
	$0.82 \times 10^3\,\text{s}^{-1}\,(\text{mm Hg})^{-1}$	216	[2.210]
	$0.55 \times 10^3\,\text{s}^{-1}\,(\text{mm Hg})^{-1}$	297	[2.210]
	$0.52 \times 10^3\,\text{s}^{-1}\,(\text{mm Hg})^{-1}$	333	[2.210]
	$0.49 \times 10^3\,\text{s}^{-1}\,(\text{mm Hg})^{-1}$	374	[2.210]
	$0.51 \times 10^3\,\text{s}^{-1}\,(\text{mm Hg})^{-1}$	427	[2.210]
	$0.61 \times 10^3\,\text{s}^{-1}\,(\text{mm Hg})^{-1}$	505	[2.210]
	$\text{HBr}(v=1) + M(v=1) \underset{k^-_{V \to V}}{\overset{k_{V \to V}}{\rightleftharpoons}} \text{HBr}(v=2) + M(v=0)$		
HBr	$4 \times 10^{-2}\,(\text{collisions})^{-1}$	300	[2.215]
	$3 \times 10^{-2}\,(\text{collisions})^{-1}$	400	[2.215]
	$1.5 \times 10^{-2}\,(\text{collisions})^{-1}$	650	[2.215]
	$k^-_{V \to V} = 1.4 \times 10^5\,\text{s}^{-1}\,(\text{mm Hg})^{-1}$	r.t.*	[2.235]
HI	$4.2 \times 10^{-3}\,(\text{collisions})^{-1}$	295	[2.223]
	$\text{HBr}(v=1) + M \rightarrow \text{HBr}(v=0) + M,\ M^v$		
DBr	$(1.96 \pm 0.2) \times 10^3\,\text{s}^{-1}\,(\text{mm Hg})^{-1}$	286	[2.231]
	$\text{HBr}(v=1) + M \xrightarrow{k_{V \to RT}} \text{HBr}(v=0) + M$		
HF	$(0.9 \pm 0.1) \times 10^4\,\text{s}^{-1}\,(\text{mm Hg})^{-1}$	295	[2.164]
	$(1.6 \pm 0.3) \times 10^4\,\text{s}^{-1}\,(\text{mm Hg})^{-1}$	300	[2.160]

*r.t. – room temperature

Table 2.3 (continued)

M	Process and kinetic data	T [K]	Reference
	$(1.1 \pm 0.2) \times 10^4 \, \text{s}^{-1} \, (\text{mm Hg})^{-1}$	350	[2.160]
DF	$(1 \pm 0.2) \times 10^4 \, \text{s}^{-1} \, (\text{mm Hg})^{-1}$	295	[2.206]
HCl	$(1.315 \pm 0.14) \times 10^3 \, \text{s}^{-1} \, (\text{mm Hg})^{-1}$	296	[2.223]
H_2	$(0.61 \pm 0.22) \times 10^3 \, \text{s}^{-1} \, (\text{mm Hg})^{-1}$	295	[2.232]
	$(0.208 \pm 0.1) \times 10^3 \, \text{s}^{-1} \, (\text{mm Hg})^{-1}$	296	[2.223]
HD	$63.5 \pm 4 \, \text{s}^{-1} \, (\text{mm Hg})^{-1}$	295	[2.233]
D_2	$40 \pm 5.5 \, \text{s}^{-1} \, (\text{mm Hg})^{-1}$	295	[2.192]

$$\text{HBr}(v=1) + M \underset{k_{V \to V}^{M-\text{HBr}}}{\overset{k_{V \to V}^{\text{HBr}-M}}{\rightleftarrows}} \text{HBr}(v=0) + M^v$$

M		T [K]	Reference
N_2	$4.5 \times 10^{-4} \, (\text{collisions})^{-1}$	295	[2.90, 2.232]
O_2	$2.5 \times 10^{-5} \, (\text{collisions})^{-1}$	295	[2.90]
CO	$1.4 \times 10^{-3} \, (\text{collisions})^{-1}$	295	[2.90]
	$7 \times 10^{-4} \, (\text{collisions})^{-1}$	295	[2.90]
CO_2	$\sigma_{V \to V}^{CO_2^* \text{HBr}} = 1.9 \, \text{Å}^2$	295	[2.90]
CH_4	$1.1 \times 10^{-3} \, (\text{collisions})^{-1}$	295	[2.232]

Relaxation of DBr

$$\text{DBr}(v=1) + M \xrightarrow{k_{V \to RT}} \text{DBr}(v=0) + M$$

M		T [K]	Reference
DBr	$(0.17 \pm 0.03) \times 10^3 \, \text{s}^{-1} \, (\text{mm Hg})^{-1}$	286	[2.231]
DF	$(0.23 \pm 0.07) \times 10^3 \, \text{s}^{-1} \, (\text{mm Hg})^{-1}$	295	[2.206]

$$\text{DBr}(v=1) + M \underset{k_{V \to V}^{M-\text{DBr}}}{\overset{k_{V \to V}^{\text{DBr}-M}}{\rightleftarrows}} \text{DBr}(v=0) + M^v$$

M		T [K]	Reference
CO_2	$\sigma_{V \to V}^{CO_2 - \text{DBr}} = 0.06 \, \text{Å}^2$	298	[2.90]

Relaxation of HI

$$\text{HI}(v=1) + M \xrightarrow{k_{V \to RT}} \text{HI}(v=0) + M$$

M		T [K]	Reference
HI	$0.375 \times 10^3 \, \text{s}^{-1} \, (\text{mm Hg})^{-1}$	295	[2.234]
	$\simeq 4 \times 10^{-4} \, (\text{collisions})^{-1}$	750	[2.175]
	$\simeq 7 \times 10^{-4} \, (\text{collisions})^{-1}$	1000	[2.175]
	$\simeq 2 \times 10^{-3} \, (\text{collisions})^{-1}$	1600	[2.175]

$$\text{HI}(v=2) + M(v=0) \xrightarrow{k_{V \to V}} \text{HI}(v=1) + M(v=1)$$

M		T [K]	Reference
HI	$(0.48 \pm 0.1) \times 10^4 \, \text{s}^{-1} \, (\text{mm Hg})^{-1}$	300	[2.160]
	$(0.32 \pm 0.1) \times 10^4 \, \text{s}^{-1} \, (\text{mm Hg})^{-1}$	350	[2.160]

$$\text{HI}(v=2) + M(v=0) \to \begin{cases} \text{HI}(v=0,1) + M(v=0) \\ \text{HI}(v=0) + M(v=1) \end{cases}$$

M		T [K]	Reference
HF	$(1.95 \pm 0.25) \times 10^4 \, \text{s}^{-1} \, (\text{mm Hg})^{-1}$	300	[2.160]
	$(1.45 \pm 0.25) \times 10^4 \, \text{s}^{-1} \, (\text{mm Hg})^{-1}$	350	[2.160]

Table 2.3 (continued)

M	Process and kinetic data	T [K]	Reference
Relaxation of DI			
CO_2	$HI(v=1)+M \underset{k_{V \to V'}^{M-HI}}{\overset{k_{V \to V'}^{HI-M}}{\rightleftarrows}} HI(v=0)+M^v$ $\sigma_{V \to V}^{CO_2-HI} = 1.5 \text{ Å}^2$	298	[2.90]
CO_2	$DI(v=1)+M \underset{k_{V \to V'}^{M-DI}}{\overset{k_{V \to V'}^{DI-M}}{\rightleftarrows}} DI(v=0)+M^v$ $\sigma_{V \to V}^{CO_2-DI} = 8.5 \times 10^{-3} \text{ Å}^2$	298	[2.90]

3. Kinetics and Numerical Analysis of Chain-Reaction Chemical Lasers (Pulsed Mode)

A great number of chemical reactions giving rise to the laser effect have been discovered since the time the first chemical laser (which depended for its operation on the chain reaction between hydrogen and chlorine, see Chap. 4) was created. Investigations of the past few years have demonstrated the possibility of developing high-power chemical lasers, both pulsed and cw (see Chaps. 4 and 5).

The desire to master and further develop chemical pumping methods have made investigations of the kinetics and population inversion mechanisms of chemical lasers a matter of top priority. Mathematical modeling of the kinetics of chemical lasers is aimed at interpreting quantitatively as well as qualitatively the processes occurring in the active laser medium in order to reveal the factors governing the efficiency of conversion of chemical energy into coherent radiation and to search for the optimum operating conditions of chemical lasers. These processes involve kinetically complex energy transformations, and so the achievement of the above goals requires the use of detailed mathematical chemical laser system models that allow for the complicated interplay of physico-chemical processes.

Section 3.1 of this chapter briefly reviews the history of the development of theoretical studies in the field of chemical lasers. Sections 3.2 and 3.3 consider in detail the general approach to the numerical modeling of the operation of chemical lasers, using by way of illustration the most efficient chain-reaction-pumped lasers.

3.1 Brief Review of the Theory

Numerical modeling of the kinetics of chemical lasers started with *Igoshin* and *Oraevsky* [3.1] and *Cohen* and co-workers [3.2], who undertook the analysis of the chain-reaction system $H_2 + Cl_2$. The simultaneous solution of the nonlinear differential equations of chemical kinetics, vibrational relaxation, and lasing [3.1] shows that oscillation ceases at a low degree of conversion as a result of relaxation processes whose rate increases in line with the accumulation of the reaction product HCl. Actually the chain character of the reaction $H_2 + Cl_2$ is not manifest in lasing, for the effective (laser) chain length v_{eff} in this system is not much in excess of unity. It has been inferred in [3.1] that it is only chain reactions involving exceptionally reactive atoms and radicals that hold promise for chemical lasers. In order for the chemical excitation process to be competitive with vibrational relaxation, the activation energy of the elementary reaction chain links

$$A + B_2 \rightarrow AB^v + B , \quad B + A_2 \rightarrow AB^v + A$$

must not exceed 2–3 kcal/mole, whereas for the reaction $Cl + H_2$ it amounts to around 5 kcal/mole and, what is more, the chain reaction cycle time in the latter case is too long. The above requirement is satisfied best of all by the chain reaction $H_2(D_2) + F_2$. The development, in 1969, of chemical lasers based on the mixtures H_2–F_2 [3.3, 4] and D_2–F_2–CO_2 [3.5], which possessed better energy characteristics because of their greater effective chain length, attracted the attention of investigators and led to the construction of chemical lasers with record-high energy performance characteristics in both pulse and cw modes.

The first numerical kinetics calculations were made in 1970–1972 for pulsed chemical lasers using H_2–F_2 [3.6–8] and D_2–F_2–CO_2 [3.6, 7, 9, 10] mixtures. The calculations were aimed primarily at:

1) revealing the relative role of the various elementary processes that take place in the active medium of the chemical laser;
2) determining the effect of the cavity on the physico-chemical processes involved in the chemical laser operation;
3) studying the effect of various parameters, such as the starting chemical composition, mixture pressure and temperature, and reaction initiation level, on the chemical laser characteristics; and
4) estimating the maximum attainable energy performance characteristics of the laser.

It was inferred on the basis of the above calculations that specific laser energies as high as 100 J/l and effective chain lengths v_{eff} in excess of 10 for the H_2–F_2 system [3.6] and in excess of 100 for the D_2–F_2–CO_2 system [3.7] were realizable. The difficulties involved in the theoretical analysis at the first stage were due to the lack of adequate data on the rate constants of the numerous competing elementary processes. By the mid-1970s, much more extensive information became available on the rate constants of these processes occurring in hydrogen-halide lasers, which made it possible to predict the characteristics of chemical lasers more accurately and formed the basis for theoretical investigations into their operation in widely differing conditions [3.11–18].

This chapter presents the results of theoretical studies of pulsed chemical laser systems. The numerical modeling of cw chemical lasers is discussed in Chap. 5. Naturally, the kinetics of elementary processes is the same for both pulsed and cw laser systems.

The most comprehensive studies of the operation of chemical lasers are based on models assuming the existence of rotational-translational equilibrium. Calculations in this approximation yield the upper limit of laser performance. As is shown by comparison between calculation and experiment [3.10–16, 3.19–22], the existing kinetic models of pulsed chemical lasers allow their performance characteristics to be calculated and predicted to an accuracy within a factor of 2–3. It is this accuracy that should be borne in mind when analyzing the calculations presented below.

The present-day development of the kinetic models of pulsed chemical lasers pursues the following objectives.

– To introduce into the models more accurate elementary process rate constants and to more adequately take account of the reaction mechanisms involved, with a view to bettering the agreement between calculated and measured laser characteristics.

– To give due consideration to deviations from true rotational and translational equilibrium conditions and transient radiative processes, so as to predict more accurately the spectral, temporal, and energy characteristics of lasers. (Naturally, giving up the assumption of rotational-translational equilibrium complicates the calculations, and as a matter of fact, the work along these lines is just beginning [3.23–25].)

– To develop simple models that include only the most important factors governing the behavior of lasers, in order to make the analysis of their energy performance more economical over a wide range of initial parameters [3.11, 12, 14, 15, 26, 27].

– To predict new modes of chemical laser operation of interest in special applications [3.11–13, 28] and to model new laser systems based on the existing approaches [3.29].

In the text below, we present the most important results of the numerical analysis of laser systems relying on the chain reactions $H_2 + F_2$ and $D_2 + F_2 + CO_2$. Attention is concentrated on those numerical results which characterize the energy capabilities of the lasers of this type. It should be noted that these capabilities have not as yet been fully realized experimentally.

3.2 H_2–F_2 System

3.2.1 Hydrogen–Fluorine Reaction Mechanism. Explosion Limits

The exothermic reaction between hydrogen and fluorine, which is described by the overall equation

$$H_2 + F_2 = 2HF$$

$$-\Delta H = 129.6 \text{ kcal/mole} = 13 \text{ kJ/g}$$

is of great interest as a source of energy for chemical lasers. The reaction goes by the chain mechanism

$$F + H_2 \rightarrow HF(v) + H, \quad -\Delta H = 31.6 \text{ kcal/mole (“cold” reaction) },$$

$$H + F_2 \rightarrow HF(v) + F, \quad -\Delta H = 98 \text{ kcal/mole (“hot” reaction) }.$$

The energy released in both stages of the chain link above goes largely to excite the vibrational levels of the HF molecules. In the "cold" reaction, the molecules are raised to the third vibrational level, to be further excited up to the eighth level in the course of the "hot" reaction. By replacing hydrogen with deuterium one can obtain vibrationally excited DF molecules.

The processes occurring in the reacting H_2–F_2 mixture are described by the following kinetic scheme which also includes, in addition to the chemical reactions, the energy transfer processes bearing on the operation of the chemical laser.

Heterogeneous chain initiation

(0) $F_2 + wall \rightarrow 2F$.

Homogeneous chain initiation

(0') $F_2 + H_2 \rightarrow F + HF + H$.

Initiation of the reaction by an external source

(0'') $F_2 + Q \rightarrow 2F$,

where Q is the electron or photon energy sufficient for the dissociation of F_2.

Chain propagation

(1) $F + H_2 \rightarrow HF(v) + H$,

(2) $H + F_2 \rightarrow HF(v) + F$.

Chain termination on oxygen

(3) $H + O_2 + M \rightarrow HO_2 + M$,

(4) $F + O_2 + M \rightarrow FO_2 + M$.

Chain termination by volume recombination of atoms

(5) $H + F + M \rightarrow HF + M$,

(6) $F + F + M \rightarrow F_2 + M$,

(7) $H + H + M \rightarrow H_2 + M$.

Chain termination on the wall

(8) $H \rightarrow wall$,

(9) $F \rightarrow wall$,

(10) $HF(v) \rightarrow wall$,

(11) $H_2(v) \rightarrow wall$,

(12) $H_2O \rightarrow wall$.

Chain branching

(13) $H_2(v) + F_2 \rightarrow HF + H + F$.

Chain restoration

(14) $HO_2 + F_2 \rightarrow HF + O_2 + F$.

Secondary processes

(15) $\mathrm{H} + \mathrm{HO}_2 \to \begin{cases} 2\mathrm{OH} , \\ \mathrm{O} + \mathrm{H}_2\mathrm{O} , \\ \mathrm{H}_2 + \mathrm{O}_2 , \end{cases}$

(16) $\mathrm{F} + \mathrm{HO}_2 \to \mathrm{HF} + \mathrm{O}_2$,

(17) $\mathrm{F} + \mathrm{OH} \to \mathrm{HF} + \mathrm{O}$,

(18) $\mathrm{OH} + \mathrm{HO}_2 \to \mathrm{H}_2\mathrm{O} + \mathrm{O}_2$,

(19) $\mathrm{OH} + \mathrm{O} \to \mathrm{H} + \mathrm{O}_2$,

(20) $\mathrm{OH} + \mathrm{H}_2 \to \mathrm{H}_2\mathrm{O} + \mathrm{H}$.

Thermal dissociation

(21) $\mathrm{F}_2 + M \to 2\mathrm{F} + M$,

(22) $\mathrm{H}_2 + M \to 2\mathrm{H} + M$,

(23) $\mathrm{HF} + M \to \mathrm{H} + \mathrm{F} + M$.

Vibrational-translational $(V \to T)$ relaxation

(24) $\mathrm{HF}(v) + M \leftrightarrow \mathrm{HF}(v-1) + M$,

(25) $\mathrm{H}_2(v) + M \leftrightarrow \mathrm{H}_2(v-1) + M$.

Vibrational-vibrational $(V \to V)$ energy transfer

(26) $\mathrm{HF}(v) + \mathrm{HF}(v) \leftrightarrow \mathrm{HF}(v+1) + \mathrm{HF}(v-1)$,

(27) $\mathrm{HF}(v) + \mathrm{H}_2(v) \leftrightarrow \mathrm{HF}(v+1) + \mathrm{H}_2(v-1)$,

(28) $\mathrm{H}_2(v) + \mathrm{H}_2(v) \leftrightarrow \mathrm{H}_2(v+1) + \mathrm{H}_2(v-1)$.

Vibrational-rotational-translational $(V \to R, T)$ energy transfer

(29) $\mathrm{HF}(v, J) + M \longleftrightarrow \mathrm{HF}(v-1, J') + M$.

Rotational $(R \to R, T)$ relaxation

(30) $\mathrm{HF}(v, J) + M \to \mathrm{HF}(v, J') + M$.

Spontaneous emission of radiation

(31) $\mathrm{HF}(v, J) \to \mathrm{HF}(v', J') + h v_{v',J'}^{v,J}$.

Stimulated emission of radiation

(32) $\mathrm{HF}(v, J) + h v_{v',J'}^{v,J} \leftrightarrow \mathrm{HF}(v', J') + 2 h v_{v',J'}^{v,J}$.

The active chain centers are initiated in processes $(0-0'')$. Processes (0) and $(0')$ are of no practical importance where the reaction is initiated by means of light, an electric discharge, or electron beam, but are essential for the stability of the H_2–F_2 mixture. The chain propagates in accordance with reactions (1) and (2) leading to the vibrational excitation of the HF molecules. The termination of the chain is due to processes (3–12). As demonstrated by *Vasil'yev* and co-workers [3.30], to calculate the chain termination rate accurately requires that due consideration should be given to secondary processes (15–20). Thermal dissociation processes (21–23) are essential where the mixture is highly heated. The reaction $H_2(D_2) + F_2$ is classed with branching chain reactions. *Semenov* and *Shilov* [3.31] have shown that branching in chain processes, particularly those involving fluorine molecules, can occur at the expense of the energy of the molecules excited in the course of the reaction. The studies of the reaction $H_2(D_2) + F_2$ [3.32–37] have revealed that energy branching here is brought about by the vibrationally excited $H_2(D_2)$ molecules [process (13)] which acquire energy from the HF(DF) molecules [process (27)].

The above scheme makes it possible to explain the causes of the explosion limits found to exist in the $H_2(D_2)$–F_2 system. Competition between processes (13) and (8–11) gives rise to the first (lower) explosion limit in the p–T coordinates (0.01 atm. and lower at room temperature). The second (upper) limit is due to competition between processes (13) and (3) and (4). *Tal'roze* [3.38] has demonstrated theoretically that in an oxygen-inhibited reaction between fluorine and hydrogen, there may exist a third, chain limit as a result of competition between chain restoration process (14) and process (12). The explosion limits depend on the mixture composition and the reactor size and wall material.

In the region above the second explosion limit, the most important chemical processes from among those numbered (0–23) are processes $(0''-3)$. Over the past few years, important information has become available on the rate constants of the elementary chemical processes occurring in the reacting H_2–F_2 mixture (see Table 3.1).

Figure 3.1 illustrates the location of the second and third explosion limits observed to exist in the H_2–F_2 mixture [3.82]. The reactor used was a passivated steel sphere of 8 l volume. To determine the third explosion limit of the stoichiometric mixture, $[H_2]/[F_2] = 1$, cooled gas mixtures H_2/He and F_2/O_2/He were leaked into a vessel at the same temperature (line *ab*). The vessel was then slowly heated until the explosion limit was reached (line *bc*). To determine the second explosion limit, an identical mixture was prepared and heated up to the point *d*. The reagents were then pumped out until explosion occurred as a result of pressure drop (line *df*).

The nature of the third explosion limit (thermal or chain explosion) is as yet unknown, but there is experimental evidence in favor of the chain character of the third explosion limit. Thus, *Chen* [3.82] has found that prior to explosion (registered by monitoring pressure and temperature changes), at both the second and third explosion limits no pre-explosion heating is necessary for thermal explosion to occur under conditions of conductive heat transfer. This, though, does not rule out the possibility of thermal explosion involving convective heat transfer.

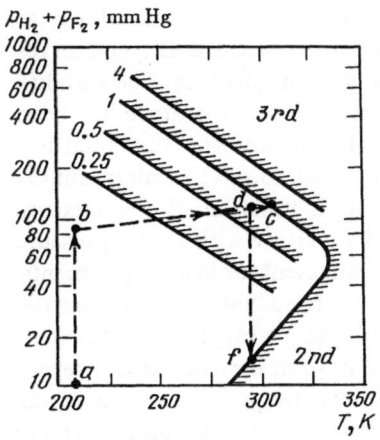

Fig. 3.1. Second and third explosion limits observed in the H_2–F_2 mixture [3.82]. The hachuring is directed towards the ignition region. The numbers on the curves denote the $[H_2]/[F_2]$ ratio. The oxygen concentration was kept at 0.2 %. The total gas pressure was maintained at 1 atm by adding He

It is exactly within the scope of the convection-controlled thermal explosion mechanism that the third explosion limit has been treated by *Bokun* and *Chaikin* [3.81]. These investigators have noted that the evolution of the chain explosion under high-pressure conditions, where, in accordance with [3.38], the existence of the third, chain explosion limit is to be expected, is prevented by the process

$$2HO_2 \rightarrow H_2O_2 + O_2 \; ,$$

whose rate can be much higher than that of reaction (12).

It has been found that with the total H_2–F_2 mixture pressure and the O_2 concentration being fixed, a mixture with a higher F_2/H_2 ratio is less stable and explodes at a lower temperature (Fig. 3.1). This can be explained qualitatively by considering the following three main effects attending the increase of the F_2 content of the H_2–F_2–O_2–He mixture [3.82]. First, the heat liberation rate increases as a result of the rising rates of the chain initiation and propagation reactions

$$F_2 \xrightarrow{\text{collisions}} 2F \; ,$$

$$H + F_2 \rightarrow HF + F \; ,$$

where the process $H + F_2$ limits the chain reaction rate. Second, the heat removal rate decreases. And third, the number of the F atoms produced in the process

$$HO_2 + F_2 \rightarrow HF + O_2 + F$$

grows larger. Thus, the increase of the proportion of F_2 in the mixture reduces its stability in the case of both thermal and chain explosion mechanisms. As reaction (13) is the main chain termination reaction, its effect on the explosion limits is very substantial. The stable mixture region between the second and third explosion limits widens as the oxygen concentration is increased.

To construct powerful pulsed lasers requires a stable active medium at a high static pressure (of the order of atmospheric pressure or higher). The exothermic chain reaction leading to the formation of vibrationally excited HF molecules gives rise to spontaneous detonation of the high-pressure H_2–F_2 mixture. Oxygen serves as a stabilizer, preventing the mixture from exploding spontaneously. The stability of multiatmospheric-pressure H_2–F_2–O_2 mixtures at room temperature has been investigated by *Truby* [3.83], who found that the explosion limits for such mixtures differ with the proportions of the components. He has also found the minimum O_2 concentration necessary to stabilize the mixture as a function of the mixture pressure and the F_2/H_2 ratio (Fig. 3.2) and obtained a mixture with a total pressure of 11 atm. that remained stable for 15 min., which is sufficient for laser experiments. The data presented in Fig. 3.2 show that with the F_2/H_2 ratio being fixed, the partial pressures of H_2 and F_2 can be increased only if the $O_2/(F_2 + H_2)$ ratio is increased as well. It can also be seen from Fig. 3.2 that more oxygen is required to prevent explosion in mixtures with higher F_2/H_2 ratios, the mixture pressures being equal. This observation agrees qualitatively with the findings of *Chen* [3.82] and *Getzinger* [3.84], who studied the explosion limits at lower pressures.

3.2.2 Mathematical Model of the Hydrogen-Fluoride Laser (HFL)

This section presents the relatively simple HFL model developed by *Igoshin*, *Nikitin*, and *Oraevsky* [3.12]. The model includes the most important physico-chemical kinetics features under the assumption of rotational-translational equilibrium and makes it possible to calculate the specific energy of coherent radiation, pulse duration, effective chain length of the reaction between fluorine and hydrogen, and chemical and engineering efficiencies of the laser. Based on this model, the above investigators have studied the effect of the main factors (mixture pressure and chemical composition, spectral conditions of stimulated emission of radiation, initiation intensity and duration) on the laser efficiency. The results obtained give an idea of the energy of coherent radiation emitted in the course of the reaction, over a wide range of initial parameters.

The characteristics of a hydrogen-fluoride laser depend on the rates of the basic processes taking place in the reactive system, namely, the rates of the production and decay of free atoms (chain active centers), the rates of chemical reactions giving

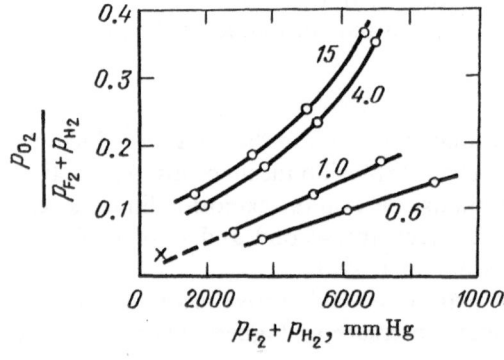

Fig. 3.2. The $[O_2]/([F_2] + [H_2])$ ratio necessary to stabilize the H_2–F_2 mixture as a function of $[H_2] + [F_2]$ for the $[F_2]/[H_2]$ values on the curves. Explosion occurs on the curves and below. The circles refer to the data of [3.83] and the cross to [3.84]

rise to vibrationally excited HF molecules, and the rate of collisional deactivation of the lasing molecules.

Below we present the kinetic scheme describing the basic processes occurring in the active HFL mixture, along with the elementary process rate constants used in the calculations of [3.12].

Chain:

(1) $F + H_2 \xrightarrow{\alpha_v^{(1)} k_1} HF(v) + H$,

$$k_1 = (1.62 \times 10^{14}) \exp(-1600/RT) \quad cm^3/mole\, s \ .$$

(2) $H + F_2 \xrightarrow{\alpha_v^{(2)} k_2} HF(v) + F$,

$$k_2 = (1.2 \times 10^{14}) \exp(-2400/RT) \quad cm^3/mole\, s \ .$$

Recombination of atoms:

(3) $F + F + M \xrightarrow{k_3} F_2 + M$,

$$k_3 = 10^{14} T^{0.5} \quad cm^6/mole^2\, s \ .$$

(4) $H + H + M \xrightarrow{k_4} H_2 + M$,

$$k_4 = 10^{18}/T \quad cm^6/mole^2\, s \ .$$

(5) $H + F + M \xrightarrow{k_5} HF + M$,

$$k_5 = 10^{18}/T \quad cm^6/mole^2\, s \ .$$

Vibrational-translational relaxation of excited HF molecules:

(6a) $HF(v) + M \xleftrightarrow{k_{6a}^M} HF(v-1) + M$,

$$k_{6a}^M = v k_{1,0}^M \ ,$$

$$k_{1,0}^F = 9 \times 10^8 T^{1/3} \quad cm^3/mole\, s \ ,$$

$$k_{1,0}^H = 64.1 T^{3.51} \exp(3946/T) \quad cm^3/mole\, s \ ,$$

$$k_{1,0}^{HF} = (5 \times 10^7 T^{1.3} + 10^{16} T^{-1.43}) \quad cm^3/mole\, s \ ,$$

$$k_{1,0}^{F_2} = 8 \times 10^{-4} T^4 \quad cm^3/mole\, s \ ,$$

$$k_{1,0}^{H_2} = 1.7 \times 10^6 T^{1.77} \quad cm^3/mole\, s \ .$$

Vibrational-vibrational energy exchange:

(6b) $HF(v) + HF(v') \xrightarrow{k_{6b}} HF(v+1) + HF(v'-1)$,

$k_{6b} = (v+1)v'k_{0,1;1,0}^{HF-HF}$,

$k_{0,1;1,0}^{HF-HF} = 1.5 \times 10^{13}$ cm^3/mole s .

(6c) $HF(v) + H_2(v') \xrightarrow{k_{6c}} HF(v+1) + H_2(v'-1)$,

$k_{6c} = (v+1)v'k_{0,1;1,0}^{HF-H_2}$,

$k_{0,1;1,0}^{HF-H_2} = (k_{1,0;0,1}^{HF-H_2})\exp[(\theta_v^{H_2} - \theta_v^{HF})/T] = 9 \times 10^{11}$ cm^3/mole s .

Vibrational-translational relaxation of excited H_2 molecules:

(6d) $H_2(v) + M \xrightarrow{k_{6d}^M} H_2(v-1) + M$,

$k_{6d}^M = vk_{1,0}^{H_2-M}$,

$k_{1,0}^{H_2-H_2} = 10^{-3} T^{4.3}$,

$k_{1,0}^{H_2-HF} = k_{1,0}^{H_2-H} = k_{1,0}^{H_2-F} = k_{1,0}^{H_2-F_2} = k_{1,0}^{H_2-Ar} = 2.5 \times 10^{-4} T^{4.3}$.

Reaction initiation:

(7) $F_2 + Q \xrightarrow{k_7} 2F$,

(8) $RF + Q \xrightarrow{k_8} R + F$,

where RF is a fluorine-containing component that can be introduced into the mixture to facilitate the initiation of the reaction, Q is the bond dissociation energy and k_7 and k_8 are functions of time dependent on the initiation technique. Chain termination on oxygen:

(9) $H + O_2 + M \xrightarrow{k_9} HO_2 + M$,

$k_9 = 10^{22}/T^2$ cm^6/mole^2 s .

Thermal dissociation:

(10) $F_2 + M \xrightarrow{k_{10}} 2F + M$,

$k_{10} = 1.38 \times 10^{13} \exp(-31120/RT)$ cm^3/mole s ,

(11) $H_2 + M \xrightarrow{k_{11}} 2H + M$,

$k_{11} = 8.1 \times 10^{16} \exp(-103240/RT)$ cm^3/mole s ,

(12) $\quad$ HF $+ M \xrightarrow{\;\;k_{12}\;\;}$ H $+$ F $+ M$,

$$k_{12} = 1.13 \times 10^{19}\, T^{-1} \exp(-13400/RT) \quad \text{cm}^3/\text{mole s} .$$

Stimulated emission of radiation:

(13) $\quad$ HF$(v) + \hbar\omega_{v,\,v-1} \leftrightarrow$ HF$(v-1) + 2\hbar\omega_{v,\,v-1}$.

Note that in the above calculations, the chain termination processes are treated in a simplified manner. The quantities k_3 to k_5 and k_9 are, generally speaking, functions of the mixture composition. In principle, to consider the mixture composition dependence of the chain termination rate presents no problem, and is most essential at mixture pressures much in excess of the atmospheric value.

$\quad$ The chain branching process

$$H_2(v) + F_2 \rightarrow H + HF + F$$

is disregarded because it is slow on the time scale of the coherent radiation pulse duration. It should be noted that, according to the more recent data [3.30, 63], the rates of branching processes involving the H$_2$ molecules excited to $v = 2$ or to higher vibrational levels are fairly fast (Table 3.1). If the branching rate is in fact so high, it is advisable to search for such an emission regime that would markedly improve the engineering efficiency of the laser (see Chap. 6).

$\quad$ Processes (1–5) and (7–12) are described by the following chemical kinetics equations:

$$\frac{d\xi_a}{dt} = -2\xi_a^2 N^2 \left\{ k_3 \left[\frac{\mu}{\mu+1} \right]^2 + k_4 \frac{1}{(\mu+1)^2} + k_5 \frac{\mu}{(\mu+1)^2} \right\} + 2k_7 \xi_{F_2}$$

$$+ k_8 \xi_{RF} - k_9 \frac{1}{\mu+1} \xi_{O_2} \xi_a N^2 + 2k_{10}\xi_{F_2} N + 2k_{11}\xi_{H_2} N + 2k_{12}\xi_{HF} N ,$$

$$\frac{d\xi_{F_2}}{dt} = -k_2 \frac{1}{\mu+1} \xi_{F_2}\xi_a N + k_3 \left[\frac{\mu}{\mu+1} \right]^2 \xi_a^2 N^2 - k_7\xi_{F_2} - k_{10}\xi_{F_2} N ,$$

$$\frac{d\xi_{H_2}}{dt} = -k_1 \frac{\mu}{\mu+1} \xi_a \xi_{H_2} N + k_4 \frac{1}{(\mu+1)^2} \xi_a^2 N^2 - k_{11}\xi_{H_2} N ,$$

$$\frac{d\xi_{HF}}{dt} = k_1 \frac{\mu}{\mu+1} \xi_a \xi_{H_2} N + k_2 \frac{1}{\mu+1} \xi_a \xi_{F_2} N + k_5 \frac{\mu}{(\mu+1)^2} \xi_a^2 N^2 - k_{12}\xi_{HF} N ,$$

$$\frac{d\xi_{HO_2}}{dt} = -k_9 \frac{1}{\mu+1} \xi_a \xi_{O_2} N^2 ,$$

$$\frac{d\xi_{RF}}{dt} = -k_8 \xi_{RF} ,$$

where N is the initial total particle concentration in the mixture, $\xi_M = [m]/N$ is the reduced concentration of the M component in the mixture, $\xi_a = \xi_F + \xi_H$ is the total reduced concentration of the atoms F and H, $\mu = (k_2 \xi_{F_2})/(k_1 \xi_{H_2})$, and k_i is the rate constant of the ith process.

The chain reaction equations are written under the assumption of quasistationary concentrations of the active centers F and H, which cuts down considerably the computer time required for numerical calculations. In this approximation, the rates of both stages (1) and (2) of the chain link are taken to be equal:

$$k_1 \xi_F \xi_{H_2} = k_2 \xi_H \xi_{F_2} , \tag{3.1}$$

i.e., a certain relationship is established between the concentrations of the atoms F and H, which depends on the ratio between the reactants F_2 and H_2 and the rate constants of the chain link stages. The time for establishment of quasistationary active center concentrations, measured over a wide range of initial parameters (pressure, mixture composition, temperature), proves to be much shorter than the time it takes for the fuel to burn out or for the active centers to decay by three-body collision, a fact that allows relation (3.1) to be used in chemical kinetics calculations.

Section 2.6 describes an effective procedure for calculating stimulated emission of radiation in multi-level molecular systems, which we apply here to the hydrogen-fluoride laser. The procedure allows one to take account of the basic factors governing the behavior of lasers operating on a vibrational-rotational population inversion under rotational equilibrium conditions.

The distribution of the HF molecules among vibrational levels in both the laser oscillator and laser amplifier modes is described by the relation

$$\xi_v^{HF} = \frac{1}{A_J} \sum_{k=1}^{v} B_J^{v-k} \Delta_{k-1}^k + B_J^v \xi_0^{HF} ,$$

where $\Delta_{v-1}^v = 1/(c\sigma_{v-1}^v \tau_{ph})$ is the threshold inversion density, σ_{v-1}^v is the cross section for stimulated emission of radiation, τ_{ph} is the photon lifetime in the cavity, $A_J = (2J-1)(T/\theta_r)\exp[-\theta_r J(J-1)/T]$, $B_J = \exp(-2Q_r J/T)$, θ_r is the characteristic rotational temperature of HF (equal to 30.2 K), and J is the rotational quantum number of the radiative transitions of the molecule. The number J is taken to be the same for all the $v \to v-1$ bands,

$$\xi_0^{HF} = \frac{B_J - 1}{B_J^{R+1} - 1} \xi_{HF} - \frac{1}{A_J} \sum_{v=1}^{R} \frac{B_J^{R+1-v} - 1}{B_J^{R+1} - 1} \Delta_{v-1}^v .$$

When calculating the laser amplification, it is necessary to put $\Delta_{v-1}^v = 0$. The stimulated emission power density is calculated by formula (2.137), which, as applied to HFL, may be represented in the form

$$P_l = \hbar \omega_l(J) \left\{ k_1 \frac{\mu}{\mu+1} \xi_a \xi_{H_2} (\varepsilon_1^{(1)} - \varepsilon_2) + k_2 \frac{1}{\mu+1} \xi_a \xi_{F_2} (\varepsilon_1^{(2)} - \varepsilon_2) - \frac{G}{N^2} \right\} N^2 ,$$

where

$$\varepsilon_1^{(1)} = \sum_{v=1}^{R} \alpha_v^{(1)} v = 2.095 \quad \text{and} \quad \varepsilon_1^{(2)} = \sum_{v=1}^{R} \alpha_v^{(2)} v = 5.435$$

are the average numbers of vibrational quanta excited in the HF molecules in the course of reactions (1) and (2), respectively; $\alpha_v^{(1)}$ and $\alpha_v^{(2)}$ are the normalization coefficients describing the distribution of energy in the elementary pumping reactions (1) and (2), respectively; R is the number of the excited $v \to v - 1$ bands, equal to nine for the (H$_2$ + F$_2$) laser; the quantity

$$\varepsilon_2 = \frac{B_J + R B_J^{R+2} - (R+1) B_J^{R+1}}{(1 - B_J)(1 - B_J^{R+1})}$$

has the meaning of the average store of vibrational quanta per HF molecule in the course of emission of radiation under conditions of saturation of the transitions $v, J - 1 \to v - 1, J$; the quantity G describes the loss of the vibrational energy of HF as a result of relaxation processes and is given by

$$G = \sum_{v=1}^{R} (n_v g_{v, v-1} - n_{v-1} g_{v-1, v}) \ ,$$

where the coefficients $g_{v, v-1}$ are equal to the total probability of the $v \to v - 1$ transition as a result of all the $V \to T$ and $V \to V$ processes, and $n_v = \xi_v^{HF} N$. Because of the additivity of the probabilities of the three collisional processes (6a), (6b), and (6c) contributing to the coefficients $g_{v, v-1}$, the quantity G may be represented as a sum of three terms:

$$G = G_{VT}^{HF-M} + G_{VV}^{HF-HF} + G_{VV}^{HF-H_2} \ .$$

Within the framework of the harmonic oscillator model, $G_{VV}^{HF-HF} = 0$, i.e., the $V \to V$ exchange between the HF molecules does not contribute to G, and the contributions from the $V \to T$ processes and the $V \to V$ exchange between HF and H$_2$ are given by

$$G_{VT}^{HF-M} = \left(\sum_M k_{1,0}^{HF-M} \xi_M N \right) \left\{ [1 - \exp(-\theta_v^{HF}/T)] \left[\xi_{HF} \varepsilon_2 N \right. \right.$$

$$\left. + \left(\frac{1}{A_J} \right) \sum_{v=1}^{R} \Delta_{v-1}^v f_v \right] - \exp(-\theta_v^{HF}/T) N [\xi_{HF} - (R+1) \xi_R^{HF}] \bigg\} \ ,$$

$$G_{VV}^{HF-H_2} = k_{0,1;1,0}^{HF-H_2} N^2 \xi_{HF} \xi_{H_2} \{ \varepsilon_{HF} (\varepsilon_{H_2} + 1) \exp[(\theta_v^{HF} - \theta_v^{H_2})/T] - \varepsilon_{H_2} (\varepsilon_{HF} + 1) \} \ ,$$

where $\theta_v^{HF} = 5725$ K, $\theta_v^{H_2} = 5991$ K; $\varepsilon_{H_2} = \Sigma_v v \xi_v^{H_2}$ and $\varepsilon_{HF} = \Sigma_v v \xi_v^{HF}$ are the numbers of vibrational quanta per H$_2$ and HF molecule, respectively, and the quantities f_v are defined by formula (2.136).

The effect of the exchange of quanta between the HF molecules manifests itself in a redistribution of the total emission power among the $v \to v - 1$ vibrational bands. In principle, the present model makes it possible to calculate the emission power in any of the excited $v \to v - 1$ bands by the formulas obtained in Sect. 2.6. But here we are interested only in the total emission power.

When calculating an HF laser oscillator, at every given moment of time we search for a rotational transition providing the maximum gain and assign the value found to the rotational quantum number J. In the course of lasing, the number J varies as a result of heating of the mixture. In the case of a laser amplifier, the number J is considered to be a specified parameter corresponding to the spectral composition of the input signal. The model permits the monitoring of the effect of the stimulated emission spectrum (the number J) on the laser efficiency. The procedure developed in Sect. 2.6 enables one to compute stimulated emission of radiation in a more general case where the rotational quantum number $J(v)$ differs between the different emitting bands $v \rightarrow v-1$, but the assumption that the number J remains constant in all the $v \rightarrow v-1$ bands gives an accurate enough idea of the specific laser characteristics.

The set of chemical kinetics equations must be supplemented with the relaxation equation for the H_2 molecule:

$$\frac{d\varepsilon_{H_2}}{dt} = -\frac{\varepsilon_{H_2} - \varepsilon_{H_2}^0}{\tau_{VT}^{H_2}} + k_{0,1;1,0}^{HF-H_2} \zeta_{HF} N \left\{ \varepsilon_{HF}(\varepsilon_{H_2} + 1) \exp[(\theta_v^{HF} - \theta_v^{H_2})/T] \right.$$

$$\left. - \varepsilon_{H_2}(\varepsilon_{HF} + 1) \right\} - (\varepsilon_{H_2}/\zeta_{H_2}) \frac{d\zeta_{H_2}}{dt} ,$$

where

$$1/\tau_{VT}^{H_2} = \sum_M k_{1,0}^{H_2-M} \zeta_M N [1 - \exp(-\theta_v^{H_2}/T)], \quad \varepsilon_{H_2}^0 = [\exp(\theta_v^{H_2}/T) - 1]^{-1} ,$$

and the heat-balance equation defining the gas temperature T:

$$N\left(\sum_M c_v^M \zeta_M\right) \frac{dT}{dt} = \sum_i q_i W_i + R_0 [\theta_v^{HF} G_{VT}^{HF-M}$$

$$+ \theta_v^{H_2} G_{VT}^{H_2-M} - (\theta_v^{H_2} - \theta_v^{HF}) G_{VV}^{HF-H_2} ,$$

where $G_{VT}^{H_2-M} = [(\varepsilon_{H_2} - \varepsilon_{H_2}^0)/\tau_{VT}^{H_2}] \zeta_{H_2} N$, c_v^M is the specific heat at constant volume of the component M, q_i the heat of the ith elementary reaction event, and R_0 the universal gas constant.

The main quantities characterizing the efficiency of a chemical laser are defined as follows.

The energy emitted per unit volume

$$\varepsilon_l = \int_{t_0}^{t_1} P_l dt ,$$

where t_0 and t_1 are the lasing (amplification) onset and quenching moments, respectively.

The chemical efficiency is

$$\eta_{chem} = \varepsilon_l/(-\Delta H) \zeta_{A_2}^0 N ,$$

where $A_2 = H_2$ at $\zeta_{H_2}^0 < \zeta_{F_2}^0$ and $A_2 = F_2$ otherwise, $(-\Delta H) = 130$ kcal/mole $= 543$ kJ/mole is the heat of the reaction $H_2 + F_2 \rightarrow 2HF$, and $\zeta_{A_2}^0$ the initial relative concentration of the component A_2 in the mixture.

The engineering efficiency is given by

$$\eta_{\text{eng}} = (\varepsilon_l / \mathscr{E}_{\text{a}} \zeta_{\text{a}}^0 N)_{\text{in}} \; ,$$

where $\mathscr{E}_{\text{a}}$ is the energy necessary to produce an active center, η_{in} the efficiency of the initiation source, and ζ_{a}^0 the relative concentration of atoms produced by the initiation source in the mixture.

The efficiency with respect to the absorbed energy (hereafter referred to as the physical laser efficiency) is

$$\eta_{\text{phys}} = \varepsilon_l / (\mathscr{E}_{\text{a}} \zeta_{\text{a}}^0 N) \; .$$

The coherent radiation quantum yield, i.e., the number of photons emitted per active center, is expressed as

$$f = \frac{\varepsilon_l}{\hbar \omega_l} \frac{1}{\zeta_{\text{a}}^0 N} \; .$$

The effective (laser) chain length, i.e., the number of HF molecules contributing to coherent radiation, per active center is

$$v_{\text{eff}} = \frac{2\varepsilon_l}{(\varepsilon_1^{(1)} + \varepsilon_1^{(2)})\hbar \omega_l} \frac{1}{\zeta_{\text{a}}^0 N} \; ,$$

where $(\varepsilon_1^{(1)} + \varepsilon_1^{(2)})\hbar \omega_l / 2$ is the average vibrational excitation energy acquired by the HF molecules in both stages of the reaction chain link.

The quantities η_{eng}, η_{phys}, f, and v_{eff} are defined above for the case where the initiating pulse is short compared to the coherent radiation pulse. If this condition is not satisfied, the quantity ζ_{a}^0 in all the above relations should be replaced by the quantity

$$\zeta_{\text{a}} = \int_0^{t_1} (2k_7 \zeta_{\text{F}_2} + k_8 \zeta_{\text{RF}}) dt$$

which determines the production of active centers by the initiating pulse in the course of oscillation (amplification). The laser characteristics considered above are not independent of one another. For instance, η_{chem} and v_{eff} are related by the relation

$$v_{\text{eff}} = \frac{1 - \Delta H}{2 \quad \hbar \omega_l} (\varepsilon_1^{(1)} + \varepsilon_1^{(2)})^{-1} \frac{\zeta_{A_2}^0}{\zeta_{\text{a}}^0} \eta_{\text{chem}} \; .$$

3.2.3 Calculation of the Effect of Basic Factors on the Energy Characteristics and Emission Dynamics of HFL in the Rotational-Translational Equilibrium Approximation

Based on the above mathematical model of HFL, we have studied the effect of the main factors on the laser efficiency and the duration of the coherent radiation pulse.

The assumed parameters were the starting chemical composition of the mixture and its concentration N, the initial temperature T_0, the time-dependent initiation process rate constants k_7 and k_8, and the spectral composition of radiation (the number J).

The temporal dependence of the initiation process rate constants was defined by

$$k_7 = \gamma_{F_2} t \exp(-t/t_0) \quad \text{and} \quad k_8 = \gamma_{RF} t \exp(-t/t_0) \; ,$$

where t_0 characterizes the initiating pulse duration; γ_{F_2} and γ_{RF} define the degree of dissociation of F_2 and RF under the action of the initiating pulse ($M = F_2$, RF):

$$\zeta_M^0 - \zeta_M^\infty = 1 - \exp\left[-\int_0^\infty \gamma_M t \exp(-t/t_0) \, dt \right] = 1 - \exp(-\gamma_M t_0^2) \; .$$

The quantities γ_M can be expressed in terms of the density of the photon flux, I (in quanta/cm^2 s), into the absorption band of the molecules M in the case of photolysis initiation, or in terms of the electron-beam current density J_b in that of electron-beam initiation, by means of the relations

$$k_7(t) = \sigma_{F_2} I(t); \quad k_8 = \sigma_{RF} I(t) \; , \quad \text{or}$$

$$2k_7 = \chi_{F_2} Q_i; \quad k_8 = \chi_{RF} Q_i \; ,$$

where χ_{F_2} (χ_{RF}) is the number of the F atoms per ion pair produced in the case of electron-beam initiation (for F_2, $\chi_{F_2} = 3$–6 [3.85–88]); Q_i [cm^{-3} s]) is the specific ionization rate of the F-atom donor:

$$Q_i = 4.744 \times 10^{21} \, J_b \,[\text{A/cm}^2] \times P_M \,[\text{atm.}] \times S_M \,[\text{ion pairs/cm mm Hg}] \; ,$$

where S_M is the total number of ion pairs generated by a fast electron per unit flight distance at a donor component pressure of $P_M = 1$ mm Hg.

The photodissociation cross section of F_2 is comparatively small: $\sigma_{F_2} = 2.3 \times 10^{-20}$ cm^2. For this reason, to increase the photolysis initiation rate, the active mixture is prepared with additions of molecules [referred to as photoinitiating components (PIC) later in the text] that dissociate to form fluorine atoms and have a large absorption cross section (e.g., WF_6 or NOF). Thus, the concentration and extent of dissociation of the F-atom donor, along with the initiating pulse duration, are the basic factors governing the reaction initiation process. It is of interest to study the effect of these factors on the operation of HFL.

Initiation Level. By "initiation level" is meant the concentration n_a of atoms produced by an external source. The relationship between the laser energy output and the initiation level is of utmost importance for chain-reaction chemical lasers. This relationship enables one to reveal the maximum attainable specific energy output and efficiency of the lasers, and also the requirements that the initiation source must meet.

Table 3.2 lists calculation results for the laser pulse duration t_l and the basic laser energy characteristics – ε_l, η_{chem}, η_{phys}, v_{eff}, and f – as a function of n_a.

Calculations were made for an energy-intensive atmospheric-pressure mixture with a specific chemical energy store of around 3 J/cm^3. More energy-intensive mixtures are difficult to prepare because of their lower stability, but give off more energy in the form of coherent radiation. The duration of the initiating pulse in the calculations was taken to be short ($t_0 = 10^{-7}$ s) compared to the laser pulse duration. The parameters γ_{F_2} and γ_{RF} were varied over the range corresponding to the change in the concentration of F$_2$ from 0.02% to 20% and in that of RF, from 0.09% to 90%. The total concentration of atoms in that case varied between the limits $\simeq 10^{14}$ and $\simeq 10^{18}$ cm^{-3}, which completely covered the initiation level range of practical interest.

It can be inferred from the analysis of the data presented in Table 3.2 that (a) the chemical efficiency η_{chem} and specific energy output ε_l of HFL rise with increasing initiation level approximately in proportion to n_a at low n_a values ($n_a < 10^{16}$ cm^{-3}) and tend to saturate at high n_a values ($n_a > 10^{17}$ cm^{-3}); (b) the effective chain length ν_{eff}, physical efficiency η_{phys}, and coherent radiation quantum yield f increase in a certain range of n_a values ($10^{15} < n_a < 10^{16}$ cm^{-3}), reaching their maximum values of $\nu_{\text{eff}} \simeq 70$, $\eta_{\text{phys}} \simeq 1000$ %, and $f \simeq 300$, and then decrease as n_a grows further (and so the choice of n_a is governed by the conflicting requirements that both ε_l and ν_{eff} should be high enough); (c) the laser pulse duration t_l decreases from $\simeq 10^{-4}$ s to $\simeq 10^{-7}$ s as n_a increases from 10^{14} cm^{-3} to 10^{18} cm^{-3} (and so a certain level of the specific energy output ε_l corresponds to a quite definite duration of the chain chemical process; this places specific demands on initiation sources, which are considered in detail in Chap. 4); (d) the initiation level range $10^{16} < n_a < 10^{17}$ cm^{-3} is optimal from the standpoint of the laser energy characteristics; in this range, the specific energy output amounts to 100–600 J/l, the physical efficiency η_{phys} in the case of electron-beam initiation ranging between 1000 and 500 %, respectively, and the laser pulse duration varying from 3×10^{-6} to 5×10^{-7} s.

Initiating Pulse Duration at Constant Pulse Energy and Concentration of the Photoinitiating Component (PIC). Table 3.3 presents calculation results for the chemical efficiency and pulse duration of FHL as a function of the initiating pulse duration for mixtures differing in chemical composition and PIC content. The calculations were made under the assumption that the photon flux into the absorption band of F$_2$ is the same as that into the RF absorption band, and hence the relationship between γ_{F_2} and γ_{RF} is determined by the absorption cross sections of F$_2$ and RF, σ_{RF} being taken at 10^{-18} cm^2 which is characteristic of polyatomic molecules such as WF$_6$ and MoF$_6$. The quantity $\gamma_{F_2} t_0^2$ corresponds to a 2 % dissociation of F$_2$ and $\gamma_{RF} t_0^2$, an 85 % dissociation of RF. The value of t_0 was varied over a range spanning four orders of magnitude – from 10^{-4} to 10^{-8} s.

It may be inferred from the data listed in the table that (a) the time it takes to produce the necessary number of active centers is a very important factor that to a large measure determines the energy output of FHL; η_{chem} rises the steepest as the initiating pulse duration is reduced to $\simeq 1$ μs and then tends to saturate; (b) the introduction of PIC into the mixture has an appreciable effect on η_{chem} only at a relatively low (< 50 mm Hg) partial pressure of F$_2$ (we do not consider here the possibility of using, with a view to increasing the efficiency of the optical radiation source used for initiation, various PIC molecules whose absorption spectra overlap the spectrum of the source); (c) mixtures with higher F_2 concentrations provide for higher η_{chem}.

Chemical Composition of the Mixture. It is of interest to find out what is the optimum (for obtaining the maximum possible specific laser energies ε_l) ratio $\zeta_{H_2}^0 : \zeta_{F_2}^0$ at a fixed total pressure of the reactants, and what is the reason for the existence, if any, of such an optimum ratio. The optimization problem may be formulated in two ways: at a fixed initiation level n_a and at a fixed initiating pulse intensity. The calculation results presented in Table 3.4 show that as the ratio $\zeta_{H_2}^0 : \zeta_{F_2}^0$ (and hence the store of chemical energy) is reduced for a fixed n_a, the quantity $\varepsilon_l \simeq \eta_{chem}\zeta_{H_2}$ remains practically unchanged up to $\zeta_{H_2}^0 : \zeta_{F_2}^0 = 1:(3$ to $4)$. But the mixture $\zeta_{H_2}^0 : \zeta_{F_2}^0 = 1:3$ is preferable compared to the stoichiometric mixture, because with the initiation level n_a set, the necessary initiating pulse intensity is inversely proportional to the concentration of the F_2 molecules. The calculations indicate that there is no need to prepare a mixture with the maximum possible chemical energy store, i.e., with $\zeta_{H_2}^0 : \zeta_{F_2}^0 = 1:1$, in order to achieve the maximum ε_l values, for more heat will be liberated in such a mixture without adding to ε_l. It also follows from the calculations that the existence of the experimentally observed optimum ratio $\zeta_{H_2}^0 : \zeta_{F_2}^0 = 1:3$ is to a large measure due to initiation conditions and cannot be explained by the chemical kinetics of the hydrogen-fluorine reaction alone.

Mixture Pressure and Oxygen Concentration. The calculation results presented in Table 3.5 demonstrate that with the degree of dissociation of F_2 and RF being fixed, the quantity η_{chem} depends but weakly on the mixture pressure in the range 1–10 atm. at a relative oxygen content of 1 %. With the oxygen content increased to 10 %, η_{chem} remains unchanged only at mixture pressures up to $\simeq 3$ atm. and then drops sharply as the pressure is further increased. This effect is explained by the decay of the active centers in the course of process (9), and sets a limit on the mixture pressure. Although the above calculations are of estimative character only, they indicate that the oxygen concentration is a factor of importance to the energy characteristics of HFL at high mixture pressures. In principle, it is desirable to carry out more comprehensive calculations on the basis of a more detailed chain termination mechanism in order to study the possibility of increasing the active medium pressure in HFL.

HFL Dynamics. Figure 3.3 illustrates the calculation of the emission dynamics of HFL. It can be seen that the emission spectrum shifts in the course of lasing towards the higher rotational states, up to $J \simeq 16$, which is explained mainly by an increase in the medium temperature beyond 1000 K during the reaction. The total laser pulse duration and the duration of lasing in a given $v, J - 1 \rightarrow v - 1, J$ transition can be controlled by varying the initiating pulse duration, as illustrated quantitatively in Fig. 3.3.

The shift of the laser emission spectrum towards higher J values (J-shifting phenomenon) materially increases the pulse duration and efficiency of HFL. At any

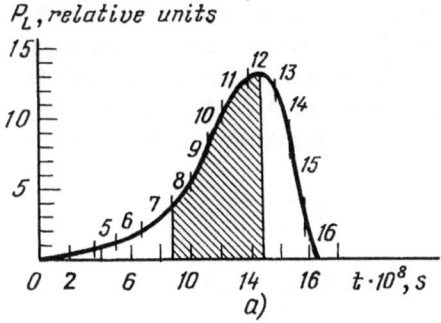

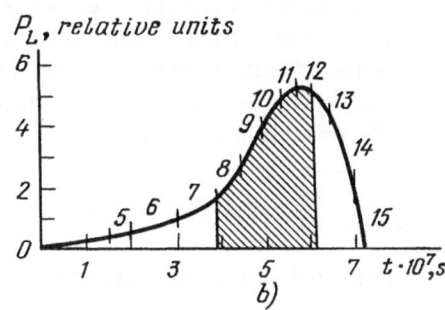

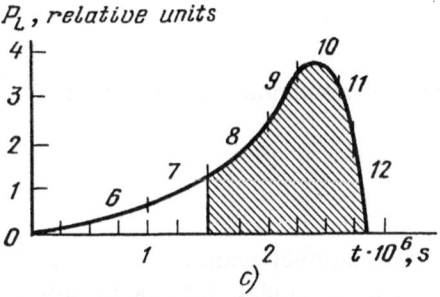

Fig. 3.3a–c. Temporal evolution of lasing in HFL operating on the $F_2:H_2:O_2:RF:He$ $=400:100:40:5:215$ mixture [3.12]: **(a)** $\gamma_{F_2} = 2.07 \times 10^{12} \text{ s}^{-2}$, $\gamma_{RF} = 9 \times 10^{13} \text{ s}^{-2}$, $t_0 = 10^{-7}$ s; **(b)** γ_{F_2} $= 2.07 \times 10^{10} \text{ s}^{-2}$, $\gamma_{RF} = 9 \times 10^{11} \text{s}^{-2}$, $t_0 = 10^{-6}$ s; **(c)** $\gamma_{F_2} = 2.07 \times 10^{8} \text{s}^{-2}$, $\gamma_{RF} = 9 \times 10^{9} \text{s}^{-2}$, t_0 $= 10^{-5}$ s. The numbers indicate lasing regions with the J values given. The shaded area defines the efficiency of lasing on transitions corresponding to the optimum J values in a master oscillator-amplifier arrangement

given moment of time, only one transition, namely that with the maximum gain, is lasing. The picture of lasing in an actual laser is more complicated owing to the violation of the rotational-translational equilibrium condition. In particular, simultaneous lasing occurs on several neighboring rotational-vibrational trans- itions. But the gradual shift of the lasing line spectrum towards the higher J values is also observed to occur experimentally.

Emission Spectrum of HFL Amplifier. The emission spectrum of a laser oscillator is formed automatically by the active medium, so that lasing occurs in transitions with the maximum gain. In a laser amplifier, the emission spectrum may be forced by the master oscillator. In that case, the question arises whether there is such a master oscillator emission spectrum that would be optimal for the maximum energy output of the associated HFL amplifier. It turns out that the vibrational-rotational transition with the maximum unsaturated gain is, generally speaking, not optimal energetically. Under rotational-translational equilibrium conditions, the use for amplification purposes of transitions whose J values exceed those of the transitions with the peak unsaturated gain allows the specific energy output of the amplifier to be substantially increased. *Igoshin* and *Oraevsky* [3.89] were the first to consider this fact. The physical cause of the effect is obvious. Lasing is quenched as a result of the competition between the rates of the reaction and relaxation processes. As the concentration of the reaction product increases, the relaxation rate grows until it exceeds the chemical pumping rate, at a certain degree of conversion of the reactants that is dependent on the initiation conditions. All other things being equal, the relaxation rate G entering into the expression for the stimulated emission power density

$$P_l = \hbar\omega_l(J)[W(\varepsilon_1 - \varepsilon_2) - G]$$

is proportional to the population of the excited vibrational levels:

$$G \simeq \sum_{v=1}^{R} n_v g_{v,v-1} \; .$$

In the presence of a strong field saturating the laser transition $v, J-1 \rightarrow v-1, J$,

$$n_v \exp[-\theta_r J(J-1)/T] = n_{v-1} \exp[-\theta_r J(J+1)/T] \quad \text{or}$$

$$n_v = n_{v-1} \exp(-2J\theta_r/T) \; .$$

It follows from this equation that the greater the rotational quantum number J of the emitting state, the smaller the total vibrational level population n_v and the lower the relaxation rate. What is more, as J increases, the number of the residual excitation quanta, ε_2, is reduced:

$$\varepsilon_2 \simeq \frac{\exp(-2J\theta_r/T)}{1 - \exp(-2J\theta_r/T)} \; .$$

All this causes the laser energy to rise with the increasing rotational quantum number J, an effect which is most important for molecules with a large vibrational constant.

The theoretically predicted dependence of the chemical efficiency of the HFL amplifier on the master oscillator emission spectrum is illustrated in Table 3.6. It can be seen from the data listed that the effect is more pronounced where initiation is weaker, η_{chem} being approximately proportional to J^2. In conditions of intense initiation, η_{chem} is almost proportional to J. Also, the increase of J from 5 to 15 enhances η_{chem} 3–10 times, depending on the initiation level. The upper bound on J is due to the weakening of the electromagnetic wave-active medium interaction at high J values as a result of depopulation of the upper sublevels:

$$n_{v,J} \simeq J \exp[-\theta_r J(J+1)/T] \; .$$

The highest rotational levels that can be used in HFL amplifiers with a reasonable active medium length can be determined by solving the radiation transfer equation. This question has been analyzed by *Basov* and co-workers [3.28], who formulated the criterion for the suitability of vibrational-rotational transitions for amplification of radiation at specified active-medium temperature and length. According to this analysis, at active-medium temperatures typical of HFL (300–1500 K) the number J is bounded above to 10–20. As the medium temperature changes in the course of the reaction, the maximum possible value of J varies as well, and the chemical efficiency of the laser oscillator-amplifier system under such conditions can be improved by making the oscillator lase at every moment of time from transitions with J values in excess of those providing for the maximum unsaturated gain in the amplifier at that moment.

The necessary time-dependent emission spectrum at the exit from the master oscillator can be obtained by the proper choice of the initiation conditions. The above analysis predicts that an HFL amplifier operating in these optimal spectral conditions will have energy characteristics approximately three times as good as those of an oscillator with the same volume. Since it is necessary that rotational-translational equilibrium should be present for the effect to be realized, it will most likely be observed in high-pressure media at moderate initiation levels, and also in media containing rotational relaxation catalysts (SF_6, C_2F_6, etc.).

Electron-Beam Initiation of HFL. Today, much consideration is being given to the initiation of chemical lasers by means of fast-electron beams from accelerators. This is due to, first, the fact that electrons, by virtue of their high penetrability, can ensure uniform initiation of large volumes of chemical laser media at sufficiently high pressures and in short times and, second, the high efficiencies of the accelerators used, which provide for high engineering efficiencies of the system as a whole. Photolysis initiation is characterized by much lower efficiencies, and electric-discharge and electroionization initiation are ineffective in the case of HFL because of the electronegativity of the compounds contained in the active mixture. Based on the mathematical model described above, *Igoshin* and co-workers [3.13] have calculated the chemical efficiency of HFL over a wide range of electron-beam current densities with a view to revealing the capabilities of the electron-beam initiation technique.

The reaction initiation rate as a function of the beam current density is determined as follows. The formation of free F atoms in the electron-beam initiated H_2–F_2 mixture basically follows the scheme [3.85]

$$F_2 + e_f \rightarrow F_2^+ + e + e_f ,$$

$$F_2 + e \rightarrow F + F^- ,$$

$$F^- + F_2^+ \rightarrow 3F ,$$

where e_f stands for a fast beam-electron and e, a slow electron. According to this scheme, each electron-ion pair gives rise to four fluorine atoms. In that case, the rate of production of fluorine atoms by the beam in a unit volume is

$$\frac{dn_F}{dt} = 4Q_i ,$$

where Q_i is the volume ionization rate (ion pairs/cm^3 s) related to the beam current density J_b by the relation $Q_i = 4.744 \times 10^{21} \times J_b$ [A/cm^2] $\times p_{F_2}$ [atm.]

$\times S_{F_2}$ [ion pairs/cm mm Hg], where S_{F_2} is the total number of ion pairs produced by a fast electron per unit flight distance at an F_2 pressure of 1 mm Hg. The quantity S_{F_2} depends on the electron energy and can be found if the energy given off by the electron beam in the medium and the energy expended to produce one fluorine atom are known. According to [3.86], $\mathscr{E}_F = 12$ eV, and so, using the data on the electron-beam absorption presented in [3.91], we find $S_{F_2} = 0.26$ ion pairs/cm^2 for electrons with an energy of 140 keV.

The calculation results are listed in Table 3.7. It is seen from these data that as the beam current density is reduced, the quantity η_{chem} decreases, while η_{phys} passes through a maximum of around 800 % at a beam current density of 1–10 A/cm^2, with η_{chem} remaining sufficiently high. We believe that this result is important because reducing the current density to around 1 A/cm^2 enables one to use commercially available electron-beam sources [3.92] that are comparatively simple and cheap. This also simplifies the problem of forming a homogeneous beam to initiate the reaction over a large area. The physical efficiency η_{phys} can be improved somewhat by shortening the initiating pulse duration in comparison with t_l [3.13].

There are experimental data available in the literature on the efficiency of an electron-beam initiated HFL operated at both high ($\simeq 10^3$ A/cm^2) and low ($\simeq 10$ A/cm^2) electron-beam current densities. In the former case, η_{phys} amounts to 130–200 % [3.93] and in the latter, 875 % [3.91], which agrees well with our calculation results. The model calculations performed by *Igoshin* and co-workers [3.13] for the exact experimental conditions reported by *Mangano* et al. [3.91] show that the model overestimates the energy output and efficiency of HFL by approximately 30 %.

It is important to note that the calculation results presented in Table 3.7 predict that more energy-intensive mixtures and more intense initiation can provide specific energy outputs several times higher than those achieved in [3.91], with η_{phys} remaining at a level of 800–900 %. This inference stimulated new experiments [3.94, 95]. *Bashkin* and co-workers [3.95] have doubled the specific energy output compared to the results of [3.91], with η_{phys} maintained at an ultimately high level. A still higher ε_l at a high η_{phys} can be expected from an oscillator-amplifier arrangement operating under the optimal spectral conditions described above.

The above numerical calculation results for the efficiency of HFL over a wide range of initial parameters point to high specific performance characteristics of such lasers. The data may be of interest in developing chemically-pumped laser oscillator-amplifier systems. It should be stressed once more that the mathematical HFL model used in the calculations is based on the assumption that the rotational levels of the HF molecules are in equilibrium. Disturbance of the rotational equilibrium conditions may impair the efficiency of HFL (see below). For this reason, the calculated specific energy characteristics of the laser should be considered as being ultimate.

3.2.4 Effect of Rotational Nonequilibrium on Laser Emission Spectrum, Dynamics, and Energy

At the present time, the questions of rotational kinetics in chemical lasers are attracting considerable interest and are the subject of much investigation in an ever increasing number of works [3.18, 21, 23–25, 90, 96–107, 124, 125]. As will be recalled (see Chap. 2), the elementary pumping reactions are characterized by an essentially non-Boltzmann energy distribution among the rotational levels of the reaction products. The rotational distribution of the molecules formed in the laser medium is determined by the competition between the chemical excitation, stimulated emission, and collisional thermalization processes. Generally speaking, the time scales of these processes may be comparable in order of magnitude.

The most pronounced qualitative discrepancy between the theory assuming the existence of rotational equilibrium and experiment manifests itself in the fact that experimentally observed laser emissions occur, as a rule, on different transitions in one and the same vibrational band that coincide either in time (pulsed systems) or in space (continuous-wave systems). In particular, lasing is frequently observed to occur in the *P*-branch transitions with high and low *J* values simultaneously. One can realize experimentally such lasing kinetics that ensure the occurrence of vibrational-rotational lines in a strict sequence typical of rotational equilibrium conditions, but this requires that the mixture should be highly diluted with an inert gas (He, N$_2$) or include good rotational relaxation catalysts (SF$_6$, C$_2$F, etc.) [3.108].

Giving up the assumption of rotational equilibrium necessary for adequate description of the operation of HFL entails a drastic increase in the number of rate equations and kinetic constants (usually unknown) that must be included in the mathematical model of the laser.

The role of rotational nonequilibrium was at first studied for the more simple, non-chain-reaction systems [3.96–103, 105]). Based on the rate equations presented in Sect. 2.5, detailed numerical kinetics modeling was carried out for the reactions $F + H_2 \rightarrow HF^v + H$ [3.96–99, 101, 105], $F + D_2 \rightarrow DF^v + D$ [3.100], and $Cl + HBr \rightarrow HCl^v + Br$ [3.102], quasiclassical lasing equations being used in [3.98]. The main results obtained may be summarized as follows.

Laser Emission Dynamics and Spectrum. Taking account of the finiteness of the rotational relaxation rate leads to simultaneous lasing on many lines, and so multilinear oscillation in chemical lasers can serve as an indication of rotational nonequilibrium. As the rotational relaxation rate decreases, the peak lasing intensity in individual lines drops, the lasing duration increases, and the lines overlap more and more. In rotational equilibrium conditions, the effect of laser line overlapping is negligible, so that the single-line lasing approximation commonly adopted under such conditions is quite adequate. All these effects are illustrated in Fig. 3.4.

It is of interest that despite the fact that rotational equilibrium is absent and lasing has a multilinear character, the position of the transition with the maximum emission intensity approximately corresponds to the lasing transition in the model assuming the existence of rotational equilibrium. This inference is useful for approximate treatment of lasing. It is only at very low pressures and slow rotational relaxation rates that the correspondence indicated above ceases to exist, so that the relation between the laser emission power and the rotational quantum number *J* is determined by the distribution of energy in an elementary pumping event. Such operating conditions are characteristic of cw lasers [3.96]. It should also be noted

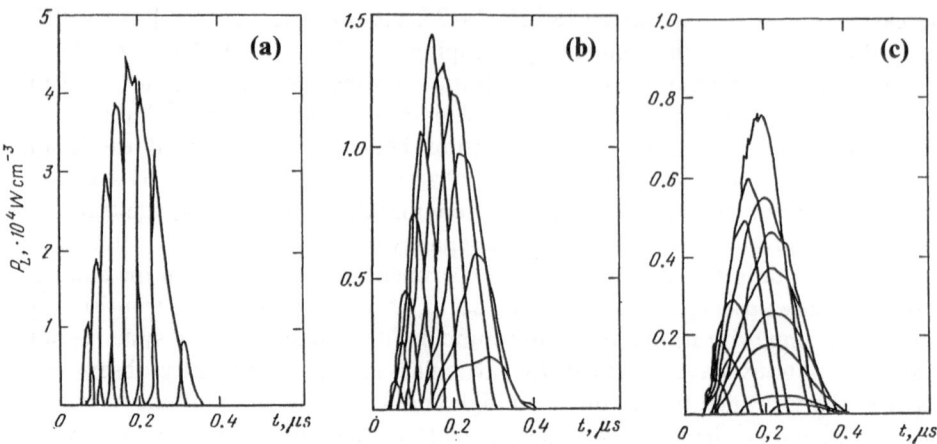

Fig. 3.4a–c. Calculated temporal dependence of the HF laser emission power on all lasing transitions in the $v = 2 \rightarrow v = 1$ band for three rotational relaxation time values [3.97]: (**a**) rotational equilibrium; (**b**) $A_r = 2 \times 10^{-8} T^{-1/2}$ s; (**c**) $A_r = 2 \times 10^{-7} T^{-1/2}$ s. The rotational relaxation time τ_r of a v, J level is related to the parameter A_r by the relation $\tau_r = A_r \exp(B\Delta E_J/kT)$, where B is the relaxation model parameter taken to be 10^{-3}, and ΔE_J is the energy difference between the adjacent levels numbered $J + 1$ and J. The numbers on the curves denote the values J of the lower transition level. Initial gas medium conditions: $H_2 : SF_6 = 1 : 5$; $T_0 = 300$ K; $p = 120$ mm Hg; $[F] : [SF_6] = 5.6\,\%$. Laser cavity parameters: $R_1 = 0.8$ m; $R_2 = 1.2$ m; $l = 50.8$ cm

that taking account of rotational nonequilibrium in non-chain-reaction systems has practically no effect on the total laser pulse duration [3.102].

Energy Characteristics of Lasing. The efficiency of lasing rises with the rotational relaxation rate. But the influence of the disturbance of rotational equilibrium conditions on the energy characteristics of non-chain-reaction systems is comparatively weak, the reduction of the emission power and energy as compared with the respective values calculated in the rotational equilibrium approximation amounting to 15–30 % [3.96, 100, 105]. The effect of rotational nonequilibrium on the energy characteristics of lasing in individual lines is more pronounced [3.96, 102].

Thus, to allow for the effect of rotational nonequilibrium in chemical lasers is of principal importance for adequate prediction of the laser emission spectrum and dynamics and also for calculation of the emission energy in individual lines. At the same time, the overall energy characteristics can be described adequately enough in the rotational equilibrium approximation.

As demonstrated in [3.23, 24, 90] and presented in the text below, in chain-reaction lasers the effect of roational nonequilibrium on the overall laser energy characteristics may be much more pronounced. Because of the involved character of chain-reaction systems, it is advisable to resort to a simplified analysis procedure. It is informative to compare the laser efficiencies in two extreme cases where (1) rotational relaxation proceeds much faster than stimulated emission and excitation processes so that rotational-translational equilibrium is maintained and (2) the rate of rotational relaxation is low in comparison with that of stimulated emission ("frozen" rotational relaxation). For this purpose, we follow [3.23] and use the expression for the stimulated emission power density of a multilevel chemical laser:

$$P_l = \hbar\omega_l [W(t)(\varepsilon_1 - \varepsilon_2) - G] \ , \tag{3.2}$$

where $\hbar\omega_l$ is the energy of the laser photon, W is the chemical reaction rate, and $\varepsilon_1 = \Sigma_v v\alpha_v$ is the average number of vibrational quanta excited in an elementary pumping event; the coefficients α_v describe the energy distribution in the course of the reaction; $\varepsilon_2 = \Sigma_v vn_v/n$ is the average number of vibrational quanta emitted per molecule under stimulated emission, $G \simeq \Sigma_v g_{v,v-1} n_v$; $g_{v,v-1}$ is the relaxation probability of the level v, and

$$n = \int_0^t W(t')dt'$$

is the total number of lasing molecules at the moment t.

Expression (3.2) is essentially a corollary to the law of conservation of energy [3.28]. It is applicable at any degree of disturbance of rotational equilibrium, unless the vibrational energy distribution function is specified.

It can be seen from (3.2) that both the quantity ε_2 determining the pumping energy fraction that cannot be converted into laser emission and the quantity G describing the rate of energy loss due to vibrational-translational relaxation are essentially determined by the distribution of the lasing molecules among vibrational levels. In the presence of a strong field, there occurs in every pair of laser energy levels the equalization of the populations of the upper and lower levels:

$$n_{v, J-1} = n_{v-1, J} \, g_{J-1}/g_J \ . \tag{3.3}$$

Using (3.3), it is not very difficult to find the molecular distribution among vibrational levels in the following two extreme cases: (1) the case of rotational equilibrium and (2) the case of "frozen" rotational relaxation. In the former case, it follows from (3.3) that the total populations of the vibrational levels are related by

$$n_v = n_{v-1} \exp(-2\theta_r J/T) \ , \quad \text{whence}$$

$$n_v = n\frac{1 - B_J}{1 - B_J^{R+1}} B_J^v, \quad B_J = \exp(-2\theta_r J/T) \ , \tag{3.4}$$

where R is the serial number of the uppermost level excited in the course of the reaction. In the latter case (ultimately nonequilibrium conditions), where each of the rotational sublevels plays the part of an isolated subsystem, the relation between the populations of the vibrational levels is found by summing (3.3) over all J's, which roughly yields

$$n_v = n_{v-1} = \ \ldots \ = n_0 = n/(R+1) \ . \tag{3.5}$$

Comparison between (3.4) and (3.5) shows that rotational equilibrium conditions are conducive to a more favorable molecular vibrational energy distribution characterized by a sparser population of the excited states in the course of lasing, hence a lower loss of vibrational energy. The quantity ε_2 for the two cases under consideration is given by the expressions

$$\varepsilon_2^{eq} = \frac{B_J + RB_J^{R+2} - (R+1)B_J^{R+1}}{(1-B_J)(1-B_J^{R+1})} \simeq \frac{B_J}{(1-B_J)}$$

and

$$\varepsilon_2^{neq} = \frac{R}{2} \; ,$$

respectively. It follows from these relationships that the difference in efficiency between the cases where rotational equilibrium is present and absent is proportional to the parameter $J\theta_r/T$.

Let us illustrate the scale of the effect using by way of example halogen-hydrogen reactions proceeding by the chain mechanism

$$A + B_2 \xrightarrow{\alpha_v^{(1)} k_1} AB(v) + B \; ,$$

$$B + A_2 \xrightarrow{\alpha_v^{(2)} k_2} AB(v) + A \; .$$

Considering that the main contribution to the relaxation of $AB(v)$ is made by the molecules AB themselves and generalizing (3.2) to the case where the lasing molecules are being excited in two concurrent reactions, we have for P_l

$$P_l = \hbar\omega_l \left[W(\varepsilon_1^{(1)} + \varepsilon_1^{(2)} - 2\varepsilon_2) - 4k_{1,0}\varepsilon_2 \left(\int_0^t W \, dt' \right)^2 \right] , \tag{3.6}$$

where W is the chain reaction rate, $\varepsilon_1^{(1)} = \Sigma_v v\alpha_v^{(1)}$, $\varepsilon_1^{(2)} = \Sigma_v v\alpha_v^{(2)}$, and $k_{1,0}$ is the rate constant of the $V \rightarrow T$ process $AB(1) + AB(0) \rightarrow 2AB(0)$. For estimation purposes, we assume that W is independent of time during the laser pulse. Then, integrating (3.6) for t going from 0 to t_l – the moment oscillation is quenched, which is determined from the condition $P(t_l) = 0$ to be

$$t_l = \left[\frac{\varepsilon_1^{(1)} + \varepsilon_1^{(2)} - 2\varepsilon_2}{4k_{1,0} W \varepsilon_2} \right]^{1/2} , \tag{3.7}$$

we get for the laser energy density

$$\varepsilon_l = \frac{1}{3}\hbar\omega_l \left[\frac{W(\varepsilon_1^{(1)} + \varepsilon_1^{(2)} - 2\varepsilon_2)^3}{k_{1,0}\varepsilon_2} \right]^{1/2} . \tag{3.8}$$

By definition, the chemical efficiency is $\eta_{chem} = \varepsilon_l/(-\Delta H)n_{X_2}$, where $(-\Delta H)$ is the heat of the chain reaction; $X_2 = B_2$ if $n_{B_2} < n_{A_2}$ and $X_2 = A_2$ otherwise. Relations (3.7) and (3.8) can be used for estimating the energy parameters and pulse duration of hydrogen halide lasers. For the $H_2 + F_2$ reaction, $\varepsilon_1^{(1)} = 2.1$ and $\varepsilon_1^{(2)} = 5.4$, and if we assume that lasing in rotational equilibrium conditions occurs on transitions with $J \simeq 10$, the ratio between the chemical efficiencies in the nonequilibrium ($\varepsilon_2^{neq} = 3.5$) and equilibrium ($\varepsilon_2^{eq} = 0.6$) conditions will be $\eta_{chem}^{neq}/\eta_{chem}^{eq} = \varepsilon_l^{neq}/\varepsilon_l^{eq} \simeq 10^{-2}$. Thus, the chemical efficiency reaches its maximum under rotational equilibrium conditions. The difference between the efficiencies of the HF laser in the above two extreme cases is so great that any excessive disturbance of rotational equilibrium must practically quench oscillation. Also, the extent to which the reactants can burn out here without making oscillation impossible is much lower. In non-chain-reaction systems, the accumulation of lasing molecules is much less intense, the rotational relaxation effects are substantially less pronounced, and so is the influence of disturbed rotational equilibrium on the laser efficiency. The order of magnitude of this influence is reflected by the change of the second term in (3.2), i.e., for non-chain-reaction systems,

$$\frac{\eta_{chem}^{eq}}{\eta_{chem}^{neq}} \simeq \frac{\varepsilon_1 - \varepsilon_2^{eq}}{\varepsilon_1 - \varepsilon_2^{neq}} . \tag{3.9}$$

For the $F + H_2 \rightarrow HF^v + H$ reaction, the values of η_{chem}^{eq} and η_{chem}^{neq} in the two extreme cases under consideration differ by no more than several fold, and in an intermediate case the effect must be much weaker, which agrees with numerical calculations.

It can be expected that the efficiency of an actual chain-reaction HF laser will be reduced by a factor of 2–3 as a result of incomplete rotational equilibrium [3.23, 24]. This question has been investigated in greater detail by *Igoshin* and co-workers [3.90].

The analysis [3.90] of the energy characteristics of chemical lasers based on self-sustained processes has revealed the existence of an ultimate laser energy output that depends on the rotational relaxation rate and cannot be exceeded by increasing the chemical process rate. What is more, the finiteness of the rotational relaxation rate imposes a principal restriction on the possible rise of the laser energy output with the increasing rotational quantum number of the emitting states in the lasing molecules under rotational equilibrium conditions. For hydrogen halide lasers, the optimum J values providing for the maximum energy output of an amplifier are in the range 10–15. The effect of rotational nonequilibrium on the laser energy output grows stronger with the increasing parameter τ_r/t_l, where τ_r is the characteristic time of relaxation from a given rotational sublevel to other rotational states (for HF at atmospheric pressure, $\tau_r \simeq 10^{-9}$) and t_l is the laser pulse duration. To take account of the effect of rotational nonequilibrium is most important where τ_r/t_l is of the order of 10^{-2}, or more. According to [3.90], rotational nonequilibrium reduces ε_l by approximately 10% for laser pulses in the microsecond duration range ($\tau_r/t_l \simeq 10^{-3}$), by a factor of 2–2.5 for those with a duration of $t_l \simeq 100$ ns ($\tau_r/t_l \simeq 10^{-2}$) and by an order of magnitude or more for pulses a few nanoseconds in duration (in an atmospheric-pressure medium). This means that the laser energy calculations made under the assumption of rotational equilibrium approximately remain valid at laser pulse durations $\geqslant 100$ ns at atmospheric pressure and $\geqslant 10$ ns at a pressure of 10 atm.

Experimental investigations into the effect of rotational nonequilibrium on the performance of HFL have been conducted by *Vasil'yev* and co-workers [3.104]. It has been found that addition of C_5F_{12} (a rotational relaxation "catalyst") to an H_2–F_2–O_2 mixture ($C_5F_{12}:H_2:F_2:O_2 = 15:22:74:37$ mm Hg) allows the laser energy output to be almost doubled. It has been demonstrated that the increase in the laser energy output observed to occur upon addition of C_5F_{12} is accompanied by a decrease in the chemical reaction rate (as a result of increased specific heat of the mixture and reduced gas temperature). The drop of the laser energy output at high concentrations of C_5F_{12} is apparently due to the high effectiveness of these molecules in chain termination:

$$H + O_2 + C_5F_{12} \rightarrow HO_2 + C_5F_{12} \ .$$

An increase in the laser energy output taking place concurrently with a decrease in the chemical reaction rate contradicts the relationship believed to exist between these laser characteristics on the basis of simple kinetic models (the two-level HFL model considered in Chap. 2 or the above-discussed multilevel HFL model assuming the existence of rotational equilibrium), but agrees quite well with what we think of the effect of rotational nonequilibrium on the energy characteristics of HFL. Experiments aimed at studying the temporal evolution of the laser spectrum have demonstrated that with C_5F_{12} added to the mixture, the laser lines belonging to a given vibrational band appear in a sequence. The body of information obtained on the effect of C_5F_{12} on the characteristics of HFL gives reason to believe that addition of this substance enhances the rotational relaxation rate.

So, the theoretical and experimental studies performed to date have shown that rotational nonequilibrium is an important factor in the operation of multilevel chemical lasers. The calculations made in the rotational equilibrium approximation fail to describe many important specific features of lasing. As shown by calculation, allowing for rotational nonequilibrium makes it possible qualitatively to explain the phenomena observed in the laser emission spectrum, dynamics, and energy characteristics. But quantitatively to predict the detailed laser emission picture, e.g., the efficiency of lasing in individual lines, is so far impossible. There is a great uncertainty as to the rates of rotational relaxation in the laser medium. More efficient calculation techniques need to be developed for complex laser systems. These problems are the subject of present-day investigations [3.127–129].

3.2.5 Modeling of HFL with Allowance Made for Rotational Nonequilibrium and Anharmonicity of Lasing Molecules

This section presents an effective technique for calculating HFL characteristics that takes account of rotational nonequilibrium and anharmonicity of the lasing molecules and allows for a more adequate quantitative prognostication of the total (summed over all lines) laser emission power and pulse duration. The model defines concretely the general approach developed by *Igoshin* and co-workers [3.127] to the calculation of stimulated emission of radiation in multilevel media under vibrational and rotational nonequilibrium conditions.

This approach is based on the possibility of introducing into consideration an equivalent two-level system to calculate the energy characteristics of a multilevel laser. As shown later in the text, a definite relationship exists between the parameters of an actual multilevel system and its conventional two-level counterpart, which makes it possible to calculate the total laser emission power within the framework of a simple two-level active medium model.

Based on the law of conservation of energy, we may write for the power density of laser radiation:

$$P_l = -\left[\frac{dl_v}{dt}\right]_{\text{rad}} - \left[\frac{dl_r}{dt}\right]_{\text{rad}} , \tag{3.10}$$

$$\frac{dl_v}{dt} = \left[\frac{dl_v}{dt}\right]_{\text{rad}} + \left[\frac{dl_v}{dt}\right]_{\text{coll}} , \tag{3.11}$$

where $[dl_v/dt]_{\text{rad}}$ and $[dl_r/dt]_{\text{rad}}$ are the rates of change of vibrational and rotational energies, respectively, occurring in a unit volume as a result of radiative processes and $[dl_v/dt]_{\text{coll}}$ is the rate of change of vibrational energy due to collisional processes (chemical reactions and relaxation). In the harmonic-oscillator approximation, for the power density of multilevel laser radiation, we have from (3.10) and (3.11)

$$P_l = \hbar\omega_l(J)\left[\sum_{k=1}^{m} W^{(k)}\varepsilon_1^{(k)} - \sum_v v\frac{dn_v}{dt} - G\right] , \tag{3.12}$$

where $\hbar\omega_l(J)$, $W^{(k)}$, $\varepsilon_1^{(k)}$, and G are the same as in (2.137).

In the case of a two-level system, we have from the balance equations

$$\frac{dn_u}{dt} = \sum_{k=1}^{m} W^{(k)}\alpha_u^{(k)} - \frac{P_l}{\hbar\omega_{ul}} - q_{u,l}n_u , \tag{3.13}$$

$$\frac{dn_l}{dt} = \sum_{k=1}^{m} W^{(k)}\alpha_l^{(k)} - \frac{P_l}{\hbar\omega_{ul}} + q_{u,l}n_u \tag{3.14}$$

the following expression for the power density of laser radiation:

$$P_l = \hbar\omega_{ul}\left[\sum_{k=1}^{m} W^{(k)}\alpha_u^{(k)} - \frac{dn_u}{dt} - q_{u,l}n_u\right] , \tag{3.15}$$

where n_u and n_l are the populations of the upper and lower laser levels, respectively, $\alpha_u^{(k)}$ and $\alpha_l^{(k)}$ are the probabilities that the lasing molecules will be formed at the upper and lower levels as a result of the kth reaction, and $q_{u,l}$ is the probability of collisional relaxation of the upper level onto the lower one. The two-level system parameters $\hbar\omega_{ul}$, $\alpha_u^{(k)}$, and $q_{u,l}$ can be chosen so as to make the powers of the energy fluxes entering into (3.15) equal to those in (3.12). In that case, the power output of the two-level laser will be equal to the total power of the multilevel laser. Also, assuming that $W^{(1)} = W^{(2)} = \ldots = W^{(m)}$ (which is realized in chain reactions), we have from the above reasoning that

$$\hbar\omega_{ul} = \hbar\omega_l \frac{\sum\limits_{k=1}^{m} \varepsilon^{(k)}}{\sum\limits_{k=1}^{m} \alpha_u^{(k)}} \,, \tag{3.16a}$$

$$q_{u,l} = q_{1,0} \,, \tag{3.16b}$$

$$n_u = \left(\sum_{v} vn_v\right) \frac{\sum\limits_{k=1}^{m} \alpha_u^{(k)}}{\sum\limits_{k=1}^{m} \varepsilon^{(k)}} \,. \tag{3.16c}$$

The fact that n_u is expressed in terms of a linear combination of n_v, on no account complicates the analysis of the two-level system because (3.16c) is only used to interpret n_u. To find n_u and n_l as a function of time requires the solution of kinetic equations which must include the basic interactions in the system and may be formulated within the scope of various approximations. To analyze chemical laser kinetics with account being taken of rotational relaxation, we resort to the rotational reservoir model [3.127]. In this model, rotational sublevels coupled by a radiative transition are singled out in the upper and lower vibrational states. All the other rotational sublevels in the states u and l are united and regarded collectively as an energy reservoir, and then consideration is given to the energy exchange between the rotational-vibrational sublevels singled out and the reservoir that takes place as a result of rotational relaxation processes.

Within the framework of the rotational reservoir model, the kinetic equations describing changes in the populations of various states may be written as

$$\frac{dn_u}{dt} = \sum_{k=1}^{m} \alpha_u^{(k)} W^{(k)} - \sigma c \varrho \varDelta - q_{ul} n_u \,, \tag{3.17a}$$

$$\frac{dn_l}{dt} = \sum_{k=1}^{m} \alpha_l^{(k)} W^{(k)} + \sigma c \varrho \varDelta + q_{ul} n_u \,, \tag{3.17b}$$

$$\frac{dn_u^{J-1}}{dt} = -\sigma c \varrho \varDelta + \frac{n_u - n_u^{J-1}}{M_{J-1}\tau_r} - \frac{n_u^{J-1}}{\tau_r} \,, \tag{3.17c}$$

$$\frac{dn_l^{J}}{dt} = \sigma c \varrho \varDelta + \frac{n_l - n_l^{J}}{M_J \tau_r} - \frac{n_l^{J}}{\tau_r} \,, \tag{3.17d}$$

$$\frac{d\varrho}{dt} = \sigma c \varrho \varDelta - \varrho/\tau_{\mathrm{ph}} \,, \tag{3.17e}$$

where $n_u = n_u^R + n_u^{J-1}$, $n_l = n_l^R + n_l^J$, n_u^R and n_l^R are the reservoir populations, n^{J-1} and n^J are the populations of the rotational sublevels singled out, $\varDelta = n_u^{J-1} - (q_{J-1}/q_J)n_l^J$, $M_J = (T/\theta_r)[1/(2J+1)]\exp[\theta_r J(J+1)/T] - 1$, and τ_r is the time of collisional transitions from a given rotational sublevel onto all the other sublevels.

This time can be determined from broadening data: $\tau_r = 1/\pi\Delta\nu_L$. The quantity $M_J\tau_r$ represents the collective relaxation time for transitions from all the reservoir levels onto the level J, which is M_J times longer than τ_r.

In the quasistationary approximation, it is not very difficult to obtain from (3.15) the following expression:

$$P_l = \hbar\omega_{ul} \frac{q_J}{q_J + q_{J-1}} \frac{n_u}{M_{J-1}\tau_r} - \frac{q_{J-1}}{q_J} \frac{n_l}{M_J\tau_r} - \frac{\Delta_{\text{thr}}}{\tau_r} .$$

Equations (3.13) and (3.14), together with the above expression for the power density of laser radiation and (3.16a) and (3.16b) for the two-level system parameters, enable one to calculate the power output of a multilevel laser with due regard for rotational nonequilibrium. These equations must be solved simultaneously with the chemical kinetics and heat balance equations. The above calculation technique allows for a simple but very important (as far as quantitative prognostication is concerned) generalization to the case of emitting anharmonic oscillator. Recent investigations have demonstrated that the relaxation rate of the HF molecules increases with v much faster than in the case of harmonic oscillator. For these molecules, the following relation holds to a good approximation:

$$g_{v,v-1} = g_{1,0}v^m ,$$

where $m \simeq 2.3$. For this reason, the harmonic oscillator model may substantially (2–3 times) overestimate the energy characteristics of a laser.

The earlier numerical calculations [3.7] of the HFL kinetics allow the ratio $n_v/n_{v-1} = \beta$ to be approximately considered as being independent of v. This approximation makes it possible to introduce into consideration an equivalent two-level model for anharmonic oscillators as well, which will allow for the increased rate of relaxation from the upper levels.

Indeed, in the general case, for the function G describing the loss of energy by relaxation, we may write

$$G = \sum_v g_{v,v-1} n_v = \sum_v g_{1,0}vn_v \frac{\sum\limits_v g_{v,v-1}\beta^v}{\sum\limits_v g_{1,0}v\beta^v} ,$$

where consideration is given to the fact that $n_v = n_0\beta^v$. Let us introduce the following notation:

$$\phi(t) = \frac{\sum\limits_v g_{v,v-1}\beta^v}{\sum\limits_v g_{1,0}v\beta^v} .$$

For the HF molecules, this quantity is given by

$$\phi(t) = \frac{\sum\limits_v v^m\beta^v}{\sum\limits_v v\beta^v} .$$

In that case, for anharmonic oscillations, instead of (3.16b) we have

$$g_{u,l} = g_{1,0}\phi(t) \ .$$

For complete formulation of the model, we have only the quantity $\beta(T)$ left to define. This quantity results automatically in a sufficiently good approximation to the integration of the basic set of equations describing the two-level system:

$$\beta(t) = n_u(t)/n_l(t) \ .$$

The above relations enable one to calculate the power output of a multilevel laser with due regard for rotational nonequilibrium and anharmonicity of the lasing molecules.

The HFL model based on this calculation technique includes the chemical kinetics equations presented in Sect. 3.2.2, relaxation equation for the H_2 molecules, gas temperature equation, and laser level population kinetics equations which define concretely (3.13) and (3.14):

$$\frac{dn_u}{dt} = k_1 n_F n_{H_2} + k_2 n_H n_{F_2} - \frac{P_l}{\hbar\omega_{ul}} + q_{ul}n_u - g_{VV}$$

$$\frac{dn_l}{dt} = \frac{P_l}{\hbar\omega_{ul}} + q_{ul}n_u - g_{VV} \ ,$$

where $n_M = \xi_M N$, $g_{VV} = G_{VV}\{2/[\varepsilon_1^{(1)} + \varepsilon_1^{(2)}]\}$; the quantity ε_{HF} entering into G_{VV} is defined as

$$\varepsilon_{HF} = \frac{\varepsilon_1^{(1)} + \varepsilon_1^{(2)}}{2}\frac{n_u}{n_{HF}} \ .$$

Note that g_{VV} describes the effect of the $V \to V$ exchange between HF and H_2 on the population of the levels u and l. The above expression for g_{VV} follows from the interpretation of n_u [formula (3.16b)]. The HFL model described here requires not very much computer time and reflects all the basic processes taking place in the laser medium: the pumping chain reaction, together with the chain termination and initiation reactions, the energy exchange between HF and H_2, the vibrational-translational relaxation of HF with allowance made for its actual dependence on the vibrational level number v, the rotational relaxatin of HF, and the self-heating of the laser mixture.

Table 3.8 presents the results of calculation by this model of the main laser characteristics. Comparison between these data and the data of Table 3.2 shows that taking account of rotational nonequilibrium and the dependence of $k_{v,v-1}^{HF-M}$ on v reduces the attainable laser power output and efficiency by a factor of 2–2.5. It should be noted that the inclusion of the rotational kinetics allows for a more exact determination of the range of J values optimal for lasing. According to our calculations, these are "moderate" J values ranging between 6 and 8. The model agrees well with the results of experiments [3.130, 131] conducted under strictly controlled reaction initiation conditions. Comparison between theory and experiment.[3.130] has made it possible to define more exactly the parameter m entering into the equation $k_{v,v-1} = k_{1,0}v^m$. Thus, at an initiation level of $n_a = 10^6 \text{ cm}^{-3}$, the specific energy output ε_l measured in the mixture F$_2$:H$_2$:O$_2$:He $= 200:48:16:500$ mm Hg is 49 ± 12 J/l. Calculation yields $\varepsilon_l = 104$ J/l for $m = 1$, 52.7 J/l for $m = 2.3$, and 28.2 J/l for $m = 2.5$. It is seen that the harmonic oscillator model sets the specific laser energy output a factor of two too high. With the parameter m taken at the most suitable value found ($m = 2.3$), the theory predicts for the experimental conditions of [3.131] a total energy output of 4.12 kJ. The experimentally obtained energy output is 4.3 kJ (F$_2$:H$_2$:O$_2$:SF$_6$ $= 208:30:62.5:35$ mm Hg). These results show that the approach developed is capable of a fairly accurate prediction of the energy characteristics of FHL.

3.3   D$_2$-F$_2$-CO$_2$ System

3.3.1   Qualitative Discussion of the Kinetic Scheme of Chemical Pumping Involving Energy Transfer

The principle of operation of chemical lasers based on the D$_2$-F$_2$-CO$_2$ mixture is as follows. The chain chemical reaction between D$_2$ and F$_2$ leads to the formation of vibrationally excited DF molecules:

$$D + F_2 \rightarrow DF^v + F \ ,$$

$$F + D_2 \rightarrow DF^v + D \ .$$

The excited DF molecules give up their energy to the CO$_2$ molecules that form the laser medium and release energy by stimulated emission of radiation:

$$DF^v + CO_2 \rightarrow DF + CO_2^v \ ,$$

$$CO_2^v + \hbar\omega_l \rightarrow CO_2 + 2\hbar\omega_l \ .$$

The above scheme gives rise to the natural question: What is the purpose of transferring energy to the CO$_2$ molecules if the DF molecules can emit it themselves? What is more, the energy of the CO$_2$ laser photon is almost a factor of 3 less than that of the vibrational quantum of the DF molecule, and so a considerable loss of potentially accessible vibrational energy is inevitable in transferring it from DF to CO$_2$ and causes the heating of the laser mixture.

The first attempts at achieving laser action by transferring energy from hydrogen halides to CO$_2$ were stimulated by the hope of avoiding, or moderating, the problem of the fast self-relaxation of the halides whose accumulation quenches oscillation at the early stages of the chain reaction. With the CO$_2$ molecules used as a laser medium, the self-relaxation problem is much less acute. Moreover, the laser level system in this molecule is rather convenient: the lower laser level is not the ground level of the molecule, and so at moderate temperatures its population is not very dense; in addition to that, it relaxes faster than the upper laser level. All this allows an inverted population to be maintained even at relatively slow pumping rates. By no means unimportant is also the fact that to produce population inversion in CO_2 by transferring energy to it from another molecular species excited in the course of chemical reaction does not require that the energy levels of this excited species should also be inversely populated. Consequently, the energy of the initially excited molecules can, in principle, be utilized even when the inverted population in them ceases to exist, provided that their store of vibrational energy remains in excess of the equilibrium value. In view of this, one may hope materially to increase the effective chain length.

However, such a scheme suffers from a number of serious disadvantages. Apart from the inevitable loss of vibrational energy owing to the smaller laser quantum inherent in CO_2, the laser efficiency may be reduced as a result of the active medium growing hot in the course of the chain reaction and thus increasing the population density of the lower laser level and enhancing the relaxation rate of the upper laser level in CO_2. In lasers operating on hydrogen halide molecules possessing large rotational constants and amplification cross sections, the heating of the active medium is not so critical, thanks to the possibility for lasing to shift to transitions with high J values, where population inversion can be maintained at high temperatures as well (if the vibrational excitation of the molecules is high enough). Another cause limiting the laser efficiency is the finiteness of the rate of energy transfer from DF to CO_2. Should the reaction proceed too fast, it may happen that there will be not enough time for energy to be transferred to the upper laser level, and this will cause an additional loss of energy as a result of a fast relaxation of the excited DF molecules. What is more, the necessary population inversion in CO_2 will in that case be especially difficult to maintain because of a strong heating of the medium due to the complete burn-out of the reactants.

Thus, the above qualitative treatment of the mechanism of the chemical laser relying on the transfer of energy from DF to CO_2 (the DF–CO_2 laser) shows that this type of laser has both advantages and disadvantages compared with HFL. The efficiency of the DF–CO_2 laser depends on many factors whose effect cannot be analyzed unless there is an adequate kinetic model of the laser and numerous data available on the rate constants of the elementary processes occurring in its active medium.

The universal interest in the DF–CO_2 laser was initially due to the success of the first experiments which demonstrated that the introduction of CO_2 into the D_2–F_2 mixture substantially improved the energy characteristics of the chemical laser operating on the $D_2 + F_2$ reaction. The experimental conditions being comparable, the introduction of CO_2 provided a ten-fold increase in the laser emission energy compared to that of the system without energy transfer [3.5]. These first experiments were conducted at low reaction initiation levels and low reagent pressures ($\simeq 1$ mm Hg). The increase achieved in both the lasing energy and duration bears witness to the fact that they actually managed largely to overcome the fast self-relaxation of DF under these experimental conditions.

In subsequent experiments, it was demonstrated that the addition of CO_2 to the D_2–F_2 mixture eliminates energetic chain branching and stabilizes the mixture at high reactant pressures. This allowed the amount of energy stored in the laser medium to be substantially increased. It was precisely the use of the D$_2$–F$_2$–CO$_2$ mixture that made it possible to create the first effective high-pressure ($\simeq 1$ atm.) chemical laser with an energy output hundreds of times the previous chemical laser outputs [3.109]. The H$_2$–F$_2$ (D$_2$–F$_2$) mixture can also be stabilized by increasing its oxygen content. But the possibility of preparing the D$_2$–F$_2$–CO$_2$ mixture free from large amounts of O$_2$ is of principal importance, for oxygen shortens the chain length of the H$_2$ + F$_2$ (D$_2$ + F$_2$) reaction, which eventually limits the engineering efficiency of the laser. As the rate of chain termination in the triple collisions

$$H + O_2 + M \rightarrow HO_2 + M$$

increases with pressure faster than the chain reaction rate, the stabilization of the laser mixture with oxygen also prevents the use of high mixture pressures. The addition of CO$_2$ stabilizes the mixture only on account of elimination of energetic chain branching and not as a result of shortening of the chemical chain length, which is essential for the realization of ultimate quantum yields of lasing.

The kinetic calculations [3.7, 16, 111, 133] and experiments [3.110, 111] conducted under a more powerful initiation and a higher laser medium pressure revealed the demerits of the DF–CO$_2$ laser qualitatively discussed above. In particular, it turned out that the energy output of the laser ceases to rise with the level of initiation much sooner than in the case of the H$_2$–F$_2$ system. Nevertheless, the laser parameter range wherein the energy characteristics of the DF–CO$_2$ system excel those of the H$_2$–F$_2$ system is very important for applications. Thus, it was exactly the capability of the DF–CO$_2$ system effectively to emit energy at slow reaction rates that made it possible to develop cw lasers with purely chemical pumping occurring in the course of mixing of NO and F$_2$, where the rate of production of the fluorine atoms as a result of the reaction

$$NO + F_2 \rightarrow NOF + F$$

could not be very high. Calculations show that the potential of the DF–CO$_2$ laser operating in the pulsed mode has not yet been fully realized. This laser can have a very high quantum yield of lasing and an engineering efficiency in excess of that of HFL [3.16].

3.3.2 Elementary Process Kinetics and Laser Performance Calculation Procedures

The mechanism of the D$_2$ + F$_2$ reaction is similar to that of the H$_2$ + F$_2$ reaction considered earlier. But the elementary process rate constants for these two systems are different. Table 3.9 lists the kinetic constants of the D$_2$–F$_2$ system that are recommended for use in calculations. These recommendations are based on the analysis of the most recent information on the kinetics involved. The addition of CO$_2$ to the reaction system necessitates consideration of the relaxation processes occurring in the energy-level ladder of CO$_2$ and the vibrational interaction between CO$_2$ and the other laser mixture components. (The chemical interactions of CO$_2$ are usually disregarded.)

The extensive investigations on CO_2 lasers with electrical, thermal, and chemical excitation have generated great interest in the relaxation processes taking place upon collisions between CO_2 and CO_2 and between CO_2 and other molecules and atoms. Information about these processes can be found in the reviews [3.118–120]. Table 3.9 also presents data on the basic relaxation processes taking place in the energy-level ladder of CO_2, along with the relaxation rate constants used in the kinetic calculations of the DF–CO_2 laser. These data refer to the interaction between a finite number of vibrational levels in CO_2: 00^00, 01^10, 02^00, 10^00, 11^10, 03^10, and 03^30, and 00^01, i.e., between the upper laser level and the levels below it. Such a restriction seems unnatural, yet it is quite justified because the higher levels with their sparce population play only a minor part over a wide range of the DF–CO_2 laser operating conditions.

Two procedures are mainly used to analyze in detail the kinetics of DF–CO_2 lasers and calculate their performance characteristics. With one of them, the relaxation processes occurring in the CO_2, D_2, and DF molecules are described by microkinetic relaxation equations, i.e., balance equations for the number of molecules at each vibrational level (level kinetics). Because it is virtually impossible with this approach to allow for the interaction of all vibrational states, the analysis is usually restricted to the interaction between the above-indicated vibrational levels in CO_2, three levels ($v = 0, 1, 2$) in D_2, and ten levels ($v = 0, 1, \ldots, 9$) in DF. This procedure has been used by *Kerber* and co-workers [3.9]. In the other calculation procedure, the relaxation processes are described by energy relaxation equations, i.e., it is assumed that within each vibrational mode there exist local quasiequilibrium conditions characterized by a Boltzmann distribution function. In that case, the CO_2 population distribution function depends on time only through the intermediary of the internal temperatures T_1, T_2, and T_3 of the CO_2 symmetric, stretching, and antisymmetric modes, respectively. The temperature T_s of the sth mode is determined by the energy store in this mode, or by the number ε_s of the vibrational quanta of the given mode per molecule:

$$\varepsilon_s = r_s/[\exp(\theta_s/T_s) - 1] \ ,$$

where θ_s is the characteristic vibrational temperature of the sth mode and r_s the degeneracy of the same mode (for CO_2, $r_1 = r_3 = 1$, $r_2 = 2$). Thus, in the case of quasiequilibrium, three balance equations are enough to describe the vibrational kinetics of CO_2, each equation defining the energy variation rate in the respective mode. The relaxation energy equations for CO_2 have been derived by *Basov* and co-workers [3.121] and then generalized to the case of an arbitrary polyatomic molecule by *Biryukov* and *Gordiets* [3.122]. By virtue of the high rate of energy exchange between the first and second modes in CO_2, the temperatures T_1 and T_2 are usually taken to be equal. Under the high-pressure conditions characteristic of the pulsed DF–CO_2 laser, the difference between the vibrational temperature of these modes and the gas temperature T is insignificant [3.7] so that one may put $T_1 = T_2 = T$ to a good approximation. In that case, the only relaxation equation for ε_3 is sufficient for the description of the nonequilibrium vibrational kinetics of CO_2. The simplified energy relaxation equations for CO_2 have been considered in detail by *Losev* [3.123].

We call the reader's attention to two assumptions that are made in deriving the energy relaxation equations for polyatomic molecules and limit their applicability. The first is that the intramode distribution function is taken to be a Boltzmann function and the second that the energy spectrum of each molecular mode is modeled by a set of harmonic oscillator equations. In the case of diatomic molecules, only the second assumption is essential, because the form of the energy relaxation equation for a harmonic oscillator is independent of the distribution function.

When calculating the laser power, the relaxation equations are supplemented with terms describing the stimulated emission of radiation in the transition under consideration (the calculation is frequently made in the quasistationary approximation as presented in Sect. 2.1).

Both the above approaches – the one based on microkinetic equations and that based on energy relaxation equations – yield compatible results over a fairly wide range of the $DF-CO_2$ laser operating conditions. The energy relaxation equations have an advantage over their microkinetic counterparts since they include fewer elementary process rate constants which are necessary for calculation. As a matter of fact, one needs to know only the rate constants of deactivation of the 00^01 and 01^10 states on the various laser mixture components, and it is precisely these kinetic constants that are at present being studied most comprehensively. The rate constants of the $V \to V$ energy exchange in the level ladders of the symmetric and stretching modes of CO_2 are studied less. There are discrepancies both in the interpretation of the experiments aimed at determining the $V \to V$ exchange rates for CO_2 and in the results of theoretical calculations of these rates. But on the whole, the state of affairs in this field has recently improved quite perceptibly.

When analyzing the operating conditions of a $DF-CO_2$ laser with a high level of initiation and a low concentration of CO_2 in the mixture, the energy relaxation equations may become inapplicable because of the rate of establishment of quasiequilibrium in the CO_2 modes becoming too slow. For this reason, to analyze and optimize the laser performance over the wide range of initial conditions of practical interest where the finiteness of the $V \to V$ process rates may become essential, use should be made of the microkinetic equations. What is more, these equations allow one to take into consideration in a most simple manner multiple-quantum relaxation processes, as well as the actual dependence of the relaxation rate on the vibrational level serial number.

For simplicity, numerical analyses of $DF-CO_2$ lasers are sometimes performed with the use of combined calculation procedures in which the kinetics of some group of vibrational states (e.g., the vibrational states of DF and D_2) is treated within the framework of the energy relaxation equations, while that of another group (the vibrational states of CO_2), within the scope of microkinetic equations. Such an approach has been used in [3.16, 111, 126].

3.3.3 Calculation of the Effect of Basic Factors on the Energy Characteristics of the $DF-CO_2$ Laser

In treating the characteristics of the $DF-CO_2$ laser, we will rely on the latest results of kinetic calculations that illustrate most comprehensively the potentialities of the

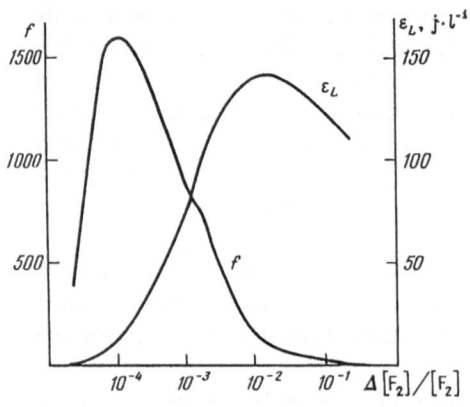

Fig. 3.5. Specific energy output and quantum lasing yield of the DF–CO$_2$ laser as a function of the initiation level [3.16]. D$_2$:F$_2$:CO$_2$:He = 1:1:4:5, $p = 1$ atm

system. These calculations are distinguished by the fact that they have been performed for energy-intensive mixtures of principal practical interest [3.16, 126].

Initiation Level. Figure 3.5 presents the results of calculation of the specific energy ε_l and the quantum yield of lasing, f, for the D$_2$: F$_2$: CO$_2$: He = 1:1:4:5 mixture at atmospheric pressure as a function of the degree of dissociation of fluorine, $\Delta[F_2]/[F_2]$ (i.e., the level of initiation). The quantum yield of lasing is defined by the formula

$$f = \varepsilon_l/\hbar\omega_l n_a \ .$$

It follows from these calculations that as the initiation level grows higher, both the specific energy output ε_l and the quantum yield of lasing, f, pass through a maximum, but what is important is the fact that these quantities do not reach their maxima under the same conditions. The specific energy output passes through a maximum at $\Delta[F_2]/[F_2] \simeq 1\,\%$. The calculated maximum value of ε_l, equal to 141 J/l, agrees well enough with the maximum value of this quantity, found experimentally to be equal to 150 ± 30 J/l [3.110][1]. At $\Delta[F_2]/[F_2]$ values in excess of 1 % the specific laser energy output drops. This effect characteristic of the DF–CO$_2$ laser has been interpreted by *Igoshin* [3.7].

At high reaction rates (high initiation levels) the finiteness of the rate of energy transfer from DF to CO$_2$ becomes manifest so that this stage becomes a bottleneck in the chemical pumping process. The rapid heating of the mixture densely populates the lower laser level 10^00 and quenches oscillation, a substantial proportion of the vibrational energy of DF being wasted. In contrast, at low reaction rates there is enough time for energy to be transferred to the upper laser level, but the rate of excitation of this level cannot compete with the relaxation rate. The combined effect of the excitation of the lower laser level and deactivation of the

---

[1] Although this value was obtained under saturating signal amplification conditions, its comparison with the calculated value is perfectly justifiable, because the calculations were made for a high mirror reflectivity, hence a low threshold inversion density.

upper one gives rise to the optimum reaction rate (and the optimum initiation level). For the given mixture, the maximum quantum yield of lasing, $f = 1.6 \times 10^3$, is reached at much lower levels of initiation: $\Delta[F_2]/[F_2] = 10^{-2}\,\%$. Where the quantity f is a maximum, the specific energy output $\varepsilon_l = 17\,\text{J/l}$, which is almost an order of magnitude lower than its maximum value. Thus, the calcultion reveals the contradictory nature of the conditions necessary to achieve the ultimate values of ε_l and f simultaneously. *Igoshin* and co-workers [3.16] have demonstrated that this can largely be overcome by concurrently optimizing the mixture composition and pressure and initiation conditions. These results are discussed in the following paragraphs.

Despite the fact that materially different levels of initiation are required to achieve the maximum values of ε_l and f, fairly high values of these quantities can be attained simultaneously at intermediate initiation levels: at $\Delta[F_2]/[F_2] = 0.3\,\%$, $\varepsilon_l = 100\,\text{J/l}$ and $f = 750$. Since the laser photon corresponds to the vibrational quantum of DF, and the average number of such vibrational quanta excited in a chain link is equal to 4.5, the effective chain length v_{eff} in this system within the range $\Delta[F_2]/[F_2] = 0.01$–$1\,\%$ varies between 300 and 30, respectively. Thus, the chain character of the pumping reaction is clearly manifest in the D_2-F_2-CO_2 system.

Initiation Duration. The rate constant k_i for the process of fluorine dissociation

$$F_2 \xrightarrow{k_i} 2F$$

under the effect of an external source of energy is defined as

$$k_i = \gamma_{F_2}\, t \exp(-t/t_0)\;,$$

where γ_{F_2} defines the initiation intensity and t_0 the initiation duration. Table 3.10 presents the results of calculation of the DF–CO_2 laser pulse energy and duration as a function of t_0 at a fixed degree of dissociation of F_2. It can be seen from these data that the DF–CO_2 laser is less sensitive to the initiation duration than HFL. The specific energy output of HFL starts to drop substantially at $t_0 \geqslant 1\,\mu\text{s}$, whereas that of the DF–CO_2 laser remains practically unchanged up to $t_0 = 10\,\mu\text{s}$.

Initial Temperature of the Laser Medium. Since the population density of the lower laser level 10^00 largely governs the laser efficiency, it is of interest to reveal the effect of the initial temperature of the medium on the operation of the DF–CO_2 laser. It can be expected that the laser energy will rise with the decreasing initial temperature T_0. The initial temperature of the DF–CO_2 laser mixture cannot be lower than 200 K because at lower temperatures carbon dioxide condenses. The effect of the initial laser medium temperature on the specific energy output and pulse duration of the DF–CO_2 laser is illustrated in Table 3.11 for various mixture compositions and initiation levels. The scale of the effect depends on the level of initiation and mixture composition. In conditions where the effect is most pronounced, the laser energy rises by 1.6–1.7 times as the initial temperature is reduced from 300 to 200 K.

Reactant Concentrations. One of the possible ways to increase the laser energy output is to raise the energy content of the mixture. It is interesting to study the possibility of enhancing the energy output by increasing the concentrations of the reactants without raising the total mixture pressure. Table 3.12 presents the results of calculation of the laser characteristics as a function of the partial pressures of D_2 and F_2 in an atmospheric-pressure laser medium at various initiation levels. These data show that increasing the energy content of the mixture at a fixed total mixture pressure enhances the laser energy over a limited scale only. Raising the partial pressures of the reactants in excess of 70 mm Hg in an atmospheric-pressure mixture adds nothing to the laser energy output. This is explained by the fact that when the total pressure (hence the heat capacity) of the mixture is fixed, its temperature rises at a certain energy release so high ($T > 2500$ K) that it no longer becomes possible to maintain a population inversion in the energy-level ladder of CO_2. At the same time, it is seen from Table 3.12 that for the more energy-intensive mixtures, the ultimate energy output ($\varepsilon_l \simeq 140$ J/l) is reached at a lower level of initiation. Thus, increasing the reactant content of the mixture can be of interest where the maximum quantum yield and engineering efficiency of the laser are to be realized.

CO_2 and He Concentrations. Calculations show that for fixed reactant partial pressures and total mixture pressure and with the initiation level specified, there are optimal relative concentrations of CO_2 and He in the mixture. Optimization of the CO_2 and He content of the mixture ensures approximately a two-fold gain in the energy output and quantum yield of the laser (Table 3.13). The optimum CO_2 : He ratio increases with rising initiation level.

Buffer Gas Pressure. The mixture temperature of the $DF-CO_2$ laser largely determines its efficiency. As the average gas temperature drops, the laser efficiency rises, which is explained by the reduction of the population densities of the lower and upper laser levels in the course of lasing and the lowering of the rate constants for the deactivation of the upper level and the reverse energy transfer from CO_2 to DF. As demonstrated by calculations, increasing the heat capacity of the mixture by raising the buffer gas pressure improves the laser efficiency very markedly (Table 3.14).

Total Mixture Pressure. The energy content of the laser mixture can be augmented by raising the total pressure of the laser mixture. It is of interest here to study the relationship between the specific energy output of the laser and its total mixture pressure in the following two cases: (1) the concentration of active centers rises in proportion to the total pressure (fixed F_2 dissociation degree $\Delta[F_2]/[F_2]$) and (2) the concentration of active centers remains constant ($\Delta[F_2]/[F_2]$ varies in inverse proportion to the pressure).

An important calculation result is that the specific energy output ε_l can be substantially increased by raising the mixture pressure while keeping constant the concentration of active centers, provided the concentration of atoms is not too low (Table 3.15a). This means that with the experiment being staged in this manner, raising the pressure of the reactants will increase not only the specific energy output ε_l, but also the quantum yield f. Physically the effect is explained by the fact that when the level of initiation is high enough, the quenching of oscillation in the DF–CO₂ laser is caused by the depletion of the reactants and heating of the mixture to high temperatures ($\simeq 2000$ K). This occus at a certain degree of conversion of the reactants that is almost independent of the initiation level. As the mixture pressure is increased, the degree of burn-out of the reactants by the moment oscillation is quenched remains unchanged, and so the laser energy output rises. At low concentrations of atoms the effect is much less pronounced (Table 3.15b), and materially to increase ε_l by raising the mixture pressure, it is necessary to keep fixed the F₂ dissociation degree $\Delta[F_2]/[F_2]$ and not the atomic concentration.

For the effect of concurrent increases of ε_l and f (predicted in [3.16]) to manifest itself most vividly, one needs to optimize the relative concentration of CO₂ (Table 3.16). It can be seen from the data of Table 3.16b that at a high – but perfectly reasonable – mixture pressure of $p \simeq 7$ atm. the value of ε_l reaches 330–740 J/l at a quantum yield of lasing of 4750–1050, which ensures a very high physical efficiency of the laser. The conditions indicated in the third column of Table 3.15b seem to be optimal for reaching high values of ε_l and f simultaneously. Under these conditions, $\varepsilon_l \simeq 330$ J/l, $f = 4750$, and the laser pulse duration $\tau_l \simeq 3$ μs. It should be noted that lasing conditions characterized by high quantum yields are sensitive to the concentration of oxygen in the mixture, which causes chain termination in the course of the trimolecular reaction $D + O_2 + M \rightarrow DO_2 + M$. As shown by calculations, the predicted conditions for lasing with a high quantum yield can be realized at a partial oxygen pressure of 10 mm Hg.

And so, the theoretical results obtained by *Igoshin* and co-workers [3.16] demonstrate the possibility of augmenting substantially both the specific energy output ε_l and quantum yield f of the DF–CO₂ laser by raising the total laser mixture pressure to a few atmospheres and optimizing at the same time the mixture composition and initiation conditions.

3.3.4 Oscillation Excitation and Quenching Mechanism

Numerical calculations have made it possible to study the relative importance of the various processes occurring in the laser medium. Figures 3.6 to 3.12 illustrate the emission dynamics and operating mechanism of the DF–CO₂ laser. The calculations refer to the energy-intensive atmospheric-pressure mixture $D_2:F_2:CO_2:He = 1:1:4:5$ at an initiation level of $\Delta[F_2]/[F_2] = 3\%$ corresponding to the maximum specific energy outputs (Fig. 3.5) but exceeding somewhat the optimum value. Let us dwell upon the main inferences from the numerical analysis of the factors governing the laser efficiency.

Oscillation continues (Fig. 3.6) until the excitation rate of the upper laser level $00^0 1$ is high enough to compensate for the effect of dissipative processes (Fig. 3.7). As a result of the joint effect of several factors oscillation is quenched.

First, at the moment oscillation is quenched the rate of deactivation of the upper laser level becomes high. As seen from Fig. 3.7, in the conditions considered, the principal part in the deactivation of the $00^0 1$ level is played by the reverse energy

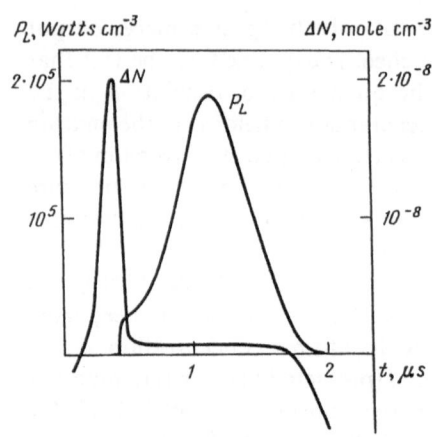

Fig. 3.6. Temporal dependence of the power output P_L and inverted population density $\Delta N = N_{00^01} - N_{10^00}$ in the DF–CO$_2$ laser

Fig. 3.7. Temporal dependence of excitation (*solid curve*) and deactivation (*dashed curves*) rates for the 00^01 upper lasing level in the DF–CO$_2$ laser ▼

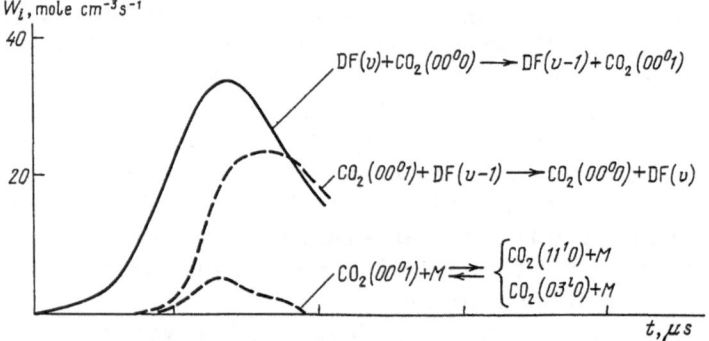

transfer from CO$_2$ to DF, the rate of which becomes more and more comparable with that of the direct energy transfer from DF to CO$_2$ as the DF molecules accumulate. Relaxation from the 00^01 level to the nm^l0 levels of the symmetric and stretching modes is less important because it is largely compensated for by the reverse processes

$$\left.\begin{array}{c}(03^10)\\(11^10)\end{array}\right\} + M \rightarrow (00^01) + M \ .$$

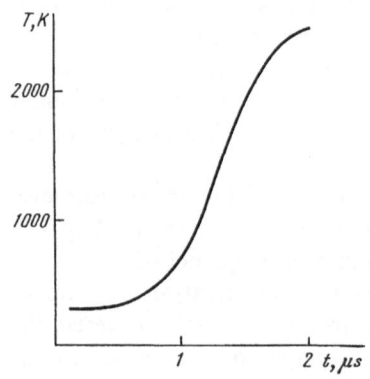

Fig. 3.8. Temporal dependence of the gas temperature in the DF–CO$_2$ laser

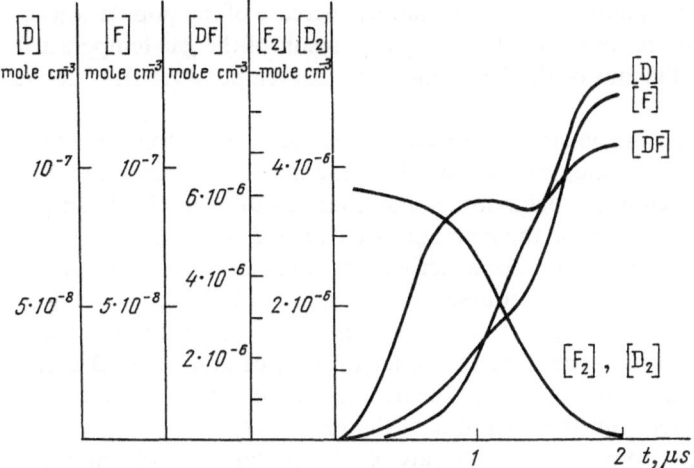

Fig. 3.9. Temporal dependence of the component concentrations in the DF–CO$_2$ laser

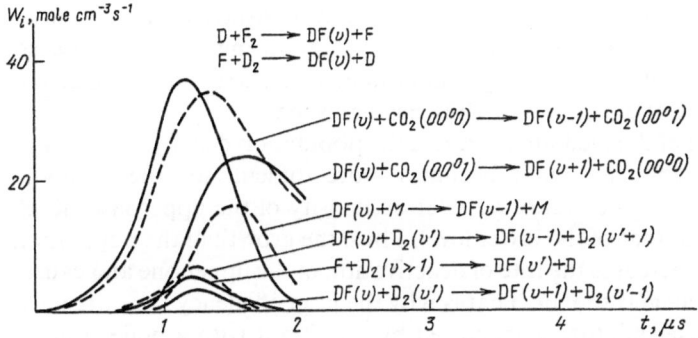

Fig. 3.10. Temporal dependence of the DF oscillation excitation (*solid curves*) and deactivation (*dashed curves*) rates in the DF–CO$_2$ laser

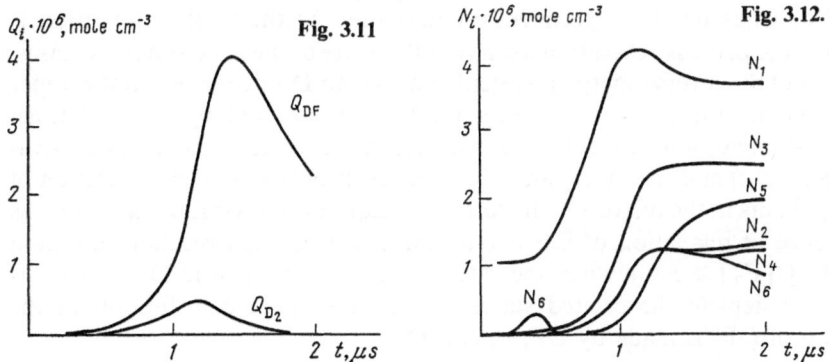

Fig. 3.11. Temporal dependence of the density of vibrational quanta stored in DF and D$_2$ in the DF–CO$_2$ laser

Fig. 3.12. Temporal dependence of the populations of the excited vibrational CO$_2$ levels in the DF–CO$_2$ laser: $N_1 = CO_2(01^10)$; $N_2 = CO_2(02^00)$; $N_3 = CO_2(02^20)$; $N_4 = CO_2(10^00)$; $N_5 = CO_2(11^10) + CO_2(03^10) + CO_2(03^30)$; $N_6 = CO_2(00^01)$

Such a compensation is explained by the substantial increase of the populations of the excited states in the first two modes of CO_2 as a result of the gas temperature growing very high. By the end of the laser pulse the gas temperature rises above 2000 K (Fig. 3.8).

Second, the depletion of the store of reactants reduces the pumping reaction rate. Figure 3.9 illustrates the kinetics of the reactant "burn-out" and DF accumulation and the temporal behavior of the concentration of the F and D atoms. By the end of the laser pulse the reactants are depleted by 90 %. Figure 3.10 shows the vibrational excitation and deactivation rates of DF. The rate of vibrational excitation of DF in the course of the reactions $F + D_2$ and $D + F_2$ passes through a maximum during the laser pulse. The rise of the excitation rate at the initial stage is due to the growth of the concentration of the F and D atoms under the action of the initiating pulse and the heating of the mixture. The drop of the excitation rate results from the burn-out of the reactants.

Third, as can be seen from Fig. 3.10, the rate of vibrational deactivation of DF has risen by the end of the laser pulse owing to the $V \rightarrow R, T$ relaxation process. The temporal dependence of the density of vibrational quanta stored in DF is illustrated in Fig. 3.11. As seen, in the intense initiation conditions considered, not all of the vibrational excitation energy of DF is transferred to CO_2 during the laser pulse, so that the vibrational quantum density Q_{DF} continues to relax to the equilibrium value, after the population inversion in CO_2 has vanished.

Fourth, the heating of the medium increases the population density of the lower laser level $10^0 0$. For this reason, it becomes more difficult for the pumping mechanism to maintain the necessary population density of the upper laser level, which is proportional to that of the lower level. Thus, the growth of the population of the lower laser level increases the rate of deactivation of the upper one and causes oscillation to come to an end sooner. That is why the laser efficiency rises when the heat capacity of the laser mixture is increased by diluting it with a buffer gas.

The relative importance of the cross-relaxation process $(00^0 1) + M \rightarrow (nm^l 0) + M$ and the reverse energy transfer from CO_2 to DF in the deactivation of the $00^0 1$ level depends on the experimental conditions. As the level of initiation is reduced, the role of cross-relaxation from the $00^0 1$ state to the $nm^l 0$ states increases, whereas that of the reverse energy transfer from CO_2 to DF decreases. At the upper laser level, the most important processes of relaxation proceeding via the channel $(00^0 1) + M \rightarrow (nm^l 0) + M$ are collisions with DF (as a result of the high deactivation rate constant) and with CO_2 and He (as a result of the high concentration of these components in the mixture). The loss of nonequilibrium vibrational energy as a consequence of relaxation of DF is only important at high reaction initiation levels $(\Delta [F_2]/[F_2] \geqslant 3 \%)$, when the rate of energy transfer from DF to CO_2 is insufficient to deplete the excited states of DF. The main contribution to the deactivation of DF^v is made by DF, F, and D.

At moderate mixture temperatures the basic mechanism responsible for the population of the lower laser level $10^0 0$ is the stimulated emission of radiation. By the time the laser pulse ceases, the decay of the $11^1 0$ and $03^l 0$ states has begun to contribute considerably to the population of the $10^0 0$ state. The $11^1 0$ and $03^l 0$ states are in turn populated as a result of deactivation of the $00^0 1$ state.

The processes

$$(10^0 0) + M \leftrightarrow \begin{cases} (02^0 0) + M, \\ (02^2 0) + M \end{cases}$$

constitute the main mechanism responsible for the depopulation of the lower laser level. The contribution of the $V \to V$ exchange process

$$(10^0 0) + (00^0 0) \leftrightarrow 2(01^1 0)$$

is smaller and, what is more, it diminishes in the course of the reaction because of the population density of the $01^1 0$ state increasing as a result of the heating of the laser mixture. The main thermalization "catalyst" for the $02^l 0$ and $01^1 0$ states is helium.

It is essential that the populations of the $10^0 0$ state and of all the states lying below it remain almost in equilibrium during the entire course of the reaction (Fig. 3.12).

3.3.5 Comparison Between the Energy Characteristics of the H_2-F_2 and $D_2-F_2-CO_2$ Systems

Disregarding the radiation wavelength, what is the main difference between the lasers operating on the H_2-F_2 and $D_2-F_2-CO_2$ mixtures? Will one of these systems always be preferable to the other in terms of energy characteristics? Table 3.17 answers these questions. It lists the results of calculation of the specific energy and duration of laser emission as a function of the reaction initiation level in $H_2-F_2-O_2-He$ and $D_2-F_2-CO_2-He$ mixtures with the same store of chemical energy. It can be seen from these data that the $DF-CO_2$ laser is preferable where the level of initiation is not very high. It is characterized by a longer laser chain, but at high initiation levels the energy transfer process in it becomes a pumping "bottleneck". The $H_2(D_2)-F_2$ system is therefore preferable where the reaction initiation process is intense and fast.

An intense and fast initiation can be realized with the use of high-current electron beams produced by accelerators, but acceleration equipment is costly. Initiation with UV radiation from a xenon lamp or an open electrical discharge is technically more simple, but its intensity is lower and duration longer. Therefore, it is advisable to use the $H_2(D_2)-F_2$ mixture in conjunction with electron-beam initiation and the $D_2-F_2-CO_2$, with photolysis initiation.

3.3.6 On the Use of Energetic Chain Branching to Initiate Laser Emission with a High Quantum Yield

In the existing chemical lasers (HF and $DF-CO_2$), the contribution of chain branching to the laser efficiency is insignificant because the duration of the chemical lasing process under studied operating conditions is much shorter than the characteristic chain-branching time. The latest investigations into the chain-branching rate in the H_2-F_2 and D_2-F_2 systems have yielded appreciably higher values than the earlier works (Table 3.1). The values obtained for the rate constant of the energetic chain-branching process

$$H_2^v(D_2^v) + F_2 \rightarrow H(D) + HF(DF) + F$$

suggest the launch of a theoretical and experimental search for laser operating conditions in which active centers would mainly be produced by chain branching.

The numerical analysis of this process in the $D_2-F_2-CO_2$ system shows that such conditions do exist (provided the data of [3.30, 63] on the chain-branching rate are correct). To initiate laser emission by chain branching, it is necessary that the mixture should contain vibrationally excited $D_2(H_2)$ molecules. In particular the following two methods can be used to produce $D_2^v(H_2^v)$:

a) The introduction of a small number of DF (HF) molecules into the mixture. They are excited by the DF (HF) laser emission to transfer their vibrational energy to the $D_2(H_2)$ molecules:

$$DF + \hbar\omega_l \rightarrow DF^v ,$$

$$DF^v + D_2 \rightarrow DF + D_2^v ;$$

b) The use of weak photolysis or some other chain-reaction initiation (to produce a small number of F atoms) giving rise to a priming amount of the DF^v (HF^v) molecules, which transfer their vibrational energy to the $D_2(H_2)$ molecules.

The chain-branching and lasing processes are impeded by the relaxation of the vibrational energy of DF^v and D_2^v, which proceeds mainly by way of the $V \rightarrow V$ exchange with CO_2. Therefore, to prevent the chain-branching process from suppression, the mixture must have sufficiently high $[F_2]/[CO_2]$ and $[D_2]/[CO_2]$ ratios.

Table 3.18 presents the results of calculation of the $DF-CO_2$ laser characteristics for two methods of producing priming centers: by means of IR and UV radiation, respectively. In both cases, gain is achieved in the ratio between the laser energy and the energy consumed to initiate the reaction, which cannot be obtained under ordinary conditions (where chain branching is suppressed). The physical laser efficiency is especially high ($\eta_{phys} = 3 \times 10^6 \%$) in the case of photolysis initiation. But initiation with IR radiation can offer some technical advantages.

Note that the condition

$$s > \sigma_{21} ,$$

where s is the chain-branching ratio and σ_{21} the relaxation probability of the upper laser level, is not satisfied at the initial reaction stage when chain branching is effected by the $D_2(v = 1)$ molecules. It is only the condition of chemical instability that is satisfied, because the calculation has been performed for an oxygen-free mixture. In the subsequent stages of chain explosion (the characteristic induction time is $10^{-4}-10^{-3}$ s), there appear D_2 molecules excited to $v = 2$, which substantially accelerate the process and initiate laser emission..

The critical oxygen concentration required to suppress chain explosion is very low ($\simeq 0.3$ mm Hg for the mixture under consideration). For this reason, to initiate experimentally laser emission by a branching-chain process it is first of all necessary to solve the problem of preparing laser mixtures extremely low in oxygen. The stringent requirement for the oxygen content of the mixture can apparently explain why the effect was never observed before.

To reach the ignition region and achieve lasing, the following method is recommended: use a mixture confined to some volume at a high pressure (above the second explosion limit for the given mixture composition) and rapidly release it via a nozzle array into a larger volume where its pressure drops below the second explosion limit, and then initiate the reaction with weak IR or UV radiation.

Table 3.1. Experimental data on the rate constants of elementary processes in the $H_2(D_2)-F_2-O_2-M$ system

T [K]	Process and rate constant	Measurement method	Reference
298–363	$H_2 + F_2 = H + HF + F$		
	$10^{12.51+0.16} \exp[-(19800 \pm 1000)/RT]$	thermal	[3.39]
298–356	$D_2 + F_2 = D + DF + F$		
	$10^{12.5(+0.14/0.21)} \exp[-(20200 \pm 1000)/RT]$	thermal	[3.40]
293	$F + H_2 = HF + H^{(a)}$		
	$(1.8 \pm 0.6) \times 10^{13}$	mass spectrometry	[3.41]
300–400	$10^{14.2} \exp(-1600/RT)$	mass spectrometry	[3.42]
	$k_{293} = 1.47 \times 10^{13}$		
195–294	$9.28 \times 10^{13} \exp[-(1080 \pm 170)/RT]$	chemical laser	[3.43, 44]
	$k_{293} = 1.03 \times 10^{13}$		
298	1.5×10^{13}	microwave discharge	[3.45]
150–300	$10^{13.74 \pm 0.38} \exp[-(510 \pm 200)/RT]$	discharge, jet,	[3.46]
	$k_{293} = 2.2 \times 10^{13}$	electron paramagnetic resonance (EPR)	
300	$10^{13.1 \pm 0.05}$	chemical laser	[3.51]
223–353	$3.85 \times 10^{14} \exp(-2145/RT)$	competing reactions	[3.49]
	$k_{293} = 10^{13}$		
	$F + D_2 = DF + D^{(a)}$		
195–294	$4.95 \times 10^{12} \exp[-(790 \pm 180)/RT]$	chemical laser	[3.43, 44]
	$k_{293} = 1.272 \times 10^{13}$		
150–300	$5.66 \times 10^{12} \exp[-(830 \pm 110)/RT]$	jet, EPR	[3.46]
	$k_{293} = 1.284 \times 10^{13}$		
170–430	$k_{H_2}/k_{D_2} = (1.04 \pm 0.02) \exp[(370 \pm 10)/RT]$	mass spectrometry	[3.47]
	$(k_{H_2}/k_{D_2})_{293} = 1.91 \pm 0.08$		
273–475	$k_{H_2}/k_{D_2} = (1.04 \pm 0.06) \exp[(382 \pm 35)/RT]$	competing reactions	[3.48]
	$(k_{H_2}/k_{D_2})_{298} = 1.94 \pm 0.04$		
77–293	$k_{H_2}/k_{D_2} = (1.48 \pm 0.22) \exp[(45 \pm 30)/RT]$	competing reactions	[3.49]
	$(k_{H_2}/k_{D_2})_{293} = 1.6 \pm 0.3$		
300	$k_{H_2}/k_{D_2} = 1.8 \pm 0.4$	isotopic	[3.50]
150–300	$k_{H_2}/k_{D_2} = 0.98 \exp[(320 \pm 290)/RT]$	jet, EPR	[3.46]
	$(k_{H_2}/k_{D_2})_{293} = 1.7$		
297	$k_{H_2}/k_{D_2} = 1.79$	chemical laser	[3.53]
	$H + F_2 = HF + F$		
288	1.8×10^{12}	thermal reaction	
	$E_a = 1.5 \pm 0.3$	$H_2 + F_2$	[3.54]
294–565	$10^{14.08 \pm 0.01} \exp[-(2400 \pm 200)/RT]$	mass spectrometry	[3.55]
	$k_{300} = 2.14 \times 10^{12}$		
300	$(2.5 \pm 2) \times 10^{12}$	EPR	[3.56]
	$E_a = 2.6$		
298	$(2.6 \pm 0.6) \times 10^{12}$	EPR	[3.57]
300	1.4×10^{12}	photochemical	[3.59]
	$D + F_2 = DF + F$		
300	$k_H/k_D = 1.4 - 1.55$	flash photolysis	[3.59]
300	$k_{D+O_2+He}/E_a k_{D+F_2} = 2.3 \pm 0.7$ cm^3/cal	flash photolysis	[3.68]

Table 3.1 (continued)

T [K]	Process and rate constant	Measurement method	Reference
	$H+O_2+M=HO_2+M$		
203–404	$10^{15.38}\exp[(473\pm92)/RT]$	flash photolysis	[360]
	$M=He,$		
	$k_{CH_4}:k_{N_2}:k_{He}:k_{Ar}=15.7:3.4:1:1$		
298	$(5.87\pm0.68)\times10^{15}$	radiolysis	[3.61]
	$M=Ar,\ k_{H_2}:k_{Ar}=2.9$		
225–1500	$1.1\times10^{16}(273/T),\ M=Ar$		[3.58–62]
300	$7.2\times10^{15(b)},\ M=He$		
293	$k_{HF}:k_{CO_2}:k_{Ar}=$	photochemical	[3.64]
	$(10.5\pm2):(3.2\pm0.5):1$		
298 ± 1	$k_{He}:k_{N_2}:k_{CO_2}:k_{Ar}=$	thermal	[3.65]
	$(1\pm0.2):(2.8\pm0.2):(4.5\pm0.2):1$		
300–600	$7.2\times10^{16},\ M=SF_6$	chemical laser	[3.66]
400	$k_{H_2}:k_{He}=(2.3\ \text{to}\ 2.8):1$	explosion limits	[3.81]
	$D+O_2+M=DO_2+M$		
300	$k_M^H=k_M^D,\ M=He,\ Ar,\ CO_2$	flash photolysis	[3.59]
300	$k_{DF}:k_{CO_2}:k_{Ar}=(6.8\pm1):(2.3\pm0.5):1$	photochemical	[3.64]
298	$k_{He}:k_{Ar}:k_{N_2}:k_{CO_2}=$	thermal	[3.67]
	$1:(1.2\pm0.2):(1.6\pm0.3):(4.5\pm0.5)$		
300	$(7.2\pm2.2)\times10^{15},\ M=He$	flash photolysis	[3.68]
	$F+O_2+M=FO_2+M$		
293 ± 2	$10^{15.38\pm0.12},\ M=He$	discharge, jet, EPR	[3.69]
293 ± 2	$10^{15.31\pm0.15},\ M=Ar$	discharge, jet, EPR	[3.69]
293 ± 2	$10^{15.68\pm0.13},\ M=N_2$	discharge, jet, EPR	[3.69]
293 ± 2	$10^{15.56},\ M=O_2$	discharge, jet, EPR	[3.70]
228–301	$(1.7\pm0.3)\times10^{14}\exp(1700/RT),\ M=Ar$	flash photolysis	[3.71]
	$k_{293}=3.1\times10^{15}$		
298	$(1.8\pm0.22)\times10^{15},\ M=F_2$	flash photolysis	[3.72]
		IR luminescence	
298	$(5.4\pm1.1)\times10^{15},\ M=O_2$	ditto	[3.72]
298	$(3\pm0.32)\times10^{15},\ M=Ar$	ditto	[3.72]
298	$(1.95\pm0.22)\times10^{15},\ M=He$	ditto	[3.72]
298	$(9\pm1.8)\times10^{17(c)},\ M=HF$	ditto	[3.72]
298	$4.5\times10^{17(c)},\ M=DF$	ditto	[3.72]
	$F+F+M=F_2+M^{(d)}$		
300–500	$(3.6\ \text{to}\ 7.2)\times10^{15},\ M=F_2$	flash photolysis	[3.76]
	$k_{F_2}:k_{Ar}:k_{He}:k_{CO_2}:k_{O_2}=40:1.3:0.7:10:1$		
293 ± 2	$7.2\times10^{14},\ M=He$	discharge, jet, EPR	[3.69]
295	$2.9\times10^{13},\ M=Ar$	jet, afterglow	[3.77]
	$k_{N_2}:k_{Ar}=10:1$		
	$H_2(v=1)+F_2=H+HF+F$		
300	1.2×10^4	explosion limits	[3.33–35]
300	$6\times10^4-2\times10^5$	explosion limits	[3.78]
330	3×10^6	flash photolysis	[3.30, 63]
	$D_2(v=1)+F_2=D+DF+F$		
330	6×10^5	flash photolysis	[3.30–63]

Table 3.1 (continued)

T [K]	Process and rate constant	Measurement method	Reference
330	$H_2(v \geqslant 2) + F_2 = H + HF + F$ 1.2×10^9	flash photolysis	[3.30–63]
330	$D_2(v \geqslant 2) + F_2 = D + DF + F$ 2.4×10^8	flash photolysis	[3.30–63]
300 300 300	$H + HO_2 = \begin{cases} (1) \ H_2 + O_2^{(e)} \\ (2) \ H_2O + O \\ (3) \ 2OH \end{cases}$ $k_1:k_2:k_3 = 1:1.2:0.11$ $k_1:k_2:k_3 = 0.38:0.34:0.28$ $k_1 + k_2 + k_3 = 6 \times 10^{13}$	 flash photolysis flash photolysis flash photolysis	 [3.79] [3.80] [3.59]
300	$D + DO_2 = \begin{cases} (1) \ D_2 + O_2 \\ (2) \ D_2O + O \\ (3) \ 2OD \end{cases}$ $k_1 + k_2 + k_3 = 6 \times 10^{13}$	 flash photolysis	 [3.59]
> 300	$2HO_2 = H_2O_2 + O_2$ $1.07 \times 10^{13} \exp(-1000/RT)$	flash photolysis	[3.80]

Notes: [a] presented are the more consistent recent experimental results; other pertinent data available can be found in reviews [3.52, 58] and also in references cited in the table;

[b] average from the data presented in [3.63];

[c] these data indicate a very high efficiency of HF and DF in the process $F + O_2 + M \rightarrow FO_2 + M$ and need to be verified; the efficiency of HF in the process $H + O_2 + M \rightarrow HO_2 + M$ is much lower;

[d] other, earlier data on the recombination rate of the F atoms can be found in review [3.52]; numerous data on the rate of recombination of H(D) $(H + H + M \rightarrow H_2 + M)$ and of H(D) with F $(H + F + M \rightarrow HF + M)$, and also on the rate of dissociation of $H_2(D_2)$, F_2, and HF (DF) are presented in reviews [3.52, 58] and reference book [3.73]; the rate constants for these processes used in laser performance calculations are given in the text of this chapter; the data on the decay rate of the F atoms on various surfaces have been obtained in [3.74, 75];

[e] other data available on the secondary processes can be found in [3.52, 58, 73].

Table 3.2. Basic energy characteristics and pulse duration of HFL as a function of the reaction initiation level n_a^*

n_a [cm^{-3}]	t_l [s]	ε_l [J/1]	η_{chem} [%]	η_{phys}^{**} [%]		v_{eff}	f
				UVI	EBI		
6.8×10^{13}	9.51×10^{-5}	0.8	0.028	3670	734	45	168
6.8×10^{14}	2.66×10^{-6}	8.3	0.29	3800	760	47	177
6.8×10^{15}	2.82×10^{-6}	123	4.3	5650	1130	69	294
6.8×10^{16}	6.1×10^{-7}	545	19	2500	500	31	116
6.8×10^{17}	1.74×10^{-7}	775	27	355	71	4.4	16.5

* Mixture $F_2 : H_2 : O_2 : RF : He = 400 : 100 : 40 : 5 : 215$ [mm Hg], $J = 10$, $t_0 = 10^{-7}$ s.
**UVI – ultraviolet initiation; EBI – electron-beam initiation.

Table 3.3. Laser chemical efficiency η_{chem} and pulse duration t_l as a function of the initiating pulse duration t_0 and concentration of the photoinitiating component RF for mixtures differing in chemical composition

		t_0 [s]					
		10^{-4}	2×10^{-5}	10^{-5}	10^{-6}	10^{-7}	10^{-8}
	γ_{F_2} [s^{-2}]	2.07×10^6	5.175×10^7	2.07×10^8	2.07×10^{10}	2.07×10^{12}	2.07×10^{14}
	γ_{RF} [s^{-2}]	9×10^7	2.25×10^9	9×10^9	9×10^{11}	9×10^{13}	9×10^{15}
1)	$p_{RF}=1$ mm Hg $\quad\eta_{chem}$[%]	0.066	0.077	1.38	3.54	3.87	3.9
	t_l[s]	2.35×10^{-5}	6.33×10^{-7}	1.54×10^{-5}	5.85×10^{-6}	1.32×10^{-6}	1.09×10^{-6}
	$p_{RF}=3$ mg Hg $\quad\eta_{chem}$[%]	0.066	0.122	3.6	6.55	7.75	7.9
	t_l[s]	1.83×10^{-5}	6.33×10^{-7}	2.68×10^{-5}	5.22×10^{-6}	1.49×10^{-6}	1.25×10^{-3}
	$p_{RF}=5$ mm Hg $\quad\eta_{chem}$[%]	0.067	0.168	4.9	11.1	14.65	15
	t_l[s]	1.56×10^{-5}	6.33×10^{-7}	2.28×10^{-5}	4.36×10^{-6}	1.36×10^{-6}	1.14×10^{-6}
2)	$p_{RF}=1$ mm Hg $\quad\eta_{chem}$[%]	2.8	10.7	13.9	23.6	29.5	31.4
	t_l[s]	2.79×10^{-5}	8.19×10^{-6}	4.54×10^{-6}	8.02×10^{-7}	1.96×10^{-7}	7.1×10^{-8}
	$p_{RF}=3$ mm Hg $\quad\eta_{chem}$[%]	3.04	11	14.4	23.6	29.7	31.6
	t_l[s]	2.66×10^{-5}	7.82×10^{-6}	4.34×10^{-6}	7.7×10^{-7}	1.88×10^{-7}	6.7×10^{-8}
	$p_{RF}=5$ mm Hg $\quad\eta_{chem}$[%]	3.05	11.3	14.5	24	30	31.9
	t_l[s]	2.54×10^{-5}	7.49×10^{-6}	4.21×10^{-6}	7.48×10^{-7}	1.82×10^{-7}	6.45×10^{-8}

Notes: 1) $F_2:H_2:O_2=50:50:40$ [mm Hg]; 2) $F_2:H_2:O_2=400:100:40$ [mm Hg]; $p_{He}=760-\sum_M p_M$[mm Hg].

Table 3.4. Effect of mixture composition on HFL characteristics

$H_2:F_2$	1:1	1:3	1:5	1:7
t_l[ns]	19.2	24.2	32	41
η_{chem}[%]	11	22.4	30	34

Notes: $p=1$ atm., $n_a=10^{18}$ cm^{-3}, $T_0=300$ K, instant initiation, $J=10$.

Table 3.5. Effect of oxygen pressure and concentration on laser chemical efficiency and pulse duration

Parameters	$p=1$ atm.	$p=3$ atm.	$p=10$ atm.
1) η_{chem}[%]	27.4	25	20.6
t_l[s]	2.76×10^{-7}	1.65×10^{-7}	1.06×10^{-7}
2) η_{chem}[%]	28.3	24.8	0.01
t_l[s]	2.9×10^{-7}	2×10^{-7}	8.33×10^{-9}

Notes: $\gamma_{F_2}=2.07 \times 10^{12}$, $\gamma_{RF}=9 \times 10^{13}$, $t_0=10^{-7}$ s, mixtures: (1) $F_2:H_2:O_2:RF:He=0.263:0.092:0.01:0.0013:0.643$ [atm.] and (2) $F_2:H_2:O_2:RF:He=0.263:0.092:0.1:0.0013:0.553$ [atm.], $J=10$.

Table 3.6. The chemical efficiency η_{chem} and pulse duration t_l of an HFL amplifier as a function of the master oscillator emission spectrum

J		1	3	5	7	9	15
1	η_{chem} [%]	0.69	9.3	17.5	23.8	28.4	36.2
	t_l [s]	1.65×10^{-7}	2.39×10^{-7}	2.64×10^{-7}	2.79×10^{-7}	2.91×10^{-7}	3.18×10^{-7}
2	η_{chem} [%]	0.77	9.2	17.3	23.2	27.7	35.3
	t_l [s]	1.19×10^{-7}	1.62×10^{-7}	1.75×10^{-7}	1.85×10^{-7}	1.93×10^{-7}	2.09×10^{-7}
3	η_{chem} [%]	0.12	1.9	4	5.9	7.3	10
	t_l [s]	8×10^{-3}	2.1×10^{-7}	2.5×10^{-7}	2.8×10^{-7}	3×10^{-7}	3.5×10^{-7}
4	η_{chem} [%]	—	0.65	3.3	6.4	10	18.3
	t_l [s]	—	1.71×10^{-6}	2.33×10^{-6}	2.65×10^{-6}	2.84×10^{-6}	3.1×10^{-6}

Notes: 1 – $F_2:H_2:O_2:RF:He = 200:70:20:1:469$ mm Hg, $\gamma_{F_2} = 2.07 \times 10^{12}$, $\gamma_{RF} = 9 \times 10^{13}$, $t_0 = 10^{-7}$ s;
2 – $F_2:H_2:O_2:RF:He = 400:100:40:1:219$ mm Hg, $\gamma_{F_2} = 2.07 \times 10^{12}$, $\gamma_{RF} = 9 \times 10^{13}$, $t_0 = 10^{-7}$ s;
3 – $F_2:H_2 = 380:380$ mm Hg, $n_F = 4.8 \times 10^{17}$ cm^{-3}, instantaneous initiation;
4 – $F_2:H_2:O_2:RF:He = 400:100:40:5:215$ mm Hg, $\gamma_{F_2} = 2.07 \times 10^8$, $\gamma_{RF} = 9 \times 10^9$, $t = 10^{-5}$ s.

Table 3.7. H_2-F_2 laser characteristics as a function of electron-beam current density

J_b[A/cm^2]	t_l [s]	η_{chem}[%]	η_{phys} [%]
10^3	2.24×10^{-7}	25.6	170
10^2	6.65×10^{-7}	19	420
10	2.11×10^{-6}	11.5	817
1	8.1×10^{-6}	4.4	811
10^{-1}	1.56×10^{-5}	0.34	340

Notes: $F_2:H_2:O_2:H_e=0.526:0.131:0.53:0.29$, $p=1$ atm

Table 3.8. Characteristics of a pulsed HFL calculated with due regard for rotational nonequilibrium and anharmonicity of emitting molecules

n_a [cm^{-3}]	t_l [s]	ε_l [J/1]	η_{chem} [%]	η_{phys} [%] (UV)	v_{eff}	f
10^{15}	1.83×10^{-6}	6.7	0.23	2210	22.5	85
10^{16}	0.83×10^{-6}	66	2.3	2180	22.3	84
10^{17}	1.32×10^{-7}	232	8	766	8	30
10^{18}	1.9×10^{-8}	340	11.7	112	1.2	4.3

Notes: $F_2 : H_2 : O_2 : He = 400 : 10 : 25 : 235$ mm Hg, instantaneous initiation, the parameter m in the equation $g_{v, v-1} = g_{1,0} v^m$ is taken at 2.3.

Table 3.9. Kinetic constants of the D_2–F_2–CO_2 system recommended for use in calculations[a]

Process	Rate constant
$F + D_2(0) = DF(v) + D$	$k(v) = g(v) \times 10^{14} \, T^{n(v)} \exp(-1960/RT)$, $g(1) = 0.1$, $g(2) = 0.35$, $g(3) = 1.2$, $g(4) = 0.7$; $n(1) = n(2) = 0$, $n(3) = n(4) = -0.1$
$D + F_2 = DF(v) + F$	$k(v) = 6.2 \times 10^{13} \, g(v) \exp(-2500/RT)$, $g(5) = 0.2$, $g(6) = 0.16$, $g(7) = 0.22$, $g(8) = 0.4$, $g(9) = 0.02$
$D + DF(v) = D_2(v') + F$	$k(v') = 10^{13} \exp(-500/RT)$, $v' = v - 4$, $v = 5, \ldots, 9$
$F_2 + M = 2F + M$	$k^M = 5 \times 10^{13} \, A_M \exp(-35\,100/RT)$, $A_{Ar} = A_{D_2} = A_{DF} = 1$, $A_{F_2} = 2.7$, $A_D = 3$, $A_F = 10$
$D_2 + M = 2D + M$	$k^{Ar} = 1.45 \times 10^{14} \exp(-93\,000/RT)$
$DF(v) + M = D + F + M$	$k^M(v) = [1.1 A_M/(n+1)] \times 10^{19} \, T^{-1}$ $\times \exp[(-137\,130 + E_v - E_0)/RT]$. $A_D = A_F = A_{DF} = 5$, $A_M = 1$ for all other M's, $v = 0, 1, \ldots, n$
$2D + M = D_2 + M$	$k_{Ar} = 10^{18} \, T^{-1}$, $k_D = 3 \times 10^{17} \, T^{-1/2}$, $k_{D_2} = 10^{17} \, T^{-0.67}$ $k_M = k_{Ar}$ for all other M's except for D, D_2
$2F + M = F_2 + M$	$k_M = 1 \times 10^{18} \, T^{-1} \, A_M$, $A_{F_2} = 1$, $k_{Ar} = 1.3$, $A_{He} = 0.7$, $A_{CO_2} = 10$, $A_F = 40$, $A_D = 20$, $A_{DF} = 2$, $A_{D_2} = 1$
$D + F + M = DF + M$	$k_M = 10^{19} \, T^{-1}$ for all M's
$D + O_2 + M = DO_2 + M$ [b]	$k_M = 2.2 \times 10^{18} \, T^{-1} \, A_M$, $A_{He} = A_{Ar} = 1$, $A_{D_2} = 2$, $A_{CO_2} = 2.5$, $A_{DF} = 7$, $A_M = 1$ for all other M's
$F + O_2 + M = FO_2 + M$ [b]	$k_M = 1.6 \times 10^{20} \, T^{-2} \, A_M$, $A_{F_2} = 1$, $A_{Ar} = 1.4$, $A_{He} = 1.2$, $A_{CO_2} = A_{O_2} = 3$, $A_{DF} = 10-10^3$, $A_M = 1$ for all other M's
$NO + F_2 = NOF + F$	$k = 3.3 \times 10 \exp(-1500/RT)$
$NOF + D = NO + DF(v)$	$k = 1.6 \times 10^{13} \, g(v) \exp(-2200/RT)$, $g(0) = 0.04$, $g(1) = 0.08$, $g(2) = 0.13$, $g(3) = 0.2$, $g(4) = 0.31$, $g(5) = 0.24$

Table 3.9 (continued)

Process	Rate constant
$DF(v) + M = DF(v-1) + M$	$k_{DF}(v) = g(v)(8 \times 10^{14} T^{-1.3} + 1.1 \times 10^4 T^{2.37})$, $g(1) = 1$, $g(2) = 6$, $g(3) = 12$, $g(4) = 20$, $g(5) = 35$, $g(6) = 60$; $k_M(v) = 7 \times 10^{-5} T^{4.3} v$, $M = Ar$, F_2; $k_{He}(v) = 4 \times 10^{-6} T^{4.7} v$; $k_{HF}(v) = g(v)(5.2 \times 10^{14} T^{-1.2} + 1.35 \, 10^2 T^3)$, $g(v) = v^{2.2}$; $k_{D_2}(v) = 56 T^3 v$; $k_{FNO}(v) = 3.5 \times 10^2 T^{2.2} v$; $k_{NO}(v) = 1.6 \times 10^{-4} T^{4.7} v$; $k_{CO_2}(v) = 3.9 \times 10^3 T^{2.2} v$; $k_{N_2}(v) = 2.3 \times 10^3 T^{2.2} v$
$DF(v) + D = DF(v') + D$	$k_D(v) = 5 \times 10^{11} g(v, v') \, 5 \times 10^{11} \exp(-2000/RT)$, $v = 1, 2, \ldots, 6$, $v' = 0, 1, \ldots, v-1$, $g(1, 0) = 1$, all other $g(v, v') = 20$
$DF(v) + F = DF(v') + F$	$k_F(v) = 4 \times 10^{12} [v/(v-v')] \exp(-3200/RT)$, $v = 1, 2, \ldots, 6$, $v' = 0, 1, \ldots, v-1$
$D_2(v) + M = D_2(v-1) + M$	$k_M(v) = 1.5 \times 10^{-6} T^{5.33} v A_M \exp(526/RT)$ $A_{D_2} = 1$, $A_F = A_{F_2} = A_{Ar} = 0.2$
$DF(v) + DF(v') = DF(v+1) + DF(v'-1)$	$k(v, v'; v+1, v'-1) = 2^{v'-v} \times 6 \times 10^{15} T^{-1}$
$DF(v) + D_2(v') = DF(v+1) + D_2(v'-1)$	$k(v, v'; v+1, v'-1) = 5 \times 10^{11} (v+1)$
$DF(v) + CO_2(00^00) = DF(v-1) + CO_2(00^01)$	$k(v) = 3.3 \times 10^{13} T^{-0.4} g(v)$, $v = 1, 2, \ldots, 9$, $g(1) = 1$, $g(2) = 1.73$, $g(3) = 4$, $g(4) = g(5)$ $= g(6) = g(7) = g(8) = g(9) = 5.52$
$D_2(v) + CO_2(00^00) = D_2(v-1) + CO_2(00^01)$	$k(v) = 5.4 \times 10^8 Tv \exp(100/RT)$, $v = 1, 2$
$CO_2(01^10) + M = CO_2(00^00) + M$	$k_{M_1} = 1.5 \times 10^2 T^{2.9} \exp(200/RT)^{(c)}$, $k_{M_2} = 1.25 \times 10^9 T \exp(680/RT)^{(c)}$, $k_{He} = 2.2 \times 10^6 T^{1.8}$
$CO_2(02^20) + M = CO_2(01^10) + M$	$k_{M_1} = 3 \times 10^2 T^{2.9} \exp(200/RT)$, $k_{M_2} = 2.5 \times 10^9 T \exp(680/RT)$, $k_{He} = 4.4 \times 10^6 T^{1.8}$
$CO_2(03^30) + M = CO_2(02^20) + M$	$k_{M_1} = 4.5 \times 10^2 T^{2.9} \exp(200/RT)$, $k_{M_2} = 3.8 \times 10^9 T \exp(680/RT)$, $k_{He} = 6.6 \times 10^6 T^{1.8}$
$CO_2(02^00) + M = CO_2(01^10) + M$	$k_{M_1} = 8.8 \times 10^2 T^{2.9} \exp(400/RT)$ $k_{M_2} = 3.6 \times 10^9 T \exp(220/RT)$, $k_{He} = 3.6 \times 10^6 T^{1.8}$
$CO_2(10^00) + M = CO_2(01^10) + M$	$k_{M_1} = 1.1 \times 10^2 T^{3.1} \exp(300/RT)$, $k_{M_2} = 1.9 \times 10^8 T^{1.3} \exp(320/RT)$, $k_{He} = 1.5 \times 10^6 T^{1.8}$
$CO_2(11^10) + M = CO_2(10^00) + M$	$k_{M_1} = 2.4 \times 10^2 T^{2.9} \exp(200/RT)$, $k_{M_2} = 2 \times 10^9 T \exp(680/RT)$, $k_{He} = 3.5 \times 10^6 T^{1.8}$

Table 3.9 (continued)

Process	Rate constant
$CO_2(11^10) + M = CO_2(02^00) + M$	$k_{M_1} = 0.17\,T^{3.4}\exp(2100/RT)$, $k_{M_2} = 3.4 \times 10^8\,T\exp(680/RT)$, $k_{He} = 5.9 \times 10^6\,T^{1.8}$
$CO_2(03^10) + M = CO_2(10^00) + M$	$k_{M_1} = 0.5\,T^{3.5}\exp(1800/RT)$, $k_{M_2} = 6 \times 10^8\,T\exp(680/RT)$, $k_{He} = 1.1 \times 10^6\,T^{1.8}$
$CO_2(03^10) + M = CO_2(02^00) + M$	$k_{M_1} = 3.6 \times 10^2\,T^{2.9}\exp(200/RT)$, $k_{M_2} = 3 \times 10^9\,T\exp(680/RT)$, $k_{He} = 5.3 \times 10^6\,T^{1.8}$
$CO_2(10^00) + M = CO_2(00^00) + M$	$k_{M_3} = 3.1 \times 10^{-7}\,T^{5.3}\exp(1700/RT)^{(c)}$, $k_{M_4} = 6 \times 10^2\,T^{2.8}\exp(970/RT)^{(c)}$
$CO_2(11^10) + M = CO_2(01^10) + M$	$k_{M_3} = 2.8 \times 10^{-7}\,T^{5.3}\exp(1700/RT)$, $k_{M_4} = 1.2 \times 10^5\,T^{2.1}$
$CO_2(03^10) + M = CO_2(01^10) + M$	$k_{M_3} = 2.2 \times 10^{-5}\,T^{4.8}\exp(1800/RT)$, $k_{M_4} = 2.6 \times 10^5\,T^{2.1}$
$CO_2(10^00) + M = CO_2(02^00) + M$	$k_{M_5} = 3.9 \times 10^8\,T^{1.5(c)}$
$CO_2(10^00) + M = CO_2(02^20) + M$	$k_{M_5} = 7.5 \times 10^8\,T^{1.5}$
$CO_2(02^20) + M = CO_2(02^00) + M$	$k_{M_5} = 8.7 \times 10^8\,T^{1.5}$
$CO_2(11^10) + M = CO_2(03^10) + M$	$k_{M_5} = 4.3 \times 10^8\,T^{1.5}$
$CO_2(11^10) + M = CO_2(03^30) + M$	$k_{M_5} = 4.2 \times 10^8\,T^{1.5}$
$CO_2(03^30) + M = CO_2(03^10) + M$	$k_{M_5} = 6.1 \times 10^8\,T^{1.5}$
$CO_2(00^01) + M = CO_2(11^10) + M$	$k_{M_6} = 0.61\,T^{3.7}\exp(640/RT)^{(c)}$ $k_{DF} = 1.45 \times 10^{14}\,T^{-1(d)}$
$CO_2(00^01) + M = CO_2(03^10) + M$	$k_{M_6} = 1.2 \times 10^{-11}\,T^{6.85}\exp(3080/RT)$ $k_{DF} = 1.1 \times 10^{11}\,T^{-0.2}$ $k_{F_2} = 3.7 \times 10^{10(e)}$
$CO_2(00^01) + M = CO_2(10^00) + M$	$k_{M_6} = 1.5 \times 10^{-16}\,T^{8.63}\exp(3940/RT)$ $k_{HF} = 5 \times 10^{11}\,T^{0.05}\exp(100/RT)$
$CO_2(00^01) + M = CO_2(02^00) + M$	$k_{M_6} = 8.6 \times 10^{-14}\,T^{7.74}\exp(1560/RT)$
$CO_2(00^01) + M = CO_2(01^10) + M$	$k_{M_6} = 1.4 \times 10^8\,T^{1.04}\exp(1320/RT)$
$N_2(v) + M = N_2(v-1) + M$	$k_{M_7} = 23\,T^{3.3}\exp(700/RT)^{(c)}$ $k_{M_8} = 1.2 \times 10^8\exp(-12.3/RT)^{(c)}$
$CO_2(00^00) + CO_2(00^01) =$ $CO_2(10^00) + CO_2(01^10)$	$k = 7.3 \times 10^{-3}\,T^{4.35}\exp(1200/RT)$
$CO_2(00^00) + CO_2(10^00) =$ $CO_2(01^10) + CO_2(01^10)$	$k = 10^6\,T^{1.4}\exp(200/RT)$
$CO_2(00^00) + CO_2(02^20) =$ $CO_2(01^10) + CO_2(01^10)$	$k = 10^8\,T^{1.4}\exp(200/RT)$
$CO_2(00^00) + CO_2(11^10) =$ $CO_2(01^10) + CO_2(02^00)$	$k = 1.6 \times 10^8\,T^{1.4}\exp(200/RT)$
$CO_2(00^00) + CO_2(11^10) =$ $CO_2(01^10) + CO_2(10^00)$	$k = 3 \times 10^6\,T^{1.4}\exp(200/RT)$

Table 3.9 (continued)

Process	Rate constant
$CO_2(00^00) + CO_2(11^10) =$ $\quad CO_2(01^10) + CO_2(02^20)$	$k = 9.1 \times 10^7 \, T^{1.4} \exp(200/RT)$
$CO_2(00^00) + CO_2(02^00) =$ $\quad CO_2(01^10) + CO_2(01^10)$	$k = 10^8 \, T^{1.4} \exp(200/RT)$
$CO_2(00^00) + CO_2(03^30) =$ $\quad CO_2(01^10) + CO_2(02^20)$	$k = 1.5 \times 10^8 \, T^{1.4} \exp(200/RT)$
$CO_2(01^10) + CO_2(02^00) =$ $\quad CO_2(02^20) + CO_2(01^10)$	$k = 2 \times 10^8 \, T^{1.4} \exp(200/RT)$
$CO_2(03^10) + CO_2(01^10) =$ $\quad CO_2(02^20) + CO_2(02^20)$	$k = 3 \times 10^8 \, T^{1.4} \exp(200/RT)$
$CO_2(03^10) + CO_2(01^10) =$ $\quad CO_2(02^00) + CO_2(02^00)$	$k = 4.2 \times 10^8 \, T^{1.4} \exp(200/RT)$
$CO_2(03^10) + CO_2(01^10) =$ $\quad CO_2(02^20) + CO_2(02^00)$	$k = 3 \times 10^8 \, T^{1.4} \exp(200/RT)$
$CO_2(03^30) + CO_2(01^10) =$ $\quad CO_2(02^20) + CO_2(02^20)$	$k = 3 \times 10^8 \, T^{1.4} \exp(200/RT)$
$CO_2(03^30) + CO_2(01^10) =$ $\quad CO_2(02^20) + CO_2(02^20)$	$k = 3 \times 10^8 \, T^{1.4} \exp(200/RT)$
$CO_2(11^10) + CO_2(01^10) =$ $\quad CO_2(02^20) + CO_2(10^00)$	$k = 10^8 \, T^{1.4} \exp(200/RT)$
$CO_2(11^10) + CO_2(01^10) =$ $\quad CO_2(02^00) + CO_2(10^00)$	$k = 10^8 \, T^{1.4} \exp(200/RT)$

Notes: [a] the kinetic constant values listed in the table are borrowed from [3.112] for processes occurring in the D_2–F_2 system and from [3.113] for relaxation processes taking place in CO_2 and for some chemical and relaxation processes in DF not considered in [3.112]; the kinetic constants are measured in [cm^3/mole s]; the rate constant for the inverse relaxation process is expressed in terms of the constant listed in the table and the equilibrium constant; the table lists the rate constants of all the chemical processes (forward and back) considered; the temperature T is measured in kelvins, and R = 1.987 cal/mole deg.;

[b] taken on the basis of the analysis of the data of Table 3.1;

[c] $M_1 = CO_2 + F_2 + F + NO + 0.7N_2 + 0.06Ar$;
    $M_2 = D_2 + 12D + 0.6DF + 0.9FNO + 0.6HF$;
    $M_3 = CO_2 + 4N_2 + F_2 + F$;
    $M_4 = DF + 0.9O_2 + 18D + 0.3He$;
    $M_5 = CO_2 + 0.5N_2 + 0.3DF + 0.3HF + 0.5He$ (the effectiveness of He is taken on the basis of the data of [3.114]; according to [3.115], the effectiveness of He in this process is not very much lower than that of CO_2 and N_2: $k_{He} = (0.8 \pm 0.3) \times 10^5 s^{-1}$ (mm Hg)$^{-1}$, $k_{CO_2} = k_{N_2} = (3 \pm 1) \times 10^5 s^{-1}$ (mm Hg)$^{-1}$);
    $M_6 = CO_2 + 0.4N_2 + 2NO + 0.2He + D_2 + 160FNO$;
    $M_7 = DF + 2HF$;
    $M_8 = CO_2 + N_2 + NO + FNO + D_2$;

[d] borrowed from [3.116];

[e] according to the data of [3.117] at $T = 300$ K; the relaxation channel is singled out here conventionally.

Table 3.10. Effect of initiation duration on characteristics of the DF–CO$_2$ laser

t_0 [s]	ε_l [J/l]	t_l [s]
10^{-8}	126	10^{-6}
10^{-7}	128	1.2×10^{-6}
10^{-6}	140	2.6×10^{-6}
10^{-5}	132	9×10^{-6}

Notes: $\Delta[F_2]/[F_2] = \gamma_{F_2} t_0^2 = 3\%$; $p = 1$ atm.; $T_0 = 300$ K; D$_2$: F$_2$: CO$_2$: He $= 1:1:4:5$.

Table 3.11. Effect of the initial laser medium temperature on characteristics of the DF–CO$_2$ laser

Mixture	T_0 [K]	ε_l [J/l]	t_l [s]
I	200	147	2×10^{-6}
	300	135	2×10^{-6}
II	200	146	1×10^{-5}
	300	88	8×10^{-6}

Notes: I – D$_2$: F$_2$: CO$_2$: He $= 1:1:4:5$, $p = 1$ atm., $\Delta[F_2]/[F_2] = 3\%$; II – D$_2$: F$_2$: CO$_2$: He $= 1:3.7:4:3.7$, $p = 1$ atm., $\Delta[F_2]/[F_2] = 0.025\%$.

Table 3.12. Effect of reactant concentrations on characteristics of the DF–CO$_2$ laser

Mixture composition: D$_2$: F$_2$: CO$_2$: He, mm Hg	Initiation level			
	$\Delta[F_2]/[F_2] = 1\%$		$\Delta[F_2]/[F_2] = 0.1\%$	
	ε_l [J/l]	t_l [s]	ε_l [J/l]	t_l [s]
50 : 50 : 200 : 360	100	5.4×10^{-6}	36	2.3×10^{-5}
70 : 70 : 270 : 340	141	3.3×10^{-6}	70	1.25×10^{-5}
140 : 140 : 270 : 200	130	1.4×10^{-6}	137	5.3×10^{-6}

Table 3.13. Effect of the CO$_2$ and He content of the mixture on characteristics of the DF–CO$_2$ laser

Relative mixture composition D$_2$: F$_2$: CO$_2$: He	ε_l [J/l]	t_l [s]
1 : 1 : 2 : 7	85.4	1.8×10^{-6}
1 : 1 : 3 : 6	117	1.9×10^{-6}
1 : 1 : 4 : 5	137	2×10^{-6}
1 : 1 : 5 : 4	151	2.2×10^{-6}
1 : 1 : 6 : 3	158	2.3×10^{-6}
1 : 1 : 7 : 2	160	2.4×10^{-6}
1 : 1 : 8 : 1	92.6	2.5×10^{-6}

Notes: $T_0 = 300$ K, $p = 1$ atm., $\Delta[F_2]/[F_2] = 3\%$.

Table 3.14. Effect of buffer gas pressure on characteristics of the DF–CO_2 laser

n	ε_l [J/l]	t_l [s]
5	137	0.8×10^{-6}
10	164	2.3×10^{-6}
15	188	2.5×10^{-6}
20	207	2.8×10^{-6}
25	222	3×10^{-6}

Notes: $D_2:F_2:CO_2:He = 1:1:4:n$, $T_0 = 300$ K, $p = 1$ atm at $n = 5$, $\Delta[F_2]/[F_2] = 3\%$.

Table 3.15. Effect of the total mixture pressure on characteristics of the DF–CO_2 laser

Mixture	p [atm.]	t_l [s]	ε_l [J/l]	f
(a)	1	1.2×10^{-6}	85	121
	3	10^{-6}	234	333
	10	10^{-6}	740	1050
(b)	1	1.9×10^{-5}	49	1160
	5	1.1×10^{-5}	61	1460
	10	8×10^{-6}	38	912
	25	5.6×10^{-6}	4	97

Notes: [a] $D_2:F_2:CO_2:He = 1:3.8:3:4.8$, $n_a = 3.7 \times 10^{16}$ cm^{-3}, $t_0 = 10^{-7}$ s.
[b] $D_2:F_2:CO_2:He = 1:1:4:5$, $n_a = 2.2 \times 10^{15}$ cm^{-3}, $t_0 = 0.5 \times 10^{-6}$ s.

Table 3.16. Effect of the total pressure and CO_2 content of the mixture on characteristics of the DF–CO_2 laser as a function of the initiation level

(a) $D_2:F_2:CO_2:He = 61:228:182:289$ mm Hg

$\Delta[F_2]/[F_2]$	2.5×10^{-5}	2.5×10^{-4}	2.5×10^{-3}
n_F [cm^{-3}]	3.68×10^{14}	3.68×10^{15}	3.68×10^{16}
ε_l [J/l]	26.6	95	85
f	3800	1360	122

(b) $D_2:F_2:CO_2:He = 610:2280:P_{CO_2}:2280$ mm Hg

P_{CO_2} [mm Hg]	2432	1216	2432	2432	2432
$\Delta[F_2]/[F_2]$	2.5×10^{-5}	2.5×10^{-5}	2.5×10^{-5}	2.5×10^{-4}	2.5×10^{-3}
n_F [cm^{-3}]	3.68×10^{15}	3.68×10^{15}	3.68×10^{15}	3.68×10^{16}	3.68×10^{17}
t_l [µs]	3.7	4.5	3.3	1.1	0.24
ε_l [J/l]	176	323	333	737	494
η_{phys} [%]:					
EBI*	2550	4600	4750	1050	71
UVI**	15200	28000	28000	6350	430
f	2520	4600	4750	1050	71

*EBI – electron-beam initiation.
**UVI – ultraviolet initiation.

Table 3.17. Comparison between characteristics of lasers operating on the H_2–F_2 and D_2–F_2–CO_2 mixtures

Laser mixture composition and characteristic		Initiation level, $\Delta[F_2]/[F_2][\%]$		
		0.1	1	10
$H_2:F_2:O_2:He = 50:50:40:620$ mm Hg	$\varepsilon_l[J/l]$	1.5	27	370
	$t_l[\mu s]$	1.3	1.97	0.6
$D_2:F_2:CO_2:He = 50:50:200:360$ mm Hg	$\varepsilon_l[J/l]$	36	100	96
	$t_l[\mu s]$	23	5.4	1.1

Table 3.18. DF–CO_2 laser characteristics under energetic chain-branching conditions

Mixture composition and pressure	$D_2:F_2:CO_2:He = 1:3.8:0.4:7$ mm Hg $p = 760$ mm Hg, $p_{DF}(0) = 3$ mm Hg		

(a) Infrared initiation

ε_{DF}^*	1	0.2	0.05
$\varepsilon_l^{**}[J/l]$	5.54	1.11	2.77×10^{-1}
$\varepsilon_l[J/l]$	36	37	36.6
$t_l[s]$	3.8×10^{-4}	9×10^{-4}	1.6×10^{-3}

(b) Ultraviolet initiation

$n_F[cm^{-3}]$	3×10^{13}	3×10^{12}
$\varepsilon_i[J/l]$	0.96×10^{-2}	0.96×10^{-3}
$\varepsilon_l[J/l]$	42.7	28.9
$t_l[s]$	1.2×10^{-4}	3.1×10^{-4}
$\eta_{phys}[\%]$	0.445×10^6	0.3×10^7
f	0.74×10^5	0.5×10^6

* ε_{DF} is the number of quanta per DF molecule excited by IR radiation.
** ε_i is the initiation energy.

4. Pulsed Chemical Lasers

Pulsed chemical lasers (PCL's) occupy a special position compared to other types of pulsed lasers whose pumping is effected exclusively by an external source of energy. Pumping in PCL's takes place on account of the internal chemical energy liberated in the course of a chemical reaction. But while the laser mixture is being prepared, it must be stabilized by some means or other. Hence to disturb it from the equilibrium condition, and to initiate the reaction one has to expend energy from an external source to produce chemically active centers (atoms or radicals) in the mixture. The main goal here is to minimize the external energy spent on initiating the reaction.

The use of the hydrogen fluorination chain reaction for which the rate of the chain chemical pumping process is much higher than that of deactivation processes has helped to make a certain progress in this direction. The basic feature of this reaction is that the number of vibrationally excited molecules produced in its course is much greater than that of the priming atoms formed during initiation. This circumstance offers opportunities for reaching high engineering efficiencies in lasers relying on this reaction. What is more, the fact that the laser energy can be drawn from the energy stored in the active F_2–H_2 mixture allows one to expect to obtain high laser energies per unit active medium volume, for the demands on the specific energies to be deposited in a dense gas medium by the external initiation source in this case prove less stringent. The emphasis in this chapter will therefore be on the lasers operating on the hydrogen fluorination chain reaction, though we will also discuss pulsed chemical lasers based on other reactions.

Inasmuch as the initiation system is an important part of a PCL, we will consider the physical processes used to produce active centers in the laser medium with various initiation techniques: flash photolysis, electrical discharge, electron-beam, and electroionization. We will also dwell upon the requirements for the intensity of initiation sources of various types and the limitations due to the physical principles underlying the initiation techniques used. We will compare the above initiation techniques with a view to establishing their relative advantages and shortcomings.

The quality of the active medium is an important factor affecting the performance of the lasers based on the hydrogen-fluorine chain reaction. For this reason, we will devote an individual section to the problems involved in the preparation of the laser mixture.

Much attention in this chapter will be given to the basic laws governing the operation of PCL's relying on the hydrogen-fluorine chain reaction. We will consider their spectral and temporal emission characteristics and the effect of various factors on their energy performance and discuss their operation under amplification conditions and when extracting laser energy not only at the fundamental frequency in transitions between adjacent vibrational levels, but also at overtones, in transitions proceeding in accordance with the selection rules $\Delta v \geqslant 2$.

On the whole, the chapter shows the achievements in the basic energy characteristics of pulsed hydrogen-fluoride lasers as of the close of 1980. The results that appeared after 1980 are considered in the last section of the chapter. They have not distorted the general picture of the trends in the behavior of the basic parameters of such lasers, but have given a deeper insight into the behavior of the energy characteristics of the lasers as a function of the initiation source energy, because methods have been developed for determining the initial atomic fluorine concentration produced by various external initiation sources. It have proved possible to raise the specific energy deposited by the initiation source in the laser mixture by using beams of relativistic electrons and additions of such fluorine-containing gases that are good donors of atomic fluorine and which only weakly deactivate the excited molecules. This has made it possible to achieve record high specific energy characteristics (up to a few tenths of a joule per cubic centimeter) for hydrogen-fluoride lasers, which exceed the respective characteristics of the other types of gas laser. Optimization of the conditions of deposition of the photo-initiation source energy in the pulsed hydrogen-fluoride laser mixture has succeeded in attaining high engineering efficiencies (up to 50%) without reducing the specific laser energy output. On the whole, this progress has not disproved the understanding gained by 1980, but on the contrary, has substantiated the trends revealed.

4.1 Requirements for the Parameters of Chemically-Pumped Pulsed Laser Systems

The purpose of creating pulsed lasers (chemical lasers included) is to obtain laser emissions of very high power within relatively short periods of time. This implies the emission of a comparatively large amount of energy during the short time the laser pulse lasts, hence the logical conclusion that the reaction rate in chemical lasers should also be very high. To minimize the size of the laser system requires that the laser energy output per unit laser medium volume should be as high as possible, which is only attainable at comparatively high reactant pressures. The problem of developing a pulsed chemical laser can therefore be formulated as follows: to prepare within a very short period of time a rapidly reacting mixture at a comparatively high pressure. Obviously the mixture preparation time must be of the order of or shorter than the required laser pulse duration. Let it be $t_l \simeq 10^{-4}$ s. To prepare in so short a time and in a reasonable volume a mixture with partial reactant pressures even as low as a few mm Hg is impossible. Thus, the problem as stated above cannot be solved. But it can be solved by preparing a stable reactant mixture and then initiating the reaction therein.

As already noted in Chap. 2, population inversion can only be produced as a result of fast reactions involving free atoms or chemically active radicals. To form chemically active centers in the laser mixture (reaction initiation) requires expenditure of energy from some external source. Obviously the lower the ratio between the energy spent to initiate the reaction in a given chemical laser and its output energy, the more interesting is the laser. Hence the desire to use in pulsed chemical lasers chain reactions with long effective (laser) chain lengths, with each chemically active center involved in a great number (equal to ν_{eff}) of transformations contributing to laser emission.

Note once more the difference between the chemical and laser chain lengths. The laser chain length is defined by relations of the form of (2.35c) and (2.40c), which include the ratio of the chain propagation rate to the lasing molecule relaxation rate. The hydrogen-chlorine and hydrogen-fluorine reactions are both chain reactions with sufficiently great chemical chain lengths. But the ratio between the chain propagation rate and the relaxation rate of the excited molecules in the H_2-F_2 mixture is much higher than in the H_2-Cl_2 mixture. From the standpoint of the laser chain length, it actually turns out that the hydrogen-fluorine reaction is a chain type ($\nu_{eff} > 1$), whereas the hydrogen-chlorine reaction is not a chain type ($\nu_{eff} < 1$). That is why lasers operating on the H_2-Cl_2 mixture have not attracted widespread attention, despite the fact that the first experimental chemical laser used precisely this mixture [4.1].

The $D_2-F_2-CO_2-He$ mixture possesses even a greater laser chain length compared to the H_2-F_2 mixture. The emitters in the former are the CO_2 molecules whose relaxation rate is slower than that of the HF molecules which are the emitters in the H_2-F_2 mixture. The chain propagation rates in these two mixtures are approximately equal. There is another mixture, $H_2 (D_2)-ClF$, whose laser chain length appreciably exceeds unity.

The detailed numerical calculations discussed in Chap. 3 and experimental investigations show that the effective (laser) chain length reaches about 30–100 for the H_2-F_2 system, 100–400 for $D_2-F_2-CO_2$, and 10 for H_2-ClF. The reactions taking place in the other mixtures used in chemical lasers are not a chain type as far as the lasing process is concerned.

It is clear from the above that a pulsed chemical laser must consist of a reactor into which the reactant mixture is injected and an initiation device which starts the reaction by producing chemically active centers in the mixture. The reactor must be provided with windows to couple out radiation, or with shutters in the case of a windowless approach [4.2, 3], and placed in an optical cavity, which in the simplest case comprises two mirrors. Where it is necessary to neutralize the exhausted gases, use is made of various traps (liquid nitrogen, carbon, etc.). Such a laser is illustrated schematically in Fig. 4.1.

The reaction can be initiated either photolytically or by electron impact. What are the conditions that the initiation device must satisfy? The answer to this question can be found by analyzing the basic kinetic laws governing the evolution of the chain process in the chemical laser.

It follows from the analysis performed in Sect. 2.1 that

1) there exists a threshold reaction rate, hence a threshold concentration of chemically active centers (atoms, radicals), below which the laser will not be excited (see formulas (2.9) and (2.13));

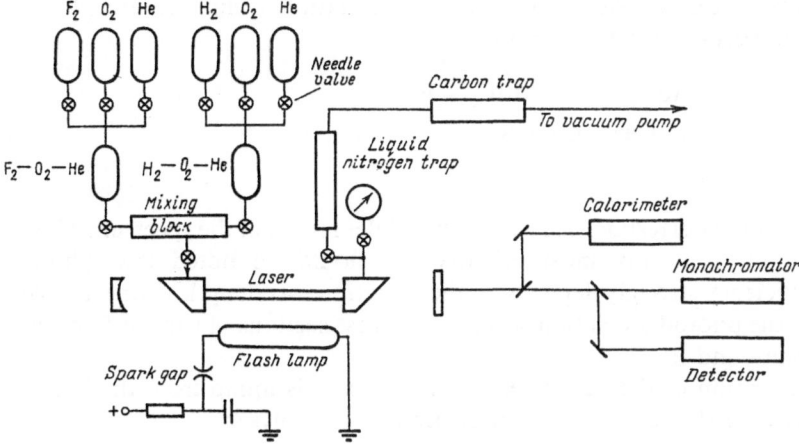

Fig. 4.1. Schematic diagram of a flash-photolysis initiated H_2–F_2 pulsed chemical laser

2) the laser pulse duration is inversely proportional to the total laser mixture pressure (see (2.35a));
3) the specific energy output of a chemical laser is directly proportional to the mixture pressure (see (2.35b) and (2.35c)) and is related to the laser pulse duration so that increasing the specific energy output reduces the pulse duration.

The above conclusions enable one to determine the requirements that the initiation source and the laser mixture pressure must meet for the laser to have a specified energy output. Here we can draw on the results of the analysis carried out in Chap. 3 for the energy characteristics of pulsed chemical lasers using the hydrogen fluorination chain reaction. Investigations into such lasers have attracted particular interest because of their high engineering efficiency. It follows from Table 3.2 that for an H_2–F_2 mixture at atmospheric pressure, a specific energy output of around 50–100 J/1 atm. is reached at an initiating pulse duration of about 2–3 μs and an initial active center concentration of $(3 \text{ to } 6) \times 10^{15}$ cm^{-3}. For the D_2–F_2–CO_2 mixture, the initiating pulse duration can be two or three times as long, the active center concentration and specific energy output being the same.

4.2 Initiation of Pulsed Chemical Lasers

4.2.1 Initiation Techniques

In this section, we will consider the physical processes used to produce active centers (atoms in the present case) in the active medium of PCL's with various initiation techniques. Attention will also be given to the requirements for the intensity of initiation sources of various types, and the limitations due to the physical principles underlying the initiation techniques used.

Flash-Photolysis Initiation. At the root of this initiation technique are photo-dissociation processes of the following type:

$$Cl_2 + h\nu \rightarrow 2Cl , \quad MoF_6 + h\nu \rightarrow MoF_5 + F ,$$
$$F_2 + h\nu \rightarrow 2F , \quad CS_2 + h\nu \rightarrow CS + S ,$$
$$O_3 + h\nu \rightarrow O(^1D) + O_2 .$$

Here $h\nu$ is a quantum corresponding to the ultraviolet region of the spectrum where photodissociation is usually most effective. It should be noted that photo-dissociation efficiency strongly depends on the radiation wavelength. (For example, Fig. 4.2 shows the photodissociation spectra and cross sections of various fluorine-containing compounds.)

The concentration of the active centers in this case is found from the kinetic equation describing the rate of their production by photolysis:

$$\frac{dN}{dt} = \Phi \sigma I_0 N_0 , \tag{4.1}$$

where N is the active center concentration, Φ the quantum yield of photo-dissociation, I_0 the number of photons emitted per unit surface area of the source per unit time, σ the photodissociation cross section, and N_0 the concentration of the molecules being dissociated. If the photon flux from the source remains constant with time and the operating time of the source is τ_i, we have from (4.1) the following expression for the required photon flux intensity:

$$I_0 = N/\Phi \sigma N_0 \tau_i . \tag{4.2}$$

Hence we can estimate the range of photon flux intensities necessary to produce the required atomic fluorine concentration $[F] \simeq 10^{-15}-10^{16}$ cm^{-3} by the photo-dissociation of F_2 ($\sigma_{eff} \simeq 10^{-20}$ cm^2, $\Phi \simeq 2$, $[F_2] \simeq 5 \times 10^{18}$ cm^{-3}) during the characteristic times the laser pulse lasts, $t_i \simeq (1$ to $5) \times 10^{-6}$ s:

$$I_0 = (0.4 \text{ to } 20) \times 10^{22} \text{ photons/cm}^2 \text{ s} .$$

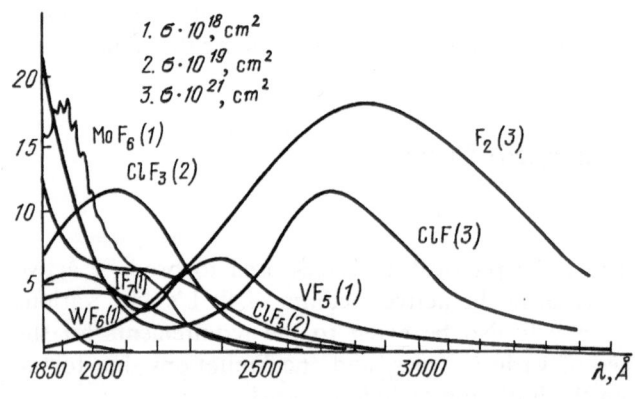

Fig. 4.2. Photodissociation spectra of F_2, ClF, ClF$_3$, ClF$_5$, VF$_5$, WF$_6$, MoF$_6$, and IF$_7$ [4.4]

Electron–Impact Initiation. The decomposition of molecules by electron impact can proceed by a number of mechanisms differing in nature. The complicated sequence of processes occurring in the active medium of electron-impact initiated hydrogen-fluoride lasers may be exemplified by the various mechanisms of formation of atomic fluorine and hydrogen:

$$F_2 + e \rightarrow F + F^- , \qquad \Delta H = -1.9 \text{ eV} , \tag{4.3}$$

$$SF_6 + e + SF_5^- + F , \tag{4.4}$$

$$SF_6 + e \rightarrow SF_5 + F + e , \tag{4.4'}$$

$$SF_6 + e \rightarrow SF_5^+ + F + 2e , \tag{4.4''}$$

$$He + e \rightarrow He^* + e , \qquad \Delta H = 19.7 \text{ eV} ,$$

$$He^* + F_2 \rightarrow He + 2F , \qquad \Delta H = -18.2 \text{ eV} ,$$

$$H_2 + e \rightarrow 2H + e , \qquad \Delta H = 8.9 \text{ eV} , \tag{4.5}$$

$$H_2 + e \rightarrow H_2^- , \qquad \Delta H = 3.9 \text{ eV} ,$$

$$F_2 + e \rightarrow F^- + F^+ + e , \tag{4.6}$$

$$F_2 + e \rightarrow F_2^+ + 2e , \tag{4.7}$$

$$Ar + e \rightarrow Ar^+ + 2e . \tag{4.8}$$

Figure 4.3 shows the cross sections of processes (4.3–8) as a function of the electron energy E_e. *De Corpo* et al. [4.5] have measured relative cross-sections for F_2, but there are sufficient data to enable us to estimate the absolute values of σ_{F_2}. Proceeding from the rate constants k for the process of dissociative electron capture by fluorine (4.3) found experimentally for electrons with an energy of $E_e \simeq 3 \times 10^{-2}$ eV ($k \simeq 3.1 \times 10^{-9}$ cm^3 s^{-1}) [4.6] and $E_e \simeq 1$ eV ($k \simeq 2.3 \times 10^{-9}$ cm^3 s^{-1}) [4.7], we can estimate the cross section σ_{F_2} of this process to be about (1 to 5)

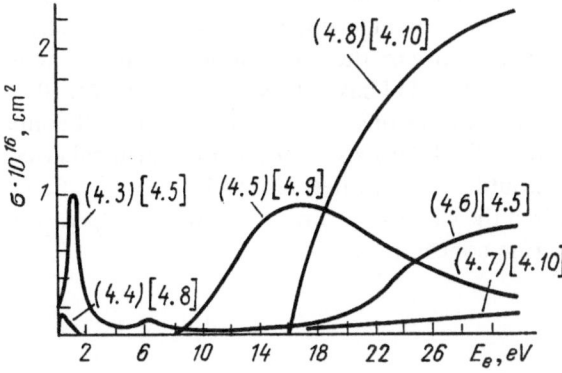

Fig. 4.3. Cross sections of various processes occurring in HF laser mixtures under the effect of electron impact

$\times 10^{16}$ cm^2 for electrons with an energy of 1.5–2 eV. The cross sections of this process for other fluorine-containing gases are of the same order of magnitude. In addition to the above processes, atomic fluorine is also formed as a result of the reaction

$$F_2^+ + F^- = 3F \tag{4.9}$$

which goes fast enough, thanks to its high rate constant ($k \simeq 10^{-8}$ cm^3 s^{-1} [4.7]).

There are two ways to produce free electrons in the active medium of a PCL: (1) to generate secondary electrons by ionizing the gas with a beam of fast electrons and (2) to generate free electrons by effecting an electrical discharge in the gas. The physical principles underlying these two techniques for initiating PCL's is considered in more detail in the following paragraphs.

It can be inferred from the analysis of the above elementary-process scheme that when initiating the H$_2$–F$_2$ mixture with a beam of fast electrons, atomic fluorine is formed mainly as a result of processes (4.3) and (4.9). The characteristic time of process (4.3) is of the order of 10^{-10} s (at a molecular fluorine concentration of around 10^{19} cm^{-3}) and that of process (4.9), 10^{-7} s (even at a secondary electron concentration as low as around 10^{15} cm^{-3}), both times are much shorter than the characteristic laser pulse duration under te same conditions (around 3 µs). Consequently, the active center concentration can be considered to be four times the secondary electron concentration n_e:

$$[F] = 4n_e = 4qt_e , \tag{4.10}$$

where t_e is the electron-beam pulse duration and q the volume ionisation rate (in ion pairs/cm^3 s) related to the electron-beam current density J_b (in A/cm^2) by the relation [4.11]

$$q = 6.2 \times 10^{18} p_{F_2} s_{F_2} J_b . \tag{4.11}$$

Here p_{F_2} is the fluorine gas pressure (in mm Hg) and s_{F_2} (in ion pairs/cm (mm Hg)) the total number of ion pairs produced by a fast electron per unit flight distance for a fluorine gas pressure of 1 mm Hg. The quantity s_{F_2} depends on the electron energy and is defined by the relation $s_{F_2} = (dU/dx)w^{-1}$, where dU/dx is the loss per unit distance of energy of the beam-electron and w the energy used per collision to produce an ion pair in the gas. This energy is independent of the beam-electron energy and differs but slightly from gas to gas; the characteristic value of w is 25–35 eV per ion pair [4.12]. *Hofland* et al. [4.11] have measured s_{F_2} at 1.29×10^{-2} ion pairs/cm (mm Hg) for a beam of fast electrons 1.7–2 MeV in energy. If some other gas, not fluorine, constitutes the bulk of the active medium of a chemical laser, the values of p and s for this gas should be substituted into (4.11).

Using (4.10) and (4.11), we can find the electron-beam current density necessary to obtain the required atomic fluorine concentration:

$$J_b = \frac{4 \times 10^{-20} [F]}{p_{F_2} s_{F_2} t_e} .$$

Taking $[F] \simeq 10^{15}-10^{16} \text{ cm}^{-3}$, $s_{F_2} = 10^{-1}-10^{-2}$ ion pairs/cm (mm Hg)[1], $p_{F_2} = 200$ mm Hg, and $t_e \simeq 10^{-7}$ s to be typical values, we have the following range of the required electron-beam current densities: $J_b \simeq 10-10^3 \text{ A/cm}^2$.

To make similar estimates for the electrical-discharge initiation technique is very difficult because this case is characterized by an extremely complicated interplay of a variety of simultaneous processes due to the wide electron energy distribution resulting from the application of a dc field across the laser medium. The determination of the form of this distribution proves a very involved task. What is more, the literature data for the cross-sections for many of the processes are either known to a poor degree of accuracy or are altogether lacking. All this makes precise calculations difficult.

For qualitative consideration of the basic laws governing electrical discharge initiation, we write the kinetic equation for the electron concentration n_e in the following approximate form:

$$\frac{dn_e}{dt} = q + \sum_i \sigma'_e p_i n'_e v'_e - \sigma_{F_2} n_e v_e [F_2] \; , \qquad (4.12)$$

where q is the rate of electron production by an external source, σ' the cross section for an ionization process of the type (4.7) for a mixture component at a pressure of p_i, v_e and v'_e are the velocities of electrons with different energies, and n_e and n'_e the respective electron concentrations. The second term in equation (4.12) describes the production of electrons as a result of ionization by collision in processes of the type (4.7) and (4.8) and the last term characterizes the loss of electrons in the dissociative capture process (4.3). In the case of an electron-beam ionized CO_2 laser, this equation is usually supplemented with a term allowing for dissociative electron-ion recombination. But where the medium contains a highly electronegative gas, the rate of this process is negligible compared to that of dissociative electron capture.

According to the method producing free electrons, there may be two types of electrical discharge: self-sustained and non-self-sustained. In a self-sustained discharge, electrons are produced as a result of collisional ionization processes taking place when the strength of the applied electric field exceeds the breakdown value. In that case, $q = 0$ and $\sum_i \sigma'_e p_i n'_e v'_e > \sigma_{F_2} n_e v_e [F_2]$, which results naturally from the maximum of the electron energy distribution being shifted towards higher energy values, with the "hot tail" of the distribution extending up to about 20 eV.

In a non-self-sustained discharge, free electrons are generated by an external source, the main proportion of the energy deposited in the active medium being provided by a dc field with a strength below the breakdown value. Photoionization and generation of secondary electrons by means of an electron beam passing through the active medium are most widely used today for producing conduction electrons in non-self-sustained-discharge initiated lasers. The condition for a discharge to be non-self-sustained is as follows:

$$q \gg \sum_i \sigma'_e p_i n'_e v'_e \; .$$

---

[1] The s_{F_2} values in excess of about 10^{-1} ion pairs/cm (mm Hg) are typical of beams of electrons several hundred keV in energy.

In that case, the maximum of the electron energy distribution is shifted towards the low energy values and there is no "hot tail" in the distribution. The presence of a dc field can materially enhance the contribution of processes (4.4'), (4.4''), and (4.5–8) to the formation of active centers and thus improve the utilization efficiency of every conduction electron. As the average electronic energy is increased, the electron attachment rate is reduced, whereas the rates of the processes indicated above grow higher. The utilization efficiency of every conduction electron can be qualitatively evaluated from the probability w_e that a zero-energy electron will acquire in the field an energy sufficient for it to leave the energy region 0–2 eV (where the process of dissociative electron capture by fluorine proceeds most effectively) and thus will not be captured by a fluorine molecule. The rate at which an electron acquires energy in a dc field with a strength of E is estimated by the formula

$$\phi = e^2 E^2 / m_e v_e \; ,$$

where m_e and v_e are the mass and elastic collision frequency of the electron, respectively. In that case, the time it takes for the electron to leave the capture energy interval will be $(2\,\text{eV})/\phi$. Considering that the characteristic electron attachment time is $(\sigma_{F_2} v_e [F_2])^{-1}$, we may find the desired probability w_e from the ratio between these times:

$$w_e = e^2 E^2 / 2 m_e v_e \sigma_{F_2} v_e [F_2] \; . \tag{4.13}$$

At a pressure of $p = 1$ atm., $v_e \simeq 2.5 \times 10^7$ cm/s ($E_e \simeq 1$ eV), $[F_2] \simeq 10^{19}$ cm^{-3}, and $E \simeq 5$ V/cm, we have $w_e \simeq 10$. This value of the probability of utilization of a single conduction electron is much lower than that for the electron-beam ionized CO_2 laser ($w > 10^3$). Of course, it should be noted that the above calculations are estimates only. Nevertheless, they graphically demonstrate the complications arising with this chemical laser initiation technique. As yet, it is difficult to perform more rigorous and reliable calculations, for the study of the electroionization technique for initiating chemical lasers is still the subject investigations. Certainly, great interest is being expressed in the further development of this initiation technique because it allows the electron-beam current density necessary to produce free electrons to be reduced substantially.

Similar reasoning can also refer to chemical lasers operating on other mixtures (SF_6–H_2, CS_2–O_2, H_2–Cl_2, etc.). A common feature of all these lasers is the presence of highly electronegative gases in their active medium, and so the above considerations are applicable to them to some extent or other.

Thermal Initiation. Fluorine gas and some fluorine-bearing molecules effectively dissociate to yield fluorine atoms at relatively low temperatures (800–1500 K). This process can be used to initiate hydrogen-fluoride chemical lasers through thermal dissociation caused by a pulsed heating of the laser mixture to a high temperature. In that case, it is possible to utilize the internal energy of the mixture, which can be released as a result of thermal explosion. Thermal explosion is a self-sustained reaction. In this respect, it is similar to branching-chain reactions, but offers a much wider choice of suitable compounds.

At present, two methods are used to initiate thermal explosion in chemical lasers (1) photoinitiation by means of xenon-filled flashlamps and (2) detonation-wave initiation. For instance, *Jensen* and *Rice* [4.13] have effected the thermal dissociation of SF_6 and NF_3 using the thermal explosion of ClN_3 initiated with a xenon-filled flashlamp. The amount of heat liberated in the explosion proved sufficient for the initiation of SF_6–H_2 and NF_3–H_2 lasers. The best results were obtained with a $ClN_3:NF_3: H_2 = 1: 1: 2$ mixture at a pressure of $p = 12$–24 mm Hg. *Gross* et al. [4.14] have obtained laser emission from HF molecules behind the front of an overdriven detonation wave in an F_2O–H_2–Ar mixture. The stability of the reaction zone was achieved by accelerating the detonation wave to a velocity of 1.8–2 km/s which exceeded the Chapman-Jougnet critical value. A certain problem here was the high diffraction loss in the cavity due to the small size (typically much less than 1 cm) of the reaction zone governed by both the high reaction rate at a temperature in excess of 2000 K behind the wave front, and the difficulty of obtaining population inversion at such high medium temperatures.

Thermal initiation can, in principle, be realized not only in hydrogen-fluoride lasers. For example, *Akulintsev* et al. [4.15] have considered theoretically the possibility of obtaining laser emission from CO molecules behind the front of an overdriven detonation wave in a CS_2–O_2 mixture. The calculated unsaturated gain of around 10^{-2} cm^{-1} points to the feasibility of such a laser, yet it would suffer from the same drawbacks as the HF laser using the same initiation technique.

4.2.2 Engineering Aspects of Initiation

The parameters of pulsed chemical lasers are largely determined by those of the initiation sources used. In the preceding section, we have considered the requirements for the energy characteristics of such sources. To eliminate the laser efficiency loss due to the initiating pulse duration τ_i exceeding the laser pulse duration t_l, the condition $\tau_i \leqslant t_l$ should be satisfied, which sets the upper limit on the initiating pulse duration. It follows from the results presented in Chap. 3 that the initiating pulse duration should not exceed a few microseconds. As will be demonstrated below, to achieve so short an initiating pulse duration is a fairly complex engineering problem.

In addition to the intensity and duration of the initiating pulse another very important characteristic of the initiation source is its efficiency. Indeed, the engineering efficiency of a PCL depends directly on the efficiency of the initiation source used, which is the ratio between the source energy that can be spent on producing active centers in the laser medium and the total energy consumed by the power-supply system of the source. For this reason, where it is required that a PCL should have a maximum engineering efficiency, use should be made of the most efficient initiation source from among the available selection of suitable sources.

Photoinitiation. The operation of flash photosources is based on the phenomenon of the luminous glow of plasma under the effect of a large pulse of current flowing through it. In wide use at present are initiation sources employing the following two main types of high-current luminous discharge: the walled low-pressure inert gas discharge (quartz flashlamps) and free discharge in the laser medium.

With flashlamps, the quartz tube separates the laser medium from the medium in which the luminous discharge occurs. This circumstance is advantageous, for the compositions of these media can be selected independently. But such lamps are not free from shortcomings of their own: the quartz tube walls have poor transmission (if any) in the vacuum ultraviolet region of the spectrum and are liable to damage and even destruction. This implies that neither type of high-current luminous discharge has indisputable advantages over the other, and so both are utilized as required by particular circumstances. The specific features of such discharges will be considered below.

Let us first formulate the main laws governing the operation of flash light sources under conditions necessary for successful operation of PCL's.

1) Operation at discharge durations of a few microseconds imposes special requirements upon the discharge circuit parameters. They may be defined from the following approximate relations for matched discharge:

$$E_b = CV_b^2/2 \simeq I^2 R \tau \; ,$$
$$I = jS, \quad R = \varrho l/S, \varrho \sim j^{-1/2}, \tau \simeq \pi (LC)^{1/2} = RC \; ,$$

where E_b is the energy stored in a bank of capacitors and deposited in the discharge plasma having a resistance of R, τ is the characteristic discharge time, S and l are the cross-sectional area and length of the discharge, respectively, L is the discharge circuit inductance, C and V_b are the capacitance and voltage of the capacitor bank, respectively, and j is the current density in the luminous plasma. It should be noted once more that the above relations are merely approximations, if only for the fact that they do not consider, for example, the variations of the discharge parameters with time and the extent to which the discharge channel is filled at the early and late stages of the discharge evolution. This is especially true of the existence of the matched discharge condition which has been analyzed in sufficient detail by *Markiewicz* and *Emmett* [4.16]. Nevertheless, these relations allow one to obtain the following important qualitative expressions:

$$j \sim P_d^{2/3}, \quad V_b = \frac{l^2}{S} \left(\frac{C}{L} \right)^{1/2}, \quad L \simeq \frac{\tau^2 V_b^2}{2\pi^2 E_b} \; ,$$

where P_d is the specific power deposited in the discharge. The above expressions make it possible to establish the relationship between the discharge current density and the specific power deposition and demonstrate that the discharge can be matched by varying its length and voltage and the capacitance of the storage capacitor bank. What is more, using the last expression, one can estimate the total inductance the discharge circuit must have in order to achieve a discharge duration of the order of 1 μs, using a capacitor bank round 1 kJ in energy store and 30 kV in voltage. The estimated inductance proves to be very low (about 50 nH). Hence the storage capacitors, the discharger, the feeders, and the flashlamp itself must have an extremely low inductance.

2) It is known that the brightness temperature T_{br} of a source varies as $j^{1/2}$. Consequently, an increase in the discharge current density will enhance the radiation flux intensity of the discharge, its radiation characteristics approaching those of an ideal black body. The radiation flux intensity in the UV region of the spectrum is approximately linear in the discharge current density, provided the latter is sufficiently high. It is therefore desirable to raise the discharge current density where the initial concentration of active centers in a PCL is to be increased. Under ordinary operating conditions, the brightness temperature of flashlamps is $T_{br} \simeq 10\,000$ K (discharge current density up to 10 kA/cm^2) [4.17], and so they can only provide a radiation flux intensity of $I_0 \simeq 5 \times 10^{21}$–$10^{22}$ photons/cm^2 s in the dissociation band of fluorine. The radiation flux intensiy can be somewhat increased by using reflectors to focus the flashlamp radiation into the region occupied by the laser medium.

3) But a flashlamp operated in severe discharge conditions is exposed to the serious danger of its quartz tube walls being damaged. It is precisely this danger that limits the specific power deposition in the discharge and thus sets an upper bound on the discharge current density. One of the most important causes of the destruction of the quartz tube walls in flashlamps is the nonuniform plasma distribution throughout the tube volume, giving rise to a shock wave. This wave can, in principle, be damped by employing the stabilizing action of a reverse current flowing through a conductive housing holding the quartz tube with the discharge plasma. The effect used here is the repulsion of two conductors carrying current in opposite directons. The following two engineering solutions are possible in this case.

1) Where it is necessary to initiate a chemical reaction in a large volume, it is advisable to employ standard tubular xenon-filled flashlamps with the return conductors run along the tubular envelope and pressed tightly against it. Use should preferably be made of low-pressure (a few tens of mm Hg) flashlamps. Where the laser cell is made of metal, the lamp should be placed in the active medium. Using such a lamp, *Volkov* et al. [4.18] have achieved an initiating pulse duration of $\tau_i \simeq 3$ μs and a discharge brightness temperature of $T_{br} \simeq 22\,000$ K in the range 2400–3000 Å at a storage capacitor energy of 1.8 kJ.

2) Even shorter initiating pulse durations ($\tau_i \simeq 1$ μs) can be achieved with the use of what is known as coaxial flashlamps [4.19–21] in which discharge occurs in the space between two coaxial quartz tubes, with the smaller diameter tube containing the laser mixture. An external reflector in the form of, say, aluminum foil wrapped around the lamp doubles as a return conductor. Using this technique, *Volkov* and *Tarasov* [4.21] have achieved an initiating pulse duration of $\tau_i \simeq 1.2$ μs ($j \simeq 2 \times 10^5$ A/cm^2) and a discharge brightness temperature of 50 000 K at a storage capacitor energy of around 6.8 kJ.

Similar principles underlie high-current luminous discharges occurring directly in the laser mixture. In this case, the discharge evolution and radiation characteristics may be affected by the mixture composition, and the absence of confining walls provides the possibility for the discharge channel to spread until the kinetic plasma pressure is balanced by the magnetic pressure of the discharge current filament and discharge constriction occurs. The initial, thin hot plasma channel can be produced as a result of the bursting of or surface breakdown on thin metal wires carrying heavy currents [4.22, 23].

A greater discharge surface can be obtained by using thin metal foils instead of the wires. With another technique (using what is known as creeping or surface discharges), hot plasma is produced by an electrical discharge occurring in material evaporated from the surface of a dielectric [4.22, 23], with the return conductor being placed beneath the dielectric surface and as close to it as possible. In both cases, it is possible to reduce materially the inductance of the discharge circuit and thus shorten the duration of the discharge, its brightness temperature remaining high.

It should be noted that the evolution of an open luminous discharge gives rise to shock waves. The laser cell walls must therefore be sufficiently strong.

A bursting wire emits a light pulse with a characteristic duration of (1 to 2) $\times 10^{-5}$ s and a brightness temperature of $T_{br} \simeq 30\,000$ K [4.22, 23]. Shorter initiating pulse durations can be obtained naturally with surface discharges, their discharge circuit inductance being lower. In that case, it is possible to achieve discharge durations shorter than 10 μs at brightness temperatures in excess of 30 000 K. Such high brightness temperatures provide radiation flux intensities as high as (3 to 10) $\times 10^{23}$ photons/cm^2 s in the photodissociation band of fluorine.

So far we have focused our attention on the energy characteristics of photo-initiation sources and touched upon the question of their efficiency only in passing, when emphasizing the necessity for making the duration of the initiating pulse shorter than that of the laser pulse. The possibility of improving the efficiency of flashlamps operating in conditions of short discharge durations (1–5 μs) by optimizing the discharge geometry, filling gas pressure and composition, and energy deposition [4.25] is still to be studied. Similar investigations concerning flashlamps with long flash-pulse durations have been performed by *Marshak* et al. [4.17] and *Belostotskii* et al. [4.26]. The efficiencies of modern short-pulse flashlamps amount to a few percent. And although the efficiencies of bursting wires and creeping discharges are somewhat higher (5–10%), the actual engineering efficiency of PCL's using them may prove lower because of the initiating pulse durations exceeding the laser pulse durations.

Electrical-Discharge Initiation. When experimenting at laser mixture pressures of a few mm Hg, it is advisable to use ordinary longitudinal electrical discharge schemes. But raising the mixture pressure causes glow-to-arc transition, and in that case, as with electrical-discharge CO_2 lasers (see, for example, the review presented in [4.27], use is made of various transverse-discharge arrangements, to improve the uniformity of discharge in the active media of chemical lasers. Specific features of the most popular arrangements will be considered now.

The most simple way to obtain a self-sustained discharge is to apply an overvoltage pulse across two electrodes, one of which is a one- or two-dimensional array of pointed pins evenly spaced at 6–10 mm intervals (Fig. 4.4a). The other

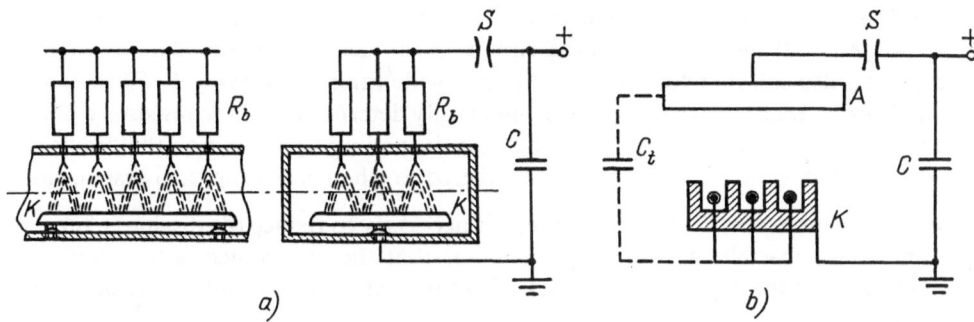

Fig. 4.4. Schematic diagrams of transverse discharge arrangements in PCL's: (**a**) pin-electrode arrangement; (**b**) double-discharge arrangement. A – anode; K – cathode; R_b – ballast resistor; S – spark gap; C_s – storage capacitor bank; C_t – trigger capacitor

electrode is a round bar or a polished strip of steel, brass, aluminum, or copper. To avoid edge effects, the strip must be much wider than the pin-electrode array, or must be shaped to have a special profile (the Rogowski profile is most popular). To prevent arcing, each pin electrode is supplied through a ballast resistor serving as a spark suppressor. As the current through one of the pins increases as a result of an arc discharge starting to develop at its pointed tip, the voltage drop across its associated ballast resistor grows higher, thus reducing the discharge voltage at the tip and hence the discharge current through the pin. The further evolution of the arc discharge is thus suppressed. Some investigators (see, for example, [4.28]) replaced the set of individual ballast resistors by an electrolyte solution (e.g., cuprous sulfate) serving as a distributed ballast resistor. The resistance of the resistor could in that case be varied by changing the electrolyte solution concentration. The ballast resistors cause a substantial proportion of the energy stored in the capacitor bank to be lost and thus reduce the engineering efficiency of the laser that employs them.

Double-discharge systems using no ballast resistors are also very popular. In these systems, initial electrons which improve the stability of a self-sustained discharge are produced by some suitable means near the discharge cathode or in the active volume of the laser. One such system (Fig. 4.4b) uses a cathode provided with a deep longitudinal slots accommodating glass or quartz tubes with thin wires inside. Voltage is supplied to the wires from the anode via a trigger capacitor forming part of a capacitive voltage divider made up of the capacitance C_t of the trigger capacitor and the circuit capacitance between the wires and the cathode. As the circuit capacitance differs from one particular cathode arrangement to another, the tirgger capacitor value is selected experimentally. The presence of the trigger capacitor is essential to high-inductance circuits. Its importance diminishes as the discharge circuit inductance is reduced, so that low-inductance circuits may use no trigger capacitor at all.

To prevent the discharge current from oscillating and optimize the deposition of energy in the discharge, the storage capacitor bank must be matched to the load. In that case, the leading edge of the discharge current pulse will be defined by the approximate relation $L/R(t)$ and its trailing edge, by the relation $C_s R(t)$, where L is the total discharge circuit inductance, C_s the capacitance of the storage capacitor bank, and $R(t)$ the total discharge circuit resistance. The shape of the discharge current pulse can therefore be varied over a wide range by varying L, C_s, and, to a much lesser extent, $R(t)$, the latter being mainly determined by the resistance of the discharge gap in the laser mixture and thus depending somewhat on the discharge voltage. Hence the general rules to be adhered to when designing such systems are:

1) to prevent a glow-to-arc transition, it is necessary to reduce the discharge time as much as possible, which in turn requires that the inductance of the discharge circuit and the capacitance of the storage capacitor bank should be made as low as possible;
2) with the energy deposition in the discharge being specified, reducing the capacitance of the storage capacitor bank requires that its voltage should be raised;
3) the main discharge circuit parameters must obey the relation $R(t) = \pi(L/C_s)^{1/2}$, referred to as the matched discharge condition.

For example, in the double-discharge arrangement [4.29] described above, the discharge circuit inductance was around 180 nH at a storage capacitor capacitance of 0.02 µF and voltage of 95 kV when the measured discharge gap resistance was around 5 Ω. These discharge circuit parameters provided for a discharge duration of approximately 100 ns. To reduce the total discharge circuit inductance, the storage capacitor was connected to the discharge electrodes in the laser cell by means of flat copper busbars. Of course, such fast discharges cannot be achieved with pin-electrode arrangements where the resistance of the ballast resistors must far exceed that of the discharge gap.

A system with an auxiliary discharge causing volume photoionization (Fig. 4.5a) has proved to be very good. In this system, the main discharge occurs between two identical Rogowski-shaped electrodes spaced 1.5–2 cm apart. A thin tungsten wire serving as a trigger electrode is stretched parallel to the main electrodes and at some

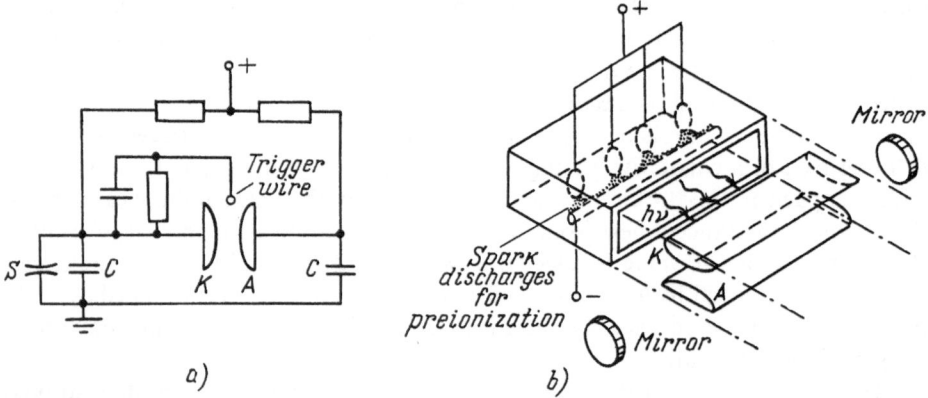

Fig. 4.5. Schematic diagrams of transverse discharge arrangements in photopreionized-electrical-discharge initiated PCL's using (a) an extended auxiliary discharge and (b) an array of auxiliary discharges

distance from them. The wire is connected to the cathode via a coupling capacitor and when a high-voltage pulse is applied across the main electrodes, a distributed weak corona discharge occurs between the wire and the anode. The coupling capacitor value (usually 50–200 pF) is selected so as to obtain a suitable auxiliary discharge. Such a discharge, whose energy is limited by the coupling capacitor, initiates a uniform main discharge. It is believed that the photoionization mechanism in this case depends either on the photoelectron emission from the cathode surface or on the effect of UV radiation emitted by the auxiliary discharge. The production of initial electrons improves the uniformity of the discharge at high pressures of the laser mixture and reduces its resistance, which allows the discharge duration to be shortened still more. Such systems helped to obtain excellent performance characteristics from both electrical-discharge CO_2 lasers [4.30] and chemical lasers [4.31, 32]. Arrays of auxiliary spark discharges (Fig. 4.5b) are also sometimes used as photoionization sources.

Voignier and *Gastaud* [4.31] have reported on some specific techniques for obtaining more uniform discharges. One of them consists in adding a small amount of C_2H_6 to the laser mixture. The above investigators have ascribed the effect obtained with this substance to the fact that (1) the ionization potential of C_2H_6 (12.8 eV) is lower than that of H_2 (15.6 eV) and (2) the C_2H_6 molecule, like C_2H_4, probably has a much greater photoionization cross section than H_2 (50 $\times 10^{-18}$ cm^2 for C_2H_4 and 7×10^{-18} cm^2 for H_2). Besides, it was noticed that the presence of helium in an optimum ratio of He: $SF_6 = 1.2:1$ added to the discharge uniformity. At a partial H_2 pressure in excess of 20 mm Hg the discharge was observed to become nonuniform. Although no data have been reported in [4.31] for the total discharge duration, the figure given there for the main current pulse rise time (about 30 ns) and evidence that the discharge in that case was matched allow us to estimate the characteristic time of energy deposition in the discharge at around 60–100 ns. As in the work reported by *Bagratashvili* et al. [4.33], such short discharge durations made it possible to operate a laser at high SF_6 pressures (up to 250 mm Hg).

It should be noted that attempts were made to operate with raised SF_6 pressures by adding low ionization potential gaseous organic substances (xylene and toluene) to the laser mixture [4.34]. Better results were obtained with xylene (ionization potential 8.56 eV). No increase in the laser power output was observed at mixture pressures below 200 mm Hg ($SF_6:H_2 = 6:1$). The maximum (60%) increase in the power output was obtained at a mixture pressure of round 300 mm Hg. The initiation system was similar to that of Fig. 4.4a. For this reason, it would be very interesting to use low ionization potential organic gas additives in conjunction with the initiation system illustrated in Fig. 4.5a where their properties could be used completely. Unfortunately, it is questionable whether it is possible to mix such organic substances with fluorine.

Electroionization Initiation. With this technique, in contrast to the overvoltage self-sustained discharge initiation considered above, energy is deposited in the active medium of chemical and other gas lasers alike from a dc field with a strength below the breakdown value with the aid of conduction electrons produced in the medium by some means or other (an electron beam or a photoionization source). The details of this initiation technique can be drawn from the papers by *Basov* et al. [4, 35] and *Wood* [4.36] on nonself-sustained discharge CO_2 lasers. For the reasons indicated in the preceding section, this technique has not as yet found wide application in chemical lasers. Nevertheless, there are already several papers available [4.37, 38, 39] that can help get some idea of the capabilities of the technique.

There are two schemes for injecting the electron beam into the dc field region. With one of them, the vector of the field which accelerates the conduction electrons produced by the beam is directed along the beam. In the other scheme (Fig. 4.6), the field vector is perpendicular to the beam. The optical axis of the laser in both these schemes is at right angles to the field and the electron beam as well.

The distance between the dc field electrodes can be varied. However, as the potential difference between the electrodes must rise with this distance, the electrode separation is bounded above by the electrode voltage at which autoelectronic emission starts from the cathode. In other words, it is inadvisable to raise the voltage across the electrodes above around 100–150 kV. Consequently, with the pre-breakdown electric field strength in atmospheric-pressure mixtures being approximately equal to 15–30 kV/cm atm., the maximum attainable electrode separation can be estimated to be 5–10 cm.

The dc field electrodes may differ in design, depending on the geometry of the experiment. When the field vector is parallel to the electron beam, the cathode must let the beam pass through, and so it is usually made in the form of a metal grid arranged parallel to the beam entrance window foil in the laser cell. In this case, the electric arcs developing at the late stages of the nonself-sustained discharge do not break the foil. To prevent such arcs, the anode is Rogowski-shaped. When the field and beam are at right angles (Fig. 4.6), both electrodes should be Rogowski-shaped.

The necessity to use an electron accelerator and the presence of such a critical component as the e-beam entrance window foil considerably complicate the experiment. For this reason, alternative ways to produce conduction electrons in the laser medium are now being sought. Most interesting in this respect is to produce such electrons by photoionization, as in the case of electrical discharge initiation, and also by means of a short overvoltage pulse [4.38, 39], the duration of which should be such as to prevent the evolution of instabilities in the overvolted

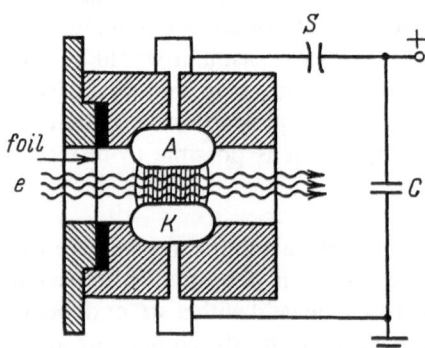

Fig. 4.6. Schematic diagram of electroionization initiation of a PCL [4.37]: A – anode; K – Cathode; e – electron beam; C – storage capacitor bank; S – spark gap

discharge during the time it lasts. With both methods, the major proportion of electrical energy is deposited in the medium by a dc field with a pre-breakdown strength. No special requirements are placed upon the discharge circuit to produce such a field, provided the time it takes for the circuit to deposit energy in the laser medium does not exceed the laser pulse duration.

Electron-Beam Initiation. Electron-beam sources depend for their operation on the emission of electrons from the cathode surface of a thermionic diode and the acceleration of these electrons through a high potential difference between the cathode and anode of the diode [4.40]. The anode is designed so as to let the electrons pass through it (e.g. use can be made of a metal grid). The electron beam thus formed in the evacuated volume of the diode is then injected into the laser medium through a window of aluminum or stainless steel foil a few tens or hundreds of microns in thickness. So great a window foil thickness is necessitated by the requirement that the foil should be strong enough to withstand the considerable increase of the medium temperature and hence pressure possible in chain-reaction chemical lasers. Generally speaking, to enhance the foil thickness is inadvisable, for this raises the e-beam energy loss in the foil. This loss can be cut down by using a protector in the form of one or two metal flanges with a large number of perforations. Where a single perforated flange is used, it is placed next to the window foil on the diode side. With two flanges, the foil is clamped between them.

The high voltage necessary to apply across the anode and cathode of the thermionic diode is produced by a pulse generator, usually a Marx generator (Fig. 4.7) or a high-voltage storage capacitor with a pulse transformer. The low inductance of such a generator allows it to be connected directly to the diode in accelerators with e-beam pulse durations of a few microseconds (sometimes even somewhat less than a microsecond [4.41]). If the e-beam pulse duration should be

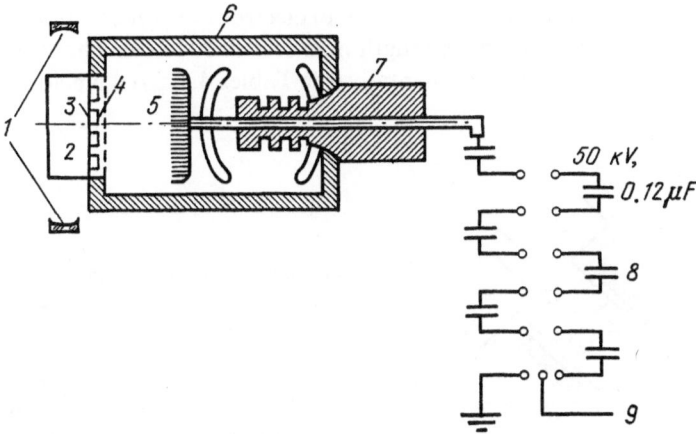

Fig. 4.7. Schematic diagram of transverse e-beam initiation of a PCL using a Marx generator [4.41]; *1* – cavity mirrors; *2* – laser cell; *3* – foil; *4* – protector; *5* – cold cathode; *6* – vacuum chamber; *7* – high-voltage bushing; *8* – six-stage Marx generator; *9* – trigger electrode

shorter than 100 ns, the pulse generator should then be supplemented with a shaper circuit to produce high-voltage pulses of the required duration.

The application of other types of accelerators with long pulse durations (of several tens and even hundreds of microseconds), e.g., those employed in e-beam preionized CO_2 lasers, is inadvisable because the characteristic laser pulse duration in chemical lasers of practical interest is less than 10 μs.

To achieve suitable e-beam pulse durations, use is made of cold pin cathodes relying for their operation on explosive autoelectronic emission under the effect of a high-strength electric field [4.42]. The parameters of electron accelerators utilizing such cathodes are governed by the physics of this phenomenon and are typically as follows: electron energy from 100 keV to 2 MeV, e-beam pulse duration from 5 ns to a few microseconds, e-beam current density from a few amperes to several kiloamperes per square centimeter, and the total e-beam current from several hundreds of amperes to a few megaamperes.

Modern electron accelerators are very complex and costly, and to deposit e-beam energy in the laser medium is as difficult as producing high-current beams of relativistic electrons. The passage of an e-beam through a dense gas is an involved and poorly studied process [4.43, 44]. It has been established experimentally that a narrow, weak, unfocused e-beam spreads to occupy an almost spherical volume in the gas [4.45]. If the beam current is increased to several tens of kiloamperes, the beam undergoes self-deceleration as a result of reflection at the dense plasma layer produced under its effect, and the electron path length is thus drastically reduced compared with that typical of low-current beams.

There is, however, a practical way to confine the beam and carry it through the dense gas (Fig. 4.8). It consists of establishing an external axial magnetic field much stronger than the magnetic field of the beam, which provides an approximately rectilinear beam propagation. If the gyromagnetic radius of the beam electrons is much smaller than the beam radius, the rate of beam spreading will be reduced. The magnetic field strength required to confine a given beam depends on the beam radius and energy. For example, a 10-cm-radius beam of electrons around 1 MeV in energy will require a confining field with a strength in excess of 1000 Oe. That this approach is valid has been confirmed experimentally (see Tables 4.3 and 4.6 below).

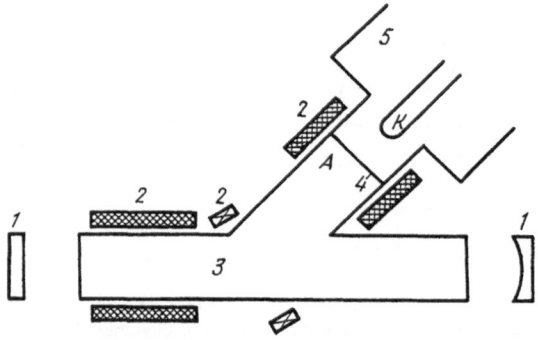

Fig. 4.8. Schematic diagram of e-beam initiation of a PCL using an external axial magnetic field to confine the beam and carry it through a dense gas: *1* – cavity mirrors; *2* – solenoid coils; *3* – laser cell; *4* – foil; *5* – electron accelerator; A – anode; K – cathode

4.2.3 Questions Relating to the Homogeneity of Initiation and the Scaling of the Laser Medium Size

To obtain the high coherent radiation energies necessary for various practical purposes, one should consider, in addition to the problem of improving the specific parameters of pulsed chemical lasers, also the possibility of effecting homogeneous initiation of their active media in large volumes. In this section, the primary emphasis will be on the homogeneity of initiation in the direction across the laser beam, for the attainment of homogeneous initiation along the beam is in most cases a purely engineering problem. We are going to discuss the principal factors limiting the depth of a sufficiently homogeneous initiation in the transverse direction. The question is very important because this depth, as a rule, characterizes the radial dimension of the laser volume that can be effectively initiated and hence determines the possibility of scaling the active medium size in the transverse direction. What is more, inhomogeneous initiation drastically increases the laser beam divergence as a result of transverse laser medium refractivity and unsaturated gain gradients (see Sect. 4.8).

In the case of flash-photolysis initiation, the depth of penetration of photons into the laser medium is

$$l \simeq (\sigma N_0)^{-1} \ . \tag{4.14}$$

Hence, with the partial fluorine pressure ranging between 50 and 200 mm Hg, the range commonly used in experiments, we have $l \simeq 12$–50 cm ($\sigma_{eff} \simeq 10^{-20}$ cm^2). Consequently, it can be expected that flash-photolysis initiation will be homogeneous enough only if the transverse dimension of the laser medium does not exceed the l values indicated above. It may not be out of place to make another remark in connection with the cylindrical geometry of photoinitiation sources a few centimeters in diameter. The radiation intensity of such a source decreases with distance as r_s/r, r_s being the radius of the source and r the distance from it. The distance dependence of the radiation intensity of the source can be greatly weakened, and the homogeneity of initiation provided by it can thus be improved, by using properly shaped reflectors.

The above values of l indicate that a good enough initiation homogeneity can be achieved in a laser medium measuring merely a few centimeters across. In that case, however, not only the energy output of the laser, but also its efficiency will be low, because only a small proportion of the available photoinitiation source radiation can be absorbed in a laser medium with so small a diametrical dimension. Therefore, to make the best of the source requires that either the cross section of the laser medium should be increased or use should be made of extremely effective reflectors. Experiments with pulsed chemical lasers have shown that the former method is preferable for improving the engineering efficiency of the laser (this point will be discussed in greater detail later in the text). Thus, it can be inferred that in flash-photolysis initiated PLC's, scaling of the active medium in the transverse direction can help improve not only the beam divergence, but also the engineering efficiency.

Now consider the characteristic depth of homogeneous initiation provided by fast electron beams. If we assume that use is made of a beam of fast electrons for which scattering is prevented by means of an external magnetic field, and that the electrons lose their energy mainly through the ionization of the gas they penetrate, the penetration depth of an electron into the gas can be estimated from the ratio between its initial energy U and the energy it loses while traversing a distance 1 cm long:

$$l \simeq U/w p_{F_2} s_{F_2} \,. \tag{4.15}$$

Substituting into this expression, as in Sect. 4.1, the typical parameters $w = 35$ eV/(ion pair), $p_{F_2} = 200$ mm Hg, and $s_{F_2} = 1.3 \times 10^{-2}$ (ion pair)/cm (mm Hg), we get $l \simeq 200$ m for $U = 2$ MeV. As the electron energy U is reduced, the quantity l will decrease not only as a result of this, but also because of an increase in s_{F_2}. Hence it follows that a beam of fast electrons with an energy of $U \simeq 300$ keV can be used to initiate transversely a laser medium measuring around 1–3 m across. It should be emphasized, however, that formula (4.15) holds only if the beam is prevented against spreading by an external uniform magnetic field, and to set up such a field in a large volume is a fairly complicated engineering problem.

As pointed out above, the use of an electrical discharge to effect homogeneous initiation of sufficiently large volumes of laser medium also produces considerable difficulties, due mainly to the fact that the optimum laser mixture composition for uniform operation of the discharge is not always the same as that for the maximum laser energy output. The presence in the medium of highly electronegative gases increases its breakdown voltage so that a self-sustained discharge, even at low gas pressures, soon transits into an arc discharge, thus excluding the possibility of volume-dominated initiation, and makes the use of double-transverse-discharge schemes usually employed in other gas lasers to combat glow-to-arc transition inefficient. That this is true has been clearly demonstrated, for example, by *Poehler* and *Walker* [4.46] who indicated specifically that while it proved possible to maintain a homogeneous volume-dominated double transverse discharge in a D_2–CO_2–He mixture over a wide pressure range, the addition to the mixture of a small amount (around 2%) of fluorine made the discharge transit into an arc at pressures in excess of 250 mm Hg, the electrode separation being no more than 2 cm. Such an interelectrode distance is typical enough for one to get an idea of the pumping depth scale attainable with the electrical-discharge initiation technique.

Bagratashvili and co-workers [4.33] have managed to increase significantly the total mixture pressure (up to 5.5 atm.) and the content of the electronegative gas SF_6 (up to 0.22 atm.) in their HF laser using an SF_6–H_2–He mixture excited by a nanosecond high-voltage self-sustained discharge between electrodes spaced the same distance apart as above. The success can be explained by the fact that the discharge, with its duration being so short, had not enough time to transit into an arc. This is clearly confirmed by comparison with the earlier work [4.47] by the same group of investigators. In this work, they used microsecond discharge durations and failed to raise the total pressure of the SF_6–H_2–He mixture above 0.9 atm. Nevertheless, from the engineering standpoint a nanosecond self-sustained discharge is, despite its obvious advantages, as difficult to realize for all practical purposes as an electron beam.

Deeper pumping in the case of electrical-discharge initiation can, in principle, be achieved with nonself-sustained discharges. This technique, however, involves difficulties of its own. First of all, the presence of an electronegative gas again drastically reduces the efficiency of utilization of conduction electrons in the initiation process (see Sect. 4.2.1). Expression (4.13) shows that the adverse effect of the dissociative electron attachment process can be diminished by increasing the dc electric field strength E. Qualitatively, the maximum value of E is defined by the condition

$$\sigma_{F_2} v_e n_e [F_2] = \sum_i \sigma'_e p_i n'_e v'_e .$$

In that case, the efficiency of utilization of every conduction electron will be a maximum. Hence it naturally follows that the strength of the dc field set up in the active laser medium must be just below the breakdown value. Thus we are faced with a contradictory situation: to make the most of the conduction electrons, it is necessary that electron multiplication processes (4.4″), (4.7), and (4.8) should become important, and this makes the discharge look more and more like a self-sustained discharge prone to all sorts of instabilities. As mentioned in Sect. 4.2.1, investigations into the capabilities of the electroionization initiation technique are far from completion, and so it is as yet difficult to draw any definite conclusions.

To provide homogeneous thermal initiation for large-volume chemical lasers is also very difficult. Indeed, even the slightest inhomogeneity in thermal explosion initiation is sufficient for detonation waves to develop in the laser medium. And detonation is characterized by the presence in the medium of local temperature and density inhomogeneities and alternate zones of intact mixture and combustion products. On the other hand, the use of "overdriven" detonation waves featuring much fewer inhomogeneities leads to a very thin reaction zone (see Sect. 4.2.1), and so to seek for ways of scaling chemical lasers relying on this initiation technique is practically hopeless.

4.2.4 Comparison Between Flash-Photolysis and Electron-Beam Initiation Techniques

Now that we have considered the basic physical and engineering aspects of the various initiation techniques, we can attempt to find out whether there is some objective physical criterion by which to compare them. Indeed, the analysis of reports on record high specific energy outputs from chemical and other lasers shows that these records are frequently achieved at the sacrifice of the depth of homogeneous initiation with all the negative consequences discussed above. Hence it follows that the specific energy output is in itself not an absolute criterion. To clarify this matter, we refer to equation (4.1) written for the case of flash-photolysis initiation. If we assume for simplicity that the photon flux intensity provided by the initiation source is time-invariable and that the operating time (i.e., flash duration) of the source is τ_i, then considering (4.14), equation (4.1) may be transformed to

$$Nl \simeq \Phi I_0 \tau_i \; ,$$

or in the case of fluorine dissociation where $\Phi = 2$ [4.48],

$$[F]l \simeq 2I_0 \tau_i \; . \tag{4.16}$$

Since the specific laser energy output is $\varepsilon_l \simeq \beta h \nu_l N v_{eff}$, we have

$$\varepsilon_l l \simeq \Phi I_0 \tau_i \beta h \nu_l v_{eff} \; ,$$

where β is the average number of photons emitted by an excited molecule. Hence it can be inferred that the most objective criterion by which to compare flash-photolysis initiated (pumped) gas lasers is the product of the specific laser energy output by the pumping depth. For nonchemically pumped gas lasers, $v_{eff} \lesssim 1$. At the same time, there are reports in the literature on H_2–F_2 lasers with $v_{eff} \gtrsim 100$. This means that the quantity εl in the H_2–F_2 laser is much greater than in lasers with nonchemical pumping. Consequently, with the specific energy output of such a laser being the same as in nonchemically pumped lasers, its pumping depth can be much greater and so can its engineering efficiency, provided the efficiency of the photosource depends weakly on the flash duration.

A relation similar to (4.16) can also be derived for the case of fast-electron-beam initiation of chemical lasers. We will do this for the H_2–F_2 laser so as to be able to make use of the appropriate expressions derived earlier. To this end it is necessary to multiply equations (4.10) and (4.15):

$$[F]l = 2.5 \times 10^{19} \, t_e J_b \frac{U}{w} \; .$$

In this example, the role of the photon flux intensity I_0 is played by the expression $10^{19} J_b U/w$. With the characteristic parameters being $U = 2$ MeV and $J_b \simeq 10^3$ A/cm^2 ($t_e \leqslant 10^{-7}$ s) for high-current electron beams and $U = 300$ keV and $J_b \simeq 10$ A/cm^2 ($t_e = 10^{-7}$ s to a few microseconds) for low-current beams, the value of this expression comes to 10^{27} and 10^{24} cm^{-2} s^{-1}, respectively. So, the product of the specific laser energy output and the pumping depth in the case of high-current electron beams may exceed by more than an order of magnitude that for the most powerful photosources ($I_0 \leqslant 10^{24}$ photons/cm^2 s, $\tau_i < 10^{-5}$ s), the two being comparable in the case of low-current beams with an operating time (i.e., e-beam pulse duration) of a few microseconds.

However, it should be pointed out once more that increasing the specific energy output in pulsed chemical lasers reduces their lasing pulse duration. Here again the possibility of cutting the operating times of electron accelerators down to $t_e \leqslant 10^{-7}$ s proves to be very attractive. After all, the use of high-power initiation sources can reduce laser pulse durations to a few hundreds and even tens of nanoseconds, and so the application of electron beams under such initiation conditions can help eliminate the loss of laser engineering efficiency due to the excess of the initiating pulse duration over the laser pulse duration, which from the engineering standpoint is very difficult to attain with powerful photolysis. But the picture would be incomplete unless we emphasize again that success here can be achieved only at a high cost, electron accelerators being as complex and expensive as they are. For this reason, no absolute preference can be given to either flash-photolysis or electron-beam initiation. In each particular case, either technique may prove more suitable, depending on the requirements the given chemical laser must meet (especially as regards its lasing pulse duration).

4.3 Laser Mixture Preparation Problems

To prepare a mixture for a non-chain-reaction laser or a mixture containing a low-activity oxidizer presents no difficulties. Examples are SF_6–H_2 and NF_3–H_2 mixtures. Mixtures for branching-chain-reaction lasers are quite another thing. In such lasers, the chain-branching process may under certain conditions gain in importance so that the mixture will become capable of self-ignition [4.49], i.e., unstable. In general, the stability of a mixture can be analyzed with reference to the p–T diagram (Fig. 4.9) typical of branching-chain reactions. If the room-temperature isotherm passes to the left of the tip of the self-ignition peninsula (isotherm 1), the mixture formed in the course of mixing of the reactants will remain stable up to the very high pressures corresponding to the third explosion limit. Such a situation is characteristic of CS_2–O_2 and H_2–Cl_2 mixtures. Consequently, these mixtures at room temperature are stable enough over a wide pressure range.

But if the initial-temperature isotherm cuts across the self-ignition peninsula (isotherm 2), the mixture will only be stable within the pressure intervals $0 < p < p_1$ and $p_2 < p < p_3$. In that case (typical of hydrogen-fluorine mixtures), the preparation

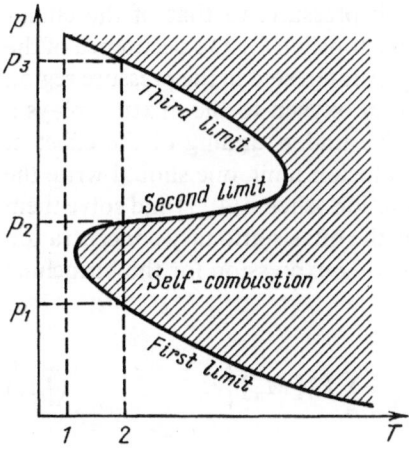

Fig. 4.9. p–T diagram characterizing the position of explosion limits for mixtures reacting by a branching-chain mechanism

of a mixture at the initial temperature meets with serious difficulties. Let us consider strictly qualitatively the reasons of the various explosion limits in the p–T diagram of the hydrogen-fluorine mixture, because it is precisely this mixture that is now being used most widely in pulsed chemical lasers. The self-ignition of the mixture is due to the fact that the hydrogen-fluorine chain reaction (4.17) below involves an energetic branching process (6) providing for the possibility of active center multiplication:

(1) $F + H_2 \rightarrow HF^* + H$,

(2) $H + F_2 \rightarrow HF^* + F$,

(3) $H + O_2 + M \rightarrow HO_2 + M$,

   $F + O_2 + M \rightarrow FO_2 + M$,   (4.17)

(4) $HF^* + H_2 \rightarrow H_2^* + HF$,

(5) $HF^* + M \rightarrow HF + M$,

(6) $H_2^* + F_2 \rightarrow HF + H + F$,

(7) $H_2^* + M \rightarrow H_2 + M$.

At pressures below the first explosion limit (less than 0.01 atm.) the particle diffusion rate is high and therefore the active centers (F, H, H*, HF*) decay or are deactivated much more effectively on the laser cell walls than by collisions with other particles. The condition of equality between the rate of disappearance of the active centers as a result of diffusion and the rate of their multiplication determines the position of the first explosion limit. When the chain-branching process rate exceeds the rate of diffusion of the active centers to the laser cell walls, the mixture undergoes self-ignition.

As the mixture pressure grows higher, a moment comes when the rate of decay of hydrogen and fluorine atoms in the three-particle recombination process (3) becomes equal, thanks to its being quadratic in pressure, to that of the chain-branching process (6). It is this moment that is responsible for the occurence of the second explosion limit. In practice, we are naturally interested in the pressure region above the second explosion limit, and addition of oxygen to the mixture plays a decisive role in reaching it. To gain a qualitative understanding of the effect of various factors on the location of the second explosion limit, one should write the kinetic equations corresponding to the above reaction scheme (4.17) and solve them (with the exception of the equation for [H_2^*]) by the quasistationary concentration method [4.50, 51]. Then one will have the following expression for the branching ratio:

$$s = \frac{14 k_1 k_6 [F_2]^2}{\sum_i k_3^i [O_2][M_i]} \left\{ \frac{1}{1 + \sum_j k_5^j [M_j]/k_4 [H_2]} - \sum_n k_7^n [M_n] \right\} .$$ (4.18)

Hence the equation for the second explosion limit (the condition $s=0$) is

$$\frac{14k_1 k_6 [F_2]^2}{\sum_n k_7^n [M_n] \sum_i k_3^i [O_2][M_i]} \frac{1}{1+\sum_j k_5^j [M_j]/k_4 [H_2]} = 1 \ . \tag{4.19}$$

The analysis of the latter expression allows the following qualitative conclusions to be drawn:

1) the higher the fluorine content of a constant-pressure mixture, the lower its stability;
2) an increase in the concentration of hydrogen in a constant-pressure mixture is not so harmful to the mixture stability as that of fluorine;
3) raising the percent concentration of oxygen relative to that of fluorine without changing the total mixture pressure allows one to prepare a mixture containing more fluorine;
4) adding to the mixture such good HF* "relaxants" as HF or CO_2 makes it possible to lower the second explosion limit. To illustrate, for $[CO_2]/[H_2]=6$, the second term in (4.19) is less than 0.1, which points to the possibility of preparing in that case a stable mixture containing much less oxygen;
5) considering the temperature dependence of the rate constants entering into (4.19), it can be inferred that lowering the temperature of the mixture is beneficial to its stability;
6) when increasing the fluorine content of a mixture, it is advantageous to raise (to a certain limit) the total mixture pressure.

It should be particularly emphasized that lowering the temperature of the mixture or adding CO_2 to it cannot eliminate completely the need for the addition of oxygen, but merely reduces its necessary concentration.

The seemingly effective method to stabilize the mixture by increasing its total pressure produces the required results only until the third explosion limit is reached. The nature of this limit is still not completely understood. Its possible cause is either thermal explosion or the occurrence of a new process preventing the termination of the chain. For example, Tal'roze [4.52] has suggested that the process

$$HO_2 + F_2 \rightarrow HF + O_2 + F$$

plays an important part in the regeneration of the active centers. But as pointed out by Bokun and Chaikin [4.53], this process is effectively counterbalanced by the HO_2 radical decay process

$$2HO_2 \rightarrow H_2O_2 + O_2 \ ,$$

and the observed significant pre-explosion heating of the mixture and the existence of the ultimate reaction rate above which the reaction turns into explosion indicate that the third explosion limit is of thermal nature. By the way, the position of the third explosion limit has proved to depend substantially on the free convection process due to the heating of the mixture by the heat of the exothermic chain reaction, and so it must lower with the increasing diameter of the cylindrical laser cell. The third explosion limit also lowers as the fluorine concentration is increased without changing the $[F_2]/[O_2]$ ratio and grows higher as the relative oxygen concentration is raised [4.54]. Thus, by drastically increasing the oxygen concentration, *Gerber* et al. [4.55] have managed substantially to shift the third explosion limit upward and prepare a mixture with a total pressure of up to 11 atm. Figure 3.2 shows the necessary relative oxygen concentration $[O_2]/([F_2]+[H_2])$ as a function of the total reactant concentration $[F_2]+[H_2]$. So, it can be concluded that the stability region of the hydrogen-fluorine mixture lying between its second and third explosion limits widens as the concentration of O_2, CO_2, or HF is increased, or as the initial mixture temperature is lowered, and narrows with increasing cylindrical laser cell diameter, initial mixture temperature, and fluorine concentration, or with decreasing diluent concentration.

It should be noted that the location of the second and third explosion limits can be influenced by other factors as well. A mixture that must in principle be stable may nevertheless ignite spontaneously on account of some causes difficult to control reliably: in the course of mixing of the reagents, there arise for a short time spots at which the mixture may enter the self-ignition region; the fluorine gas may react chemically with the inner surfaces of the laser cell and the gas bleed-in system to produce atomic fluorine; the gas bleed-in system, if faulty or poorly designed, may let specks of dust or fine particles of foreign matter enter the laser cell with the mixture component gases, where they burn away to give off heat that brings some mixture into the self-ignition region; the presence in the gases of organic or combustible impurities may also cause self-ignition of the mixture. For this reason, the actual behavior of the second and third explosion limits may differ considerably between particular laser cells. The prime object of the experimenter is to eliminate as far as possible the factors interfering with the preparation of a stable mixture at a sufficiently high pressure. Let us consider the experimental techniques employed to this end.

1) First of all, the reactant mixing procedure is developed. Separate premixers are used individually to prepare the F_2–O_2–He and H_2–O_2–He premixes (see Fig. 4.1) which are then mixed. This latter operation may be performed in one of the following three main ways:

a) by admitting into the laser cell first one and then the other premix (in most cases, it is the F_2–O_2–He premix that is admitted first [4.55–4.57][2]);

b) by preparing the mixture in an individual mixer at a pressure known to be above the second explosion limit and then admitting it into the laser cell [4.59]; and

c) by flow mixing the premixes [4.37, 54, 60].

In all cases, it is desirable that the pressure of the F_2–O_2–He premix and its relative oxygen content $[O_2]/[F_2]$ should be such as to ensure that the mixing can be carried out at pressures above the second explosion limit for fluorine-rich

---

[2] In [4.58], the premixes were admitted in the reverse order.

mixtures, for it is precisely such mixtures that are formed in the boundary region when admitting the H_2-O_2-He premix.

Where the laser is operated at not very high pressures close to the second explosion limit, the second way is preferable. It allows the mixture to be prepared at pressures far above the second explosion limit and then admitted into the laser cell at the required pressure [4.59]. But if the cell has been preliminarily evacuated, the pressure of the mixture can at first drop so that it will enter the self-ignition region. For this reason, one must first fill the laser cell with some inert diluent and then displace the diluent with the mixture being admitted into the cell. The diluent pressure must be above the second explosion limit of the mixture. The stability of the mixture can be much improved if it is cooled, as well as the premixes, down to around 200 K [4.61] or even 100 K [4.62].

2) To prevent chemical interaction between fluorine and the laser cell surface, the latter is passivated. This is done by filling the cell for several hours with fluorine. The operation is sometimes repeated until the change in the pressure of the freshly admitted fluorine charge becomes negligible. Passivation results in the formation of a protective coating of metal fluorides and hence prevents nucleation on the laser cell walls. It is advisable to passivate the walls at an increased temperature if possible. The use in the laser cell of materials that cannot be passivated (e.g., quartz, glass) should be avoided as far as possible. *Agroskin* et al. [4.63] have observed an autocatalytic acceleration of the dark reaction on the surface of a quartz laser cell leading to the burn-out of a freshly prepared mixture. Therefore, where the experimental laser cell cannot do without quartz or glass (e.g., flash-photolysis initiated lasers use quartz for both the cell walls and flashlamps), it is good practice not to store the mixture in the cell for any prolonged period of time.

3) The combustible particles or harmful impurities entering the laser cell may be contained in the starting gases. Whereas the purity of commercial-grade hydrogen and helium is high enough (around 99.9%), that of fluorine frequently proves much lower (around 98%). Consequently, provision should be made in the gas bleed-in system for the purification of fluorine, first of all to free it from hydrogen fluoride, by passing the gas through liquid-nitrogen-cooled or sodium-fluoride-filled traps [4.64]. Commercial-grade carbon dioxide requires especially thorough purification. Fine particles of foreign matter may form in the gas bleed-in system if it is not clean enough or uses unsuitable materials that fail to develop a protective coating but interact instead with fluorine at spots of local heating to produce such fine mechanical impurities. All the component parts of the gas bleed-in system must therefore be manufactured from fluorine-resistant materials, such as stainless steel, copper, aluminum, nickel, and monel metal. Vacuum-tight joints between the components must use copper, aluminum, or fluoroplastic gaskets. The use of glass and quartz is also permissible, but with the reservations indicated above. No organic substances easy to react with fluorine should be employed. Pressures can be measured with standard beryllium-bronze Bourdon-tube pressure and vacuum gauges or U-tube manometers filled with fluorinated oil.

If none of the above measures proves effective enough, so that combustible particles, for example, find their way into the laser cell, there will occur local microexplosions giving rise to unexcited HF molecules. But even if the mixture somewhere in the cell has locally entered the self-ignition region, this does not mean that the entire mixture charge will ignite. For this to happen takes a certain characteristic time referred to as the induction period. During this period the reaction accelerates to reach a critical temperature leading irreversibly to the ignition of the mixture. If the mixture is close to the second explosion limit, the induction period τ may be evaluated approximately from the condition $1/\tau = s$. In that case, the smaller the difference in the expression (4.18) for the branching ratio s, the longer the induction period. It follows from the analysis of (4.18) that the induction period increases with the pressure of the mixture, especially with that of the diluent, oxygen, and HF. Therefore, as long as the critical temperature is not reached, the joint action of the oxygen contained in the mixture, the HF molecules being locally formed, and the heat transfer process (not to mention the dynamics of the mixing flows of gases) may localize and suppress the microexplosions. Thus, insofar as the position of the second explosion limit depends, according to (4.19), on the HF concentration, there may exist conditions under which the mixture will be self-stabilized should it exceed this limit [4.65]. In that case, some mixture will have reacted to yield a quantity of unexcited HF molecules. It will be demonstrated later that this process is harmful as it may drastically worsen the laser parameters. At first glance it would seem that a boundless increase of the oxygen content of the mixture must help surely to suppress its self-ignition. But, as will also be shown later, in certain conditions this may drastically reduce the laser energy output as well.

Where no reasonable method can help to stabilize the mixture, the laser is operated with the reactants being flow-mixed directly at the entrance to the laser cell and the mixture pumped continuously through it. In that case, the time it takes to change the mixture in the cell must be shorter than the induction period. This holds for the ClF_3-H_2 [4.66] and Cl_2-HCl [4.67] systems.

What is the general conclusion forced upon one analyzing the above problems involved in the preparation of hydrogen-fluorine mixtures at pressures above the second explosion limit? First of all, the job of mixture preparation is an intricate experimental art depending on the construction of the gas bleed-in system and laser cell and the procedures used to prepare the system for operation and mix the premixes. The compositions and pressures of the mixtures used in particular experiments are listed in Tables 4.4–6 which are discussed more fully later. The situation is further complicated by the fact that the experimenters frequently fail to monitor adequately the degree of the mixture burn-out prior to initiation so as to know exactly the number of unexcited HF molecules in the mixture. This is one of the causes of the wide spread in the data obtained not only in different experiments but often even in one and the same experiment.

4.4 Non-Chain-Reaction Pulsed Hydrogen-Fluoride Lasers

Tables 4.1–3 list the main characteristics of hydrogen-fluoride pulsed chemical lasers based on non-chain reactions proceeding in various mixtures. These characteristics include the engineering efficiency η_{eng}, the physical efficiency η_{phys}, the chemical efficiency η_{chem}, the specific energy output ε_l, the total energy output E_l, the spectral composition of the emission, the mixture pressure and composition, the laser cell volume V_l or dimensions, the characteristic initiating pulse duration τ_i, the characteristic laser pulse duration t_l, the initiation system parameters, and the initiation ratio[3] $\Delta[RF]_0/[RF]_0$ (i.e., the relative change of the initial RF concentration due to decomposition under the action of the initiation source). In the tables, lasers are grouped according to the initiation technique they use. As a rule, the tables give the maximum values of the characteristic laser parameters obtained to date with optimum mixture compositions, initiation energies, and so on.

As demonstrated above, the PCL's based on the chain hydrogen fluorination reaction show the best energy performance. Nevertheless, the non-chain-reaction hydrogen-fluoride lasers have played an important part in the investigation and development of chemical lasers. What is more, there are practical applications where these lasers prove most suitable. For example, one such application is laser chemistry which requires laser emissions of moderate energies (of the order of a few joules) in the range 2.6–3.3 µm. Let us therefore consider the main characteristics of such lasers without going into details.

The exothermic reactions used for chemical pumping in non-chain-reaction PCL's are substitution reactions of the type

$$F + RH \rightarrow HF + R - \Delta H \; ,$$

where R is a suitable atom or radical. The exothermicity $-\Delta H$ of such reactions depends on the R–H bond energy in the RH molecule and is equal to 37.5, 31.5, 32.1, 46.3, and 69.5 kcal/mole for C_2H_6, H_2, HCl, HBr, and HI, respectively. Special investigations have shown that the exothermicity fraction that goes into the vibrational excitation of the HF molecule formed in the reactions involving the molecules indicated above amounts respectively to 62% [4.96], 66% [4.97, 98], 59% [4.99, 100], 55% [4.100], and 62% [4.100]. These percentages are at the same time the maximum chemical efficiency values of HF lasers which operate on mixtures containing these RH molecules.

The indicated exothermicity values are high enough to excite the HF molecule to $v=3$ in the reactions involving the first three molecules, $v=4$ in that involving the fourth molecule, and $v=6$ in the one involving the last molecule. Of course, due consideration should be given to the fact that these are the maximum possible values v_{max}. In that case, the emission spectrum of such lasers can be governed by transitions from all vibrational levels in the range $1 \leqslant v \leqslant v_{max}$. In experiments with C_2H_6 and H_2, transitions are as a rule observed to occur in the vibrational bands $v=3 \rightarrow v=2$, $v=2 \rightarrow v=1$, and $v=1 \rightarrow v=0$. Laser emissions due to transitions from higher vibrational levels have been detected in experiments with such hydrogen-

---

[3] This term has been introduced to reflect more unambiguously the physical essence of the process of producing active centers in the laser mixture and enable one to compare experiments conducted at different RF concentrations.

bearing molecules as HBr ($v_{max} = 4$) [4.101] and especially HI ($v_{max} = 6$) [4.102, 103]. The emission intensity in each vibrational-rotational transition depends on the initial distribution of the energy going into the vibrational excitation of the HF molecule produced by the pumping reaction among its vibrational levels, the vibrational relaxation rate, and the rate of establishment of vibrational equilibrium within the limits of each vibrational band. The form of this energy distribution is in turn dependent on the hydrogen-bearing species RH, the vibrational relaxation and vibrational equilibrium establishment rates being governed by both the molecular species used to prepare the laser mixture and the mixture pressure (see Chap. 2). In conditions of perfect rotational equilibrium, each individual vibrational-rotational line occurs in a strict sequence characterized by the increase with time of the rotational quantum number of the lasing transition within each vibrational band. In non-chain-reaction PCL's, rotational equilibrium has been attained by strongly diluting the laser mixture with inert gases or complex fluorine compounds (e.g., SF_6, C_2F_6) [4.69, 84].

Non-chain-reaction PCL's can be initiated by any of the initiation techniques described in Sect. 4.2. Flash-photolysis initiation has been mainly used for purely research purposes, viz., searching for new active media and establishing the inversion mechanism in them, studying the spectral composition of laser emissions, and so on (see Table 4.1). The better energy characteristics have been obtained with electron-impact initiation of mixtures containing sulfur hexafluoride (SF_6) as an atomic fluorine donor and H_2 or C_2H_6 as a hydrogen-bearing species. Let us consider the main results obtained with these mixtures.

It follows from Tables 4.1–3 that the attained laser energy outputs depend on the intensity of the initiation source used, initiating pulse duration, and the possibility of efficient deposition of the initiation source energy in the active medium of the lasers. It has been noticed that the total and specific laser energy outputs grow higher with the laser mixture pressure so long as the pressure rise does not prevent homogeneous initiation of the mixture. It should be noted in this connection that Table 4.2 does not list specific energy outputs reduced to atmospheric pressure because electrical discharge can only ensure sufficiently homogeneous initiation at mixture pressures below atmospheric value. In that case, the notion of reduced specific energy output becomes physically meaningless.

The ratio between SF_6 and the hydrogen-bearing mixture component in the various experiments ranged from 5 to about 20, a ratio in excess of 10 being optimal for electrical-discharge initiation and below 10, for electron-beam initiation.

The highest energy outputs (up to 380 J [4.94]) from non-chain-reaction PCL's have been obtained with electron-beam initiation, for this technique makes it possible to deposit in the laser medium a very large amount of energy in a very short time. It is of interest to note, however, that although electrical-discharge initiation can deposit a materially smaller total amount of energy in the mixture and hence provide for a much lower laser energy output (up to 11 J [4.28]), it can ensure, if adequately designed, approximately the same specific energy output (up to 16 J/l [4.31]) as high-power e-beam initiation (up to 15 J/l [4.94]). The pulse duration of non-chain-reaction PCL's is determined by the initiating pulse duration and, as a rule, does not exceed it, no matter what its value (up to several tens of nanoseconds).

The ultimate physical efficiency of the SF$_6$–H$_2$ laser comes to some 22%. This value has been estimated proceeding from the measured energy it takes for a fast-electron beam passing through the laser medium to detach a single fluorine atom from the SF$_6$ molecule (4.5 eV [4.91]) and the energy fraction going into the vibrational excitation of the HF molecule in the reaction F + H$_2$ (0.66 × 31.5 kcal/mole [4.97]). Similar estimates show the ultimate physical efficiency of the SF$_6$–C$_2$H$_6$ laser to be around 25%. The experimentally measured maximum physical efficiency values amount to 11% for the SF$_6$–H$_2$ laser [4.95] and 14% for the SF$_6$–C$_2$H$_6$ laser [4.94], i.e., about 50% of the ultimate values indicated above, and correspond to one quantum emitted per fluorine atom produced upon initiation. The output energy deficit in such lasers may be due to the mixture temperature rising as a result of energy liberation in the chemical pumping reaction and the adverse effect of relaxation processes. It is the joint action of these two factors that causes premature quenching of the lasing process. This also explains why the experimentally measured chemical efficiencies of these lasers are a factor of 2–3 lower than the ultimate values.

Knowing the physical efficiency η_{phys} of a laser and the efficiency η_i of the initiation source used, one can calculate the engineering efficiency of the laser by the formula $\eta_{eng} = \eta_{phys}\eta_i$. Considering the fact that the efficiency of modern electron accelerators reaches 20–50%, the engineering efficiency of electron-beam-initiated non-chain-reaction PCL's can be estimated at 2–5% for the SF$_6$–H$_2$ laser and 3–7% for the SF$_6$–C$_2$H$_2$ laser. Similar engineering efficiency values have also been achieved with electrical-discharge initiation of these lasers (see Table 4.2).

4.5 H$_2$(D$_2$)–F$_2$ Lasers

The most important parameters of PCL's based on the hydrogen fluorination chain reaction are listed in Tables 4.4–6. These parameters (engineering, physical, and chemical efficiencies, laser pulse duration, specific and total energy outputs, and spectral composition of the laser emission) adequately characterize the results achieved in the field since such lasers have made their appearance [4.101, 105] and up to the time the manuscript of this monograph has been written. The tables list mainly the maximum parameter values reported in the respective references cited. They also include information on experimental conditions, such as the optimum composition and total pressure of the mixture, laser cell dimensions, initiation scheme, initiation source parameters, and the initiation ratio $\Delta[F_2]_0/[F_2]_0$.

To characterize any H$_2$–F$_2$ laser comprehensively, it is essential to know the relationships between its main parameters and the mixture composition and pressure, the energy deposited in the mixture to initiate the pumping reaction, and the like. The reason is that the optimum conditions for the different laser parameters are far from being identical, and the knowledge of these conditions helps not only understand the physical and chemical aspects of the processes occurring in the laser, but also find ways of improving its performance. This aspect will be given special consideration in the present section.

The characteristics of H_2–F_2 lasers depend in a large measure on the mixture preparation procedures. Unfortunately, we have to note that the overwhelming majority of the experiments known from the literature fail to control the quality of the prepared mixture. What is more, they frequently report on the H_2–F_2 laser parameters as a function of the total energy available from the initiation source and not as that of the energy deposited in the mixture to initiate the reaction, and as demonstrated above, these two are not the same. Also, it should be noted once more that the extremely high gain of the hydrogen-fluoride laser makes it especially liable to parasitic oscillation. In that case, much of the laser energy is lost uselessly and the recorded laser energy parameters are thus reduced. All this shows how difficult it is to compare different experiments. But comparison between the results of a great many of such experiments allows one to describe the principal H_2–F_2 laser parameter relations reliably enough.

4.5.1 Spectral and Temporal Characteristics

The behavior of the H_2–F_2 laser emission spectrum is governed first of all by the initial relative distribution of the excited HF molecules produced in the chemical pumping reactions (1) and (2) of (4.17). This distribution is usually characterized by the relative constants $k(v)$ for the rates of population of the individual vibrational levels v. Many experiments have been conducted to establish these rate constants. For example, *Anlauf* et al. [4.125] have obtained $k(1):k(2):k(3)=0.31:1:0.48$ for reaction (1) and *Polanyi* and *Hiroshi* [4.126], $k(1):k(2):k(3):k(4):k(5):k(6):k(7):k(8):k(9)=0.12:0.13:0.25:0.35:0.78:1:0.4:0.26:0.16$ for reaction (2). It can thus be inferred that the vibrational energy distribution of the HF molecules features two maxima, one at $v=2$ being due to reaction (1) and the other at $v=6$ due to reaction (2). This feature is also generally reflected by the experimentally measured H_2–F_2 laser energy output distribution among vibrational transitions [4.127] (Fig. 4.10). As an example, Table 4.7 presents the lines observed within the limits of each vibrational band and their relative energies [4.128]. The total emission bandwidth of the HF molecules in the H_2–F_2 lasers extends from 2.67 μm to 3.33 μm.

An analysis of various experimental results shows that the HF-laser emission spectrum depends not only on the vibrational distribution of the energy released in the chemical pumping reactions, but also on a number of other factors, such as (1) the presence of unexcited ($v=0$) HF molecules in the mixture, which leads to the quenching of lasing in the $1\rightarrow0$ transition [4.129], (2) the quality factor of the laser cavity (to observe the complete set of transitions from $1\rightarrow0$ to $6\rightarrow5$ requires a high-quality cavity [4.127]), (3) the mixture composition, and (4) the level of initiation. It should be noted that increasing the initiation energy causes the laser emission spectrum to extend toward higher values of v and J, as illustrated by the data of [4.125] (see Table 4.4) and [4.122] (see Table 4.6). Hence it is clear why the measured distribution of the H_2–F_2 laser output energy among vibrational bands differs from experiment to experiment. Thus in [4.127, 128], most (65–80%) of the laser output came from the first three vibrational bands, in [4.20] the first two

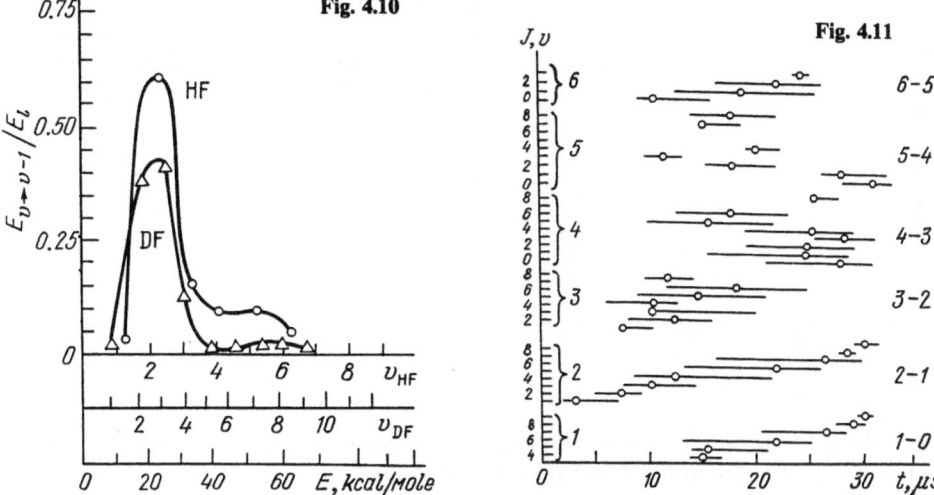

Fig. 4.10. Relative energy $E_{v \to v-1}/E_l$ of various vibrational bands of the H_2-F_2 and D_2-F_2 laser emissions as a function of the vibrational quantum number v of the upper laser level [4.127]

Fig. 4.11. Temporal sequence of transitions observed in the flash-photolysis initiated $H_2:F_2:He = 0.5:1:40$ mixture ($p = 50$ mm Hg) [4.128]. The vibrational band is indicated on the right. The circles mark the intensity peaks

vibrational bands yielded about 90% of the output energy, and in [4.129] the same proportion of the energy was obtained in the $1 \to 0$ band.

Figure 4.11 illustrates a typical temporal behavior of the H_2-F_2 laser emission spectrum. It is characterized by simultaneous lasing in several rotational transitions within the limits of each vibrational band, which certainly bears witness to the fact that the lasing HF molecules are not fully equilibrated rotationally. As discussed in Sect. 3.2, this may substantially reduce the laser energy output [4.130, 131]. Another characteristic feature is the increase with time of the rotational quantum number of the lasing transition within each vibrational band (Fig. 4.11). This results from the laser medium temperature growing higher in the course of the chemical pumping reaction and points to the fact that rotational equilibrium is only partially absent. And so, the composition of the HF laser mixture, especially the presence in it of species conducive to a rapid establishment of rotational equilibrium, must materially affect the laser emission spectrum. Experimental measurements have demonstrated that such species may be both the HF molecules themselves [4.132, 133] and complex fluorine compounds, such as C_5F_{12} [4.133].

As can be seen from Fig. 4.10, the measured output energy distribution of the D_2-F_2 laser among vibrational transitions is similar to that of the H_2-F_2 laser, the only difference being in the number of the populated vibrational levels. Table 4.8 presents the observed D_2-F_2 laser lines and their relative energies [4.127]. It should be emphasized that all the spectral measurement results listed in Tables 4.7 and 4.8 were obtained with axial pulse-discharge initiation of flowing $H_2(D_2)-F_2$ mixtures.

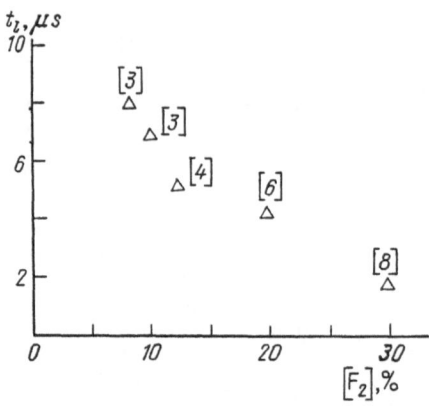

Fig. 4.12. Laser pulse duration as a function of the fluorine content of an H_2–F_2–O_2–He atmospheric-pressure mixture ($[F_2]/[O_2]=25$) [4.60]. The figures in brackets indicate the H_2 percentage

Since the gas pressure in such a discharge cannot be too high, the reactant pressures in these experiments did not exceed a few mm Hg.

It follows from above that studies into the spectral and temporal emission characteristics of the $H_2(D_2)$–F_2 lasers are essential both to the understanding of the physico-chemical processes occurring in the laser media and to the application of such lasers. It is especially important to know what the emission spectrum is like and to be able to control it when developing laser oscillator-amplifier systems where the matching of the spectral characteristics of the master oscillator and the amplifier is most critical (see Sect. 4.10).

Let us now consider the behavior of the total H_2–F_2 laser pulse duration as a function of the composition and pressure of the laser mixture and the initiation energy deposited in it. An analysis of the results presented in Fig. 4.12 leads to the following conclusion: the laser pulse duration decreases not only with increasing pressure of the fluorine and hydrogen gases kept in a fixed proportion (i.e., with the increasing energy content of the mixture), but also with the increasing relative concentration of fluorine, the hydrogen content of the mixture being kept fixed. The latter circumstance is obviously due in a large measure to a greater amount of initiation energy being deposited in the mixture. The relationship between the laser pulse duration and this energy has not been directly studied. But its behavior can be inferred reliably enough from the e-beam initiation results listed in Table 4.6. For instance, *Mangano* et al. [4.60] and *Gerber* et al. [4.122] used comparable fluorine and hydrogen pressures, but entirely different initiation ratios, and the laser pulse durations measured in these two experiments differ by more than an order of magnitude. Hence it follows that the H_2–F_2 laser pulse duration decreases not only with the increasing energy content of the mixture, but also with rising initiation ratio.

4.5.2 Effect of Oxygen and Mixture Preparation Techniques on Laser Performance Characteristics

It has been shown in Sect. 4.3 that to make a hydrogen-fluorine mixture stable, it is necessary to add oxygen to it. Oxygen prevents the mixture not only from self-ignition, but also from slow burn-out. By slowing down the dark reaction, oxygen reduces the production of unexcited HF ($v=0$) molecules in the laser medium prior to the start of the initiating pulse. There can be no doubt that the presence of such molecules in the active medium may adversely affect the energy characteristics of the laser by preventing lasing in the $v=1 \to v=0$ transition. Moreover, this accelerates the $V \to T$ relaxation of the excited HF molecules and vibrational energy redistribution in $V \to V$ processes. These effects are especially important in conditions of weak initiation. Indeed, addition of 0.1% HF to the mixture in the experiments [4.129] using flashlamp initiation quenched lasing in the $v=1 \to v=0$ transition and materially reduced the laser energy output.

But it would be wrong to stabilize any hydrogen-fluorine mixture by adding oxygen to it without restriction (which might seem a good way out of all the engineering difficulties involved in the mixture preparation), for the presence of oxygen in the active medium of a laser may reduce its energy output and efficiency. A qualitative explanation of this can be given by considering chain reaction mechanism (4.17) including chain termination process (3). Considering (2.40a)[4], the condition for the chain termination process to influence the laser energy characteristics in the case of instantaneous initiation, $t_l \gg (k_3[O_2]M)^{-1}$, may qualitatively be expressed as

$$\frac{[F]}{[F_2]} \gg \frac{k_3^2(\alpha^* - \alpha)M^2}{4k_2k_5}\frac{[O_2]^2}{[F_2]^2} . \tag{4.20}$$

This inequality may hold in conditions of flashlamp or low-current e-beam initiation where the initiation ratio is not very high. Then the role of oxygen in worsening the laser parameters becomes significant. This is well illustrated by Fig. 4.13 which shows that addition of 2 mm Hg of oxygen to the $H_2:F_2 = 1:1$

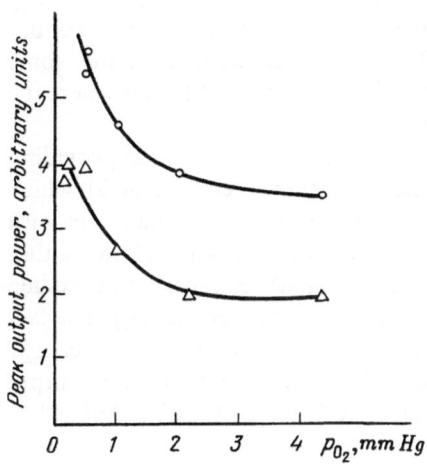

Fig. 4.13. Peak output energy of a H_2–F_2 laser as a function of the partial O_2 pressure ($p_{F_2} = p_{H_2}$ = 10 mm Hg, $T = 97$ K) [4.61]. The top curve was plotted for a higher initiation energy

---

[4]In (2.40a), α^* and α are the probabilities of the HF molecules being formed in an excited and the ground state, respectively.

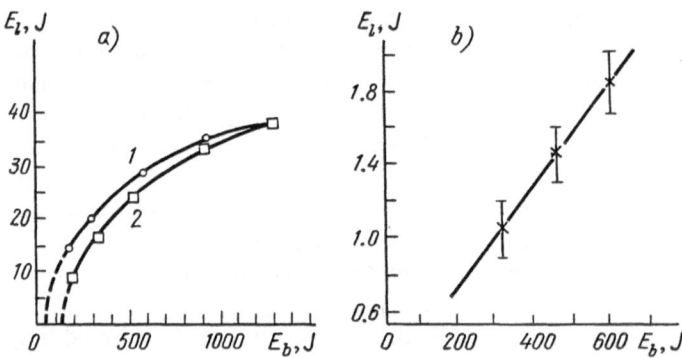

Fig. 4.14. Output energy of an H_2-F_2 laser as a function of photoinitiation energy: (**a**) tubular flashlamp initiation (*curve 1*—$F_2:H_2:O_2 = 10:3:5$ mixture, $p=250$ mm Hg; *curve 2*—oxygen content of the mixture increased from 70 to 85 mm Hg) [4.114]; (**b**) coaxial flashlamp initiation ($F_2:H_2:He = 0.04:0.12:0.84$, $p=1.1$ atm., $[F_2]/[O_2]=20$) [4.20]

mixture ($p=20$ mm Hg) reduced the peak laser energy by nearly one half in the case of weak initiation and by approximately a factor of 1.5 under stronger initiation. Comparison between curves 1 and 2 in Fig. 4.14a reveals the same tendency: increasing the partial oxygen pressure from 70 to 85 mm Hg reduced the laser energy output.

Thus, the oxygen content of the laser mixture is very critical where the initiation ratio is low. To prevent inequality (4.20) from being satisfied, the concentration ratio $[F_2]/[O_2]$ must be very high, and this may cause the mixture to ignite spontaneously, the particular self-ignition limits depending on the mixture preparation procedures employed. The maximum experimentally attained values of the $[F_2]/[O_2]$ ratio range between 12 and 40 [4.3, 20, 60, 123]. With higher ratios, the mixture was not stable enough, and with lower ones, the laser energy output was reduced. The difficulty of preparing a stable hydrogen-fluorine mixture with a low concentration of oxygen relative to that of fluorine made some investigators [4.32, 37, 60] operate in continuous flow-mixing conditions where the burn-out of a low-oxygen mixture is insignificant.

The situation is much simpler where use is made of initiation sources providing for high initiation ratios. That this is actually the case can be seen from (4.20) which shows that the maximum permissible oxygen content of the mixture increases with the initiation ratio. Qualitatively this effect is illustrated in Fig. 4.14a showing that the difference in output energy between lasers using mixtures differing in oxygen content can go down to zero as the initiation energy is increased [4.114]. The fact that strong initiation materially weakens the effect of oxygen on the energy characteristics of the H_2-F_2 laser made it possible to achieve good laser output parameters even with the $[F_2]/[O_2]$ ratio reduced to 2.5 (see [4.122] in Table 4.6).

4.5.3 Energy Characteristics

Before giving the maximum parameters of H_2-F_2 lasers achieved in practice, we will dwell upon the main energy characteristics. The energy characteristics of hydrogen-fluoride lasers depend not only on the mixture preparation procedures, but also, as is the case with any other laser, on the method of coupling emission out of the laser cavity. H_2–F_2 lasers are noted for their extremely high gain reaching a few tenths of a cm^{-1} even under weak initiation. This greatly eases the problem of coupling energy out of the laser cavity. Figure 4.15 shows an experimental output energy vs. coupling fraction curve for an H_2–F_2 laser initiated with a coaxial flashlamp. The maximum output energy was achieved with the output mirror in the form of a plane-parallel sapphire plate (about 88% output coupling). Accordingly, the cavity volume here is more densely filled with radiating modes than in other lasers with a lower gain. For this reason, H_2–F_2 laser cavities commonly use one or two plane-parallel plates of sapphire or CaF_2 for output coupling.

Let us now consider the effect of mixture composition and pressure on the output energy and the physical and chemical efficiencies of H_2–F_2 lasers. Figures 4.16–18 show the above parameters as a function of mixture composition for various initiation techniques and ratios. What conclusions can be drawn from comparison between these figures? First, in the case of weak initiation (be it flash-photolysis or a low-current c-beam initiation), the laser energy output and physical efficiency increase with the $[H_2]/[F_2]$ ratio (varied by changing the partial pressure of H_2 with that of F_2 remaining fixed and the total mixture pressure thus varying but little), as long as this ratio remains below 0.3–0.5, but then it ceases to rise. On the other hand, under strong initiation these laser parameters increase continuously with the $[H_2]/[F_2]$ ratio growing at least to unity. The cause of this difference in behaviour is not clear. Most likely it is due to a difference in initiation ratio. Second, the chemical efficiency of the lasers decreases rather strongly with increasing $[H_2]/[F_2]$ ratio, no matter what the initiation ratio, except for the region $[H_2]/[F_2] < 0.25$ in the case of weak initiation (Fig. 4.17). The decrease of the laser energy output in this mixture parameter region is due to the fact that the active medium gain in this region is close to the threshold value.

To reduce the harmful effect of the excessive heating of the active laser medium during the chemical pumping reaction and to ease the preparation of the H_2–F_2 mixture, the latter is diluted with substances of high heat capacity which do not

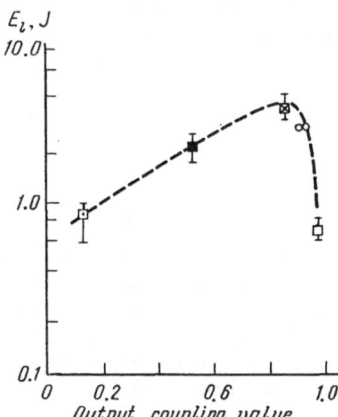

Fig. 4.15. Output energy of an H_2–F_2 laser as a function of the relative transmissivity of the output cavity mirror. For more detail, see [4.20]

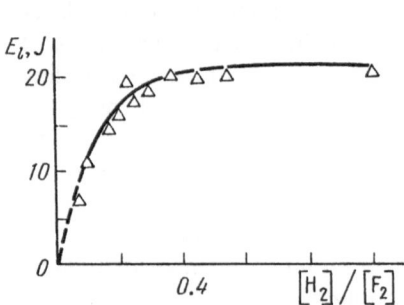

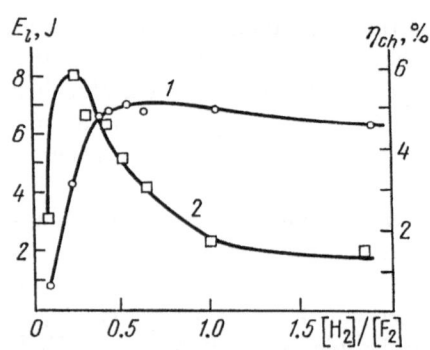

Fig. 4.16. Output energy of a flash-photolysis initiated H_2–F_2 laser as a function of the $[H_2]/[F_2]$ ratio at $p_{F_2} = 74$ mm Hg and $p_{O_2} = 37$ mm Hg [4.114]

Fig. 4.17. Output energy (*1*) and chemical efficiency (*2*) of a low-current e-beam initiated H_2–F_2 laser as a function of the $[H_2]/[F_2]$ ratio in the F_2:H_2:O_2:He = 100: $[H_2]$:8:333 mm Hg laser mixture [4.123]

affect adversely the laser energy characteristics. Inert gases or nitrogen are most commonly used for this purpose. Typical output energy vs. total mixture pressure curves for H_2–F_2 lasers with low and high initiation levels are shown in Fig. 4.19 and 4.20, respectively. As can be seen, the laser energy is linear for mixture pressures up to atmospheric, no matter what the initiation level. As the pressure is further raised, the relation starts to deviate from the linear law. Particularly strong deviations (up to saturation) are observed when adding 100 mm Hg of SF_6 to the mixture. But the causes of this deviation in the case of weak and strong initiation are different. In the former case, the deviation is explained by the deposition of the e-beam energy along the propagation direction becoming nonuniform as the mixture pressure is increased. The point is that with the e-beam current being small, the electron energy usually amounts to some 200–300 keV, and with the electron energy being so low, an increase in the mixture pressure may make the electron flight distance shorter than the laser cell length, thus giving rise to a longitudinal initiation inhomogeneity. The nonlinearity in the case of strong initiation is due to the laser pulse delay time and duration becoming shorter than the initiation time so that a rise of the mixture pressure is accompanied by a decrease in the rate of growth

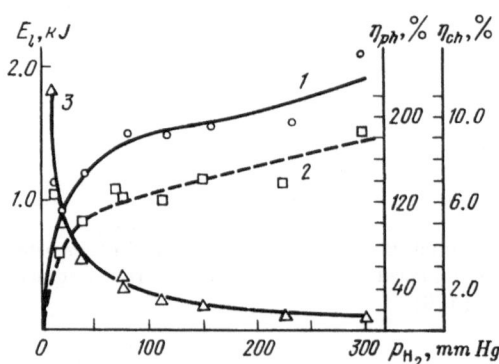

Fig. 4.18. Output energy (*1*) and physical (*2*) and chemical (*3*) efficiencies of a high-current e-beam initiated H_2–F_2 laser as a function of the partial H_2 pressure with the partial F_2 and O_2 pressures being fixed at 300 and 90 mm Hg, respectively [4.57]

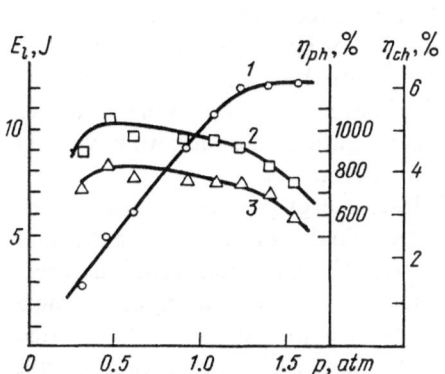

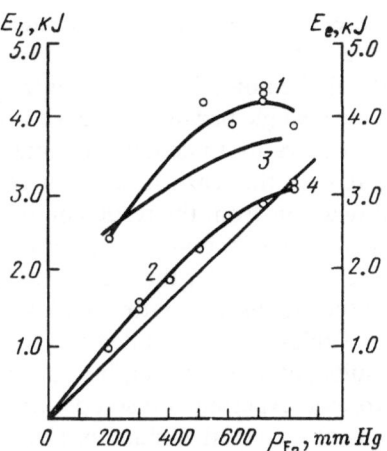

Fig. 4.19. Output energy (*1*) and physical (*2*) and chemical (*3*) efficiencies of a low-current e-beam initiated H_2–F_2 laser as a function of the total pressure of the $H_2:F_2:O_2:He=6:15:1.2:50$ laser mixture [4.123]

Fig. 4.20. Output energy (*curves 1* and *2*) and e-beam energy deposition (*curves 3* and *4*) of a high-current e-beam initiated H_2–F_2 laser as a function of the partial F_2 pressure in the $F_2:H_2:O_2 = 1:0.25:0.3$ laser mixture [4.57]: *curves 1* and *3*—with 100 mm Hg of SF_6 added to the mixture; *curves 2* and *4*—without SF_6

of the initiation ratio and hence of the laser energy. It can thus be inferred that insofar as the output energy of the H_2–F_2 laser is proportional to the mixture pressure (i.e., to both the amount of initiation energy deposited in the mixture and the amount of chemical energy stored in it), provided the pressure is not very much in excess of the atmospheric value, the physical and chemical efficiencies of the laser must depend but little on the mixture pressure within the limits indicated, regardless of the initiation ratio. This is clearly demonstrated in Fig. 4.19 for the case of weak initiation.

The energy parameters of the H_2–F_2 laser also strongly depend on the amount of energy spent to initiate the chain pumping reaction or, in other words, on the initiation ratio. At first glance, the relationships shown in Fig. 4.14 between the laser energy output and the amount of electrical energy stored in the initiation capacitor bank for the cases of initiation with a coaxial and a tubular flashlamp seem to differ in character. However, it has been found that in the case of initiation with the coaxial flashlamp, the atomic fluorine concentration is approximately proportional to the amount of energy stored in the capacitor bank. But the energy output of the H_2–F_2 laser can in this case be considered to be proportional to the initiation ratio, at least until the latter reaches 0.4% (see [4.20] in Table 4.4). In the case of initiation with the tubular flashlamp, no attempt was made to find the relation between the initiation ratio and the amount of energy stored in the capacitor bank, and this relation is in fact not necessarily linear. Moreover, the discordance may also be due to the differences between the mixture pressures and compositions used in these two cases. Thus, the above disagreement is most likely apparent than real. It follows from Fig. 4.14b that the physical efficiency of the

H_2–F_2 laser must be independent of the initiation ratio (up to around 1%), whereas its chemical efficiency must be proportional to it.

It should be noted in conclusion that all the above relationships were obtained by varying a single parameter, be it the composition or pressure of the mixture, or else initiation ratio. In actual experiments, it frequently happens that two or three are variable. In that case, it becomes very difficult, even if at all possible, to derive general relations from the results of different investigations. Besides, we have only used the results of the most adequately staged experiments.

Let us now consider the maximum H_2–F_2 laser energy parameters attained by the time the manuscript of this book went to press. To do this, we refer to the data listed in Tables 4.3–6 compiled in accordance with the initiation technique used. Two results obtained with flash-photolysis initiation attract our attention. Using a small-volume experimental setup with an atmospheric-pressure mixture initiated by means of a coaxial flashlamp, *Chen* and co-workers [4.20] achieved a specific energy output of about 80 J/l at an engineering efficiency of some 1.3% and a chemical efficiency around 4%. As demonstrated above, one can only hope to attain higher engineering efficiencies if the size of the setup is enlarged so as to "intercept" most of the radiant energy of the photosource. And indeed, using a U-shaped flashlamp in a setup 12.8 l in active volume and about 10 cm in active region diameter, *Nickols* et al. [4.3] managed to obtain high engineering and chemical efficiency values ($\approx$ 20–30% and 8%, respectively), but the specific energy output of their laser was not very high (around 23 J/l). Roughly the same parameters were also achieved by *Bashkin* et al. [4.117].

Electrical-discharge initiation helped to reach higher engineering efficiencies (100% [4.32] and even 144% [4.118]), but the specific energy outputs achieved with this initiation technique remained rather low because of the difficulties involved in increasing the energy content of the laser mixture.

Fast electron-beam initiation naturally allowed high specific energy outputs to be obtained along with high physical efficiencies. When initiating a fluorine-rich mixture at nearly atmospheric pressure with a high-current beam of fast electrons, *Gerber* and co-workers [4.122] obtained at least 100 J/l in specific energy output at a physical efficiency of 180% and a chemical efficiency of 3.3%. It was difficult to evaluate the output more accurately, the active laser medium volume being indefinite because of the lack of information on the scattering of the electron beam upon its passage through the medium. *Bashkin* et al. [4.123] achieved a high enough specific energy output of some 91 J/l at a physical efficiency of 940% and a chemical efficiency of 4.7% by using electron beams of lower current densities.

The last result illustrates vividly the still untapped possibilities of improving the parameters of pulsed H_2–F_2 lasers. Indeed, bearing in mind the fact that to produce a single fluorine atom takes around 8–12 eV of energy in the case of e-beam initiation [4.11, 60, 134, 135] and only some 2 eV under photoinitiation and that the actual efficiency of photoinitiation sources amounts to a few percent, the potential engineering efficiency of the flash-photolysis initiated hydrogen-fluoride laser can be estimated to be at least 100% (and at a high specific energy output at that), which is much in excess of what has been reached to date. This means that particular emphasis should be placed on matching the characteristics of the photoinitiation source to those of the laser medium. It may then prove possible to obtain specific energy outputs as high as 100 J/l at an engineering efficiency in the neighborhood of 100% even with so simple an initiation technique (compared to e-beam initiation) as flash photolysis, provided there are reliable photoinitiation sources available possessing the necessary efficiency and capable of operation at a flash duration of 1–2 μs.

Another as yet incompletely utilized possibility of improving chemical laser parameters is the effective use of the electroionization initiation technique. Setting up a dc field close to the breakdown value in strength in an SF_6–H_2 mixture raised the output energy of the laser 2–3 times [4.91, 95]. A similar improvement was also observed with an F_2:H_2:He:Ar = 6:3:54:37 mixture at atmospheric pressure: the establishment of a dc field of about 11 kV/cm in it increased the specific energy output four times (up to 42 J/l) at an engineering efficiency of about 150% and a chemical efficiency of some 6.3% [4.37]. It is interesting to note that the physical efficiency of this laser was about 850% in the absence of the d.c. field. But the ultimate capabilities of the electroionization initiated H_2–F_2 laser are as yet unknown because of the involved character of the processes occurring in its active medium. This question requires further investigations to clarify.

The D_2–F_2 lasers have been studied much less than their H_2–F_2 counterparts. But in principle, there is no reason to believe that the main relationships in them will behave differently. As to their energy parameters, these were usually found to be a factor of 2–2.5 lower under similar experimental conditions (see [4.3] in Table 4.4 and [4.124] in Table 4.6).

4.6 D_2–F_2–CO_2 Lasers

Since the D_2–F_2–CO_2 pulsed chemical lasers made their appearance [4.136], they have received much less attention than the H_2–F_2 lasers. It is therefore too early to consider their behavior to be completely understood. Nevertheless, the information available on such lasers using flash-photolysis initiation allows us to make some relevant remarks on them.

In the case of flash-photolysis initiation, the initiation ratio is only determined by the flash intensity and duration, no matter what the mixture composition and pressure, provided the laser pulse duration exceeds the flash duration.

The optimum laser mixture composition can be determined from the analysis of Figs. 4.21–23. The maximum at $[CO_2]/[D_2] \simeq 5$ in Fig. 4.21 can be explained by the double role the CO_2 molecules play: the rise of the partial CO_2 pressure is first accompanied by an increase in the rate of energy transfer from the DF to CO_2 molecules and the corresponding growth of the laser energy output, but then, as the pressure is raised further, there may come into play a number of adverse factors, such as an increase in the threshold population density of the upper laser level, a decrease in gain as a result of the lasing transition linewidth growing wider, an increase in the collisional relaxation rate of the molecules at the upper laser level, a decrease in the chemical pumping reaction rate due to the enhancement of the heat

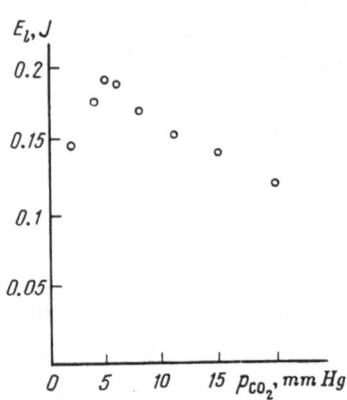

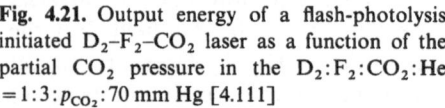

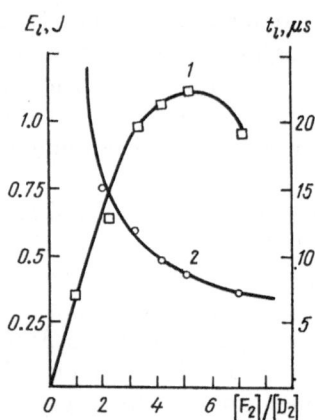

Fig. 4.21. Output energy of a flash-photolysis initiated D_2–F_2–CO_2 laser as a function of the partial CO_2 pressure in the D_2:F_2:CO_2:He $=1:3:p_{CO_2}:70$ mm Hg [4.111]

Fig. 4.22. Output energy and pulse duration of a flash-photolysis initiated D_2–F_2–CO_2 laser as a function of the $[F_2]/[D_2]$ ratio at $[CO_2]/[D_2]$ $=6$ and $p_{D_2}=10$ mm Hg [4.111]

capacity of the mixture, and an increased chain termination rate with $M \equiv CO_2$. *Agroskin* and co-workers [4.112] suggested that the last factor was predominant, their experiments demonstrating the specific laser energy output at constant initial reaction and chain termination rates (achieved by varying the initial oxygen concentration and initiation energy) to grow steadily with CO_2 concentration up to saturation without any drop until He was completely replaced by CO_2.

A maximum is also observed in the plot of the laser energy output against the relative fluorine content of the mixture (see Fig. 4.22). The laser energy grows linearly up to $[F_2]/[D_2] \simeq 3$, which can be explained by the reaction rate increasing at a constant initiation ratio. But as the $[F_2]/[D_2]$ ratio grows further, the laser energy output first slows its growth and then drops. This drop is most likely due to experimental difficulties involved in the preparation of a stable fluorine-rich mixture. An increase in the rate of relaxation of the excited CO_2 molecules on fluorine molecules is also possible.

As the heating of the active laser medium in the course of the chemical pumping reaction reduces the population inversion between the laser levels of the CO_2 molecule, the D_2–F_2–CO_2 laser mixture is usually diluted with some substances of a high heat capacity. Figure 4.23 presents experimental relationships between the laser energy output and the partial pressures of the inert diluents He and Ar. These curves also feature a maximum at $[He]/[D_2]=60$. The use of such a mixture (D_2:F_2:CO_2:He$=1:3:6:60$, $p=70$ mm Hg) yielded a very high chemical efficiency of 6.4% for the laser medium volume occupied by the radiating mode (or around 1.6% for the total laser medium volume). It should be noted, however, that such mixtures with their low energy content will naturally provide only low specific energy outputs. Oxygen added to the D_2–F_2–CO_2 laser mixture must in principle have the same effect on the energy characteristics of the laser as in H_2–F_2 lasers. And though the stabilizing effect of the CO_2 molecule is fairly high, to prepare a D_2–F_2–CO_2 mixture at atmospheric pressure ($T \simeq 300$ K), one has to add to it

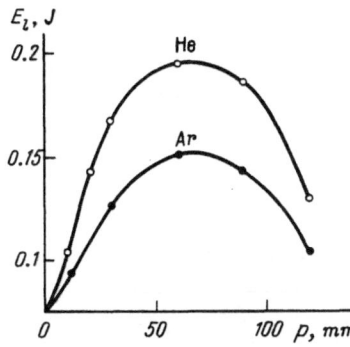

Fig. 4.23. Output energy of a flash-photolysis initiated D_2–F_2–CO_2 laser as a function of the partial He (Ar) pressure in the D_2:F_2:CO_2:He (Ar) = 1:3:6:$p_{He(Ar)}$ mixture at p_{D_2} = 1 mm Hg [4.111]

$\geqslant 5\%$ oxygen relative to the fluorine concentration, which amounts to a few mm Hg in terms of partial pressure [4.110]. Inasmuch as the pulse duration of the D_2–F_2–CO_2 laser is longer than that of the H_2–F_2 laser under similar initial conditions (see [4.124] in Table 4.6), addition of oxygen to the D_2–F_2–CO_2 laser mixture can also be expected to have an adverse effect. Unfortunately, no special quantitative investigations have been conducted in this respect.

The behavior of the energy output and pulse duration of the D_2–F_2–CO_2 laser as a function of the total laser mixture pressure and initiation energy is in principle the same as in the H_2–F_2 laser. In other words, increasing the initiation energy or mixture pressure, or else the partial pressure of fluorine (see Fig. 4.22), reduces the pulse duration of the laser but raises its energy output. However, the main characteristics of the D_2–F_2–CO_2 laser are still not clearly understood. Thus, the character of the mixture pressure dependence of the laser energy output differs from experiment to experiment. For example, in [4.107] this dependence is linear almost up to atmospheric pressure (Fig. 4.24a), whereas in [4.111] it starts deviating from the linear law at a pressure of some 300 mm Hg (Fig. 4.24b). A similar discrepancy is observed between the dependences of the laser energy output on the photoinitiation source energy obtained in different experiments. In [4.107], this dependence tends

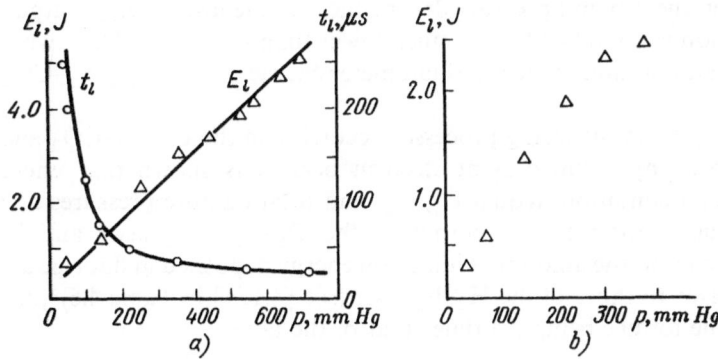

Fig. 4.24. Output energy of a flash-photolysis initiated D_2–F_2–CO_2 laser as a function of the total mixture pressure for (**a**) D_2:F_2:CO_2:He = 1:1:6:19 [4.107] and (**b**) D_2:F_2:CO_2 = 1:3:5 [4.111] mixtures. The plot for (**a**) also includes the laser pulse duration *vs.* total mixture pressure curve

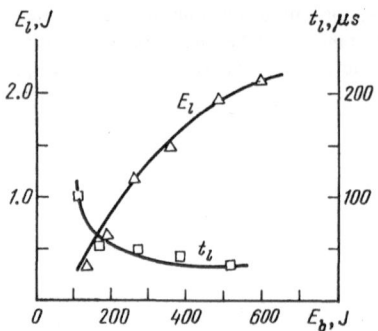

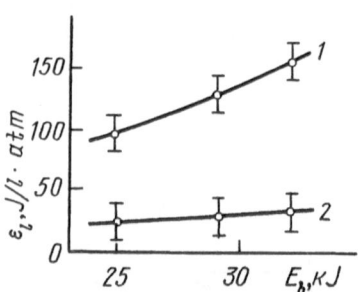

Fig. 4.25. Output energy and pulse duration of a flash-photolysis initiated D_2–F_2–CO_2 laser operating on the $D_2:F_2:CO_2:He=1:1:6:19$ ($p=250$ mm Hg) mixture as a function of the photoinitiation source energy [4.107]

Fig. 4.26. Specific energy output of a creeping-discharge initiated D_2–F_2–Co_2 atmospheric-pressure laser amplifier as a function of the photoinitiation source energy [4.110]: 1 — $D_2:F_2:CO_2:He=1:1:4:5$ mixture; 2 — $D_2:F_2:Fe:CO_2:He=1:1:6:15$ mixture

towards saturation (Fig. 4.25), with the laser pulse duration exceeding the initiation time (flash duration) irrespective of the experimental conditions. At the same time, in [4.110], for a mixture of almost the same composition, no saturation was observed even at a much higher photoinitiation source energy (Fig. 4.26). Unfortunately, these two studies used different types of photoinitiation source and failed to report on the relationship between the initiation ratio and the source energy. It therefore remains to be seen whether the above observations actually disagree. However, an explanation of the cause of such a disagreement can be offered by an analysis of the results reported in [4.112] where the specific energy output yielded by a mixture of a certain composition was observed to grow with initiation ratio, no matter what the mixture pressure, first monotonically and then with a tendency to saturate at high initiation ratio values, the deviation from the linear law starting sooner at lower mixture pressures. The mixture pressure in [4.107] was substantially lower than in [4.110]. It may be likewise with the discordance between the mixture pressure dependences of the laser energy output. After all, the initiation ratio in [4.11] was much lower than in [4.107]. Thus, what we see here is the manifestation of the multiparameter character of the D_2–F_2–CO_2 laser system.

A comparative study of the lasing processes occurring in the D_2–F_2–CO_2 and H_2–F_2 lasers initiated by a low-current electron beam has shown that under identical experimental conditions (equal F_2, H_2, and total mixture pressures and initiation ratios) the specific energy output of the D_2–F_2–CO_2 laser and its efficiency with reference to the amount of initiation energy deposited in fluorine are in practice not inferior to those of the H_2–F_2 laser (see [4.124] in Table 4.6)[5], the pulse duration of the former being 2.5 times that of the latter.

---

[5]A similar study was made earlier in [4.112] for flash-photolysis initiation, and the results of this work disagree with those of [4.124].

As a rule, the emission spectrum of the D_2–F_2–CO_2 laser consists of several P-branch lines, the major part of the emission being concentrated in the $P(16)$, $P(18)$, and $P(20)$ lines (see, for example, [4.137, 112]). It is precisely in the $P(20)$ transition that the maximum unsaturated gain (around $0.07\ cm^{-1}$) was achieved in a D_2–F_2–CO_2 laser initiated by a high-power creeping discharge [4.110]. A slightly lower gain (0.04–$0.06\ cm^{-1}$) was obtained with a coaxial flashlamp initiation [4.138].

Let us now dwell upon the question of attaining the maximum chemical efficiency and specific energy output in D_2–F_2–CO_2 lasers. As the measured unsaturated gain values in the flash-photolysis initiated lasers of this type are usually not very high (0.02–$0.03\ cm^{-1}$ [4.107, 109]), use has to be made of cavities with a sufficiently high Q-factor, in which the radiating mode volume only accounts for part of the total cavity volume. For this reason, many investigators frequently state their experimentally obtained laser parameters in terms of the cavity volume occupied by the radiating mode. With this approach, the chemical efficiency would have reached around 16% in [4.111] and even somewhat over 20% in [4.107], while the theoretical ultimate chemical efficiency of the D_2–F_2–CO_2 laser is in the neighborhood of 20%. If such an emitting-mode volume correction of chemical efficiency (restored to in [4.107]) is considered legitimate, a similar correction of the specific energy output reported in [4.111] (see Table 4.4) will yield no less than 120 J/l atm. But in another experiment [4.110] using much more intense initiation of a mixture of almost the same composition in a laser amplifier with the cavity volume wholly occupied by the emitting modes, the specific energy output was less than 50 J/l atm. All this highlights the need to exercise extreme caution in deciding on the legitimacy of such corrections.

It can be seen from Table 4.4 that the maximum specific energy output obtained to date from a D_2–F_2–CO_2 laser initiated by a high-power creeping discharge amounts to some 150 J/l atm. in the laser amplifier mode ($\eta_{chem} \simeq 7\%$) and 70 J/l atm. in the laser oscillator mode [4.110]. A sufficiently high specific energy output ($\simeq 70$ J/l atm.) was achieved in a D_2–F_2–CO_2 laser initiated with a coaxial xenon-filled flashlamp [4.112]. In the case of flash-photolysis initiation, as with the H_2–F_2 laser, conducting experiments in sufficiently large volumes made it possible to achieve an engineering efficiency of approximately 12% [4.2]. The engineering efficiency of the flash-photolysis initiated D_2–F_2–CO_2 laser can further be improved, as evidenced by the fact that the efficiency of such a laser measured with reference to the amount of energy deposited in fluorine is in practice no worse than that of the H_2–F_2 laser. And so, all the considerations given in this regard in Sect. 4.5 are applicable to this case as well.

4.7 Effect of Additives Accelerating the Hydrogen Fluorination Chain Reaction

It follows from the preceding sections that the energy characteristics of the hydrogen-fluoride lasers based on the hydrogen fluorination chain reaction depend on both the partial fluorine pressure and initiation ratio. And given the mixture composition, the initiation ratio can be raised by increasing the initiation source energy, but this is not always technically feasible. There is another way to accelerate the chain reaction. It consists in adding to the hydrogen-fluorine mixture special substances serving as atomic fluorine donors. In other words, these substances yield additional fluorine atoms under the action of the initiation source used (e.g., an e-beam or a flashlamp). In that case, the concentration of fluorine atoms has no direct relation to that of molecular fluorine, which makes it possible to vary the laser initiation ratio over a wide range without changing the initiation source energy.

These atomic fluorine donors may act by different mechanisms, depending on what initiation technique (e-beam or flash-photolysis) a given chemical laser uses. Inasmuch as the electron-impact initiation of fluorine-bearing mixtures relies on dissociative electron attachment processes leading to the loss of electrons, the use of fluorine donors is only advisable where the initiating electron beam has surplus energy. Such experiments were conducted with SF_6 added to the H_2-F_2 mixture of an e-beam initiated chain-reaction laser [4.57]. The positive effect of introducing SF_6 molecules into the laser mixture is well illustrated in Fig. 4.20.

When analyzing the effect of the above substances in the case of flash-photolysis initiation, one has to consider two cases: one where the absorption band of the fluorine donor coincides with that of fluorine itself and the other where the absorption bands of the two do not coincide. The former case is in principle analogous to the case of e-beam initiation. Adding the donor in the latter case may be useful from the standpoint of raising the engineering efficiency of the laser by improving the "interception" of the photoinitiation source emission; in this case, it is as if the spectral range of the source quanta absorbed by the laser medium is broadened. The fluorine donor substances used in the case of flash-photolysis initiation are sometimes referred to as photoinitiating components (PIC's). According to (4.1), the contribution from the addition of a PIC to the laser mixture will be appreciable if the condition

$$(\Phi\sigma I_0 N_0)_{\text{PIC}} \geqslant (\Phi\sigma I_0 N_0)_{F_2} \tag{4.21}$$

is satisfied. To make the analysis more comprehensive, the above inequality should be supplemented with the relation between the depths of homogeneous initiation in fluorine and the PIC. If $l_{F_2} \gg l_{\text{PIC}} \gg l_l$ (l_l being the transverse dimension of the laser cell), then $(\sigma N)_{\text{PIC}} \gg (\sigma N)_{F_2}$, and so inequality (4.21) obviously holds because the quantum yield usually has little variation (from 0.5 to 2[4.103]). To state it differently, where the partial fluorine pressure and the transverse dimension of the laser cell are not very great, the role of a PIC will be significant irrespective of whether the absorption bands of fluorine and the PIC coincide or not.

To make quantitative estimates, we reduce (4.21) to the form

$$\frac{N_{F_2}}{N_{\text{PIC}}} \leqslant \frac{(\Phi\sigma I_0)_{\text{PIC}}}{\Phi\sigma I_0)_{F_2}}$$

and use the relative values of the product $\Phi\sigma I_0$ (Table 4.9) reported in [4.4] for a particular xenon-filled flashlamp and some fluorides that can be used as PIC's. The data listed in Table 4.9 characterize the necessary relative concentrations of these PIC's.

The situation changes where $l_{F_2} = l_{PIC} = l_l$. In that case, $(\sigma N)_{PIC} = (\sigma N)_{F_2}$ and inequality (4.21) holds with difficulty, if at all. Hence it can be inferred that where the partial pressure of fluorine and the transverse dimension of the laser cell are great enough, the effect of PIC's is drastically reduced, and their use is only justifiable if their absorption band does not coincide with that of fluorine.

The MoF_6 molecule has been the most widely studied experimentally in the capacity of a PIC. Using a $F_2:H_2:MoF_6:He = 1:1:0.25:40$ mixture with a total pressure of 600 mm Hg, *Hess* [4.113] obtained a laser energy of 65 mJ. The same mixture yielded only 16 mJ when deprived of MoF_6 and a mere 9 mJ when deprived of F_2. Thus, adding MoF_6 to the above hydrogen-fluoride laser mixture increased the laser pulse energy some 2.5 times compared to the total energy of the lasers using F_2-H_2-He and MoF_6-H_2-He mixtures. It was emphasized in this study that as the mixture pressure was reduced, the effect of the PIC became more pronounced.

Basov and co-workers [4.108, 109] also added MoF_6 to a mixture made up of F_2 (45 mm Hg), D_2 (45 mm Hg), CO_2 (187 mm Hg), and He (480 mm Hg), initiated by a bursting wire emission. Figure 4.27 shows the specific energy output of this laser as a function of the partial pressure of MoF_6. The maximum specific energy output (somewhat over 20 J/l atm.) was obtained at a MoF_6 pressure of 1 mm Hg. At higher MoF_6 pressures the results were difficult to reproduce because of fog forming in the laser cell. The cause of this fogging is not clear. One possible cause may be complexing involving the MoF_6 molecules or their photolysis products.

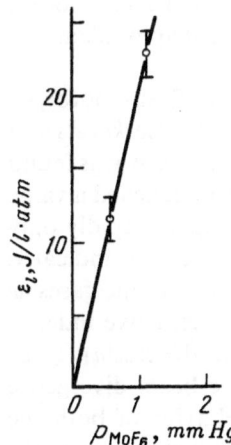

Fig. 4.27. Specific energy output of a bursting-wire initiated $D_2-F_2-CO_2$ atmospheric-pressure laser as a function of the partial pressure of the PIC MoF_6 [4.109]

4.8 Beam Divergence in Hydrogen-Fluoride Lasers

As is the case with other types of laser, the beam divergence of hydrogen-fluoride lasers depends mainly on the following two factors: the type of the optical cavity used and the variation of the refractive index of the laser medium in the transverse direction. In this respect the results of *Gerber* and *Patterson* [4.57] are typical. They investigated an H_2–F_2 laser using three different types of cavity: a plane-parallel cavity, a concave-mirror cavity, and an unstable cavity. The widest beam divergence (35×10^{-3} rad) was observed with the first two types of cavity. As with other lasers featuring a high unsaturated gain, the use of the unstable cavity improved the beam divergence by almost an order of magnitude, narrowing it down to 4×10^{-3} rad. The search for optical cavity systems most suitable for hydrogen-fluoride lasers is thus a high-priority task.

What might be advantageous here is the use of the laser amplifier mode, no cavity being required in this case. It would then be possible to improve the laser beam divergence up to the ultimate limit set by the inhomogeneity of laser medium initiation in the transverse direction, leading to variations in the refractive index of the medium. These variations may be due to (1) a nonuniform heating of the active medium, (2) changes in the polarizability of the active medium as a result of the formation of new molecules and the decay of the initial ones, and (3) gas-dynamic displacements of the active medium as a result of the development in it of regions differing in pressure. But in principle, short pulse durations (up to a few microseconds) in pulsed hydrogen-fluoride lasers must prevent large gas-dynamic displacements of their active medium. The effect of the first two causes is as yet imperfectly understood. The inhomogeneity of initiation also gives rise to a transverse gain gradient which in turn may substantially affect the laser beam divergence. The use of the laser amplifier mode here should also help weaken this effect [4.139].

The variation of the refractive index of the active medium in flash-photolysis initiated hydrogen-fluoride lasers was experimentally studied by *Belotserkovets* and *Zykov* et al. [4.140, 141]. The greater part (> 85–95%) of the laser energy was found to be released before the moment the refractive index variation $|\Delta n|$ reached a value of $\leqslant 2 \times 10^{-7}$, the refractive index being observed to start changing rapidly only near the end of the laser pulse, owing most likely to the first two causes indicated above. It was also found experimentally that addition of such polyatomic gases as CO_2 and SF_6 to the laser medium reduced the variation of its refractive index.

The laser beam divergence θ estimated by the formula $\theta = 2(2\Delta n)^{1/2}$ for $|\Delta n| = 2 \times 10^{-7}$ amounts to 10^{-3} rad. Actually, to calculate the beam divergence more accurately, it is necessary to establish the temporal behavior of both the refractive index variation Δn and the laser energy generation and then, knowing the relation between Δn and the laser energy at any moment, average Δn over time. For this reason, the beam divergence $\theta = 10^{-3}$ rad. estimated above is but an upper limit. There is a lot of work to do to elucidate this matter and find ways to reduce the beam divergence of PCL's and maximize at the same time their energy characteristics. Much depends here on the engineering aspects of the initiation system design.

4.9 Pulsed Chemical Lasers Operating on Vibrational Overtones of HF and DF Molecules

The capabilities of pulsed chemical lasers will undoubtedly be materially extended if their wavelength is shortened. One possible way to attain this end is to use lasing on vibrational overtones of various active molecules. In the case of HF lasers, for example, this will make it possible to use glass as a material for making laser optical elements instead of quartz or salts. In this respect, halving the laser output wavelength seems to be of principal importance. Shortening laser wavelengths is also essential for narrowing the laser beam divergence. Moreover, laser emissions at wavelengths corresponding to the vibrational-rotational transitions with $\Delta v = 1$ in HF are characterized by a higher atmospheric absorption (except for narrow atmospheric bands) compared to the shorter-wavelength transitions with $\Delta v > 1$. Besides, the usefulness of coherent emissions covering new spectral regions is beyond any doubt.

In connection with the above considerations, it is certainly interesting to study lasers relying on the overtone transitions of the HF molecule. That such lasers are feasible can be seen from Table 4.10 which lists calculated relative intensities of various transitions in HF and DF. The ratio between the relative intensities of transitions with $\Delta v = 1$ and $\Delta v = f$ is determined by the ratio between the respective dipole moment matrix elements of these transitions:

$$\frac{|m_v^{v+f}|^2}{|m_0^1|^2} = \frac{X_e^{f-1}(v+f)!}{f^2 v} \quad \text{for} \quad (2v+f) < X_e^{-1} \; ,$$

where X_e is the spectroscopic constant of the molecule ($X_e = 0.0218$ for HF [4.142] and $X_e = 0.01575$ for DF [4.143]). The experimentally measured values of the unsaturated gain α in lasers using lean mixtures range between 10^{-1} and 10^{-2} cm^{-1} ($\Delta v = 1$). For example, *Hess* [4.64] measured α at around 3×10^{-2} cm^{-1} for the flash-photolysis initiated F_2: H_2: $He = 3:1:60$ mixture at a total pressure of 128 mm Hg. Hence, according to Table 4.10, one can expect to achieve $\alpha \simeq 10^{-3} - 10^{-4}$ cm^{-1} for transitions with $\Delta v = 2$. This means that with the laser medium length L being equal to 1 m, lasing on transitions with $\Delta v = 2$ can be effected using 90–98% reflecting mirrors in the cavity. The development of a laser depending for its operation on transition with $\Delta v = 3$ and over is more problematic.

To use chemical pumping reactions for which the peak of the vibrational energy distribution of the excited products falls at high vibrational levels (see Sect. 4.5) seems to be an extremely simple way to obtain first-overtone laser emissions from hydrogen halide molecules. Such a laser emission was experimentally produced by *Hon* and *Novak* [4.144] using a pin-cathode transverse discharge in an SF$_6$–HBr mixture and by *Bashkin* and co-workers [4.145] using longitudinal discharge in an SF$_6$–H$_2$S (HBr) mixture. In both cases, lasing was observed to occur in the P_{3-1} and P_{4-2} transitions at an output mirror reflectivity of 99.5% at $\lambda \simeq 1.3$ μm. It is interesting to note that the ratio obtained in [4.145] between the laser energy on the overtone and that on the fundamental harmonic under similar experimental conditions amounted to some 15%. This figure cannot be considered the ultimate, because the laser cavity was optimized when operating on the fundamental harmonic, but no cavity optimization was made when operating the laser on the overtone.

Lasing on the first onvertone of the DF molecule was experimentally observed by *Suchard* and *Pimentel* [4.81] using a flash-photolysis initiated N_2F_4–CD_4 mixture in a cavity incorporating a selective filter to lower the gain of the fundamental-harmonic emission. It should be noted that this was the first reported observation of overtone-laser emission from a hydrogen halide molecule.

It follows from the experiments described above that suppression of lasing in transitions with $\Delta v = 1$ is essential to the operation of lasers based on overtone transitions. Aside from selective elements, these lasers used low energy content mixtures strongly diluted with inert gases. If resort is made to the more energy-intensive mixtures of practical interest (especially H_2–F_2 mixtures) which provide for high gains ($\alpha \geqslant 1\,\mathrm{cm}^{-1}$), the rapid development of superluminescence in transitions with $\Delta v = 1$ will impair the population inversion in transitions with $\Delta v = 2$. This problem can be solved in two ways. First of all, one can add to the laser mixture some absorbing substances which will reduce unsaturated gain in transitions with $\Delta v = 1$. These absorbents must mix well with fluorine and must not absorb laser emissions due to transitions with $\Delta v = 2$. The search for such absorbents is a very difficult task. For this reason, the use of a non-chain reaction laser master oscillator operating on an overtone transition of HF in combination with a laser power amplifier seems to be the more promising solution to the problem of extracting energy from the active medium of the H_2–F_2 laser in transitions with $\Delta v = 2$. In that case, the population inversion in the amplifier will be controlled by the emission from the master oscillator. Calculations by *Basov* and co-workers [4.146] have shown that the first-overtone laser energy produced by such a combination can reach up to 30% of the fundamental-harmonic output from the H_2–F_2 laser.

4.10 Amplifier-Mode Operation of Pulsed Hydrogen-Fluoride Lasers

The various chemical laser applications requiring high-energy emissions, for example, laser-controlled thermonuclear fusion, depend on the development of laser master oscillator-laser power amplified systems (MOPA's). What are the drawbacks of a high-power laser operated in the oscillator mode? First, such a laser requires high-quality optics capable of withstanding high radiation power densities in the cavity. Second, the inhomogeneities of the active medium refractive index due to insufficiently uniform initiation widen the divergence of the laser beam repeatedly traversing these inhomogeneous regions in the cavity. And third, in large-sized laser oscillators, the loss of energy due to superluminescence becomes important. What is more, superluminescence also adds to the laser beam divergence.

The use of MOPA's allows the above drawbacks to be eliminated or at least moderated substantially. The requirements for the optical elements of the amplifier are not very demanding, there being no need for an optical cavity. The influence of refractive index inhomogeneities in the amplifier is minimal because the laser beam traverses the gain medium only once. The master oscillator emission can materially reduce the population inversion level in the amplifier medium, thus impeding the development of superluminescence.

The study of MOPA's is necessary not only to solve practical application problems, but also to gain an insight into the ultimate capabilities of chemical lasers. It is precisely the amplifier-mode operation of lasers, especially chemical lasers with a low unsaturated gain (e.g., D_2–F_2–CO_2 types), that allows their energy characteristics to be truly maximized and not simply claimed to be excellent in some questionable terms. In fact, with the amplifier, there is no problem with the optimum output coupling (a most critical issue indeed for chemical lasers with their spatially nonuniform and rapidly varying unsaturated gain), which must make for the maximum extraction of the energy stored in the active medium. The absence in the amplifier of the mode field structure typical of the oscillator facilitates the estimation of the actual values of specific parameters. Finally, since amplification always lasts longer than the threshold oscillation condition, the energy output and engineering efficiency of a laser amplifier should be higher than those of a laser oscillator operating under the same initial conditions.

There are some principal differences between the H_2–F_2 and D_2–F_2–CO_2 laser amplifiers. First, as has been already indicated above, the number of lines in the emission spectrum of the D_2–F_2–CO_2 laser is small (two or three) compared to that of the H_2–F_2 laser (a few tens). Second, the unsaturated gain in the H_2–F_2 laser is at least an order of magnitude higher than in the D_2–F_2–CO_2 laser, which greatly facilitates the possible development of parasitic oscillations in the H_2–F_2 laser amplifier.

For estimation, we use the formula $\alpha \leqslant -m(\ln R)/2L$ for the threshold unsaturated gain along the optical axis of the amplifier (Fig. 4.28). In this formula, m is the number of beam reflections necessary to restore the initial wave front, R the reflectivity of a single surface in the path of the beam (for simplicity, all the surfaces in the beam path are assumed to have the same reflectivity), and L the length of the gain region. In that case, for $m = 6$, $R = 5\%$, and $L = 1$ m we have $\alpha \leqslant 0.1$ cm^{-1}. For a metal-wall amplifier of the same size, the requirement for the maximum value of the gain α will be even more stringent. With the D_2–F_2–CO_2 laser amplifier, the problem of suppression of parasitic oscillations is therefore not so critical. Besides, it is in principle always possible here to add a small amount of an absorbent to the amplifier medium in order to lower its unsaturated gain (even bring it down to zero,

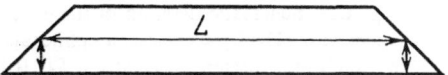

Fig. 4.28. Path of light rays in a laser amplifier in the case of self-excitation in a direction along the optical axis

if necessary). In contrast, parasitic oscillations are a major problem with the H_2-F_2 laser amplifier with its unsaturated gain α characteristically in excess of 1 cm^{-1}.

In both types of amplifier, parasitic oscillations can be suppressed by making the power density produced in the active medium of the amplifier by the master oscillator sufficiently high, for this greatly lowers the required population inversion level in the medium and thus impairs the conditions necessary for such oscillations to develop. That this can be effective in the case of a large-size $D_2-F_2-CO_2$ laser amplifier is quite obvious (the population inversion level needs to be reduced by a factor of ten to several tens). The H_2-F_2 laser amplifier on the other hand requires a much greater reduction of its population inversion level, and in a much greater number of transitions and also under partial rotational equilibrium conditions.

Basov et al. [4.110] experimentally studied a $D_2-F_2-CO_2$ laser amplifier using a powerful creeping-discharge UV-source to initiate the reaction. The master oscillator used was either a $D_2-F_2-CO_2$ chemical laser or a TEA–CO$_2$ laser. In the former case, use was made of part of the laser cell to create the $D_2-F_2-CO_2$ laser amplifier under study (Fig. 4.29). The oscillator and the amplifier parts of the cell were arranged symmetrically about the initiation source, which made the oscillator and amplifier active media identical. It was established experimentally that the chemical-laser-based master oscillator emitted in the $P(20)$ and $P(22)$ transitions of the CO$_2$ molecule. For this reason, when using the TEA-CO$_2$-laser-based master oscillator, its emission was tuned to resonance with the $P(20)$ and $P(22)$ transitions by means of a selective cavity. With the energy output of the both types of master oscillator being the same, the output energies yielded by the amplifier were practically equal.

The maximum specific energy output of the amplifier was achieved with the TEA–CO$_2$ laser oscillator emission subjected to preliminary amplification. The preamplifier was that part of the large laser cell which in other experiments (Fig. 4.29) served as a master oscillator. The preamplification of the TEA–CO$_2$ laser oscillator emission was necessary to reach a radiation power density of no less than 150 kW/cm^2 at the input of the chemical laser amplifier. So high an input radiation power density made it possible to obtain a record specific energy output of 150 $\pm$ 30 J/l atm. from the amplifier using the $D_2:F_2:CO_2:He = 1:1:4:5$ mixture. The unsaturated gain in that case was 0.07 cm^{-1}. The specific energy output from the $D_2-F_2-CO_2$ oscillator under the same conditions (the same mixture, the same

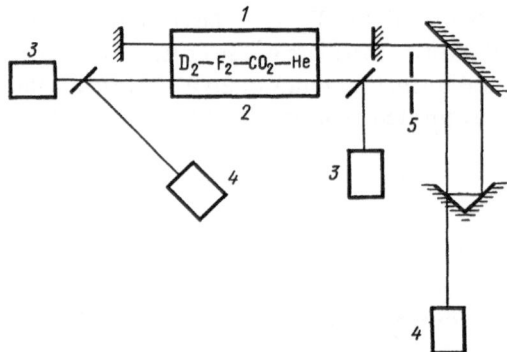

Fig. 4.29. Schematic diagram of the experiment with a $D_2-F_2-CO_2$ laser amplifier and a built-in master oscillator [4.110]: *1*–chemical laser oscillator; *2*–chemical laser amplifier; *3*–calorimeters; *4*–photodetectors; *5*–aperture stop

initiation regime) did not exceed 70 J/l atm (see also Sect. 4.6). The high radiation power densities necessary to obtain high specific energy outputs apparently caused damage to the cavity mirrors used when operating the system in the oscillator mode. Examination of the mirrors in the course of the experiments revealed damaged areas on their surface. This belief is confirmed by other experiments. *Agroskin* et al. [4.138] studied a D_2–F_2–CO_2 chemical laser using a less powerful initiation source. These investigators achieved a specific energy output of 40 J/l atm. and found their system to have the same energy characteristics when operated in the oscillator and the amplifier mode.

The amplifier-mode operation of non-chain-reaction HF (DF) lasers (SF_6:C_2H_6 = 10:1, $p \simeq 0.5$ atm.) was studied by *Getzinger* and co-workers [4.147] using axial e-beam initiation. The master oscillator was pin-discharge initiated and operated on the same active mixture components (SF_6 and C_2H_6). These investigators managed to obtain 11.6 J of energy from a double-pass amplifier, the radiation power density at its input being equal to some 40 kW/cm². The same amplifier yielded a mere 7.7 J when operating under superluminescence conditions. Unfortunately, no data were reported on the energy the laser could have given if operated in the oscillator mode.

Tisone and *Hoffman* [4.148] carried out more comprehensive investigations of an HF laser amplifier relying on the chain reaction in an axial e-beam initiated H_2–F_2 mixture. The amplifier was the F_2–H_2 laser used earlier by *Gerber* and *Patterson* [4.57] as an oscillator (see [4.57] in Table 4.6). The master oscillator was a self-sustained-discharge initiated SF_6–H_2 chemical laser. Its radiation power density (summed over 17 lines – $P_1(3)$–$P_1(8)$, $P_2(3)$–$P_2(8)$, and $P_3(3)$–$P_3(7)$) at the input of the amplifier was around 20 kW/cm². It is seen from Fig. 4.30 that the power gain of the amplifier reached a value of some 500.

An important result of this investigation was the measurement of the superluminescence intensity in the backward direction. The peak value obtained (3.3×10^5 W/cm²) was about a factor of 30 lower than the radiation intensity in the forward direction at the amplifier output. This, in principle, attests to the validity of the method considered here for combating superluminescence, but the energy output from the amplifier was obviously insufficient. It did not exceed some 110 J, but it reached 500 J under superluminescence conditions and 1800 J in oscillator-mode operation. This means that the chemical-energy-to-coherent-radiation con-

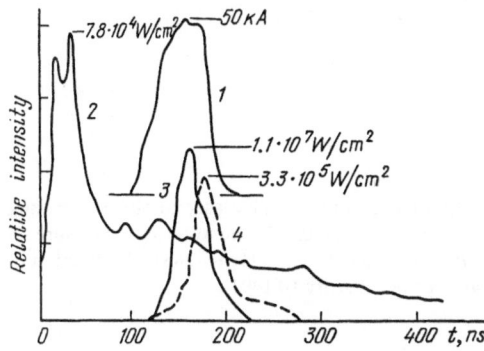

Fig. 4.30. Temporal behavior of e-beam current density (*curve 1*), power density of the SF_6–H_2 master oscillator signal (*curve 2*), and power density of the H_2–F_2 amplifier emission in the forward and backward directions (*curves 3* and *4*, respectively) [4.148]

version efficiency in the amplifier-mode operation was only about 6% of that in the oscillator-mode operation. The reason for so low results is most likely a poor matching of the spectral characteristics of the oscillator and the amplifier, though the possibility cannot be ruled out that the radiation power density at the input of the amplifier was insufficient as well.

Further progress in the understanding of the main principles of operation of the H_2–F_2 laser amplifier was made by *Basov* and co-workers [4.149]. The master oscillator used by these investigators was a flash-photolysis initiated H_2–F_2 laser, and the amplifier itself was initiated by means of an axial electron beam. It can be seen from Fig. 4.31 showing the amplifier output energy *vs.* input signal power density curves for two different laser mixture pressure values that particular attention should be given to the radiation power density at the amplifier input.

At the lower mixture pressure, saturation was reached with an input radiation power density of about 0.5 kW/cm². Under these conditions, the chemical-to-laser energy conversion efficiency of the laser in the amplifier mode amounted to 35–45% of that in the oscillator mode. But with the mixture pressure merely doubled, saturation was not reached even with trebled input radiation power density. Apart from using identical mixtures in the master oscillator and amplifer, no special measures were taken in these experiments to match their spectral characteristics. What is more, difficulties were encountered in time-locking the oscillator and amplifier.

To check on the importance of these factors, use was made in [4.149] of the method for ensuring an almost perfect spectral matching and time locking of the master oscillator and amplifier which proved a success in the studies of the D_2–F_2–CO_2 laser amplifier [4.110]. In this method, the master oscillator was made part of the active medium of the amplifier (see Fig. 4.29). Under these conditions, the power-extraction efficiency of the H_2–F_2 laser amplifier was around 73%, which once more bore out the need for a thorough matching of the spectral and temporal characteristics of any master oscillator and amplifier in a combination operating on many transitions.

In order to find out whether it is possible to suppress superluminescence in the H_2–F_2 laser amplifier, *Basov* et al. [4.149] studied the temporal behavior of the

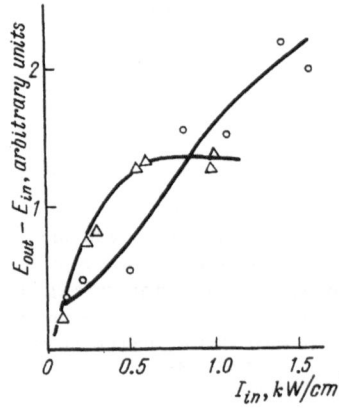

Fig. 4.31. Output energy of an H_2–F_2 laser amplifier operating on the F_2:H_2:O_2:He$=3$:1:0.6:10 mixture at a pressure of 230 mm Hg (*triangles*) and 450 mm Hg (*circles*) as a function of the input signal power density [4.149]

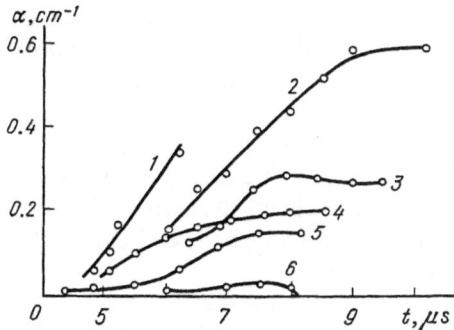

Fig. 4.32. Temporal behavior of the unsaturated gain of an H_2–F_2 laser amplifier in the absence of the master oscillator signal, α_0 (*curve 1*), and in the presence of this signal, α_s (*curves 2* through *6*), with the amplifier active region length L being 3 cm (*curve 2*), 5 cm (*curve 3*), 10 cm (*curve 4*), 15 cm (*curve 5*), 55 cm (*curve 6*) [4.149].

unsaturated gain α in the absence of the master oscillator emission, i.e., $\alpha_0(t)$, and in the presence of this emission, $\alpha_s(t)$. The value of $\alpha_s(t)$ in the amplifier was calculated by the formula $I_{out} = I_{in}\exp(\alpha_s L)$ using the input and output laser pulse oscillograms (I_{out} and I_{in} are the output and input signal intensities, respectively and L is the active region length of the amplifier). The influence of the master oscillator emission on the active medium gain $\alpha_s(t)$ is illustrated in Fig. 4.32. As the gain region length and the master oscillator emission power density are increased, the sharp decline of α_s becomes more and more manifest. Such a considerable decrease of the amplifier gain in the presence of the master oscillator emission suggests that superluminescence in the H_2–F_2 laser amplifier can be suppressed effectively by appropriate selection of the input radiation power density.

   The above investigations are only the initial stage of detailed research into the H_2–F_2 laser amplifier. A number of questions are still to be answered for us to understand completely the main principles of its operation. The more important among these questions are the following.

1) What is the power-extraction efficiency of the amplifier for individual vibrational-rotational transitions?
2) What are the possibilities of increasing the rotational relaxation rate?
3) Can the power-extraction efficiency of the amplifier be improved by utilizing transitions with large rotational quantum numbers J (see Sect. 4.2)?
4) What are the particular conditions required to suppress superluminescence in large-volume H_2–F_2 amplifiers completely?

4.11 High-Repetition-Rate Operation of Chemical Lasers

To achieve lasing in high-repetition-rate chemical lasers, it is necessary that the laser mixture be continuously pumped through the laser cell. To develop such lasers on the basis of non-chain reactions is not very difficult. For example, *Aprahamian* et al. [4.150] achieved a laser-pulse repetition rate of 5 Hz using e-beam initiated SF_6–H_2 and SF_6–D_2 mixtures (1.5:1, $p=600$ mm Hg). The maximum specific energy output per pulse amounted to around 1 J/1 for the SF_6–H_2 mixture and was half as much as that for the SF_6–D_2 mixture. A still higher repetition rate (up to 1100 Hz) was obtained by *Jacobson* et al. [4.151] with an electrical-discharge initiated SF_6–C_3H_8 mixture ($p = 50$ mm Hg) flowing at a velocity of about 1.5 m/s ($\eta_{eng} \simeq 0.7\%$, average power 7.2 W). Also, there have already been developed transverse electrical-discharge initiated closed-cycle high-repetition-rate HF and DF lasers [4.152] using SF_6–H_2 and SF_6–D_2 mixtures (specific energy output 3 J/1, $\eta_{eng} \simeq 2\%$). The HF molecules produced in the course of chemical reactions in these lasers are removed by means of an absorber (sodium hydroxide and zeolite) with the gases circulating continuously around a closed circuit.

It is much more difficult to develop such lasers on the basis of the branching-chain hydrogen-fluorine reaction, where much heat is evolved in the reaction zone, giving rise to the ignition of the laser mixture. Depending on the concentration of fluorine, hydrogen, and oxygen, the resulting flame may propagate with velocities from a few meters a second to supersonic values of the order of 2–3 km/s. High-repetition-rate operation can be realized if the mixture flow velocity is higher than the flame propagation velocity. In that case the flame will be extinguished by the continuous flow of the cold gases.

Flame propagation conditions in hydrogen-fluorine mixtures were experimentally studied by *Chen* and co-workers [4.54]. The stable F_2–O_2–He and H_2–He premixes were mixed in the laser cell region upstream of the optical cavity (Fig. 4.33). The continuously flowing laser mixture was photolysis–initiated with a flashlamp. Investigations into the mixture composition dependence of the flame propagation velocity revealed the existence of two regions differing in the composition and pressure of the starting gases. These regions are separated by a distinct boundary whose position can be altered by varying the oxygen concentration (Fig. 4.34). The regions on either side of the boundary differ greatly in flame propagation velocity: in low-concentration mixtures, flames propagate with velocities equal to the burning rates, i.e., a few tens of meters per second, while in high-concentration mixtures, flames propagate with the shock velocity. Thus, for example, the flame

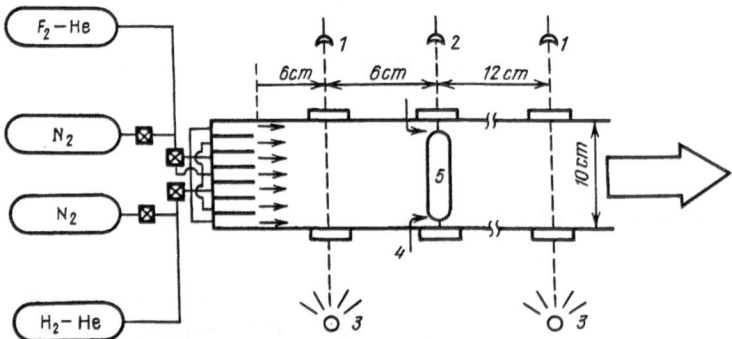

Fig. 4.33. Schematic diagram of the experimental setup for studying flame propagation velocity in the flow of mixed F_2–O_2–He and H_2–He premixes as a function of the mixture composition [4.117]: *1* – photomultiplier for measuring the F_2 concentration in the gas flow: *2* – IR laser radiation detector *3* – mercury lamp (2537 Å); *4* – injection of He to produce a boundary layer; *5* – laser initiation flashlamp

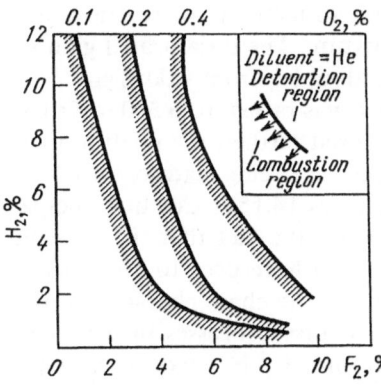

Fig. 4.34. Position of the boundary between the two regions differing in flame propagation velocity in an $H_2-F_2-O_2-He$ mixture as a function of the mixture composition [4.117]

propagation velocity in a mixture composed of 4% F_2, 4% H_2, 0.4% O_2, and 91.6% He at a total pressure of 1 atm. (specific energy output around 10 J/1) turned out to be lower than the gas flow velocity (35 m/s). These investigators have already operated their laser at a pulse repetition rate of 10 Hz, the value predicted for the near future being 100 Hz.

Operation on high-concentration mixtures providing for good specific laser parameters calls for supersonic laser mixture flow velocities. These velocities can in principle be achieved, but will greatly complicate the laser design. There is a more expedient method involving modulation of the hydrogen gas flow. The main problem with this method is the development of a reliable high-speed H_2-flow controller. *Woodroffe* and *Limpaecher* [4.153] managed to build such a controller capable of operating at a frequency of 370 Hz. The atmospheric-pressure laser mixture used by these investigators contained 11.6% F_2, 8% H_2, 0.5% O_2, and 79.9% He.

The above summary of the results already achieved in the field show the first steps in the development of high-repetition-rate hydrogen-fluoride lasers to be a success. But a number of important problems still remain to be solved to allow the energy output, pulse repetition rate, and physical size of such lasers to be increased.

4.12 Pulsed Chemical Lasers Based on Chain Reactions Other than the Hydrogen-Fluorine Reaction

In addition to the hydrogen-fluorine chain reaction, there are other chain reactions that can give rise to population inversion. First of all, we would like to mention the reaction in the H_2-Cl_2 mixture, for it was precisely this mixture that was used back in 1965 by *Kasper* and *Pimentel* [4.1] to create the first chemical laser. But the reaction $Cl_2 + H_2 \rightarrow HCl + H$ is endothermic, its activation energy being fairly high ($\simeq 5$ kcal/mole). That is why the experimentally observed laser chain length in the H_2-Cl_2 laser is small ($\leqslant 1$). The reaction can be somewhat accelerated by raising the

$[H_2]/[Cl_2]$ ratio to over 10 [4.154]. Another way is to increase the reaction rate constant by raising the temperature of the laser mixture. The unsaturated gain in the H_2–Cl_2 laser was experimentally observed to double – from 0.0058 cm^{-1} to 0.012 cm^{-1} – when the initial mixture temperature T was raised from 300 to 510 K [4.154]. But if the initial mixture temperature is raised further, the mixture may ignite spontaneously. This naturally prevents the most use being made of the effect.

The reaction in the IF_5–H_2 and IF_7–H_2 mixtures [4.156] can in principle proceed by a chain mechanism as well. For instance, the fact that the e–beam initiated IF_7–H_2 laser showed a 100% efficiency with respect to the energy deposited in the electron beam undoubtedly points to the chain character of the reaction in this mixutre. The sequence of the elementary processes in the chain mechanism of the reaction occurring in the IF_5–H_2 and IF_7–H_2 mixtures is as yet unknown.

There is more information available on the chain mechanism of the reactions betweeen chlorofluorides and hydrogen. And inasmuch as chlorofluoride molecules contain both fluorine and chlorine atoms, lasing in this case can take place on vibrational transitions of the HF and HCl molecules simultaneously. This phenomenon was observed with all the chlorofluoride-hydrogen mixtures: ClF–H_2 [4.66], ClF_3–H_2 [4.66, 157–161], and ClF_5–H_2 [4.161, 162]. There is no point in considering in detail the mechanisms of the reactions occurring in all the above mixtures, for their prospects differ greatly as a result of the different rates of relaxation of the vibrationally excited HF^* molecules on the initial chlorofluoride molecules. The respective rate constants were measured by *Hancock* and *Green* [4.163], *Bott* and *Cohen* [4.164], and *Chebotarev* et al. [4.165]. Proceeding from the values of these constants ($5 \times 10^{-12} \text{ cm}^3 \text{ s}^{-1}$ for ClF_5 and ClF_3 and $1.5 \times 10^{-13} \text{ cm}^3 \text{ s}^{-1}$ for ClF) found in [4.165], it can be inferred that the mixture pressure can be raised high enough only in the case of the ClF–H_2 mixture. *Chebotarev* et al. [4.161] suggested that the reaction in this mixture proceeds by a chain mechanism. The kinetic model of the ClF–H_2 chemical laser developed by *Igoshin* [4.166] demonstrates that the effective chain length in it can reach about 10. In addition to the possibility of lasing using the HCl molecule, there is another point in favor of the ClF–H_2 mixture: its higher stability compared to the H_2–F_2 mixture. This is quite understandable because the chain-branching reaction similar to reaction (6) in reaction mechanism (4.17) is here impeded as a result of the high bond energy of the ClF molecule, while the chain termination process is the same as in the H_2–F_2 mixture. While experimenting with the e-beam initiated $ClF:H_2:He = 2.5:1:2.5$ mixture ($p = 360 \text{ mm Hg}$), *Bashkin* and co-workers [4.167] obtained around 22 J/1 of specific laser energy output, the output from the lasing HCl molecules amounting to 24% of the total. The physical efficiency of the ClF–H_2 laser was around a factor of 4 lower than that of the H_2–F_2 laser operated under identical initial conditions.

Much was expected from the use of the reaction taking place in the CS_2–O_2 mixture because it was demonstrated, as far back as the 1930's, to proceed by a branching-chain mechanism [4.168], and lasing in a flash-photolysis initiated CS_2–O_2 mixture was achieved in 1966 [4.169]. At present, a number of investigators (e.g., [4.170, 171]) have established that the chain mechanism of the reaction is ensured by the following main processes:

1) $S + O_2 \rightarrow SO + O$, $\Delta H = -5.6$ kcal/mole;

2) $SO + O_2 \rightarrow SO_2 + O$, $\Delta H = -12.8$ kcal/mole;

3) $O + CS_2 \rightarrow CS + SO$, $\Delta H = -31.1$ kcal/mole;

4) $O + CS \rightarrow CO^* + S$, $\Delta H = -85$ kcal/mole.

It is seen that process (4) gives rise to vibrational population inversion in CO. The initiation of the reaction, i.e., the production of atomic oxygen or sulfur, can be effected by the electrical-discharge dissociation of O_2 or CS_2 or by the flash photolysis of CS_2. Lasing usually occurs in a very great number of vibrational-rotational transitions in CO. For example, Gregg and Thomas [4.172] reported on the observation of lasing on 270 P– and R–branch transitions from P_{16} to P_1 (5.805–4.745 μm) and from R_{15} to R_9 (5.374–5.109 μm). As a rule, lasing starts on the 13–12 and 11–10 band transitions, and then transitions in the 10–9, 9–8, 8–7, etc. bands come into play consecutively. This circumstance suggests a high selectivity of population of high-lying vibrational levels in the CO molecule and a cascade lasing mechanism in the CS_2–O_2 chemical laser. The investigations into the chemical CO laser by Suart et al. [4.173] and Pilloff et al. [4.174] have shown that small amounts of CO, N_2O, and OCS added to the laser mixture increase the laser output energy. The positive role of these substances is believed to be due to acceleration of the relaxation of the lower vibrational levels in the CO molecule, which leads to an increase in the extent of its population inversion and in the time this inversion lasts. All the above molecules have their fundamental vibration frequencies close to the transition frequencies between the lower vibrational levels of CO.

Its chain reaction mechanism and laser wavelength range apart, the attractiveness of utilizing the CS_2–O_2 mixture in chemical lasers is physically explained by two factors. First, the reaction in this mixture is highly exothermic and, as can be seen from the reaction scheme given above, a sizeable proportion of its exothermicity is accounted for by the process (4) giving rise to the excitation of the vibrational degrees of freedom of the lasing CO molecule. According to the estimates by various investigators [4.175–177], from 66 to 90% of the total 84.7 kcal/mole of energy released in this process go with a high selectivity into the vibrational excitation of CO. Second, the rate constant for the vibrational self-relaxation of the lasing CO molecules is relatively low [4.178–180], which gives reason to expect that it will be possible to effect lasing under high mixture pressures where the best use can be made of the advantages of the chain reaction mechanism.

Lasing in the CS_2–O_2 mixture was achieved in numerous experiments using flash-photolysis [4.169, 172, 181–183], longitudinal electrical-dischage [4.184–186], and transverse electrical-discharge [4.187, 188] initiation techniques. But in no experiment done the optimum CS_2 pressure exceed a few mm Hg. Gordon et al. [4.181] tried to explain this effect by the competition between the three-particle atomic sulfur recombination process

$$S + CS_2 + M \rightarrow CS_3 + M$$

and the process (1) responsible for the production of atomic oxygen. But the rate constant used by these investigators for process (1) turned out to be almost three orders of magnitude smaller than the more accurate figure established later, and so their suggestion was wrong. More likely explanations are the high rate of relaxation of the excited CO molecules on CS_2 molecules (for $v=1$, $k=3.5 \times 10^{-13}$ cm^3 s^{-1} at $T=296$ K [4.189]), which, in addition, increases with the number v, and the homogeneous flash-photolysis initiation depth being reduced as a result of the very large absorption cross section (over 10^{-17} cm^2) of CS_2 in its photodissociation band from 1800 to 2200 Å [4.190].

The difficulties involved in raising the partial pressure of CS_2 are aggravated by the fact that the engineering efficiency of the electrical-discharge initiated CS_2–O_2 laser does not exceed a few percent. This is evidence that the chain reaction mechanism is not effective enough as far as the laser process is concerned. The main obstacle here is the slow rate of process (2). The rate constant of this process, $k=6 \times 10^{-3} \exp(-6500/RT)$ cm^3 s^{-1} [4.191], is so small that the contribution to the laser energy made by the chain mechanism during the laser pulse is indeed negligible.

The search for more efficient oxidizers resulted in the use of ozone instead of oxygen. Using the flash–photolysis initiated $CS_2:O_3:N_2O:SF_6=3:5:6:15$ mixture with a total pressure of 58 mm Hg, *Bashkin* et al. [4.182] obtained around 0.28 J of laser energy at a specific energy output of 4.5 J/1 which is twice that of their CS_2–O_2 laser. The low efficiency of the CS_2–O_3 laser, measured at around 14% with respect to the initiation energy deposited in the laser mixture, suggests an inefficient chain reaction mechanism (if any) in the CS_2–O_3 mixture. The doubled specific energy output of the CS_2–O_3 laser can in that case be directly explained by the additional production of atomic oxygen as a result of photodissociation of ozone. As the photodissociation regions of ozone and carbon disulfide do not coincide, there occurs simultaneous production of atomic oxygen and CS radicals, which increases both the power output and efficiency of the laser. Oxygen can also be replaced by NO_2 [4.192], SO_2 [4.193], and SO_3 [4.194]. But for all practical purposes, none of these substances can appreciably improve the energy characteristics of the chemical CO lasers.

It seems interesting to try to realize a chain reaction in a mixture containing CS radicals and molecular oxygen. That such a reaction is possible follows from the analysis of processes (1) and (4) in the above chain reaction scheme. Unfortunately, the CS radical is unstable, and so such a mixture can only be prepared in continuous gas flow conditions. The potentialities of such a laser are now being explored successfully [4.195].

To date, lasing on the CO molecules has been achieved with a number of other chemical reactions involving atomic oxygen:

$$O + CSe \rightarrow CO^* + Se \ , \qquad \Delta H = -118 \text{ kcal/mole [4.193, 4.196]};$$

$$O + CN \rightarrow CO^* + N \ , \qquad \Delta H = -74 \text{ kcal/mole [4.197]};$$

$$O + CH_2 \rightarrow CO^* + 2H \ , \qquad \Delta H = -75 \text{ kcal/mole [4.197]};$$

$$O + CH \rightarrow CO^* + H \ , \qquad \Delta H = -176 \text{ kcal/mole [4.197]};$$

$$O(^1D) + C_3O_2 \rightarrow 3CO^* \ , \qquad \Delta H = -160 \text{ kcal/mole [4.197]}.$$

In all the above experiments, the necessary atomic oxygen has been produced either through electrical-discharge dissociation of O_2 [4.196] or through photolysis of O_3 or SO_2 [4.193, 197]. No chain mechanism has been found for these reactions, and the parameters of all these lasers are inferior to those of the CS_2–O_2 laser.

There is another mixture, H_2–O_3, which has been believed to hold promise for lasing on the basis of the chain reaction

$$H + O_3 \rightarrow OH^* + O \ , \quad \Delta H = -78 \text{ kcal/mole},$$

$$OH + H_2 \rightarrow H_2O + H \ .$$

This reaction is also attractive because its products are absolutely nontoxic.

The flash photolysis of the H_2–O_3 mixture gives rise to atomic oxygen in the ^1D state, and so involves the following process as well:

$$O(^1D) + H \rightarrow OH^* + H \ , \quad \Delta H = -44 \text{ kcal/mole} \ .$$

But despite the perceptible exothermicity of the above reactions, in practice, lasing has been very weak. This might be explained by a rapid decay of the vibrational population inversion in the OH radical as a result of either relaxation or the following chemical reaction:

$$OH^* + O_3 \rightarrow \text{products}.$$

If that is the case, the efficiency of the laser could be improved by utilizing, by analogy with the D_2–F_2–CO_2 laser, vibrational energy transfer to the CO_2 molecule which relaxes more slowly [4.198]. The mechanism by which population inversion in this case is produced in CO_2 can be perceived from the following scheme of the basic processes occurring in the D_2–O_3–CO_2 mixture under flash-photolysis initiation:

0) $O_3 + h\nu + O(^1D) + O_2$,

1) $O(^1D) + D_2 \rightarrow OD^* + D$,

2) $D + O_3 \rightarrow OD^* + O_2$,

3) $OD^* + CO_2 \rightarrow OD + CO_2^*$,

4) $O(^1D) + M \rightarrow O(^3P) + M$,

5) $OD^* + O_3 \rightarrow \text{products}$,

6) $O(^1D) + O_3 \rightarrow \text{products}$,

7) $OD^* + D_2 \rightarrow OD + D_2$.

Because the rate constants for processes (3), (5), and (7) were unknown, a special laser procedure has been developed to measure them [4.199]. Experimental optimization of the parameters of this laser [4.200] has made it possible to obtain 1 J/1 of laser energy from the $D_2:O_3:CO_2 = 24:1:10$ mixture at a total pressure of 0.5 atm. Replacing deuterium by hydrogen has somewhat reduced the laser energy, apparently as a result of the slower rate of energy transfer from OH* to CO_2. Lasing has also been achieved with vibrational energy transferred from OD* to N_2O [4.200]. Unfortunately, the chain reaction mechanism in that case has also proved to be extremely inefficient from the standpoint of the laser process. The attempts at extending the chain length in excess of unity by adding CO to the mixture to effect the reduction of atomic deuterium through the reaction $CO + OD \rightarrow CO_2 + D$ have also proved unsuccessful, for the stability of the mixture deteriorated, and the laser energy output was reduced [4.200].

4.13 Current Status of Pulsed Chemical Lasers

This section deals with the main trends in the development of PCL's and the results obtained since 1980. As far as non-chain-reaction hydrogen-fluoride lasers are concerned, it must be said that very little work has appeared. Some progress was made in lasers of this type using electron-beam initiation. *Bashkin* et al. [4.202] established that the reduced energy output from the SF_6–H_2 laser at increased mixture pressures is due to enhanced initiation inhomogeneity in the direction along the electron beam. They also demonstrated that the energy output of this laser can, in principle, be progressively increased by raising the mixture pressure at least up to 2.5 atm, provided the electron-beam energy is high enough to ensure homogeneous initiation. Thus, a record level of specific energy output ($\varepsilon_l = 45 \pm 5$ mJ/cm^3) was obtained from the $SF_6:H_2 = 20:1$ mixture at a sufficiently high pressure (up to 1.5 atm.) and an increased initial atomic fluorine concentration ($[F]_0 \simeq 7 \times 10^{17}$ cm^{-3}) produced by an electron beam propagating at right angles to the optical axis. What set those experiments apart from the ordinary was that the laser volume was strictly defined by means of an external magnetic field preventing the electron beam from scattering and by using suitable restricting diaphragms. Reliability of definition was checked against the trace left by the beam on a sensitive plastic (Astrolon). As a result, the accuracy of laser volume definition was no worse than 10%. With the ratio $SF_6:H_2$ increased up to 100:1, the highest ever chemical efficiency ($\eta_{chem} = 37\%$) was obtained. Some time later, roughly the same chemical efficiency value (39%) was achieved by *Kannari* and co-workers [4.203].

Recent studies of chain-reaction pulsed hydrogen-fluoride lasers have not changed the existing understanding of the general laws governing their operation. The importance of the results obtained is that the experiments have been conducted in conditions of reliable control over the initial mixture composition and atomic fluorine concentration produced by various types of initiation sources. This has improved the reproducibility of experimental results and allowed a better understanding to be gained into a number of the results by way of comparison with theoretical calculations. It has been mentioned in Sect. 4.5.3 that the relationships reported in a number of works between laser energy characteristics and the energy stored in the storage capacitor bank of the initiation source have no comprehensive physical meaning. Indeed, the atomic fluorine concentration $[F]_0$ produced with one and the same source energy may differ depending on particular design features of the source used. It is clear from the foregoing sections that the energy characteristics of a laser are governed by the concentration of atomic fluorine and the temporal behavior of its production by the initiation source.

The methods of determining $[F]_0$ used to date in the case of flash-photolysis initiation were largely indirect, prone to measurement error, and required absolute calibration, an extremely involved procedure. The initial fluorine concentration could be measured only where the initiation source was some laser (excimer [4.204], ruby [4.62, 205, 206], or neodymium glass [4.207]) allowing for the initiation energy E_i to be measured directly. In that case, with the energy spent on the photolysis of F_2 at a given wavelength known, the concentration $[F]_0$ of fluorine atoms produced could be found.

The method suggested by *Chegodayev* [4.208], who studied the main processes involved in the photolysis of fluorine-oxygen mixtures, has greatly simplified the measurement of $[F]_0$ produced by flash photolysis. It was further developed by *Bashkin* et al. [4.117, 206] and *Porodinkov* [4.209] for conditions typical of flash-photolysis initiated H_2–F_2 lasers. The method is as follows. The laser cell is first filled with an F_2–O_2 mixture with the partial F_2 pressure equal to that in the chemical laser under study. Then part of the molecular fluorine is dissociated by flash photolysis, leading to the fast reaction

$$F + O_2 + M \rightarrow FO_2 + M$$

to yield metastable FO_2 radicals whose concentration is equal, under conditions found in [4.206], to $[F]_0$ accurately enough. The concentration of FO_2 is measured from changes in the optical density of the mixture in the 190–230 nm region, where the radical has a continuous absorption spectrum. Unfortunately, the method is inapplicable to electron-beam initiation, as it requires a great excess of O_2 over F_2 ($[O_2]/[F_2] \geqslant 3$) distorting the true picture of the electron beam-F_2 interaction.

The concentration of atomic fluorine produced by a beam of fast electrons passing through a medium containing F_2 can be found by the formula

$$[F]_0^{eb} = E_i^{eb} / \varepsilon_F^{eb} V^{eb} \, ,$$

where E_i^{eb} is the measured e-beam energy deposited in the medium, ε_F^{eb} the e-beam energy required to produce a single fluorine atom, and V^{eb} the effective e-beam initiated laser volume. The ε_F^{eb} values recommended in various works are from 6 eV [4.135] to 12 eV [4.134]. *Bashkin* et al. [4.210] have suggested a method allowing for a more accurate determination of ε_F^{eb}. They compared the energy output of an H_2–F_2 laser initiated first with an electron beam and then with the second-harmonic output from a ruby laser. In the latter case, the initial concentration of fluorine atoms can easily be found by measuring the absorbed ruby laser energy E_i^{hv}, because here it is practically completely expended in the photodissociation of molecular fluorine:

$$[F]_0^{hv} = E_i^{hv}/\varepsilon_F^{hv}\, V^{hv}\ ,$$

where ε_F^{hv} is the energy spent to produce a single fluorine atom by laser photolysis and V^{hv} the laser medium volume involved in the absorption of the second-harmonic ruby laser energy. With the laser mixture composition being the same in both cases, the laser energy outputs were made equal by varying E_i^{eb} and E_i^{hv}. The equality of the energy outputs in these conditions means the equality $[F]_0^{eb} = [F]_0^{hv}$. From this equality, with $V^{eb} = V^{hv}$, it is not very difficult to find

$$\varepsilon_F^{eb} = \varepsilon_F^{hv} E_i^{eb}/E_i^{hv}\ .$$

The experiments in [4.210] have been conducted with the typical $F_2:H_2:O_2:He = 50:12:5:125$ laser mixtures with total pressures of 192 and 576 mm Hg. The lasing parameters have proved to be identical at $E_i^{eb}/E_i^{hv} = 2.3\text{--}2.5$. Since ε_F^{hv} is in this case equal to 1.7 eV, $\varepsilon_F^{eb} = 4.3 \pm 0.9$ eV. This value is substantially lower than those recommended earlier.

Using the above methods of determining the initial fluorine concentration, *Bashkin* and co-workers [4.211, 212] measured the energy characteristics of the H_2–F_2 laser as a function of $[F]_0$. In these experiments, the initiating electron beam entered the laser cell at right angles to the optical axis and was prevented from scattering by means of an external magnetic field. As a result, the effective laser volume was 65 mm long, 5.5 mm high, and 25 mm wide. The total active region volume was 9 ± 1 cm^3. The e-beam current density could be varied to ensure the production of atomic fluorine in concentrations from 5×10^{15} to 4.5×10^{16} cm^{-3} in the $F_2:H_2:O_2:He = 200:48:16:500$ mixture [4.211]. At $[F]_0 = 1.5 \times 10^{16}$ cm^{-3}, the specific energy output ε_l of the laser was some 90 mJ/cm^3 at a physical efficiency of around 900% and a chemical efficiency of around 6.5%. It follows from the theory that in conditions where the excited HF* molecules are quenched mainly by their unexcited HF counterparts, $\varepsilon_l \sim [F]_0^{1/2}$ (see Chap. 2 and also [4.213]). Such a relation is typical of the case where the oscillation threshold is considerably exceeded. In the experiment [4.211], the relation $\varepsilon_1 \sim [F]_0^{1/2}$ was observed to hold for $[F]_0 > 1.5 \times 10^{16}$ cm^{-3}.

To further increase $[F]_0$, *Bashkin* et al. [4.212] replaced some helium in the laser mixture by sulfur hexafluoride which played the part of a source of atomic fluorine, the total mixture pressure being maintained at atmospheric value. With 400 mm Hg of SF_6 substituted for He in this way, the specific e-beam energy deposition amounted to 265 mJ/cm^3 and $[F]_0$ increased to 2.8×10^{17} cm^{-3}. Under these conditions, the specific energy output and chemical efficiency of the H_2–F_2 laser reached record-high levels of 370 mJ/cm^3 and 24%, respectively, but its physical efficiency dropped to 140%. Thus, the specific energy output and chemical efficiency here also varied approximately as $[F]_0^{1/2}$.

Such high values of ε_l and especially of η_{chem} offer a good reason for further development of the suggestion by *Cooper* [4.214] that the H_2–F_2 laser be used as a highly efficient driver for laser-controlled thermonuclear fusion purposes [4.215].

An important kinetic feature of the H_2–F_2 laser, which plays a key part in interpreting the relationships between the laser energy output and various laser parameters, is to be noted. The hydrogen-fluorine chain reaction rate grows progressively higher as the temperature rises to 1000–1200 K, and then its growth slows down. The $V-T$ deactivation rate of HF* first drops as temperature rises to around 800 K and then starts to increase. For this reason, the ratio between the rate of production of the excited molecules and that of their deactivation has a maximum in the neighborhood of 800–1000 K. This is the most favorable temperature regime for the H_2–F_2 laser.

Under weak initiation conditions, the total energy release is not very high, and so the mixture temperature may be below 1000 K for most of the time the laser pulse lasts.

Where initiation is intense, the course of lasing rapidly heats the active mixture to temperatures above 1000 K, so that the process mostly occurs at increased temperatures, i.e., in unfavorable conditions. Consequently, under intense initiation, it is desirable to have a mixture with a higher heat capacity than in the case of weak initiation. Addition of SF_6 to the mixture with a view to increasing the initial concentration of atomic fluorine is also advantageous from the standpoint of increasing the heat capacity of the mixture, the heat capacity of this substance being high, and thus preventing the mixture from overheating. SF_6 also accelerates the rotational relaxation of the HF molecules, which in turn increases the energy output of the H_2–F_2 laser [4.216].

Thus, SF_6 has a many-sided effect on the lasing process in the H_2–F_2 laser [4.217]. Nevertheless, experience shows that at sufficiently high initial fluorine concentrations $[F]_0$ obtained with the addition of SF_6 to the mixture, the energy produced by the laser obeys the law $\varepsilon_l \sim [F]_0^{1/2}$. The same is true of the H_2–F_2 laser initiated with the third-harmonic output of a neodymium glass laser, where no addition of large amounts of SF_6 is required to produce high $[F]_0$ concentrations [4.216].

When studying an e-beam initiated H_2–F_2 laser, *Amimoto* et al. [4.218] have obtained 79 mJ/cm³ of laser energy at a physical efficiency of $\eta_{phys} = 380\%$ from the $F_2:H_2:O_2:SF_6:He = 160:64:24:200:352$ mm Hg mixture. These results correlate quite well with the data of [4.211, 212]. Indeed, the initial concentration $[F]_0$ in [4.212] was two times that in [4.218]. This should have raised the specific energy output 1.4 times. The actual value obtained in [4.212] ($\varepsilon_l = 156$ mJ/cm³) is two times that in [4.218]. But there is no contradiction here because the authors of [4.212] used a mixture containing a factor of 4 less oxygen than the mixture used by the authors of [4.218]. As has been already noted, oxygen reduces the laser chain length and hence the energy output. Calculations show that, due to the effect of oxygen alone, the specific laser energy output in [4.218] must be about a factor of 1.5 lower than that in [4.212], which in the final analysis sets up a correspondence between the results of [4.212] and [4.218].

The results reported in [4.212] have also been confirmed by the recent work of *Fujioka* et al. [4.219] who have achieved a very high specific energy output of $\varepsilon_l = 244$ J/l using the $F_2:H_2:O_2:SF_6 = 200:45:60:84$ mm Hg mixture at a specific e-beam energy deposition of 127 J/cm³. As can be seen, the F_2 and H_2 concentrations here are the same as in [4.212], but the specific e-beam energy deposition is only half as high, and this agrees well with the factor of 1.5 lower value of ε_l. The fact that the Japanese scientists have obtained their result using a laser cell with a large active volume (18 l) is encouraging, for it is evidence that the results obtained in [4.212] with a much smaller active laser volume can be scaled up.

Thus, experiments with the e-beam initiated H_2–F_2 laser have shown that it can operate in two different regimes: at high initiation ratios providing for high specific energy outputs and chemical efficiencies but low physical efficiencies, and vice versa. The same relationships have also been found to hold true in the case of the flashlamp-initiated H_2–F_2 laser [4.209, 220]. These experiments have been conducted on the modernized setup (6 l in active volume) described in [4.117] (see Table 4.4). The transverse dimension (around 13 cm) of the active region allowed for the flashlamp radiation to be effectively absorbed (intercepted) by molecular fluorine at a pressure of 150–200 mm Hg. With the partial F_2 pressure kept at 150 mm Hg, the initial atomic fluorine concentration produced by flash photolysis ranged from 4.5×10^{15} cm^{-3} to 1.8×10^{16} cm^{-3}. Optimization of the cavity, flashlamp, and mixture parameters, the initiation ratio, and mixture preparation procedures allowed the energy characteristics of the laser to be improved in comparison to those achieved in [4.117]. When operating the laser on the $F_2:H_2:O_2:He = 150:75:15:520$ mm Hg mixture at a minimal initiation level ($[F]_0 = 4.5 \times 10^{15}$ cm^{-3}), the specific energy output of the laser was at its minimum of 32 J/l, its physical and engineering efficiencies reaching their maximum values of 2060% and 50%, respectively. With the initiation level being at its maximum ($[F]_0 = 1.8 \times 10^{16}$ cm^{-3}), the specific energy output was also maximal (46 J/l), whereas the physical and engineering efficiencies were minimal (720% and 17%, respectively). The specific energy output and engineering efficiency values obtained in [4.209, 220] are the best as of now for flash-photolysis initiated H_2–F_2 lasers. Somewhat worse but fairly close results confirming the possibility of achieving physical efficiency values in excess of 1000% have been obtained by *Gordon* et al. [4.221] with an H_2–F_2 laser initiated by the radiation of a XeCl excimer laser. The initiating radiation propagated along the optical axis of the H_2–F_2 laser, and for this reason, the length of the laser cell was always made greater than the effective depth of initiation which in these conditions was essentially inhomogeneous. For example, in the case of the $F_2:O_2:H_2:SF_6 = 17:7.5:25:35$ mm Hg mixture, the value of $[F]_0$ varied over the laser cell length ($l = 25$ cm) from 7.5×10^{15} to 2×10^{15} cm^{-3}. Under these conditions, the physical efficiency of the laser amounted to 1100% at a specific laser energy output of 20 J/l. These investigators even managed to increase the physical efficiency of their laser to 1500% by raising its initiation level slightly.

When analyzing in Sect. 4.6 the results of studies into the flash-photolysis initiated D_2–F_2–CO_2 laser obtained before 1980, we have noted that the maximum engineering efficiency value (12%) achieved with this laser at that time [4.2] (see Table 4.4) is not the limit and can further be increased, especially in view of the established fact that the physical efficiency of the D_2–F_2–CO_2 laser is no worse than that of its H_2–F_2 counterpart [4.124]. And indeed, recent research on this type of laser [4.209, 220] carried out with the $D_2:F_2:CO_2:He:O_2 = 36:122:122:$

450:5 mm Hg mixture in the same 6-liter experimental setup has shown that this laser can also be optimized to have energy characteristics similar to those of the H_2-F_2 laser. What is more, this laser has also been found to feature two entirely different regimes, one providing for maximum specific energy outputs and the other, for maximum engineering (as well as physical) efficiencies: a specific energy output of 51 mJ/cm^3 has been obtained at $[F]_0 = 1.6 \times 10^{16}$ cm^{-3}, the physical and engineering efficiencies being 900% and 20%, respectively, whereas physical and engineering efficiencies of 2800% and 65%, respectively, have been achieved at $[F]_0 = 2.1 \times 10^{15}$ cm^{-3}, the specific energy output of the laser being in this case equal to 21 mJ/cm^3. Thus, at moderate initiation levels, the energy characteristics of the $D_2-F_2-CO_2$ laser are, in fact, close to those of the H_2-F_2 laser.

Certain progress has been made in the H_2-F_2 lasers using electrical discharge initiation, both longitudinal and transverse. Of unquestionable interest is the laser design [4.222] utilizing a longitudinal discharge set up in a cell 9.2 cm in diameter and about 120 cm long ($V = 8$ l). Sufficiently homogeneous discharges have been achieved with the $F_2:O_2:SF_6:H_2 = 3:1:3:1$ mixture at a total pressure of up to 24 mm Hg. The engineering efficiency of the laser has reached 63% at a specific energy output of around 2.5 J/l, which is close to the results reported in [4.32] (see Table 4.5), but the active region volume here has been almost 700 times that in [4.32]. The total laser energy level ($\simeq 20$ J) is quite sufficient for many practical applications. Good enough results have also been obtained with the transverse-discharge initiated H_2-F_2 laser [4.223] using some SF_6 to increase the initiation ratio. The original double-discharge system of the laser has allowed for a sufficiently homogeneous main discharge to be set up in a volume of about 0.3 l. And though the engineering efficiency achieved (23%) is inferior to the data of Table 4.5, the specific energy output is substantially higher (17.5 J/l).

Some headway have in recent years been made in investigations aimed at improving the beam divergence of H_2-F_2 lasers. Experiments in this field have been staged in two ways: to study the influence of optical inhomogeneities in the active medium of a laser on the divergence of an external laser beam passing through it and to study the laser beam divergence directly. The first approach has been successfully realized by *Patterson* et al. [4.224] who studied changes in the divergence of a beam passing through the active medium of a 7-cm-dia. H_2-F_2 laser amplifier transversely initiated with a high-current electron beam. The mixture used was energy-intensive: $F_2:H_2:O_2 = 1:0.25:0.3$ at a total pressure of 1240 mm Hg. The initiation ratio was such that the specific energy output of the amplifier operated in the oscillator mode reached 90 J/l, i.e., the active medium of the amplifier provided for relatively high laser energy parameters. Measurements of the field distribution in the far field demonstrated that the beam divergence, which was a factor of 1.8 worse than the diffraction-limited value at the entrance to the amplifier, became a factor of 2.7 worse than the diffraction-limited value after passing through the amplifier medium.

These experiments have shown that the optical quality of the H_2-F_2 laser medium can, in principle, be made good. But this was done in conditions of fairly high initiation homogeneity. Indeed, the 1.5–MeV electrons in the initiating beam used ensured a great initiation depth (up to a few meters) in the given mixture.

Hence it is clear that with the e-beam path length in the laser cell being around 7 cm, only a small proportion of the electron energy was absorbed in the laser medium. In principle, increasing this proportion may worsen the optical quality of the active medium. But this increase is most interesting because it may help reach high engineering efficiency values.

With the second approach, the beam divergence of the H_2-F_2 laser has been studied directly. The importance of selecting a suitable unstable cavity (UC) to solve the problem of improving the laser beam divergence has been demonstrated by *Borisov* et al. [4.225] who studied the possibility of achieving the diffraction-limited divergence in a flash-photolysis initiated D_2-F_2 laser. The experiments have been performed with the $F_2:D_2:SF_6:O_2 = 22:7:22:7$ mm Hg low-energy-content mixture having an effective initiation depth of $l_{eff} \simeq (\sigma_{eff}[F_2])^{-1} = 120$ cm for an effective F_2 photodissociation cross section of $\sigma_{eff} \pm 10^{-20}$ cm, which was much greater than the laser cell diameter (7.2 cm). What is more, the laser mixture was strongly diluted with SF_6 smoothing out temperature inhomogeneities in the active medium. All this substantially reduced refractive index variations in the transverse direction. In these conditions, the laser beam divergence was determined only by the cavity parameters and the quality of the optics used. The experiments were carried out using a Fabry–Perot cavity (FPC) and a telescopic UC with three magnification values: $M = 2.5, 4$, and 8. The beam divergence values measured in terms of the half-maximum energy level amounted to 1.2, 0.16, 0.09, and 0.07 mrad, respectively. The latter figure practically coincides with the diffraction-limited divergence value. The energy loss in that case came to a mere 40% compared to that with the FPC.

In [4.225], the low energy content of the mixture and operation with an effective initiation depth much in excess of the laser cell diameter ($l_{eff} \gg d$) were responsible for the low values of specific energy output ($\simeq 2.4$ mJ/cm^3) and engineering efficiency ($\simeq 1.1\%$) achieved with the FPC. But it has remained unclear whether it is possible to attain a beam divergence close to the diffraction-limited value in hydrogen-fluoride lasers operating with high-energy-content mixtures and with efficient utilization of the initiation source energy, i.e., with $l_{eff} \simeq d$. It is precisely in this case that the refractive characteristics of the active medium vary most strongly as a result of initiation inhomogeneity in the transverse direction. Such investigations have been conducted by *Bashkin* and co-workers [4.226] using the same flash-photolysis initiated laser setup they experimented with previously when studying the H_2-F_2 and $D_2-F_2-CO_2$ lasers [4.209, 220]. The only differences here have been that the laser cell windows were tilted at an angle of $5°$ with respect to the optical axis in order to prevent the formation of a combination optical cavity, that the partial fluorine pressure in the main experiments was lowered somewhat, and that hydrogen was replaced by deuterium to exclude atmospheric absorption when taking beam divergence measurements (mixture composition $F_2:D_2:O_2:He = 113:38:11:600$ mm Hg).

Two magnification values, $M = 2.1$ and 3.1, were chosen in these experiments to suit the measured unsaturated gain values ($\alpha_0 = 0.05$ cm^{-1}). The beam divergence of the D_2-F_2 laser with a FPC, measured in terms of the half-maximum energy level, has proved to be almost a factor of 50 worse than the diffraction-limited divergence. With the initiation level at its maximum ($[F]_0 = 1.6 \times 10^{16}$ cm^{-3}), the specific energy output and engineering efficiency of the laser have amounted to 12 mJ/cm^3 and 4.4%, respectively. The beam divergence of the laser using an UC with $M=2.1$ was only a factor of 1.7 worse than the diffraction-limited divergence at a specific energy output of 8 mJ/cm^3 and an engineering efficiency of 3%. The use of a more energy-intensive mixture ($F_2:D_2:O_2:He=150:53:15:540$ mm Hg) has made it possible to reach $\varepsilon_l=13$ mJ/cm^3 and $\eta_{eng}=5\%$ at $[F]_0=1.8\times10^{16}$ cm^{-3}, account being taken of the energy coupled out of the UC with $M=2.1$ upon reflections at the laser cell windows. No deterioration of the laser beam divergence has been observed in that case. The use of a UC with $M=3.1$ had a small effect on the result. Thus, these experiments have demonstrated that the specific energy output and engineering efficiency of the D_2–F_2 laser can be made fairly high by appropriately selecting unstable cavity parameters in conditions where the initiation depth does not exceed the transverse dimension of the active medium.

Table 4.1. Flash-photolysis initiated non-chain reaction HF (DF) lasers

Mixture	Pressure [mm Hg]	Radiation spectrum	t_l [µs]	Reference
$UF_6-(H_2, D_2, HD)$	2.25	$P_2(3)-P_2(8)$	10	[4.68, 69]
$UF_6-(CH_4, CD_4, CH_3F,$ $C_nH_{2n+2}, CH_2F_2, CHF_3,$ $CH_3Cl, CH_2Cl_2, CHCl_3)$	2.25	$P_2(4)-P_2(18)$	4–8	[4.70]
UF_6-H_2	2–5			[4.71]
SbF_5-H_2	6		1	[4.72]
UF_6-CH_3Cl		P_1, P_2, P_3		[4.73]
WF_6-H_2	6.5	$P_1(3)-P_1(8),$ $P_2(2)-P_2(8)$ $P_3(1)-P_3(5)$	1–2	[4.74]
$(UF_6, MoF_6)-H_2$	2–45	$P_2(4)-P_2(7)$	7–10	[4.75]
$WF_6-(H_2, D_2, CH_4,$ $C_4H_{10}, HCl)$				[4.76]
MoF_6-H_2				[4.77]
$XeF_4-(H_2, CH_4)$	2.5–10		1	[4.72]
F_2O-H_2	100		5	[4.78]
$N_2F_4-(HCl, CH_4, CH_3F,$ $CH_2F_2, CH_3Br, C_2H_6,$ $C_2H_5F, C_2H_5I)$	1–24	$P_1(6)-P_1(9),$ $P_2(4)-P_2(8)$	4	[4.79]
$N_2F_4-CH_3I$	12	$P(2)$	20	[4.80]
$N_2F_4-CD_4$	80	Second harmonic of transitions in DF	20	[4.81]
CF_3I-HI				[4.82]
$CF_3I-(C_2H_2, C_2H_4, C_2H_6, CH_4)$				[4.83]
CF_3I-CH_3I	5–40	$P_1(3)-P_1(10)$ $P_2(3)-P_2(5)$		[4.84]
$CH_3F-C_2H_2O$				[4.85]
$O_3-CH_nX_{4-n}$ $(X=F, Cl; n=1, 2, 3)$	30	$P_1(6)-P_1(10)$ $P_2(4)-P_2(6)$	2	[4.86]

Table 4.2. Main parameters of electrical-discharge initiated non-chain reaction pulsed chemical HF lasers

Mixture composition and pressure	Discharge region size [cm³]	Discharge circuit design, U_b, E_b, τ_i	$\dfrac{\Delta[F_2]_0}{[F_2]_0}$ [%]	t_i [µs]	E_i [J]	Radiation spectrum	ε_i [J/l]	η_{eng} [%]	η_{chem} [%]	Reference
SF₆:H₂=24:1, p=62.5 mm Hg	3×3×200	Fig. 4.4a; 25 kV, 155 J, 1 µs	—	0.2	1.2	$P_1(4)-P_1(17)$, $P_2(3)-P_2(13)$, $P_3(2)-P_3(9)$		0.8		[4.87]
SF₆:H₂=24:1, p=150 mm Hg	4×5×300	Fig. 4.4a; 120 kV, 260J, 0.4 µs		0.15	11			3.8		[4.28]
SF₆:C₄H₁₀=35:1, p=50 mm Hg	2.5×2.4×51	Fig. 4.4b; 80 kV, 16 J		0.5	0.1		0.3	0.6	0.25	[4.29]
SF₆:H₂=5.5:1, p=80 mm Hg	2.5×2.4×51	Fig. 4.4b; 80 kV, 16 J		0.5	0.09	$P_1(4)-P_1(7)$, $P_2(3)-P_2(8)$, $P_3(3)-P_3(6)$	0.3	0.6	0.25	[4.29]
SF₆:C₂H₆=21:1, p=50-150 mm Hg	2.5×2.4×51	Fig. 4.4b; 66 kV, 50 J, 0.1 µs; 95 kV, 200 J		0.1	1.4		4.2	3.1		[4.88]
					3.5		10.5	1.8		
SF₆:H₂:He =12:1:14, p=400–600 mm Hg	2×1.5×20	Fig. 4.5a; 40 kV, 25 J, 60–100 ns	4.8	0.06	1	$P_1(8)$, $P_2(3)-P_2(9)$, $P_3(3)-P_3(5)$	16.5	4		[4.31]
p_{SF_6} = 0.22 atm., p_{H_2} = 0.056 atm., $0 < p_{He} \leqslant 5.2$ atm.		250 kV, 20 J, 14 ns								[4.33]

Table 4.3. Main parameters of electron-beam initiated non-chain reaction pulsed chemical HF lasers

Mixture composition and pressure	V_l [1]	U, I, J_b, τ_e, cathode size*	$\dfrac{\Delta[\mathrm{RF}]_0}{[\mathrm{RF}]_0}$ [%]	t_l [ns]	E_l [J]	Radiation spectrum	ε_l [J/l]	$\dfrac{\varepsilon_l/p}{\left[\dfrac{\mathrm{J}}{\mathrm{l\,atm}}\right]}$	η_{phys} [%]	η_{chem} [%]	Reference
N$_2$F$_4$–(H$_2$, B$_2$H$_6$), NF$_3$–H$_2$ $p=$10–250 mm Hg	0.0125	1.4 MeV, 400 J, 50 ns, $\perp$		10		$P_1(4)$–(6), $P_2(3)$–(6), $P_3(3)$–(4)					[4.89]
SF$_6$:C$_2$H$_6$=6:1, $p=$140 mm Hg		60 keV, 5 kJ 10–15 kOe, $\parallel$		250	60						[4.91]
SF$_6$:H$_2$=11:1, $p=$450 mm Hg	1	120 keV, 4.65 kA, 26 ns, 3.5×35 cm^2, $\perp$	4×10^{-4}	20–30	0.2		0.2	0.35	5.2		[4.90]
SF$_6$:C$_2$H$_6$=5:1 $p=$24 mm Hg	1.4	300 keV, 25 kA, 30 ns, ϕ10 cm, 6 kOe, $\angle$	4.7	45	5.5	$P_1(3)$–(9), $P_2(2)$–(9), $P_3(3)$–(9)	4	120	4		[4.92]
SF$_6$:C$_2$H$_6$=10:1 $p=$440 mm Hg	22	2 MeV, 55 kA, 70 ns, ϕ15.2 cm, 2 kOe, $\parallel$	1–2	55	228		10.5	18	8		[4.93]
SF$_6$:C$_2$H$_6$=10:1, $p=$300–400 mm Hg	26	2 MeV, 55 kA, 70 ns, ϕ14.9 cm, 2.5 kOe, $\angle$	1–2	23	380	$P_1(3)$–(9), $P_2(2)$–(9), $P_3(3)$–(9)	15	30	14	28	[4.94]
SF$_6$:H$_2$=7:1, $p=$0.7 atm.	8×10^{-3}	50 keV, 75 A, 100 ns, 7 A/cm^2, $\perp$			10^{-2}		1.25	1.7	11		[4.95]

* Indicated here are also the strength of the confining magnetic field and e-beam injection geometry relative to the optical axis of the laser cell $\parallel$ – longitudinal, $\perp$ – transverse, and $\angle$ – oblique.

Table 4.4. Main parameters of flash-photolysis initiated chain-reaction pulsed chemical HF, DF, DF–CO$_2$ lasers

Mixture composition and pressure	V_i [l]	Initiation source, E_b, τ_i	$\dfrac{\Delta[F_2]_0}{[F_0]}$ [%]	t_i [μs]	E_i^* [J]	Radiation spectrum	ε_i [J/l]	$\varepsilon_i^\dagger/p$ $\left[\dfrac{J}{1\,\text{atm.}}\right]$	η_{eng} [%]	$\eta_{\text{chem}}^\dagger$ [%]	Reference
F$_2$:D$_2$:O$_2$:CO$_2$:He =2:1:0.1:5:10, p=300 mm Hg	10	Tubular Xe-filled flashlamp			189		19	50	12	3.7	[4.2]
F$_2$:D$_2$:CO$_2$:He =3:1:24:30, p=0.5 atm.	0.29 (0.059)	Tubular Xe-filled flashlamp, 2400 J, 50 μs		17	2.8		10	20	0.1	5.1 (15.3)	[4.106]
F$_2$:D$_2$:CO$_2$:He =1:1:6:19, p=1 atm.	0.23	Tubular Xe-filled flashlamp, 350 J, 25 μs	1	45	5 (0.03)		20	20	1.5	3.2 (>20)	[4.107]
F$_2$:D$_2$:CO$_2$:He =1:1:4:11, p_{MoF_2}=1–1.5 mm Hg, p=1 atm.		Bursting wire, 13.5 kJ, >10 μs	2–3 (estimate)	7–10	10 (≥0.024)		>20	>20			[4.108, 109]
F$_2$:D$_2$:CO$_2$:He =1:1:4:5, p=1 atm.		Creeping discharge, 50 kJ, 5–10 μs	3–7 (estimate)	8–12	(0.07)	P(20), P(22)	150	150		7	[4.110]
F$_2$:D$_2$:CO$_2$ =3:1:5, p=0.5 atm.	0.1 (0.04)	Tubular Xe-filled flashlamp, 790 J, 13 μs	0.13	4	2.5		25	50		2 (5)	[4.111]

Table 4.4 (continued)

Mixture composition and pressure	V_i [l]	Initiation source, E_b, τ_i	$\dfrac{\Delta[F_2]_0}{[F_0]}$ [%]	t_i [µs]	E_i^* [J]	Radiation spectrum	ε_i [J/l]	$\varepsilon_i^\dagger/p$ $\left[\dfrac{J}{1\,\text{atm}}\right]$	η_{eng} [%]	$\eta_{\text{chem}}^\dagger$ [%]	Reference
$F_2:D_2:O_2:CO_2$ $\simeq 3:1:0.1:6$, $p=1$ atm.	0.6	Coaxial Xe-filled flashlamp, 20 kJ, 7 µs	0.5–0.9	1	42	$P(14)$: :$P(16)$: :$P(18)$: :$P(20)$: :$P(22)$: :$P(24)\simeq$ 1:8:8:80: :2:1	70	70	0.2	3.5	[4.112]
$F_2:H_2:MoF_6:He$ $=1:1:0.25:40$, $p=600$ mm Hg	0.02	Tubular Xe-filled flashlamp, 3 µs		5	65		3.3	4		0.8	[4.113]
$F_2:H_2:O_2$ $=10:3:5$, $p=0.3$ atm.	4.5	Tubular Xe-filled flashlamp, 940 J			37		8	24	3.9	0.64	[4.114]
$F_2:H_2:O_2:He$ $\simeq 3:1:0.5:20$, $p=1$ atm.	0.6	Coaxial Xe-filled flashlamp, 10 kJ, 7 µs	0.15	1.5–2	9		15	15	0.1	1.5	[4.112, 115]
$F_2:H_2:O_2$ $=1:1:0.05$, $T=213$ K, $p=114$ mm Hg	0.0057	Tubular Xe-filled flashlamp, 500 J, $\geqslant 10$ µs	0.75	0.4	0.035	$P_1(3)$–(8), $P_2(3)$–(10), $P_3(4)$–(8), $P_4(3)^{*\dagger}$	6	40 (77)		0.5	[4.61]
$F_2:H_2:O_2:Ar$ $\simeq 6.4:1.8:1:0.8$, $p=560$ mm Hg	0.04	Tubular Xe-filled flashlamp	0.035	0.25	0.83	$P_1(5)$–(10), $P_2(4)$–(10), $P_3(4)$–(9), $P_4(5)$–(6), $P_4(8)$, $P_5(5)$–(6)	20	30		0.8	[4.116]

F$_2$:H$_2$:O$_2$:He =1:2:0.025:9.5, p=1.1 atm.	0.1	Coaxial Xe-filled flashlamp, 600 J, 1–1.5 µs	0.4±0.1	2	8	P$_1$(3)–(9), P$_2$(3)–(16), P$_3$(4)–(8), P$_4$(4)–(7), P$_5$(3)–(6), P$_6$(1)–(5)	80	73	1.3	4	[4.20]
F$_2$:H$_2$:O$_2$:Ar =10:1.2:0.7:88.1, p=1 atm.	12.8	U-type flashlamps, 1–1.5 kJ, 3 µs		2.7	292	P$_1$, P$_2$, P$_3$	23	23	20–30	8	[4.3]
F$_2$:D$_2$:O$_2$:Ar =10:1.2:0.7:88.1, p=1 atm.	12.8	U-type flashlamps, 1–1.5 kJ, 3 µs		3.8	144		11	11	14	4	[4.3]
F$_2$:H$_2$:O$_2$:He =20:3:2:75, p=1 atm.	7.3	Tubular Xe-filled flashlamp, 650 J, 2.5 µs	≃10^{-2}	3.5	194		26	26	30	3	[4.117]

*Figures in parentheses indicate unsaturated gain in cm^{-1}.

†Figures in parentheses refer to the case where the laser volume was taken to be the cavity volume occupied by the radiating modes.

*†Observed under other experimental conditions were weak P$_1$(9)–(10), P$_2$(11)–(13), P$_4$(5)–(8), P$_5$(5)–(8), and P$_6$(5)–(7) lines.

Table 4.5. Main parameters of electrical-discharge initiated chain-reaction pulsed chemical HF, DF, DF-CO$_2$ lasers

Mixture composition and pressure	Discharge region size [cm^3]	Discharge circuit design U_b, E_b, τ_i	$\dfrac{\Delta[F_2]_0}{[F_2]_0}$ [%]	t_i [µs]	E_i [J]	Radiation spectrum	ε_i [J/l]	η_{eng} [%]	η_{chem} [%]	Reference
F$_2$:H$_2$:O$_2$:He = 1:0.23:0.08:12, $p=240$ mm Hg	ϕ 2.5×130	Fig. 4.4a, 25 kV, 0.1 J, 0.1 µs	0.2	40	0.15		0.25	144		[4.118]
F$_2$:H$_2$:O$_2$:He = 1:1:0.25 (0.06):10, $p=36$-120 mm Hg	0.8×1×15	Fig. 4.5b, 3–5 kV, 0.016–0.12 J, 0.2–07 µs	0.2–2	<4	(1 to 1.6)× 10^{-2}	$P_1(3)$–(14), $P_2(3)$–(13), $P_3(3)$–(9), $P_4(5)$–(9), $P_5(5)$–(8), $P_6(5)$–(6)	2–3	25–100	1–2	[4.32]
D$_2$:F$_2$:O$_2$:CO$_2$:He = 2:4.5:6:82, $p=300$-760 mm Hg	2.2×1.3×43	Fig. 4.5a, 50 kV, 7 J, 0.15 µs		15		$P(18, 20, 26, 34, 42, 50)$, $P_1(5)$–(6)				[4.119]
D$_2$:F$_2$:O$_2$:He = 2:4.5:6:88, $p=300$-760 mm Hg	2.2×1.3×43	Fig. 4.5a, 50 kV, 7 J, 0.15 µs				$P_2(9)$, $P_3(4)$–(7), $P_4(5)$–(6)				[4.119]
D$_2$:F$_2$:CO$_2$:He = 1:1:3:20, $p=250$ mm Hg	ϕ 1.2×65	Fig. 4.4a, 30 kV, 2.5 J, 0.3 µs		50	0.5		8	20	2.4	[4.45]

Table 4.6. Main parameters of electron-beam initiated chain-reaction pulsed chemical HF, DF, DF–CO$_2$ lasers

Mixture composition and pressure	V_l [l]	U, I, J_b, τ_e, cathode size*	$\dfrac{\Delta[F_2]_0}{[F_2]_0}$ [%]	t_l [ns]	E_l [J]	Radiation spectrum	ε_l [J/l]	$\dfrac{\varepsilon_l}{p}\left[\dfrac{\text{J}}{\text{1 atm}}\right]$	η_{phys} [%]	η_{chem} [%]	Reference
F_2:H_2=1:1, p=200–660 mm Hg, $T \simeq 100$ K	0.011	2 MeV, 4 kA, 6×10^{-8} s, 5×2.25 cm^2, $\perp$		2×10^3	3.5×10^{-2}		3.5		150–180		[4.120]
F_2:H_2:SF_6 = 1:1:1.5, p=420 mm Hg	0.236	300 keV, 3 kA, 1 μs, 3.8×20.3 cm^2, $\perp$		500	11		46	84	6		[4.41]
F_2:H_2:$O_2 \simeq$ 4.4:4.8:0.8, p=620 mm Hg	0.67	2 MeV, 1.5 kA, 80 ns, ϕ 2.4 cm, 7 kOe, $\perp$		60	13		20	25	100	0.25	[4.121]
F_2:H_2:SF_6:$O_2 \simeq$ 5.1:1.4:1.4:2, p=700 mm Hg	<28	2 MeV, 50 kA, 70 ns, ϕ 14.9 cm, 790 Oe, $\parallel$		35	2.3×10^3	$P_1(4)$–(13), $P_1(15)$–(18), $P_1(21)$, $P_2(4)$–(14), $P_2(16)$, $P_3(4)$–(12), $P_4(7)$–(10), $P_5(4)$, $P_5(7)$–(10), $P_5(13)$–(15), $P_5(17)$, $P_6(3)$, (5), $P_6(10)$	$\geqslant 100$	$\geqslant 100$	180	3.3	[4.122]
F_2:H_2:O_2:SF_6 = 1:0.25:0.3:0.125, p=1300 mm Hg	<32	2 MeV, 50 kA, 70 ns, ϕ 14.9 cm, 3 kOe, $\angle$		20	4.2×10^3	P_1, P_2, P_3, P_4, P_5, P_6	$\geqslant 130$	$\geqslant 80$	110	3.5	[4.57]

Table 4.6 (continued)

Mixture composition and pressure	V_l [l]	U, I, J_b, τ_e, cathode size	$\frac{\Delta[F_2]_0}{[F_2]_0}$ [%]	t_l [ns]	E_l [J]	Radiation spectrum	ε_l [J/l]	ε_l/p $\left[\frac{J}{1\,atm}\right]$	η_{phys} [%]	η_{chem} [%]	Reference
$F_2:H_2:O_2:He = 0.3:0.08:0.01:0.61$, $p = 1$ atm.	0.1 (ϕ 2.3 cm)	140 keV, 200 ns, 10×25 cm², 20 A/cm² (12 A/cm² in cell), $\perp$	5×10^{-2}	2×10^3	5.1		51	51	875	2.8	[4.60]
$F_2:H_2:O_2:SF_6 = 10:3:4:10$, $p = 400$ mm Hg	0.008	50 keV, 75 A, 100 ns, 7 A/cm², $\perp$			1.1×10^{-1}		14 / 28†	26 / 52†	200 / 400†		[4.95]
$F_2:H_2:O_2:He = 15:6:1.2:50$, $p = 720$ mm Hg	0.12	200 keV, 800 A, 35 ns, ϕ 2 cm, 6 kOe, $\perp$	0.1	$\simeq 10^3$	11		91	96	940	4.7	[4.123]
$F_2:D_2:O_2:CO_2:He = 150:60:12:150:375$ mm Hg	0.12	200 keV, 800 A, 35 ns, ϕ 2 cm, 6 kOe, $\perp$	0.08	3.5×10^3	7.2		60	60	330	3.5	[4.124]
$F_2:D_2:O_2:CO_2:He = 50:20:4:50:125$ mm Hg	0.12	200 keV, 800 A, 35 ns, ϕ 2 cm, 6 kOe, $\perp$	0.08	9×10^3	2.6		22	66	360	3.5	[4.124]
$F_2:D_2(H_2):O_2:He = 50:20:4:125$ mm Hg	0.12	200 keV, 800 A, 35 ns, ϕ 2 cm, 6 kOe, $\perp$	0.08 / 0.08†	3.5×10^3 / 3.5×10^3†	1.2 / 2.88†		10 / 24†	38 / 91†	375 / 900†	1.6 / 3.7†	[4.124]

*Indicated here are also the strength of the confining magnetic field and e-beam injection geometry relative to the optical axis of the laser cell: ||—longitudinal, $\perp$—transverse, and $\angle$—oblique.

†Estimated values for the optimum laser cavity output coupling.

Table 4.7. Lines observed in the radiation spectrum of the H_2–F_2 chemical laser [4.128]

Measured wavelength [μm]	Identification		Calculated wavelength [μm]	Relative peak power
	Band	Transition J		
2.6721	1–0	5	2.672	1
2.707		6	2.7068	2190
2.7432		7	2.7434	6910
2.7817		8	2.782	4390
2.8224		9	2.8224	781
2.865		10	2.865	89
2.6955	2–1	2	2.6956	872
2.7267		3	2.7267	1405
2.7599		4	2.7597	2985
2.7946		5	2.7945	3850
2.8312		6	2.8311	6100
2.8696		7	2.8697	4340
2.91		8	2.9104	536
2.9529		9	2.9532	449
2.8531	3–2	3	2.8534	622
2.8881		4	2.8881	1195
2.9249		5	2.9248	1650
2.9635		6	2.9636	496
3.0044		7	3.0044	1785
3.0473		8	3.0473	3750
3.0926		9	3.0926	573
2.9216	4–3	1	2.9212	2920
2.9542		2	2.9539	3710
2.989		3	2.9887	5980
3.0257		4	3.0254	3125
3.0644		5	3.0642	2225
3.1053		6	3.1052	684
3.1483		7	3.1484	1024
3.2416		9	3.2419	295
3.0628	5–4	1	3.0625	1120
3.0973		2	3.0972	1725
3.1342		3	3.1339	500
3.1733		4	3.1728	148
3.2142		5	3.2139	177
3.352		8	3.3516	377
3.4031		9	3.4026	527
3.2155	6–5	1	3.2151	467
3.2523		2	3.2518	995
3.2913		3	3.2908	458
3.3325		4	3.332	584

Table 4.8. Lines observed in the radiation spectrum of the D_2–F_2 chemical laser [4.127]

Measured frequency [cm^{-1}]	Identification		Calculated frequency [cm^{-1}]	Relative line energy
	Band	Transition		
2665.2	1–0	P(10)	2665.2	1
2727.3	2–1	P(4)	2727.28	0.04
2704		P(5)	2703.97	0.26
2680.1		P(6)	2680.14	0.94
2655.8		P(7)	2655.82	1.36
2631		P(8)	2631.02	2.37
2605.8		P(9)	2605.76	2.68
2580		P(10)	2580.04	4.24
2553.9		P(11)	2553.9	3.46
2617.3	3–2	P(5)	2617.32	0.41
2594.1		P(6)	2594.1	1.14
2570.4		P(7)	2570.45	1.79
2546.3		P(8)	2546.3	2.74
2521.7		P(9)	2521.7	5.38
2496.6		P(10)	2496.65	3.35
2471.2		P(11)	2471.18	2.26
2445.3		P(12)	2445.3	0.29
2419		P(13)	2419.02	
2392.4		P(14)	2392.36	
2509.8	4–3	P(6)	2509.82	0.38
2486.7		P(7)	2486.77	0.91
2463.2		P(8)	2463.25	1.08
2439.3		P(9)	2439.29	1.78
2414.9		P(10)	2414.89	0.65
2390.1		P(11)	2390.07	0.19
2404.6	5–4	P(7)	2404.63	0.016
2381.7		P(8)	2381.73	0.065
2334.6		P(10)	2334.63	
2310.4		P(11)	2310.45	
2285.9		P(12)	2285.88	
2260.9		P(13)	2260.92	
2388	6–5	P(4)	2388.02	0.003
2323.9		P(7)	2323.89	0.045
2301.6		P(8)	2301.6	0.09
2278.9		P(9)	2278.87	0.19
2255.7		P(10)	2255.71	0.07
2232.1		P(11)	2232.15	
2286.5	7–6	P(5)	2286.45	0.016
2265.6		P(6)	2265.65	0.11
2244.4		P(7)	2244.38	0.06
2222.7		P(8)	2222.68	0.19
2200.5		P(9)	2200.54	
2178		P(10)	2177.99	
2155		P(11)	2155.03	
2131.7		P(12)	2131.68	

Table 4.8. (continued)

Measured frequency [cm^{-1}]	Identification		Calculated frequency [cm^{-1}]	Relative line energy
	Band	Transition		
2206.9	8–7	$P(5)$	2206.87	0.01
2186.6		$P(6)$	2186.63	0.08
2165.9		$P(7)$	2165.93	0.18
2144.8		$P(8)$	2144.8	0.11
2123.2		$P(9)$	2123.24	0.03
2101.3		$P(10)$	2101.27	
2056.1		$P(12)$	2056.14	
2033		$P(13)$	2033.01	
2108.5	9–8	$P(6)$	2108.48	0.01
2088.3		$P(7)$	2088.34	0.06
2067.8		$P(8)$	2067.76	0.08
2046.8		$P(9)$	2046.77	0.0008
2025.4		$P(10)$	2025.36	
2003.6		$P(11)$	2003.56	
1981.4		$P(12)$	1981.38	

Table 4.9. Efficiency of photolysis, $\Phi\sigma I_0$, of some fluorides relative to that of fluorine [4.4]

Substance	F_2	ClF	ClF$_5$	IF$_7$	MoF$_6$	VF$_5$	WF$_6$
$\Phi\sigma I_0$	1	0.25	15	64	64	225	5

Table 4.10. Ratio between dipole moment matrix elements of various transitions in HF and DF

Transition $v+f\rightarrow v$	$\lambda[\mu m]$ (band center)	$(|m_v^{v+f}|^2)/(|m|_0^1)^2$	Transition $v+f\rightarrow v$	$\lambda[\mu m]$ (band center)	$(|m_v^{v+f}|^2)/(|m_0^1|^2)$
	HF			DF	
1–0	2.52	1	1–0	3.44	1
2–1	2.64	2	2–1	3.55	2
3–2	2.76	3	3–2	3.67	3
4–3	2.89	4	4–3	3.79	4
5–4	3.04	5	5–4	3.92	5
6–5	3.18	6	6–5	4.05	6
2–0	1.29	1.1×10^{-2}	2–0	1.75	0.79×10^{-2}
3–1	1.35	3.3×10^{-2}	3–1	1.8	2.4×10^{-2}
4–2	1.41	6.5×10^{-2}	4–2	1.86	4.7×10^{-2}
5–3	1.48	10.9×10^{-2}	5–3	1.93	7.9×10^{-2}
6–4	1.55	16.3×10^{-2}	6–4	1.99	11.8×10^{-2}
3–0	0.88	3.2×10^{-4}			
4–1	0.92	12.6×10^{-4}			
5–2	0.963	31.5×10^{-4}			
6–3	1.01	63×10^{-4}			

5. Continuous-Wave Chemical Lasers

The practical interest in continuous-wave chemical lasers (CWCL's) is due first of all to the possibility of realizing independent operation without the need for any additional power-supply source. This circumstance distinguishes CWCL's from other types of lasers, such as electrical-discharge ones.

The present chapter consists of seven sections. Section 5.1 deals with the physical principles underlying the operation of CWCL's and considers their main structural components.

Section 5.2 is devoted to the purely chemical subsonic $DF-CO_2$ laser. This system has embodied two ideas: the use of energy transfer from the excited molecules produced in the course of the chemical pumping reaction to the "cold" lasing molecules and the use of supplementary reactants to initiate the reaction. The section also discusses the design features of the laser and expounds a simple theoretical model of the kinetic processes responsible for the formation of its active medium.

The supersonic chemical HF (DF) lasers are described in Sect. 5.3. According to the type of initiation of the chemical pumping reaction, these lasers are divided into two classes: lasers using an external initiation source (usually an arc discharge) and fully self-contained laser systems in which initiation is effected by an auxiliary exothermic reaction. For the latter type, physical principles are formulated, to be adhered to when composing the starting fuel mixtures.

The power performance characteristics of supersonic CWCL's are considered in Sect. 5.4. The section tentatively formulates problems for theoretical calculation, including various physico-chemical processes that occur in chemical lasers relying on the diffusion mixing of the reactants.

Section 5.5 discusses the outlook for the development of the $DF-CO_2$ and HF CWCL's. Considered here are possibilities for raising the reactant pressure at the exit from the nozzle bank and creating supersonic chain-reaction HF lasers with a cylinderical nozzle bank geometry.

Section 5.6 briefly describes CWCL's other than the HF and $DF-CO_2$ types. These are first of all the CO lasers based on the branching-chain carbon disulfide oxidation reaction and $OD-CO_2$ lasers distinguished by nontoxic final products. The material concerning one of the most interesting laser systems developed in recent years, namely, the oxygen-iodine laser, is presented in an individual chapter. The section ends with a summary in the form of a table listing the main types of CWCL's.

Finally, Sect. 5.7 considers some CWCL versions that extend the engineering capabilities of chemical lasers, such as the standing detonation wave HF CWCL, a

DF–CO$_2$ CWCL operating in the 16-µm region, and an HF CWCL based on optical resonant energy transfer.

5.1 Physical Principles of Operation of CWCL's

Continuous-wave operation of a chemical laser is based on the flow-mixing of the reactants undergoing the chemical pumping reaction. The reaction must proceed very rapidly for it to compete successfully with the relaxation processes which quench the excited states of the molecules formed. These requirements are met by the atom (radical)–molecule exchange reactions considered in Chap. 1. As is the case with pulsed chemical lasers, population inversion in CWCL's is produced either (1) directly in the course of the chemical pumping reaction (hydrogen-halide and CO lasers) or (2) indirectly, by way of energy transfer from the molecules excited chemically in the course of the pumping reaction to the lasing molecules (hydrogen-halide-CO$_2$ lasers).

To maintain lasing at a constant level continuously, provision must be made for a fast change of the reactants in the reactor. Obviously the reactants must be changed within a time interval shorter than the time it takes for the excited states to be quenched. For typical chemical laser mixtures, the quenching time τ is of the order of $10^{-4} - 10^{-5}$ s even at relatively low reactant pressures (from a few mm Hg to a few tens of mm Hg). Such short deactivation times specify the necessary requirements for the pumping rates of the gas mixture through the reactor. With the minimum linear dimension l of the reactor being of the order of 1 cm, the required flow velocities are $u \sim l/\tau \simeq 10^4$–$10^5$ cm/s. Thus, to create a CWCL, use should be made of sonic or supersonic reactant flow velocities, even if the reactant pressures are not very high.

Another physically clear requirement dictated by the very nature of the CWCL is that the initial flows of the chemically active reactants should be mixed rapidly enough. Under laminar flow mixing conditions, the characteristic diffusion time is estimated by the relation $\Delta t \sim \delta^2/D$, where D is the diffusion coefficient and δ the depth of mixing in a direction perpendicular to the velocity of the gas flows being mixed. In the most favorable case (H$_2$–He mixture), $D \simeq (10^3/p)(T/300)^{3/2}$. The coefficient D here is expressed in cm^2/s, the gas pressure p in mm Hg, the mixing depth δ in cm, and the temperature T in K, so that at $T \simeq 300$ K, $p \simeq 1$–5 mm Hg, and $\delta \simeq 0.2$ cm, for example, we have $\Delta t \simeq 4 \times 10^{-5} - 2 \times 10^{-4}$ s. It is seen that the characteristic diffusion time is of the same order of magnitude as or even longer than the deactivation time of the excited states of the lasing molecules.

Thus, the main requirements to be met when constructing efficient CWCL's are as follows: (1) to use fast chemical reactions, (2) to effect fast flows of the active gas mixtures, and (3) to ensure fast and deep intermixing of the reactant gas flows. Before proceeding to the description of the designs of CWCL's and the analysis of their operating conditions, let us briefly dwell upon some characteristic features of such systems that are predetermined by the requirements formulated above.

A CWCL includes the following main elements: a gas generator wherein active centers are produced, a gas-dynamic duct wherein the starting gas mixture is accelerated to the necessary velocities, a gas injection system, an optical cavity, a diffuser, and an exhaust system (a vacuum pump or an ejector in high-power lasers). The diffuser serves to raise the static gas pressure upstream of the ejector in order to improve its efficiency. An important advantage of CWCL's is that the inherent fast change of the reactants in the reaction zone allows them to utilize components that require no external initiation to undergo reaction, which greatly simplifies the problem of creating a purely chemical laser. What is more, the specific features of continuous-wave operation make it possible to extend the range of gas reactants suitable for lasing.

As is the case with other flowing-gas lasers, the use in CWCL's of supersonic gas flows allows one to effect a rapid convective heat removal from the laser interaction region, which is of principal importance in developing high-efficiency continuous-wave laser systems. In CWCL's, the initial conditions in the optical cavity region are controlled by the conditions produced in the gas generator. In supersonic CWCL's, the necessary composition of the chemical reactants (and particularly the required active center concentration) can be "frozen" in the course of the rapid expansion of the gas flow in the supersonic nozzles. In contrast to pulsed lasers, to obtain low initial gas temperatures, no cryogenic cooling systems are required, the gas cooling process being inseparably connected with the formation of the supersonic gas flow.

The relatively low static gas pressures in the optical cavity region of CWCL's favor the development of laser systems producing high-quality emissions, for they make it in principle possible to achieve a beam divergence close to the diffraction-limited value. At the same time, the low gas pressures make it difficult to construct high-power CWCL's. First of all, there is the engineering problem of exhausting the waste gases. Insufficient gas pressure in the cavity makes the exhaust system complex and bulky. Raising the static pressure of the supersonic gas flow by means of a diffuser allows the gas pressure in the exhaust flow to be increased to the atmospheric value, which greatly simplifies the exhaust system.

Improving the efficiency of CWCL's requires that the starting gas flows should be rapidly intermixed. In practice, this important engineering problem is solved by dividing the gas flows into extremely narrow jets by means of an array of small-size nozzles with an effective throat section depth of the order of a few tenths or even hundredths of a millimeter.

To design an optimum CWCL and determine the optimum gas mixture composition, one has to find answers to a number of questions. The most important among them are as follows. What is the maximum static reactant pressure in the laser cavity that provides for the maximum power output? How does the chemical efficiency of the laser depend on the system's parameters? What is the optimum partial mixture composition from the standpoint of maximizing the power output? What is the optimum mutual arrangement of the gas injection systems and the laser cavity? The answer to the first question is important in designing the vacuum (exhaust) systems. Improving the chemical efficiency of the laser cuts down the expenditure of reactants per unit laser emission power. And the optimization of the system design parameters and mixture composition makes it possible to achieve the necessary power output with the minimum size of the system.

Continuous-wave amplification of radiation due to chemical reactions was observed for the first time by *Anlauf* and co-workers [5.1]. The experiment was conducted at extremely low reactant pressures (10^{-2}–10^{-3} mm Hg) and flow rates, so that the reactants could be changed using ordinary laboratory vacuum pumping equipment. Excited molecules were produced by the reactions

$$H + Cl_2(Br_2) \rightarrow HCl^*(HBr^*) + Cl(Br) ,$$

$$Cl + HI \rightarrow HCl^* + I. \tag{5.1}$$

Atomic hydrogen or chlorine was obtained by flowing the respective molecules through a discharge gap. The atomic reactant flow was then mixed with the secondary reactant flow, which gave rise to an active gain medium through reactions (5.1). It is not very difficult to estimate that if attempts had been made to achieve lasing under the experimental conditions of [5.1], the laser power obtained would not have been more than a few fractions of a milliwatt per cubic centimeter of the reactor volume. The abbove work [5.1] was of a purely research character and was really the first to prove directly the possibility of attaining continuous stimulated emission of radiation through the use of chemically active gas flows.

At present, the following two types of CWCL are being developed most intensively: the purely chemical DF–CO$_2$ laser and the supersonic HF laser. The success in the development and improvement of these continuous-wave laser systems is due first of all to two factors: (1) the large chemical energy store of the gas mixtures used and (2) the high rate of the elementary chemical processes transforming chemical bond energy into the excitation energy of the lasing molecules and ensuring in the final analysis a high efficiency of chemical energy conversion into coherent laser emission energy.

In the sections below we will consider the main schemes and the design and physical features of such lasers, analyze their operating principles and conditions, and discuss their further progress. Consideration will also be given to other chemically pumped continuous-wave laser systems.

5.2 Purely Chemical Subsonic DF–CO$_2$ Lasers

5.2.1 Basic Schemes. Experimental Situation

The first purely chemical continuous-wave laser was created by *Cool* et al. [5.2–4]. The system developed by these investigators embodied two ideas put forward earlier: the possibility of using the transfer of energy from the excited molecules produced in the course of a chemical reaction to the cold lasing molecules [5.5] and the possibility of using supplementary (priming) reactants to initiate the reaction [5.6].

The laser was based on the deuterium fluorination reaction to produce vibrationally excited deuterium fluoride molecules whose vibrational excitation quanta were then transferred to CO_2 molecules. As noted earlier, this reaction features the greatest known laser chain length. To "ignite" the reaction, use was made of a supplementary reactant — nitrogen oxide existing as a stable radical at room temperature. These radicals react with fluorine molecules to yield atomic fluorine:

$$NO + F_2 \overset{k_0}{\rightarrow} NOF + F \quad (\Delta H = -18 \text{ kcal/mole}), \tag{5.2}$$

the atomic fluorine serving as the very active center starting the chemical-laser chain in the $D_2 + F_2 + CO_2$ mixture:

$$F + D_2 \rightarrow DF^* + D \ ,$$

$$D + F_2 \rightarrow DF^* + F \ , \tag{5.3}$$

$$DF^* + CO_2(00^00) \rightarrow DF + CO_2(00^01) \ ,$$

$$CO_2(00^01) + nh\nu \rightarrow CO_2(10^00) + (n+1)h\nu \ . \tag{5.4}$$

From the standpoint of chemical and vibrational kinetics, once atomic fluorine is produced, the process is identical to the one considered in the preceding chapter. The specific features of the process here that place certain requirements upon the laser design are associated with the need to inject and mix the reactant flows in a definite sequence.

First, the NO and F_2 flows are to be mixed, it being convenient to feed NO premixed with He, the need for which is dictated by the same considerations as in the case of pulsed laser systems. The subsequent injection of the CO_2 and D_2 flows completes the set of components necessary for lasing. All this is illustrated schematically in Fig. 5.1.

The first purely chemical continuous-wave DF–CO_2 laser [5.2, 3] had a longitudinal arrangement, i.e., the axis of the cavity coincided with the axis of the flow tube (Fig. 5.2). Atomic fluorine was produced in a 45 cm long tube,

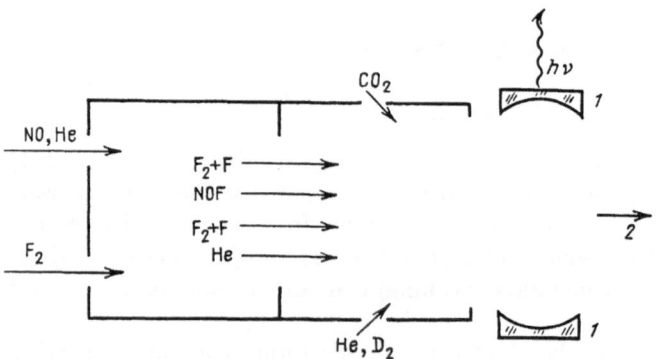

Fig. 5.1. Schematic diagram of a purely chemical laser operating on an $NO + F_2 + D_2 + CO_2 + He$ mixtures: *1* – cavity mirrors; *2* – to exhaust

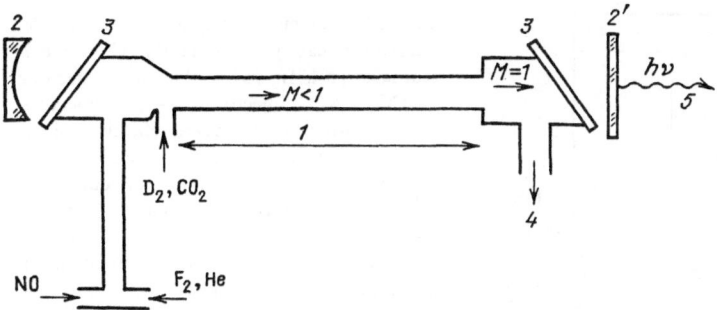

Fig. 5.2. Schematic diagram of a longitudinal–cavity purely chemical subsonic DF–CO$_2$ laser [5.2, 3]: *1*–mixing and reaction zone; *2*–99.4% reflecting mirror 2 m in radius of curvature; *2'*–99% reflecting flat mirror; *3*–NaCl windows; *4*–to exhaust; *5*–laser emission ($\lambda = 10.6$ μm)

and the chemical pumping reaction was sustained in a Teflon tube 21 cm long and 9 mm in inside diameter. The gas pressure at the entrance to the reaction tube (the left-hand end of the reaction zone in the figure) was kept at 19 mm Hg, and the mean flow velocity was 400 m/s. A laser power of 8 W [5.2, 3] (around 45 W s/g in terms of specific power) was obtained from the F$_2$:NO:D$_2$: CO$_2$:He $= 0.39:0.019:0.36:1.57:3.83$ mixture at a total reactant flow rate of 0.0062 mole/s, the reactant proportions being close to the optimum.

The longitudinal laser arrangement was tested in a number of other works [5.7–9], but found no wide recognition. In practice, the transverse arrangement in which the axis of the flow tube is at right angles to the axis of the cavity is preferable, for it ensures a faster and more homogeneous mixing of the reactants. The arrangement is also especially amenable to scaling up.

Substantial power outputs were obtained by *Cool* et al. [5.4, 10, 13], *Brunet* and *Mabru* [5.11], *Falk* [5.12], and *Tregay* et al. [5.14] from transversely arranged purely chemical CW lasers operating on the D$_2$ + F$_2$ + CO$_2$ + He mixture "ignited" by means of a supplementary reactant. These authors studied the effect of the molar mixture composition and gas flow pressure on the laser power characteristics.

Cool and co-workers [5.4, 10, 13] reported on the development of a laser with a total power output of 160 W. A schematic diagram of this laser [5.4, 13] is presented in Fig. 5.3, and its main parameters are listed in Table 5.1. The laser used a rectangular flow channel 1 × 15 cm^2 in cross-sectional area. The reactants were injected through three rows of tubes with a number of small holes each. The tubes in the rows were spaced at 1.5–2 mm intervals. The first row served to inject the He + F$_2$ mixture and the second, the CO$_2$ + NO mixture. The small holes in these tubes were around 0.8 mm in diameter. The length of the flow zone where atomic fluorine was produced came to 30–60 cm, and the third row of tubes serving to inject D$_2$ was set up at the downstream end of this zone, the small holes in these tubes being of the order of 0.1 mm in diameter. A specific feature of the optical system of the laser [5.4, 10] is a multipass cavity allowing the energy stored in the vibrationally excited CO$_2$ molecules on account of the chemical pumping reaction to be effectively utilized.

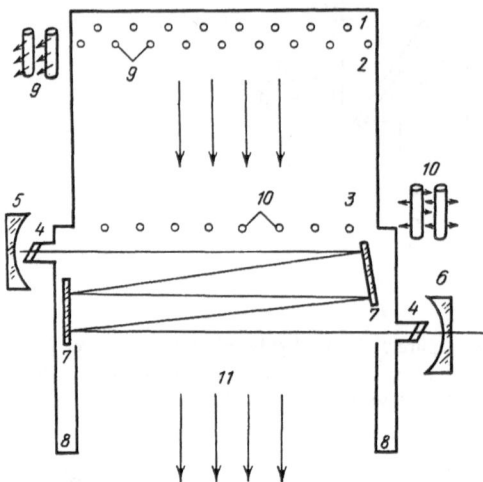

Fig. 5.3. Schematic diagram of a 160-W transverse–cavity subsonic DF–CO$_2$ chemical laser [5.4, 10, 13]: 1 – He + Fe$_2$ injectors; 2 – CO$_2$ + NO injectors; 3 – D$_2$ injectors; 4 – NaCl windows; 5 – 99.4% reflecting mirror 10 m in radius of curvature; 6 – semitransparent mirror 10 m in radius of curvature; 7 – 99.4% reflecting flat mirrors; 8 – nitrogen – purged mirror chambers; 9 – detail of CO$_2$ + NO injector; 10 – detail of D$_2$ injector; 11 – to exhaust

Falk [5.12] managed to obtain a total of 560 W from his laser. This laser differed from the one illustrated in Fig. 5.3 in the design of the cavity. The cavity was formed by two rectangular mirrors installed on both sides of the flow channel, the output mirror being provided with a large number of perforations to couple energy out of the cavity. The rectangular flow channel was 2.5×15 cm^2 in cross-sectional area. The chemical-energy-to-laser-emission conversion efficiency achieved with this laser ($\simeq 4\%$) is close to the experimental maximum efficiency for this type of laser. Falk studied the relationship between the power output of the laser and the static pressure of the reactants in the cavity. Lasing was observed to occur at reactant pressures up to around 0.1 atm., but the laser power produced at this pressure was a negligible fraction of the maximum (560 W) achieved at 17 mm Hg.

The measurements taken in [5.4, 10, 12] show that the optimum proportions of the reactants in the above mixture are roughly NO:D$_2$:F$_2$:CO$_2$:He $= 1:5:6:45:100$.

One of the most important laser parameters is the saturation intensity corresponding to the complete extraction of the energy stored by the excited particles in the flowing active medium. This intensity depends on the lasing species, mixture pressure and composition, gas flow velocity, and so on. Under the optimum lasing conditions realized in [5.4, 10, 12], the saturation intensity is of the order of 140–160 W/cm^2.

Substantial progress in the development of high-power self-contained subsonic continuous-wave DF–CO$_2$ lasers was made by *Tregay* and co-workers [5.14]. They reported a power output of 15 kW from the IRIS-I large-scale laser setup with a flow channel 10.7×76 cm^2 in cross-sectional area. The setup used original injectors allowing for an effective mixing of the reactants over a wide range of pressures. As a result, high laser powers were obtained at pressures up to 60 mm Hg (see Fig. 5.5a). The optimum gas mixture pressure corresponding to the 15 kW power output amounted to 35 mm Hg. The optical system was a confocal unstable cavity with an output coupling of 50%. The apertures of the small (convex) mirrors used were 6×9.5 cm^2 or 7.2×10 cm^2 and the optical windows were made of NaCl or KCl.

A modification of the D_2 injection system made it possible to operate the laser at supersonic flow velocities (IRIS-II setup). In this setup, the D_2 injector tubes and supersonic nozzles were incorporated in a single block. Despite the fact that a number of chemical and gasdynamic factors limited the maximum flow velocity in the cavity (the Mach number M was equal to 1.75 at a pressure differential of 6:1), it proved possible to obtain around 8 kW of laser power at a static gas pressure of $p \simeq 15$ mm Hg in the cavity. Some experimental results obtained with the IRIS-II setup are listed in Table 5.2. As can be seen, the laser was operated at increased NO/F_2 ratios, which was dictated by the need to raise the gas temperature upstream of the nozzle block. It is expected [5.14] that it will be possible to improve the performance of the setup, in particular, by optimizing the cavity. *Emanuel* et al. [5.15] constructed a supersonic DF–CO₂ laser with igniting reactant. The authors of [5.16–18] worked with a supersonic DF–CO₂ laser driven by combustion of some fuel compounds. More detailed comments are given in Sect. 5.5.

5.2.2 Simplified Theoretical Model. Discussion of Experimental Data

Figures 5.4 and 5.5 present typical relationships between the output power of the continuous-wave DF–CO₂ laser and the static mixture pressure in the laser cavity [5.12, 14]. It is seen that with the molar mixture composition being fixed, there exists an optimal mixture pressure providing maximum laser power. Figures 5.4 and 5.5b show that this pressure depends on the concentration of the initiating agent (NO): higher optimum mixture pressure values correspond to lower molar concentrations of NO. Recall for comparison that the main energy characteristics of the pulsed DF–CO₂ laser behave in a different manner: as the mixture pressure is increased (up to 1 atm.), the total laser energy output rises [5.19]. The laser energy also rises with the initiation level (the proportion of active centers relative to the F_2 concentration) increasing from 10^{-3} to 10^{-1} [5.20].

A qualitative analysis of a CW DF–CO₂ laser using an initiating agent requires consideration of a great number of physico-chemical factors affecting the formation of the active medium and its interaction with the optical cavity, such as the processes of mixing of the chemically active flows, kinetic processes, and thermal

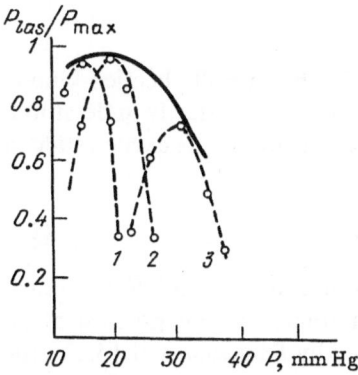

Fig. 5.4. Output power of the DF–CO₂ laser as a function of the static gas pressure in the cavity for various NO concentrations [5.12]: *1* – 0.011; *2* – 0.007; *3* – 0.002. Open circles – experimental data points. Solid curve – output power *vs.* static pressure plot for optimum NO concentration

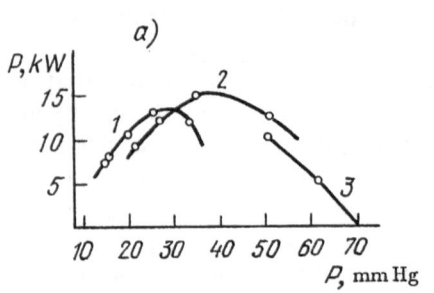
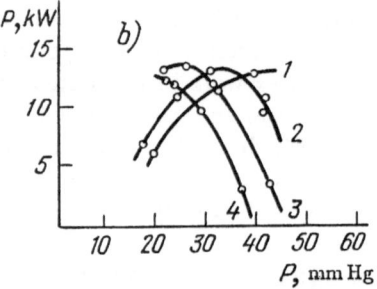

Fig. 5.5a, b. Output power of the DF–CO$_2$ laser as a function of the static gas pressure in the cavity [5.14] for various molar concentration ratios. (**a**) [CO$_2$]/[F$_2$]: *1* – 3.2; *2* – 4.3; *3* – 6.3 ([NO]/[F$_2$] fixed at 0.22)); (**b**) [NO]/[F$_2$]: *1* – 0.14; *2* – 0.22; *3* – 0.34; *4* – 0.44 ([CO$_2$]/[F$_2$] fixed at 5.8). Unstable cavity with the smaller mirror 9.5 × 6 cm^2 in aperture

and gasdynamic effects. The difficulties involved in constructing an adequate mathematical model to describe the laser system as a whole are thus evident. These difficulties are aggravated by the fact that some of the important processes to be taken into account are still not clearly understood. This is true in particular of the key process for the given type of laser, namely, the production of active centers (fluorine atoms) in that part of the flow channel where reaction (5.2) takes place along with the associated recombination reactions

$$F + NO + M \xrightarrow{k_{rec}} NOF + M \ ,$$

$$F + F + M \rightarrow F_2 + M \ . \tag{5.5}$$

Unfortunately, there are no reliable experimental data on the rate constants of the above processes. The estimate of 3×10^{-13} cm^3/s by *Kondrat'yev* [5.21] for the rate constant of reaction (5.2) is questioned by *Wolga* [5.22] who considers it to be much smaller. The author of [5.22] recommend the following values for the rate constants of reactions (5.2) and (5.5):

$$k_0 = 5.5 \times 10^{-14} \exp(-750/T) \ \text{cm}^3/\text{s} \ ,$$

$$k_{rec} = 9.1 \times 10^{-29}/T \ \text{cm}^6/\text{s} \ . \tag{5.6}$$

It is likely that NOF plays an important part in the overall chemico-kinetic process occurring in the laser cavity, but what it is is as yet imperfectly understood. The NOF molecule can in principle act as a strong deactivator of the excited state in the lasing CO$_2$ molecule in accordance with the process

$$CO_2(00^01) + NOF \rightarrow CO_2(00^00) + NOF(v_1, v_2, 0) \ , \tag{5.7}$$

where v_1 and v_2 denote the first two vibrationally excited states in NOF. Inasmuch as process (5.7) is of a quasiresonant exchange character, it can proceed at an appreciably high rate. According to the data reported by *Baumann* et al. [5.23], the rate constant of process (5.7) is of the order of 2×10^{-12} cm^3/s at room temperature, which is three times that of deactivation of the 00^{0}1 state in CO$_2$ by collision with DF. The NOF molecule can also participate in the exchange process

$$D + NOF \rightarrow NO + DF^* \tag{5.8}$$

which, though it gives rise to excited deuterium fluoride molecules, inhibits the chain reaction (5.3) because the rate of production of fluorine atoms as a result of reaction (5.2) between molecular fluorine and NO is lower than the rate of the reaction of molecular fluorine with atomic deuterium. As regards the rate constant of process (5.8), the literature data are contradictory [5.13].

The above uncertainties make it difficult to interpret experimental results on the CW DF–CO$_2$ laser using the initiating agent, and cause ambiguity in forecasting new and more effective operating conditions for this type of laser. In this connection, we think it of interest to analyze simplified theoretical models that would allow for the most important kinetic processes, which is essential to the understanding of a number of general features of the power characteristics of the laser, and make it possible to reveal its ultimate energy capabilities. In the text below, we present such a model for the mixture considered above, in which the pumping source is reaction (5.3) and the CO$_2$ molecules are excited as a result of exchange process (5.4). The basic assumptions are as follows:

1) The finiteness of the rate of mixing of the primary gas flow containing F, F$_2$, NO, NOF, CO$_2$, and He with the secondary one containing D$_2$ is disregarded.
2) No consideration is given to thermal effects in the gas flow.
3) The vibrational temperatures of the v_1 and v_2 modes of the CO$_2$ molecule are considered to be the same as the gas temperature T.
4) The kinetic processes occurring in the lasing zone are characterized by certain effective times.

Let us introduce the notation E and E_* for the densities of vibrational quanta (in quanta/cm^3) stored in the v_3 mode of CO$_2$ and in the vibrational mode of DF, respectively. The set of balance equations describing the kinetics of vibrational energy redistribution between CO$_2$ and DF in the presence of the laser field may be written as

$$u\frac{dE_*}{dx} = Q_{chem} - \frac{E_*}{\tau_1} - \frac{E_*}{\tau_2} \;, \tag{5.9}$$

$$Q_{chem} = e_1^* k_1 [F][D_2] + e_2^* k_2 [D][F_2] \;,$$

$$u\frac{dE}{dx} = \frac{E_*}{\tau_2} - \frac{E}{\tau} - gI \;, \tag{5.10}$$

where u is the initial reactant flow velocity, x the coordinate along the flow (Fig. 5.6), Q_{chem} the chemical pumping energy due to the chain process (5.3), k_1 and k_2 are the rate constants of the respective elementary processes, e_1^* and e_2^* denote the average numbers of quanta (per oscillator) acquired by the DF molecule in these elementary

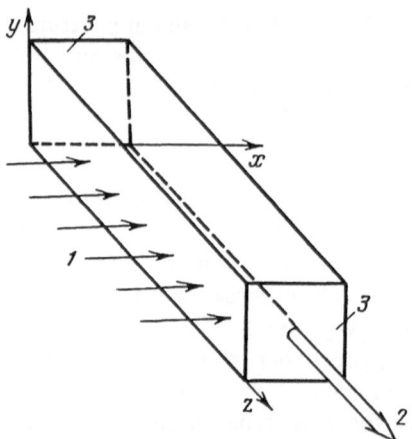

processes, respectively, τ_1 is the deactivation time of the excited DF molecules, τ_2^{-1} the rate of vibrational quantum transfer from the DF to CO_2 molecules, and τ^{-1} the deactivation rate of the ν_3 mode of CO_2. The concentrations [F], [D], [F_2], and [D_2] are expressed in particles/cm^3. The term gI, where I is the radiation intensity (in quanta/cm^2 s) and g the gain of the lasing transition $[00^01] \rightarrow [10^00]$, character-izes the loss of energy in the ν_3 mode of the lasing molecules on account of induced radiative transitions. The gain g may be expressed as

$$g = \sigma[CO_2](e_3 - e_1)(1 + e_1)^{-2}(1 + e_2)^{-2}(1 + e_3)^{-2} . \tag{5.11}$$

In (5.11), the symbols e_i stand for the average numbers of quanta (per oscillator) stored in the respective vibrational modes of CO_2 and σ is the stimulated transition cross section. Since $e_3 = E/[CO_2]$, the relation between the density E of vibrational quanta and the gain g is given by the expression

$$E = e_1[CO_2] + \frac{g}{\sigma}(1 + e_1)^2(1 + e_2)^2(1 + e_3)^2 . \tag{5.12}$$

Using the relation $k_2 \ll k_1$, which is typical of the hydrogen fluorination chain reaction, and assuming that the initial concentrations of F_2 and D_2 are equal, one can easily reduce the term responsible for the chemical pumping in (5.9) to the form

$$e_1^* k_1[F][D_2] + e_2^* k_2[D][F_2] = \frac{e_1^* + e_2^*}{\tau_0}[F_2]_0 e^{-x/u\tau_0} , \tag{5.13}$$

where $\tau_0^{-1} = k_2[F]_0$ is the evolution time of the chain process and $[F_2]_0$ and $[F]_0$ are the initial concentrations of the molecular and atomic fluorine, respectively. Considering (5.13), we find from (5.9) the following expression for the density of vibrational quanta stored in the reaction products:

$$E_* = [F_2]_0 \left[\frac{e_1^* + e_2^*}{\tau_0 \left(\dfrac{1}{\tau_1} + \dfrac{1}{\tau_2} \right) - 1} \right] \left[\exp\frac{-x}{u\tau_0} - \exp\frac{-x}{u}\left(\frac{1}{\tau_1} + \frac{1}{\tau_2} \right) \right] . \tag{5.14}$$

To determine the laser power, use can be made of the quasistationary oscillation technique [5.24, 25]. The threshold vibrational quantum density can be found from (5.12) on putting $g = g_{thr} = (1/L)\ln(1/R_0^{1/2})$ (R_0 being the reflectivity of the output cavity mirror and L the active region length coincident with the gas flow width). So,

$$E_{thr} \simeq e_1[CO_2] + (1+e_1)^2(1+e_2)^2 \frac{g_{thr}}{\sigma} . \tag{5.15}$$

According to the above technique, the output laser emission intensity is found from (5.10) to be

$$I_{out} = Lg_{thr}I = L\left(\frac{E_*}{\tau_2} - \frac{E_{thr}}{\tau}\right) . \tag{5.16}$$

The expression for the laser power is obtained by integrating (5.16) for x going from the threshold section x_0 to the section x_+ were oscillation ceases, so that the lasing zone width $x_l = x_+ - x_0$. The coordinates x_0 and x_+ are determined in the routine way. For the practically important case where the relations

$$\tau_0^{-1} < \tau^{-1} < \tau_1^{-1} + \tau_2^{-1} \tag{5.17}$$

hold true, one can easily find

$$x_0 = u\tau \ln\frac{1}{1-\alpha} , \quad x_+ = u\tau_0 \ln\frac{1}{\alpha} , \tag{5.18}$$

where

$$\alpha = \frac{E_{thr}}{E_0}\frac{\tau_0}{\tau}\left(1+\frac{\tau_2}{\tau_1}\right) , \quad E_0 = [F_2]_0(e_1^* + e_2^*) . \tag{5.19}$$

Relations (5.18) immediately yield the oscillation conditions

$$\alpha < 1 \quad \text{and} \quad \tau/\tau_0 < \frac{\ln\alpha}{\ln(1-\alpha)} .$$

The above inequalities define a domain of parameters in the (τ/τ_0, α) plane and comply with the physically clear requirements for the initial energy store, the rates of the respective processes, and the threshold oscillation level.

In the most interesting case where the threshold oscillation level is not very high, the width of the lasing zone is determined by the coordinate x_+ ($x_0 \simeq 0$ and $x_+ \simeq x_l$) so that the expression for the laser power per unit cross-sectional area of the gas flow may be written in the following simple form:

$$P \text{ (quanta/cm}^2 \text{ s)} = P_0\left[1-\alpha\left(1+\ln\frac{1}{\alpha}\right)\right] , \tag{5.20}$$

where

$$P_0 = \frac{u(e_1^* + e_2^*)[F_2]_0}{1+\tau_2/\tau_1} = \frac{uE_0}{1+\tau_2/\tau_1}$$

is the maximum laser power. Accordingly, for the chemical efficiency of the laser, we have

$$\eta_{\text{chem}} = \chi_1 \chi_2 \chi_3 \; ,$$
$$\chi_1 = h\nu_{\text{CO}_2}/h\nu_{\text{DF}}, \quad \chi_2 = h\nu_{\text{DF}}(e^* + e^*)/Q_{\text{chem}} \; , \tag{5.21}$$
$$\chi_3 = \frac{1 - \alpha[1 + \ln(1/\alpha)]}{1 + \tau_2/\tau_1} \; ,$$

where χ_1 is the ratio between the lasing transition quantum of the emitting molecule and the quantum of the molecule excited in the course of the chain pumping process, χ_2 the fraction of energy localized in the vibrational modes of the reaction products relative to the total amount of energy released (per chain reaction product particle), χ_3 the vibrational-to-laser energy conversion efficiency, and Q_{chem} the heat of reaction. As follows from (5.21), the ultimate kinetic efficiency of the laser is ($\alpha \to 0$)

$$\chi_3^0 = 1/(1 + \tau_2/\tau_1) \; .$$

Note that the energy relationships obtained also hold true, except for transformations, for the pulsed DF–CO_2 chemical laser.[1]

When analysing the energy characteristics of the CW DF–CO_2 laser, consideration should be given to the specific character of the initial data associated with the method of production of the active centers, which, as will be demonstrated below, strongly affects their behavior. Consider two possible situations:

1) A situation where recombination processes (5.5) can be disregarded.

a) If the length l of the active center production channel is relatively short ($k_0[\text{NO}]_0 l/u < 1$), the active center production rate w_F is proportional to the product of the NO and F_2 concentrations [5.24]:

$$w_F = k_0[\text{NO}]_0[F_2]_0 \; , \tag{5.22}$$

and the atomic fluorine concentration in the deuterium injection region is

$$[F] = w_F(l/u) \sim k_0[\text{NO}]_0[F_2]_0 \sim p^2 \xi_{\text{NO}}^0 \; , \tag{5.23}$$

where ξ_{NO}^0 is the molar concentration of NO.

b) If the initiation channel is long enough, the NO radicals are almost completely consumed by the moment the gas flow reaches the deuterium injection region, so that the atomic fluorine concentration is close to the initial NO concentration:

$$[F] \simeq [\text{NO}]_0 \sim p\xi_{\text{NO}}^0 \; . \tag{5.24}$$

---

[1] It is not very difficult to see that all the basic qulitative features typical of the pulsed laser mode can be explained within the scope of the simple notions expounded above: the laser pulse energy is proportional to the pressure p, average laser power to p^2, laser pulse duration to p^{-1}, and so on.

Thus, the atomic fluorine concentration is in both cases proportional to the initial molar concentration of the NO radicals and varies as the square of pressure in the former case (5.23) and is linear in pressure in the latter (5.24). Consequently, for the evolution time of the chain process in the cavity, we have

a) $\tau_0^{-1} \sim \zeta_{NO}^0 p^2$,

b) $\tau_0^{-1} \sim \zeta_{NO}^0 p$.

(5.25)

Using (5.20) and (5.25), it is not very difficult to find that the laser power increases with pressure (and linearly at that, provided p is high enough). However, it should be borne in mind that relation (5.20) holds for the optimum cavity arrangement and size. (What we mean is that the cavity is of the Fabry-Perot type and uses mirrors whose length along the flow is equal to that of the active region, i.e., the flow region where the oscillation threshold condition is satisfied.) But experimentally both the position of the cavity axis and the size of the cavity may prove to be not optimum. This circumstance, along with a poor mixing of the reactants at high pressures, may, in the situation considered here, be responsible for the laser power characteristics deteriorating with rising pressure.

If we assume that there exists an optimum chain reaction evolution time τ_0, then, according to (5.25), the molar proportion of NO must be reduced in inverse proportion to (a) the square or (b) the first power of the total gas pressure in the cavity. The existence of the optimum reaction time may be due to a number of reasons. One such reason is the optimum location of the active medium inversion maximum with respect to the cavity. It is easy to understand that the position of this maximum is governed by the rate of the chemical pumping reaction (provided the reactant flow mixing length is fixed). As τ_0 decreases, the population inversion distribution becomes more and more strongly dependent on the distance from the deuterium injector, so that the inversion maximum shifts towards the injector. For this reason, there is an optimum reaction rate for each particular position of the cavity.

2) A situation where the recombination process (5.5) accompanying reaction (5.2) plays an important part (increased pressures, high recombination rate constant values). Processes (5.2, 5) correspond to the following set of kinetic equations:

$$\frac{d}{dt}[F_2] = -k_0[NO][F_2] ,$$

$$\frac{d}{dt}[F] = k_0[NO][F_2] - k_{rec}[NO][F][M]$$

(5.26)

In accordance with the law of conservation of chemical elements, the set of equations (5.26) must be supplemented with the relations

$$[NO] + [NOF] = [NO]_0 ,$$

(5.27)

$$[F] + 2[F_2] + [NOF] = 2[F_2]_0 .$$

(5.28)

Using (5.26), one can easily find the relationship between the molecular and atomic fluorine concentrations:

$$\frac{[F]}{[F_2]_0} = \frac{1}{(q-1)} \frac{[F_2]}{[F_2]_0} \left\{1 - \left(\frac{[F_2]}{[F_2]_0}\right)^{q-1}\right\} , \tag{5.29}$$

where $q = k_{rec}[M]/k_0$. Estimates based on the values of k_{rec} and k_0 recommended in [5.22] show that at $T \simeq 300$ K, the parameter $q \simeq 1$ already at a pressure of $p \simeq 0.5$ mm Hg, and so at pressures of the order of 10 mm Hg and higher, $q \gg 1$.

Expression (5.28) shows that the concentrations of F and F_2 are rigidly related, although the atomic fluorine concentration is in this case controlled by the fast recombination process (5.5). It follows from (5.29) that the number density of the active centers is limited to

$$[F]_{max} = ([F_2]_0/q) q^{1/(1-q)} . \tag{5.30}$$

This limit can be reached when

$$[F_2]/[F_2]_0 = q^{1/(1-q)} .$$

If $q \gg 1$, $[F]_{max} = [F_2]_0/q \ll [F_2]_0$.

a) If the length of the initiation channel is not very large ($q^{-1} \leqslant k_0[NO] \times l/u < 1$), then by the moment the gas flow reaches the deuterium injection region the concentration of F_2 is reduced by the amount

$$[F_2]_0 \left[1 - \exp\left(-k_0[NO]_0 \frac{l}{u}\right)\right] \simeq k_0[F_2]_0[NO]_0 \frac{l}{u} ,$$

$2k_0[F_2]_0[NO]_0 l/u$ of the NO radicals being consumed and, correspondingly, roughly the same amount of the NOF molecules produced, and the active center concentration comes, in accordance with (5.29), to

$$[F] = \frac{[F_2]_0}{q} \left[1 - \exp\left(-k_0 q[NO]_0 \frac{l}{u}\right)\right] .$$

The quantity q is proportional to pressure; at relatively high p values

$$[F] \sim \frac{[F_2]_0}{q} \sim \xi_{F_2}^0, \quad [NOF] \sim p^2 \xi_{NO}^0 . \tag{5.31}$$

b) If the length of the initiation channel is long enough (and this is exactly what the laser design must provide), by the moment it reaches the D_2 injector the chemical mixture will be close to its steady-state condition. In this case, the concentration of the NOF molecules is proportional to the initial concentration of the NO radicals:

$$[NOF] \sim [NO]_0 \sim p\xi_{NO}^0, \quad [F] \sim [F_2]_0/q \sim \xi_{F_2}^0 . \tag{5.32}$$

As can be seen, the active center concentration, hence the chain reaction time τ_0, is in both cases almost independent of pressure. This inference is physically clear: fluorine atoms are produced at a slow rate in the course of reaction (5.2) and decay rapidly as a result of recombination process (5.5), the decay rate increasing with rising gas mixture pressure. For this reason, in the system, there is sustained a quasistationary atomic fluorine concentration independent of pressure. Referring to relations (5.18, 20, 21) and considering that $E_{thr} \sim E_0 \sim p$, $P_0 \sim p$, $\tau_1^{-1} \sim \tau_2^{-1}$, $\tau^{-1} \sim p$, and τ_0 is independent of pressure, we have, for fixed molar concentrations of the starting gas mixture components, the pressure dependences of the laser power, chemical efficiency and the lasing region width in the following functional forms:

$$P \sim p\left[1 - Ap\left(1 + \ln\frac{1}{Ap}\right)\right] ,$$

$$\eta_{chem} \sim \left\{1 - Ap\left[1 + \ln\frac{1}{Ap}\right]\right\} , \qquad (5.33)$$

$$x_+ \sim \ln\frac{1}{Ap} ,$$

where $A \sim \Sigma_i \xi_i k_i$, ξ_i are the molar concentrations of the mixture components, and k_i the respective rate constants for the deactivation of the $[00^01]$ level in CO$_2$, ξ_{NOF} being proportional to (a) $\xi_{NO}^0 p$ or (b) ξ_{NO}^0 in accordance with (5.31) or (5.32).

It follows from (5.33) that the laser power as a function of pressure has a maximum at some optimum static pressure p_{opt}. The presence of the maximum in this case is apparently not connected with the relative position of the cavity but is of a purely kinetic origin, although the cavity location is important in practice. It also follows from (5.33) that raising the gas pressure reduces the laser efficiency. The reason for the laser power and chemical efficiency dropping with rising pressure is readily attributable to the specific features of initiation of the chain reaction giving rise to excited deuterium fluoride molecules. In view of the slow rate of production of the active centers (supplementary reaction (5.2)), the recombination processes start manifesting themselves at fairly low gas mixture pressures (compared to the pressure providing for the maximum power output). As pressure is increased, the active center number density and the associated chain propagation rate cease to be pressure-dependent. On the other hand, the rate of deactivation of the excited molecules increases linearly with pressure. Therefore, once the gas mixture pressure has reached some fixed value depending on the initial molar composition of the mixture, lasing is quenched.

If we assume that the NOF molecule is a strong deactivator of the excited state in CO$_2$, then at relatively high values of ξ_{NO}^0 the quantity A will be proportional to (a) $k_{NOF}\xi_{NO}^0 p$ or (b) $k_{NOF}\xi_{NO}^0$ so that the functional pressure dependence of the laser power, (5.33), will take the form

$$P \sim \begin{cases} \text{(a) } p\{1 - A'p^2\xi_{NO}^0[1 + \ln(1/A'p_{NO}^2)]\} , \\ \text{(b) } p\{1 - A''p\xi_{NO}^0[1 + \ln(1/A''p\xi_{NO}^0)]\} . \end{cases} \qquad (5.34)$$

Correspondingly, for the lasing region width and the chemical efficiency of the laser, we have

$$x_+ \sim \begin{cases} \text{(a)} & \ln(1/A'p^2\xi_{NO}^0) \ , \\ \text{(b)} & \ln(1/A''p\xi_{NO}^0) \ , \end{cases} \tag{5.35'}$$

$$\eta_{chem} \sim \begin{cases} \text{(a)} & 1 - A'p^2\xi_{NO}^0[1 + \ln(1/A'p^2\xi_{NO}^0)] \ , \\ \text{(b)} & 1 - A''p\xi_{NO}^0[1 + \ln(1/A''p\xi_{NO}^0)] \ . \end{cases} \tag{5.35''}$$

It follows directly from (5.34) that with the initial molar concentration of the NO radicals (ξ_{NO}^0) being fixed, there exists an optimum combination $p^2\xi_{NO}^0$ (case (a)) or $p\xi_{NO}^0$ (case (b)), for which the laser power reaches its maximum. This implies that the higher the initial concentration of the NO radicals, the lower the optimum gas mixture pressure: $\xi_{NO}^0 \sim p_{opt}^{-2}$ (case (a)) or $\xi_{NO}^0 \sim p_{opt}^{-1}$ (case (b)). Thus, we have actually obtained the same relationships as in the first situation, but their physical essence is different. In the situation considered here, these relationships are governed by the strong deactivating effect of the NOF molecules on the excited CO_2 molecules (it will be recalled that according to (5.31) or (5.32), the NOF concentration in the deuterium injection region of the flow channel is proportional to (a) p^2 or (b) p).

The above simplified analysis has made it possible to reveal, subject to the initial assumptions, the relationship between the optimum gas mixture pressure and the concentration of the priming reactant NO. This relationship agrees qualitatively with the experimental data on subsonic DF–CO_2 lasers reported by *Falk* [5.12] and *Tregay* et al. [5.14]. *Falk* [5.12], who studied a small-scale laser setup, demonstrated that the molar concentration of NO was inversely proportional to the square of the optimum static gas pressure. In the experiments conducted with the IRIS-I setup [5.14], this dependence was weaker.

The above two approaches qualitatively interpret these experimental results from different positions but do not rule out the possibility of existence of some other (or a supplementary) explanation considering more comprehensively the specific features of the processes occurring both in the initiation channel and in the optical cavity region. In particular, in the gas flow region where the active centers are produced, factors which apparently may become important, under certain conditions, include the recombination of fluorine atoms on the gas channel walls, the reaction $F_2 \leftrightarrow 2F$, and the heating of the mixture (at increased NO concentrations or relatively low diluent concentrations). Process (5.8) is likely to play an important part in the cavity region.

5.2.3 Design-Basis Theoretical Model of Continuous-Wave DF–CO_2 Laser

In the preceding section, we have obtained, on the basis of a very simple model, a number of major results characteristic of the continuous-wave self-contained DF–CO_2 laser. Of course, a comprehensive description of this type of laser must be based on a much more complex, design-basis theoretical model allowing for the complete set of the gasdynamic, kinetic, and radiative processes taking place in the laser, and this inevitably implies the use of computer art.

Despite the fact that no sufficiently reliable data are available on the rate constants of a number of the kinetic processes (some primary ones included), we believe it useful to carry out the numerical modelling of the continuous-wave mode of the DF–CO$_2$ laser, first of all because theoretical calculations add to our understanding of the physico-chemical mechanisms at the root of the laser and help reveal the general trends in the behavior of the system under various conditions. On the other hand, comparison between theoretical and experimental data may provide information about the kinetic processes themselves. A complete theoretical model of a CW chemical laser considers chemical and vibrational kinetics, thermal and gasdynamic effects, and radiation kinetics, the solution of the laser problem proper being preceded by the calculation of the gasdynamic parameters of the starting gas flow, as well as the number densities of the active centers (fluorine atoms) and the other reactants.

Numerical calculations of CW DF–CO$_2$ chemical lasers were made by *Baumann* et al. [5.23], *Pimenov* and *Shcheglov* [5.26, 29], *Basov* et al. [5.27], *Zelazny* et al. [5.28], and *Vieceli* et al. [5.30]. The calculations in [5.23, 26, 28] deal with the subsonic laser version using an initiating reactant and those in [5.27, 29, 30] treat the supersonic laser version.

The chemical and vibrational kinetics of the D$_2$ + F$_2$ + CO$_2$ + He mixture were analyzed in detail by *Poehler* et al. [5.19], *Kerber* et al. [5.20, 31], and *Bashkin* et al. [5.24]. Additional information on the kinetic constants of the respective processes in the NO + D$_2$ + F$_2$ + CO$_2$ + He mixture can be found in [5.4, 10, 13, 23, 26].

In accordance with what has been said above, the set of equations modelling the continuous-wave DF–CO$_2$ chemical laser includes four interrelated groups of equations describing chemical, vibrational, and radiation kinetics and gasdynamic variations in the gas flow, respectively. The sets of chemical kinetics equations and radiation kinetics equations are written in the usual fashion (see Chap. 1), and so we do not present them here in any particular form. When describing the vibrational kinetics of the D$_2$ + F$_2$ + CO$_2$ + He mixture, use is usually made of one of the following two approaches. The first involves the analysis of a set of balance equations for the populations of the vibrational levels in the CO$_2$, D$_2$, and DF molecules [5.23, 28].

The second method for describing the vibrational kinetics of gas lasers, proposed and substantiated by *Basov* and co-workers [5.32] using as an example the N$_2$–CO$_2$ system, employs a set of balance equations for the average numbers of vibrational mode quanta, each vibrational mode of the polyatomic molecules of interest being characterized by the Boltzmann temperature of its own (the energetic approach). One of the methods of deriving such a set has been presented in Chap. 2. In the case of CW chemical lasers, it is advisable to write the set for the numbers of vibrational quantum moles, ε_i, stored in each partial mode per gram of the mixture. The quantities ε_i, which may be referred to as the mole-mass concentrations of vibrational quanta, are related to the average numbers of quanta per mode oscillator, e_i, by the relations

$$\varepsilon_i = n_i e_i \ ,$$

where n_i is the mole-mass concentration (in moles/g) of the respective mixture component. That the introduction of the quantities ε_i is convenient is obvious, for they can be treated as the concentrations of additional components taking part in the overall chemical transformation cycle.

It should be noted that the energetic approach to the description of vibrational kinetics is being widely used, thanks to its convenience and physical clearness, in calculating gasdynamic lasers with thermal [5.32–34], chemical [5.26, 27, 29, 30], and electrical [5.35] pumping and also pulsed lasers with electrical [5.36] and chemical [5.19, 24, 37] excitation.

Introducing the notation ε_1, ε_2, ε_3, ε_4, and ε_5 for the numbers of vibrational quantum moles (per gram of the laser mixture) stored in the v_1, v_2, and v_3 modes of CO_2 and in the vibrational modes of DF and D_2, respectively (allowing for degeneracy, the total energy of the bending vibrations v_2 in CO_2 is equal to $2\varepsilon_2$), we write the set of equations of vibrational kinetics for the $NO+D_2+F_2+CO_2+He$ mixture in the form

$$\frac{u}{\varrho}\frac{d}{dx}(\varepsilon_1+\varepsilon_2)=\frac{3}{2}L_{3,12}+\frac{3}{2}L_{3,2}+R_{12}+R_{\mathrm{rad}} \; ,$$

$$\frac{u}{\varrho}\frac{d}{dx}\varepsilon_3=-L_{3,12}-L_{3,2}+L_{34}+R_3-R_{\mathrm{rad}} \; ,$$

$$\frac{u}{\varrho}\frac{d}{dx}\varepsilon_4=-L_{34}-L_{45}+R_4+R_{\mathrm{chem}} \; ,$$

$$\frac{u}{\varrho}\frac{d}{dx}\varepsilon_5=-L_{35}+L_{45}+R_5 \; ,$$

(5.36)

where u, ϱ, and x are the gas flow velocity, density, and coordinate, respectively, and

$$L_{3,12}=\left[\sum_M k_{3,12}(M)n_M\right]\left\{\varepsilon_3\left[1+\frac{\varepsilon_1}{n_{CO_2}}\right]\left[1+\frac{\varepsilon_2}{n_{CO_2}}\right]\right.$$
$$\left.-\exp\left(\frac{-500}{T}\right)\frac{\varepsilon_1\varepsilon_2}{n_{CO_2}^2}(\varepsilon_3+n_{CO_2})\right\} \; ,$$

$$L_{3,2}=\left[\sum_M k_{3,2}(M)n_M\right]\left\{\varepsilon_3\left[1+\frac{\varepsilon_3}{n_{CO_2}}\right]^3\right.$$
$$\left.-\exp\left(\frac{-500}{T}\right)\left(\frac{\varepsilon_2}{n_{CO_2}}\right)^3(\varepsilon_3+n_{CO_2})\right\} \; ,$$

$$L_{34}=k_{34}\left[\varepsilon_4(\varepsilon_3+n_{CO_2})-\exp\left(\frac{-800}{T}\right)\varepsilon_3(\varepsilon_4+n_{DF})\right] \; ,$$

$$L_{35}=k_{35}\left[\varepsilon_5(\varepsilon_3+n_{CO_2})-\exp\left(\frac{-920}{T}\right)\varepsilon_3(\varepsilon_5+n_{D_2})\right] \; ,$$

$$L_{45}=k_{45}\left[\varepsilon_4(\varepsilon_5+n_{D_2})-\exp\left(\frac{-120}{T}\right)\varepsilon_5(\varepsilon_4+n_{DF})\right] \; ,$$

$$R_{12} = \left[1 - \exp\left(\frac{-960}{T}\right)\right]\left[\sum_M k_{12}(M)n_M\right]\left[\frac{n_{CO_2}}{\exp(960/T) - 1} - \varepsilon_2\right] ,$$

$$R_3 = \left[1 - \exp\left(\frac{-3380}{T}\right)\right](k_3 n_{NOF})\left[\frac{n_{CO_2}}{\exp(3380/T) - 1} - \varepsilon_3\right] ,$$

$$R_4 = \left[1 - \exp\left(\frac{-4180}{T}\right)\right]\left[\sum_M k_4(M)n_M\right]\left[\frac{n_{DF}}{\exp(4180/T) - 1} - \varepsilon_4\right] ,$$

$$R_5 = \left[1 - \exp\left(\frac{-4300}{T}\right)\right]\left[\sum_M k_5(M)n_M\right]\left[\frac{n_{D_2}}{\exp(4300/T) - 1} - \varepsilon_5\right] ,$$

$$R_{chem} = e_1^* k_1 n_F n_{D_2} + e_2^* k_2 n_D n_{F_2} ,$$

$$R_{rad} = gI/h\nu N_A \varrho^2 .$$

The terms entering into (5.36) have the following meaning:

$L_{3,12}$ and $L_{3,2}$ represent the complex deactivation channels of the ν_3 mode in CO$_3$ (i.e., $\nu_3 \rightarrow \nu_1 + \nu_2$, $\nu_3 \rightarrow 3\nu_2$);

L_{34} allows for the quasiresonant quanta exchange between the DF and CO$_2$ molecules, as a result of which the excitation energy acquired by the DF molecules in the course of the pumping chain reaction is converted to the vibrational energy of the ν_3 mode in CO$_2$;

L_{35} describes the exchange process between the ν_3 mode of CO$_2$ and the D$_2$ molecule;

L_{45} reflects the exchange of quanta between the DF and D$_2$ molecules;

R_{12}, R_3, R_4, and R_5 describe the vibrational-translational relaxation of the ν_2 and ν_3 modes of CO$_2$ and the vibrational modes of the DF and D$_2$ molecules, respectively (for simplicity, the deactivation of the ν_3 mode upon collision between CO$_2$ and NOF is considered an ordinary $V \rightarrow T$ relaxation process);

R_{chem} is equivalent to the term Q_{chem} entering into (5.9) and represents the chemical pumping process ($e_1^* = 2.5$, $e_2^* = 6.5$);

R_{rad} describes stimulated transitions between the [00^{0}1] and [10^{0}1] levels of CO$_2$ in the laser field of intensity I (W/cm^2) ($h\nu$ is the lasing transition quantum of CO$_2$ and N_A the Avogadro number; $h\nu N_A = 11.44$ J/mole).

The kinetic constants $k_{3,12}$, $k_{3,2}$, and so on are expressed in cm^3/mole s.

When modelling the continuous-wave mode of the DF–CO$_2$ laser, one can usually restrict oneself to the following one-dimensional[2] gasdynamic equations:

$$\varrho u S = G = \text{const} ,$$

$$\varrho u \frac{du}{dx} + \frac{dp}{dx} = 0 , \qquad\qquad (5.37)$$

$$\varrho u \frac{d}{dx}(h + u^2/2) = -gI .$$

---

[2] *Thoens* and *Ratliff* [5.38] calculated a supersonic DF–CO$_2$ chemical laser within the framework of a two-dimensional approximation.

These equations comply with the laws of conservation of mass, momentum, and energy. In (5.37), p and h are the pressure and specific enthalpy of the gas mixture, respectively, S is the cross-sectional area of the flow channel, and the term gI defines the loss of the active flow energy by radiation in the laser cavity. Set (5.37) is supplemented with the thermal and caloric equations of state

$$p = \frac{R}{W}\varrho T, \quad h = \sum_i n_i H_i ,$$

where W is the molecular weight of the gas mixture and H_i are the molar enthalpies of the mixture components.

In the case of a subsonic DF–CO_2 laser using an initiating reactant, the adoption of a one-dimensional gasdynamic flow model presumes that the distance it takes for the primary gas flow carrying atomic and molecular fluorine to mix completely with the gas flow containing deuterium is shorter than the width of the active (lasing) flow region. In this connection, it may not be out of place to make the following remarks [5.23].

It follows from experiments on turbulent mixing that with the gas injection method being optimal, the distance that the gas flow covers until its components are mixed completely is equal to 10–15 gauges. (By "gauge" is meant the diameter of the small holes in the tubes through which deuterium is injected.) The operational lasers [5.10, 14] use injection systems with a D_2-injector hole diameter of $d \simeq 0.01$ cm, so that the turbulent diffusion length in them is $x_D \simeq 0.1$–0.15 cm. At the same time, the characteristic active region length under typical laser conditions is some 5–10 cm. However, it should be borne in mind that while the flow covers a distance of around x_D, the mixing in it occurs only on a macroscopic level, characterized by a predominance of large-scale turbulent fluctuations [5.39, 40]. According to the general theory of turbulence [5.39], the break-down rate of large-scale turbulence elements into small-scale ones is of the order of the rate of dissipation of the turbulent kinetic energy as heat. The time scale of this dissipation is $\tau \simeq 10\lambda/u$, where λ is the scale of large vortices. Considering that the convective velocity associated with the motion of large-scale vortices is $u' \simeq u/2$, the characteristic distance that the mixture flow covers until its components are mixed on a molecular level is $x'_D \simeq u'\tau \simeq 5\lambda$. Setting λ to be equal to the distance between the injector tubes, $l_0 \simeq 0.2$–0.4 cm, we get $x'_D \simeq 1$–2 cm.

It is clear that if the characteristic width of the active flow region is of the order of $x'_D \gg x_D$, then, generally speaking, account should be taken of the effect the break-down of turbulent fluctuations has on the laser kinetics. This can be done by introducing into the pertinent kinetic equations the effective rate constants $\{k_{eff}\}$ for the processes occurring upon collisions between the components of the primary and secondary gas flows [5.23, 26, 28], it being convenient to establish the relationships between $\{k_{eff}\}$ and the ordinary rate constants $\{k\}$ characteristic of the homogeneous gas phase in the form $\{k_{eff}\} = \{k\}(1 - \exp(-x/l_D))$, where $l_D \simeq x'_D$ is the effective mixing length.

Pimenov and *Shcheglov* [5.26] have performed calculations for conditions characteristic of the experiments by *Cool* et al. [5.4] and *Shirley* et al. [5.10]. A numerical parametric analysis has shown that the $NO:F_2:CO_2:D_2:He$ $= 1.5:7:60:7:100$ mixture at a static pressure of $p \simeq 15$ mm Hg is optimal as regards the basic power characteristics of the laser. In the given case, around 2.5% of the molecular fluorine in the initiation channel is decomposed by the moment the gas flow reaches the D_2 injection region, the consumption of NO amounting to 15%. The optimal mixture composition provides for the following laser energy characteristics: chemical efficiency $\eta_{chem} \simeq 5\%$, reduced laser power $P \simeq 10$ W/cm^2, specific energy output $E \simeq 50$ J per gram of the mixture and $E_F \simeq 650$ J per gram of the starting molecular fluorine, saturation intensity $I \simeq 150$ W/cm^2, and the active region width $\Delta x_l \simeq 8$ cm. Lasing is quenched upon a 50% burn-out of the fluorine, with the gas mixture temperature reaching some 600 K.

Comparison between the calculation results reported by *Baumann* et al. [5.23] and *Zelazny* et al. [5.28] and the experimental data obtained by *Shirley* et al. [5.10], *Falk* [5.12], and *Tregay* et al. [5.14] has demonstrated a reasonable correlation between the gain distributions in the active region. The authors of [5.23] have analyzed the role of the various collisional processes responsible for the excitation and decay of the $[00^01]$ level in CO_2 under the experimental conditions of [5.12]. They have demonstrated that the population of this level depends not only on the dominant process of vibrational quanta transfer from the excited DF molecules, but to a certain extent on the transfer of quanta from D_2 as well, the contribution of this process increasing toward the downstream end of the active region. Where the mixture is diluted with helium, the most important deactivation mechanisms for the $[00^01]$ level of CO_2 are associated with the formation of $CO_2(11^01)$ throughout the active region primarily as a result of collisions between $CO_2(00^01)$ and DF, F, and D_2. In the case of dilution with nitrogen, the main part in the deactivation of the $[00^01]$ level is played by the $V \rightarrow V$ exchange between N_2 and CO_2 at the early stages and the complex deactivation channels at the subsequent stages of the process. As the gas flow approaches the end of the active region, the role of N_2 is reversed: the vibrational energy stored by nitrogen is transferred to CO_2.

5.3 Supersonic HF (DF) Lasers. A Review of Experimental Work

Shirley and *Cool* et al. [5.10, 41] achieved continuous-wave lasing on the HF and DF molecules in a laser using a purely chemical initiation by reaction (5.1). The experiment was conducted on the basis of the subsonic laser setup reported by *Cool* et al. [5.4], with the CO_2 flow being shut off. They found that the power output (10 W) obtained in that case was much lower than in the laser using the transfer of excitation to CO_2. This is explained first of all by the high threshold reaction rates necessary to excite the HF (DF) molecules and their high relaxation rates. That is why the achievement of high laser emission powers from these molecules is associated with the use of supersonic gas flows. To attain high efflux velocities of the gas, it must be heated to high temperatures, which in turn causes the molecules to dissociate and thus form active centers. It was precisely on this basis that the first supersonic HF and DF laser models were developed [5.42–44]. According to the initiation technique used, the continuous-wave hydrogen- (deuterium-) fluoride chemical lasers are divided into two classes: lasers using an external energy source, usually an arc discharge [5.42, 45–59, 62], and fully self-contained lasers in which initiation is effected by a supplementary exothermic reaction [5.17, 59–66]. To excite oscillation in these lasers, use is made of the processes

$$F + H_2 \rightarrow HF(v) + H \quad (\Delta H = -31.7 \text{ kcal/mole}) \quad \text{or}$$

$$F + D_2 \rightarrow DF(v) + D \quad (\Delta H = -31.7 \text{ kcal/mole}). \tag{5.38}$$

5.3.1 Plasma-Generator Laser Version

Figure 5.7 presents a schematic diagram of a supersonic laser using an external initiation source. The gas-generator unit of the laser consists of a primary (heating) chamber containing the initiation source (heater), whose energy is used to raise the temperature of a heat-carrying agent (diluent) injected into the chamber, and a secondary (mixing) chamber in series with the primary one, into which a chemical reactant ("oxidizer") is injected to make it undergo thermal dissociation upon mixing with the high-temperature heat-carrying agent and thus yield the active centers necessary for the operation of the laser. The heater is usually a plasma generator (an arc discharge) whose operation is considered in detail in a vast body of special literature. The secondary chamber is provided with special injectors through which additional amounts of the diluent can be injected to control the temperature and mass (heat capacity) of the gas mixture in the chamber.

The flow of the heat-carrying agent mixed with the chemically active centers is discharged through an array of supersonic nozzles from which gas jets issue at a supersonic velocity. Arranged at the exit from the nozzle bank are numerous "fuel" injectors. The fuel molecules are entrained by the supersonic gas jets issuing from the bank and enter into chemical reaction with the active centers contained in the

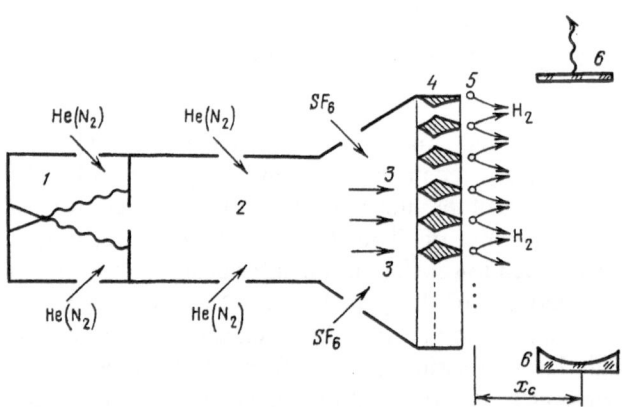

Fig. 5.7. Schematic diagram of an externally initiated supersonic chemical laser [5.50]: *1* – heating chamber; *2* – secondary (mixing) chamber; *3* – oxidizer injector; *4* – supersonic nozzle bank; *5* – fuel injector; *6* – cavity mirrors

gas. This reaction produces the excited molecules which give rise to lasing in the laser cavity downstream of the nozzle bank.

To improve the mixing of the reactants, the nozzle bank is usually made in the form of a honeycomb set of small nozzles, and fuel injectors are located in the exit plane of the nozzles. Figure 5.8 shows typical nozzle designs, and Fig. 5.9 illustrates schematically the process of mixing of a jet of heat-carrying agent containing chemically active centers with jets of fuel.

Either nitrogen or helium is used as a diluent, and sulfur hexafluoride (SF_6) usually serves as an oxidizer subject to thermal decomposition. After heating, the diluent temperature ranges between 2000 and 4000 K. When sulfur hexafluoride is mixed with the hot diluent, it dissociates into fluorine atoms, some radicals the type of SF_n ($n \leqslant 5$), and sulfur atoms, the atomic fluorine and sulfur being predominant. The fuel, H_2 or D_2, is injected at the exit face of the nozzle bank.

The composition and characteristics of the supersonic jet are determined by the parameters of the starting gas mixture in the gas generator and the nozzle-area ratio, i.e., the ratio between the nozzle-exit area A and the nozzle-throat area A^*. Table 5.3 lists typical experimental ratios between the gas-jet parameters at the nozzle exit and the gas parameters at the nozzle entrance for two heat-carrying agents–nitrogen and helium [5.50].

As has been already noted, the position of the inversion maximum and the active region width along the flow direction are important laser characteristics. *Spencer* and co-workers [5.50] determined these parameters by measuring the intensity of radiation coupled from a Fabry-Perot cavity by means of a multiple-hole coupler, a flat metal mirror with a row of small coupling holes drilled through its face and arranged in a line along the flow direction. It was found that in the case where helium was used as a diluent, the active region width Δx was greater and the inversion maximum was located farther downstream from the exit face of the nozzle bank than in the case of nitrogen. The ratio between the coordinates of the inversion maxima, $x^0_{He}/x^0_{N_2}$, approximately coincided with that between the respective

Fig. 5.8a–c   **Fig. 5.9**

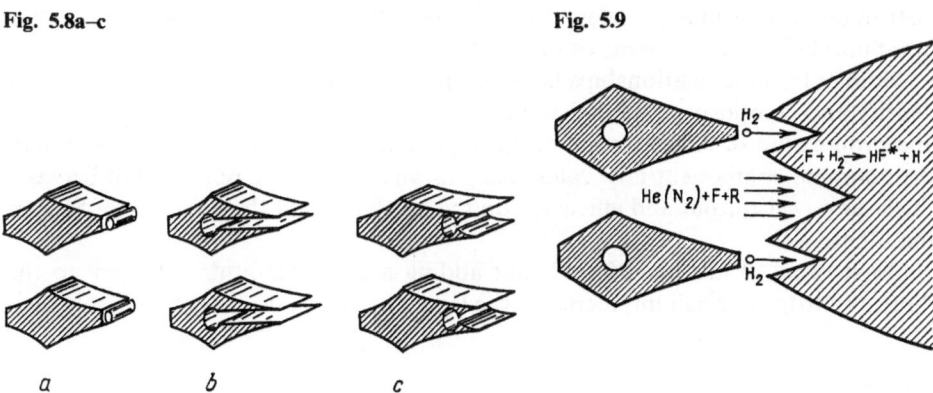

a   *b*   *c*

Fig. 5.8a–c. Details of nozzles used in supersonic HF (DF) lasers. H_2 (D_2) injected through (**a**) perforated tubes; (**b**) narrow slits; (**c**) nozzles

Fig. 5.9. Illustrating the mixing of reactants in a supersonic oxidizer jet

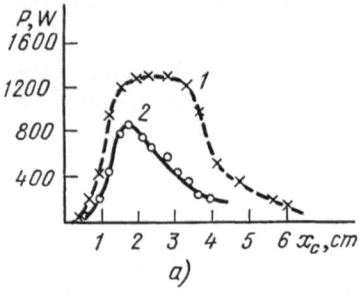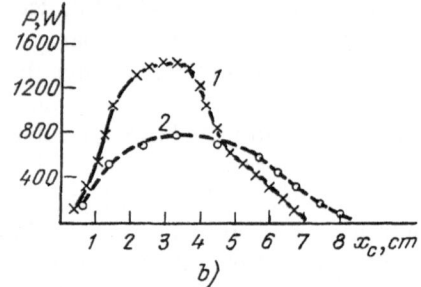

Fig. 5.10a, b. Output power of the HF laser [5.50] as a function of the distance x_c between the cavity axis and nozzle bank exit face (location plane of H_2 injectors, see Fig. 5.7). (**a**) For various diluents $1 - \dot{m}_{He}$ $= 2.06$ g/s and $2 - \dot{m}_{N_2} = 8.5$ g/s ($\dot{m}_{SF_6} = 2.2$ g/s); (**b**) for various SF_6 flow rates: $1 - \dot{m}_{SF_6} = 2.2$ g/s and $2 - \dot{m}_{SF_6} = 0.8$ g/s ($\dot{m}_{He} = 2.06$ g/s). $\dot{m}_{H_2} = 1$ g/s; $\dot{m}_{O_2} = 0.8$ g/s

inversion region widths, $\Delta x_{He}/\Delta x_{N_2}$, and corresponded to the velocity ratio u_{He}/u_{N_2}:

$$x^0_{He}/x^0_{N_2} \simeq \Delta x_{H_e}/\Delta x_{N_2} \simeq u_{He}/u_{N_2} .$$

When He was used as a diluent, the jet velocity was approximately twice that in the case of nitrogen diluent.[3]

The power distribution along the flow direction can also be studied by means of a spatial scanning cavity, i.e., a cavity with small-sized mirrors, which can be moved along the flow. Using this technique, *Spencer* et al. [5.50] studied the relationship between the power coupled out of the cavity and the distance between the cavity axis and the nozzle bank exit face, i.e., the location plane of the hydrogen injectors. The curves presented in Fig. 5.10a indicate that helium diluent provides for a wider and smoother laser power distribution along the flow direction. The maximum power in this case is 25% higher than that for nitrogen diluent, the molar flow rates of the mixture components being the same. Figure 5.10b illustrates the power distribution for various SF_6 flow rates. It is seen that raising the flow rate of sulfur hexafluoride increases the maximum power and narrows the power distribution.

Studies into the relationships between the laser characteristics and the gas flow density and composition have demonstrated the existence of optimum flow rates and proportions of the active mixture components. Let us consider the question of the optimum component flow rates, using as an example the relationship between the laser power output and efficiency on the one hand and the flow rate of SF_6 on the other.

First of all, it has been found that adding a small amount of oxygen to the diluent (nitrogen or helium) increases the laser power by 20–30% at relatively high

[3]In [5.50], the arc discharge power was varied over the range 35–50 kW. Accordingly, the stagnation temperature was $T_0 \simeq 2000$–4000 K. Under the nominal conditions $p_0 \simeq 1$ atm. and $T_0 \simeq 2000$ K, in the case of N_2, $p \simeq 3$ mm Hg, $T \simeq 400$ K, and $u \simeq 1.8 \times 10^5$ cm/s, and in the case of He, $p \simeq 1.4$ mm Hg, $T \simeq 160$ K, and $u \simeq 4.3 \times 10^5$ cm/s.

flow rates of SF_6. This is apparently [5.49, 50] due to the production of some additional atomic fluorine as a result of the reactions

$$SF_n + O_2 \rightarrow SO_2F_{n-2} + 2F \; ,$$

the process involving the SF_4 radical being considered predominant [5.49]. Such an explanation seems plausible, for as the concentration of SF_6 in the flow is increased, the amount of incompletely dissociated sulfur hexafluoride molecules grows larger. One could also suppose that as the concentration of the SF_n radicals increases and radicals of the type SO_2F_{n-2} appear in the mixture, the reactions

$$SF_n + H \rightarrow HF^* + SF_{n-1}$$

and $\hspace{6cm}$ (5.39)

$$SO_2F_{n-2} + H \rightarrow HF^* + SO_2F_{n-3}$$

gain in importance. The energy released in these reactions is appreciably higher than that resulting from reactions (5.38), and so there would be observed an increase in the chemiluminescence of the HF molecules from the $v = 3$ level. However, no such effect was noted in [5.50]: the addition of oxygen did not change the vibrational energy distribution of HF. The analysis of the emission spectrum (Fig. 5.11) also gives no reason to believe that processes of the type (5.39) play any part in the production of the active medium.

Second, the analysis of the relationship between the total laser power and the flow rate of SF_6 has shown the existence of a clearly defined maximum in the case of nitrogen diluent. In the case of helium diluent, the laser power tends to saturate so that no distinct maximum is observed (Fig. 5.12a). To explain this fact, one can assume [5.50] that in conditions of arc (plasma) heating, some helium atoms transit to the $2s^2S$ and $2s^1S$ metastable states with energies of 19.8 and 20.6 eV, respectively. These states may serve as an additional source of energy in the dissociation of SF_6 by collision. The presence of $He(2s^2S)$ and $He(2s^1S)$ was detected by adding neon to the gas flow and observing neon luminescence at a wavelength of 6328 Å resulting from the energy transfer from $He(2s^2S)$ and $He(2s^1S)$ to neon. But it proved impossible to estimate the amount of such helium atoms. The arc discharges used to heat high-pressure gases are a low-voltage type, and therefore the presence of large amounts of the metastable atoms in the hot helium can hardly be expected, their excitation energies being high. The energy of a

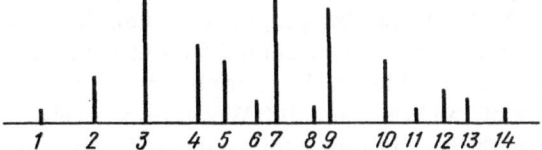

Fig. 5.11. Emission spectrum of the HF laser [5.50] using He diluent: $1-P_1(4)$; $2-P_1(5)$; $3-P_1(6)$; $4-P_1(7)$; $5-P_2(4)$ $6-P_1(8)$; $7-P_2(5)$; $8-P_1(9)$; $9-P_2(6)$; $10-P_2(7)$; $11-P_3(4)$; $12-P_2(8)$; $13-P_3(5)$; $14-P_2(9)$

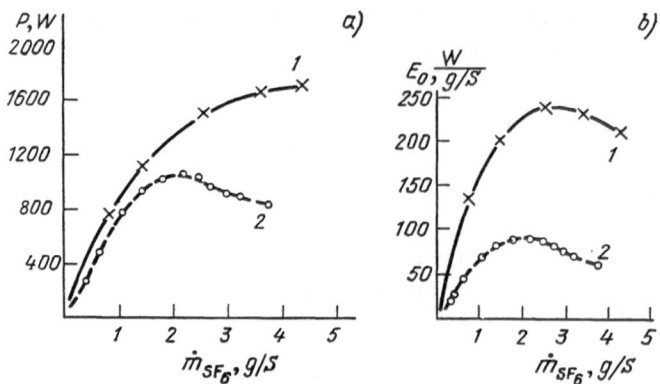

Fig. 5.12. (a) Total power P and (b) specific power E_0 of the HF laser [5.50] as a function of the flow rate of SF_6: 1 – helium diluent ($\dot{m}_{He} = 2.06$ g/s, $\dot{m}_{O_2} = 0.75$ g/s, $\dot{m}_{H_2} = 1$ g/s); 2 – nitrogen diluent ($\dot{m}_{N_2} = 0.85$ g/s, $\dot{m}_{O_2} = 0.4$ g/s, $\dot{m}_{H_2} = 1.25$ g/s)

metastable atom is 40–50 times the kinetic energy of an unexcited atom at a temperature of 3000–4000 K, so that a mere 1% of excited helium atoms makes a noticeable contribution to the overall energy balance.

Another possible explanation of the laser power behavior as a function of the flow rate of SF_6 is associated with the specific features of the heater operation. An increase in the SF_6 content of the gas flow enhances the diffusion of sulfur hexafluoride molecules into the heater, which raises the voltage and reduces the current of the arc discharge. The discharge in nitrogen operates close to the constant-power region of the volt-ampere characteristic. In this region, the rise of the discharge voltage and the attendant reduction of the discharge current occur in such a manner that their product (discharge power) remains constant. A different picture is observed in the case of helium. The discharge in helium operates in a low-voltage region where an increase in the discharge voltage raises the total electric power dissipated by the discharge, some reduction of the discharge current notwithstanding. Thus, the higher laser power provided by helium diluent compared to nitrogen diluent at high flow rates of SF_6 can be explained simply by a higher power consumed by the heater. It cannot be ruled out that both the above factors make contributions of their own to the experimental results observed.

Despite the fact that raising the flow rate of SF_6 increases the total laser power, this eventually leads to a reduction in the laser power per unit gas mixture flow rate (i.e., in the specific laser power $W/(g/s) = W\,s/g$). This inference follows from the experimental data presented in Fig. 5.12b. As can be seen, the optimum SF_6 flow rate as regards the maximum specific laser power is around 2.5 g/s in the case of helium diluent and 1.8 g/s in that of nitrogen diluent, the laser powers per unit gas mixture flow rate being of the order of 250 W s/g and 100 W s/g, respectively.

Knowing the relationship between the laser power output and the SF_6 flow rate, one can calculate the corresponding relationship for the chemical efficiency of the laser. In the present case, the chemical efficiency is understood to be the ratio of the laser power output to the total power generated by reaction (5.38) under the

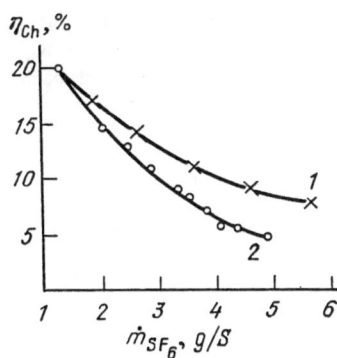

Fig. 5.13. Chemical efficiency of the HF laser [5.50] as a function of the flow rate of SF_6: 1 – helium diluent ($\dot{m}_{He}$ = 2.06 g/s, $\dot{m}_{H_2}$ = 1.25 g/s, $\dot{m}_{O_2}$ = 0.75 g/s); 2 – nitrogen diluent ($\dot{m}_{N_2}$ = 8.5 g/s, $\dot{m}_{H_2}$ = 1 g/s, $\dot{m}_{O_2}$ = 0.75 g/s)

assumption of complete dissociation of SF_6. The results of such calculations are presented in Fig. 5.13. The drop of the efficiency with the increasing flow rate of SF_6 is most likely due to a decrease in the degree of dissociation of this reactant.

When the chemical pumping reaction is initiated by an external source of energy, consideration should also be given to the engineering efficiency of the laser, which is defined as the ratio between the laser power output and the initiation source power consumed by the laser. It is obvious that the engineering efficiency of the laser will be enhanced if it is operated in the parameter variation region where the increase in the laser power output exceeds that in the power consumed to initiate the reaction.

The relationships between the laser performance characteristics and the other parameters (the flow rate of hydrogen in particular) can be studied in a similar way. Such investigations carried out by *Spencer* et al. [5.42, 45, 46, 48–50] and *Kwok* et al. [5.47] made it possible to establish the optimum proportion between the mixture components in the supersonic flow as regards the maximum engineering efficiency and specific and total power outputs of the laser. In their work reported in [5.46], the authors achieved record high specific energy characteristics for the supersonic HF laser. The pertinent data are listed in Table 5.4.

Thus, experimental results show that lasers using helium diluent have better performance characteristics than those operating on nitrogen diluent. The advantage of helium over nitrogen is especially evident in DF lasers. According to [5.50], the maximum laser power output attainable with DF is only 70% of that attainable with HF, provided the diluent used is nitrogen. At the same time, if helium diluent is used, the maximum laser power outputs attainable with DF and HF are the same for all practical purposes (Fig. 5.14).

For the partial flow densities indicated in Table 5.4, the respective power outputs per unit nozzle bank exit face area were 70 W/cm² (He diluent) and 45 W/cm² (N_2 diluent). In the case of He diluent (see Fig. 5.12a), as the flow rate of SF_6 was increased, the laser power output per unit nozzle bank exit face area rose to 80 W/cm² and the total laser power, to 1.8 kW, but the specific characteristics of the laser were worse (chemical efficiency amounted to 7.5% and specific power output, 200 W s/g). *Giedt* [5.62] reported a total power of 11.8 kW from a plasma-generator version of the supersonic HF laser.

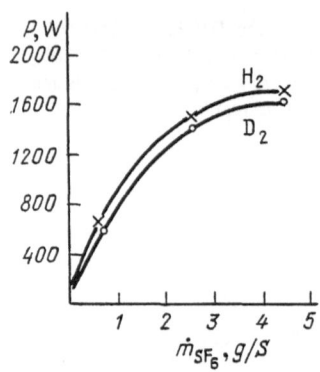

Fig. 5.14. Output power of the HF and DF lasers [5.50] as a function of the flow rate of SF_6. $\dot{m}_{He} = 2.06$ g/s, $\dot{m}_{O_2} = 0.8$ g/s, $\dot{m}_{H_2} = 0.5$ g/s. The distance between the cavity axis and the nozzle bank exit face, $x_c = 2.5$ cm

The high specific laser parameters reported in [5.50] were obtained at low operating pressures in the cavity. When the pressures are increased, in the supersonic jet there may become manifest a number of effects (shock waves, turbulization of the gas, thermal "plugs", etc.) influencing the power characteristics of the laser. In this connection, comprehensive experimental investigations allowing information to be obtained on the character of mixing processes, the distribution of gasdynamic parameters in the gas flow, and the population distribution of the lasing molecules [5.52–56, 58, 67] become particularly important.

5.3.2 Self-Contained Laser Version

The second version of the supersonic HF (DF) laser is distinguished by the fact that the production of active centers (fluorine atoms) in its energetic unit occurs as a result of burning a suitable fuel in the atmosphere of a fluorine-bearing reactant. The principle of operation of such a laser is as follows (Fig. 5.15). The fuel (H_2 or D_2) and the fluorine-bearing reactant (e.g., F_2, ClF_5) are injected into the combustion chamber, the amount of the latter exceeding what is required for the reaction. The

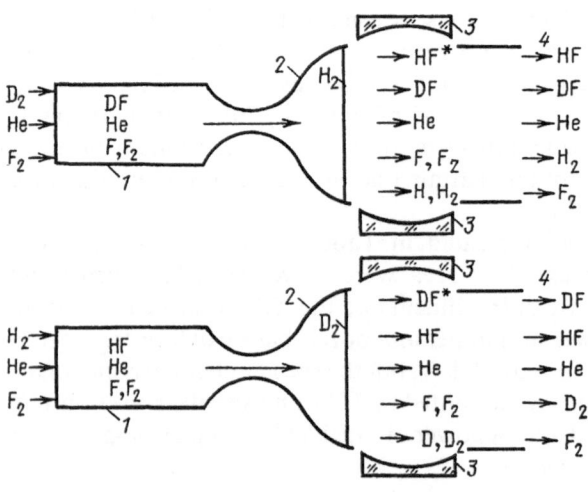

Fig. 5.15. Schematic diagram of the self–contained supersonic HF (DF) laser [5.60]: *1*–combustion chamber; *2*–nozzle bank; *3*–cavity mirrors; *4*–exhaust system

heat liberated in the reaction causes the excess fluorine-bearing reactant to dissociate and thus yield F atoms. The gas temperature is controlled as usual by injecting a diluent (He or N_2) into the mixture. The resulting gas mixture is then expanded through a bank of supersonic nozzles at the exit from which H_2 (or D_2) is injected into the jet formed. As a result, vibrationally excited HF* (or DF*) molecules are produced in the supersonic jet. These then lase in the cavity downstream of the nozzle bank. So, the supersonic HF (DF) laser version considered here requires no external power supply and is thus a completely self-contained device.

Nowadays, serious attention is being paid to the development of highly efficient self-contained supersonic hydrogen-fluoride CWCL's [5.60–72]. 5 kW HF (DF) lasers have already been built [5.60–62]. Judging from the available publications, work on the improvement of this class of lasers is extending into all directions – design and manufacture of combustion chambers and nozzle banks, vacuum pump and exhaust systems, cavity design, and design of fuel compositions facilitating better laser operating and energy characteristics. *Nagai* and co-workers [5.61] studied such a self-contained supersonic HF (DF) laser. The necessary energy parameters of the flow in the gas generator were achieved by using the reaction of HF (DF) with F_2 (typical stagnation temperature $T_0 \simeq 1600$ K). An important feature of the gas generator was the use of a regenerative cooling system. Lasing took place at a pressure of 0.01 atm. in the cavity. The efficient operation of the laser under such a pressure was ensured by the well-organized mixing process in the cavity region, owing first of all to the use of small-sized nozzles for the oxidizer flow and supersonic deuterium injectors (Fig. 5.8c). The pertinent information about the supersonic nozzle bank of the laser is collected in Table 5.5.

The regeneratively cooled small-size HF laser model [5.61] generated up to 7.3 kW of power at a specific energy output of 105 J/g, the output power per unit nozzle bank exit face area reaching 125 W/cm^2.

An important aspect of the construction of high-power CWCL's is the development of effective methods for the restoration of the supersonic flow pressure in the diffuser. A successful solution of this problem has a direct effect on the required exhaust system capacity. In practice, the problem is solved by selecting an appropriate diffuser channel (ejector) geometry. *Cavalleri* and *Laeger* [5.69] analyzed the design principles and operating characteristics of a solid-fuel powered ejector and demonstrated that with the ejector entrance-to-exit pressure ratio below 0.1, the ejection coefficient was over 20 and rose rapidly with the decreasing ejector-entrance pressure. The authors also considered a method for decontaminating toxic exhaust products of CWCL's, based on adding alkali-metal compounds to the ejector fuel.

The effective operation of the self-contained supersonic hydrogen-fluoride laser greatly depends on the composition of the fuel used in the combustion chamber. Without going into purely technical requirements for the fuel (convenience in storage, possibility of uncontrolled ignition, injection details, etc.), let us formulate the basic objects at which to aim when selecting fuel mixtures: (1) to ensure the necessary energetics in the combustion chamber, (2) to produce the maximum possible atomic fluorine concentration, and (3) to provide for a low concentration of reaction by-products which intensively deactivate the lasing molecules. With these ends in view, *Roback* and *Lynds* [5.70] and *Ratliff* et al. [5.71] analyzed a wide array of possible candidates to replace the ordinary $F_2/H_2(D_2)$ mixture. The oxidizers considered included F_2, ClF_2, N_2F_4, and NF_3. The analysis [5.70] has demonstrated that the most preferable "oxidizer + {fuel}" combinations are

$$NF_3 + \{N_2O, CS_2, (CN)_2, CO, NO, C_2H_2\} \ ,$$

$$F_2 + \{N_2O, CS_2, (CN)_2, CO, C_2H_2\} \ ,$$

$$ClF_5 + \{C_6F_6, C_6H_6, C_2H_2\} \ ,$$

$$N_2F_4 + \{C_2H_2\} \ ,$$

i.e., the circle of potential candidates is, as we see, wide enough. Obviously the validity of the recommendations of [5.70] can be reliably verified by experiment only. And such experiments are already being conducted (see, for example, [5.63]).

The development of high-power supersonic machines involves the scale-up of the system, i.e., the enlargement of the nozzle bank exit area. If the nozzle bank is flat, the scale-up entails certain difficulties in the cavity design. In this connection, *Finkleman* and *Greenberg* [5.68] have discussed a cylindrical nozzle bank geometry allowing for compact construction, toroidal-shaped mirrors being suggested for the cavity.

5.4 Principal Power Performance Features of Supersonic CWCL's

5.4.1 Introductory Remarks. Flame Front Concept

The analysis carried out in the preceding section shows that the characteristics of the supersonic-diffusion chemical laser depend on a great variety of physico-chemical processes. Let us formulate the basic problems typical of such systems.

1) Determination of the molar composition, pressure, and temperature of the products in the gas-generator unit proceeding from the initial pressure, composition, and thermodynamic parameters of the starting mixture components (and also the initiation source power in the case of external initiation). A necessary element of this analysis is the inclusion of the thermal losses which inevitably occur when the hot gas mixture comes in contact with the gas generator walls.

2) Calculation of a multicomponent, chemically nonequilibrium flow containing a dissociated oxidizer through a supersonic duct of specified profile and dimensions. Generally speaking, account should be taken here of the presence of wall boundary layers, the effect of which on flows with relatively small Reynolds numbers may be substantial.

3) Calculation of the energy characteristics of the chemical laser proceeding from the specified parameters of the primary (oxidizer) and secondary (fuel) flows and optical system parameters. As distinct from the case of pulsed laser systems, this

problem is greatly complicated by the fact that the evolution of the kinetic processes here takes place against the background of varying temperatures and densities in the mixing gas flows.

4) Calculation of an optimum ejector system – a necessary component part of the laser system.

It should be noted that whereas the approach to the solution of the first, second, and fourth problems has been studied thoroughly (see, for example, [5.73–76]), to analyze the third problem (the laser problem proper) requires, in view of its complexity, that a special method be devised. It is no accident that much of the published theoretical work on the supersonic HF chemical laser has paid much attention to this question. As far as methodology is concerned, it is advisable to divide this work into several groups.

The first group includes the papers in which the active medium is described within the scope of one-dimensional equations giving no consideration to mixing processes [5.24, 72, 77–85]. Such an approach is useful in revealing the role of various kinetic mechanisms in the formation of the active medium itself, but is hardly suitable for the description of actual laser models where the factor of mixing materially affects the energy characteristics of the laser.

The second group embraces the theoretical work based on the use of two-dimensional hydrodynamic equations in the boundary-layer approximation. Belonging to this group are the papers concerned with the study of laminar mixing conditions [5.86–88] and mixing in conditions of well-developed turbulence [5.89–95]. Generally speaking, most consistent numerical calculations must be performed using the complete set of Navier–Stokes equations (see, for example, [5.96–98]; the representation of the respective equations as applied to CWCL's with flat and cylindrical nozzle bank geometries can be found in [5.99, 100]). Note, however, that the need to integrate a great number of equations to the high precision dictated by the gaskinetic system itself entails a heavy expenditure of time, even for the highest-powered modern computers [5.100].

The latter circumstance stimulates the development of simpler methods allowing only for the key factors of the problem in hand. These methods include the one based on the quasiunidimensional approach [5.101–118]. The advantage of the method is that it considerably simplifies the mathematical aspect of the problem and at the same time makes it possible to interpret experimental results graphically, and reveal the general trends in the system's behavior in response to changes in the external parameters. Based on a flame-front model, *Broadwell* [5.102] proposed an approach to the description of the supersonic-diffusion laser. The lasing molecule was modelled by a two-level system, and no account was taken of thermal effects. *Stepanov* and *Shcheglov* [5.106] generalized this approach to include the multiple-level character of excitation of the lasing molecules and their rotational level structure (see also [5.107]), thermal effects in the flow and their influence upon gasdynamic motions. We now explain the flame front concept and its application to the supersonic-diffusion CWCL.

The principle of the approach itself is widely used in the theory of burning of unmixed gases [5.119–121], and is based on the well-known qualitative notion that the surface of a flame separates the region where there is an oxidizer but no fuel from the one where the oxidizer is absent but the fuel is present. One of the most important inferences drawn in [5.121] is as follows: when unmixed gases burn rapidly, the concentration of the products in the reaction zone turns out to be exactly the same as would have been the case if the starting gases were mixed in the stoichiometric proportion on the flame surface and the reaction was effected without any diffusive exchange.

In the CW chemical HF laser operating on the non-chain reaction $F + H_2 \rightarrow HF^* + H$, the flows of the oxidizer (F) and fuel (H_2) at the exit from the supersonic nozzles are not mixed and the burning process is of diffusive character. The hydrogen concentration usually greatly exceeds the stoichiometric value. Being the lighter component, the hydrogen molecules diffuse into the oxidizer flow, and as a result, there occurs the mixing of chemically active jets. If the burning rate is sufficiently high, chemical transformations take place in relatively narrow zones separated from nonreacted gas flow zones by flame surfaces (Fig. 5.16). In the general case, the shape and structure of the flame surface should be found from the solution of a self-consistent problem including kinetic and diffusive transfer processes (see, for example [5.86]), the spatial localization of the surface being determined by the stoichiometry condition.

Based on the flame front concept, the quasiunidimensional approach proposed by *Stepanov* and *Shcheglov* [5.106] presumes the surface of mixing to be pre-determined. Its coordinate y_f is found from additional considerations; in particular, use is made of empirical data on the character of the mixing process. The situation may be pictured to be as though a flow of premixed reactants feeds on this surface, and the reaction between the reactants is "switched on" and proceeds rapidly on the surface (Fig. 5.16).

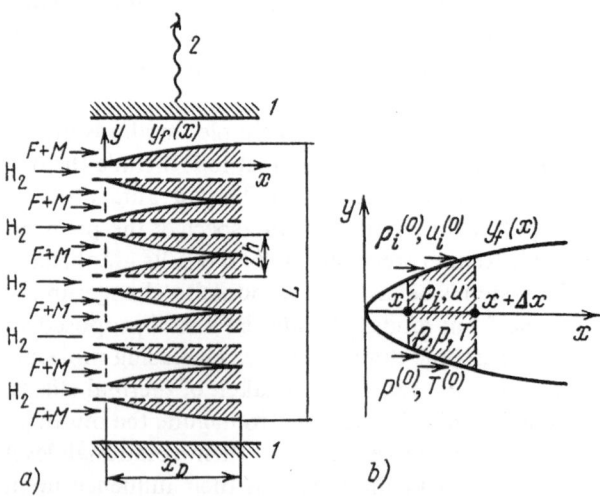

Fig. 5.16a, b. Quasiunidimensional model of the mixing of chemically active jets. (a) Alternate ($F + M$) and (H_2) jets; $y_f(x)$ – sheet flame coordinate; x – flow coordinate; x_D – mixing length; $2h$ – space period of structure; 1 – cavity mirrors; 2 – laser emission; (b) $p^{(0)}$ and $T^{(0)}$ – initial pressure and temperature, respectively ; $\rho_i^{(0)}$ and $u_i^{(0)}$ – initial component density and velocity, respectively

According to the quasiunidimensional approach, the mixing process is taken into consideration by averaging the dynamic problem variables (reactant densities ϱ_i and gasdynamic flow parameters u, ϱ, p, and T) over the flow cross section in the cavity region.

Let x be the coordinate along the flow direction, y that along the cavity axis, and y_f the mixing front coordinate (flat case). Within the framework of the model under consideration, it is not very difficult to get equations for the averaged quantities in the form

$$\bar{\phi} \equiv \phi(x) = \frac{1}{y_f(x)} \int_0^{y_f(x)} \phi(x, y) dy \ ,$$

where $\phi(x, y)$ is any of the independent variables. For example, referring to Fig. 5.16b, we write the law of conservation of the mass flow of the ith component for the section $(x, x + \Delta x)$ with account being taken of the inflow of particles through the surface y_f:

$$(\varrho_i u y_f)_{x + \Delta x} - (\varrho_i u y_x)_x = \varrho_{0i} u_{0i} \Delta y_f + y_f \Delta x \left(\frac{d\varrho}{dt} \right)_i \ . \tag{5.40}$$

The subscript "0" here stands for the values of the respective quantities on the mixing front and $(d\varrho/dt)_i$ is the rate of production of the ith species as a result of chemical transformations in the active region; for the lasing molecules, the term $(d\varrho/dt)_i$ also allows for the processes of vibrational and radiative kinetics.

The passage to the limit $\Delta x \to 0$ in (5.40) yields

$$\frac{d}{dx} \varrho_i u y_f = \varrho_{0i} u_{0i} \frac{dy_f}{dx} + y_f \left(\frac{d\varrho}{dt} \right)_i \ . \tag{5.41}$$

The equations expressing the laws of conservation of the total mass flow, momentum, and energy may be obtained in a similar way:

$$\frac{d}{dx} (\varrho u y_f) = \varrho_0 u_0 \frac{dy_f}{dx},$$

$$\frac{d}{dx} (\varrho u^2 y_f) = \varrho_0 u_0^2 \frac{dy_f}{dx} - y_f \frac{dp}{dx} \ , \tag{5.42}$$

$$\frac{d}{dx} \varrho u y_f (h + u^2/2) = \varrho_0 u_0 (h_0 + u_0^2/2) \frac{dy_f}{dx} - y_f \frac{dW}{dV} \ .$$

In the above expressions, $h = \Sigma_i (\varrho_i/\varrho) h_i$ is the specific enthalpy of the mixture, h_i are the partial enthalpies, W is the total laser power extracted from the entire active medium, $dV = 2n y_f z \, dx$ is the active region volume differential, n the number of the oxidizer (fuel) jets, and z the coordinate along an axis perpendicular to the (x, y) plane.

It is convenient to represent the term describing the stimulated emission of radiation in the energy balance equation in the form

$$\frac{dW}{dV} = \Sigma_v g_{v,v-1} I_{v,v-1} \ , \tag{5.43}$$

where $g_{v,v-1}$ is the gain in the $(v, J-1 \to v-1, J)$ vibrational-rotational transition and $I_{v,v-1}$ the respective radiation intensity.

After supplementing equations (5.41)–(5.42) with the thermal and caloric equations of state (Sect. 5.2), we get the complete set of equations describing the active gas medium in the supersonic-diffusion CWCL.

It should be noted that the quasiunidimensional approach makes it possible to take account of the nonuniform distribution of the gasdynamic flow parameters at the exit face of the nozzle bank, due in particular to the boundary layers at the surfaces of the supersonic channels [5.122, 123]. Note also that such an approach is especially convenient in calculating CWCL's with unstable cavity systems [5.114–117, 124].

5.4.2 Estimation of Scale Factors

The quasiunidimensional approach allows a fairly simple analysis of the characteristics of the supersonic-diffusion HF laser. To start with, let us estimate the problem's scale parameters – the mixing (diffusion) length x_D, the linear dimension of the reaction zone, $x_{chem} = u\tau_{chem}$, and the dimension characteristic of the collisional deactivation of the excited molecules, $x_{VT} = u\tau_{VT}$ – for typical conditions.

The characteristic reaction and deactivation lengths are estimated from the relations

$$x_{chem} = u\left(\frac{1}{[F]}\frac{d[F]}{dt}\right)^{-1} = \frac{u}{k_{chem}[H_2]_0} \ ,$$

$$\tag{5.44}$$

$$x_{VT} = u\left(\frac{1}{[HF^*]}\frac{d[HF^*]}{dt}\right)^{-1} = \frac{u}{k_{VT}[M]} \ .$$

These relations define the respective scales of the length it takes for the concentration of fluorine atoms and that of excited hydrogen fluoride molecules to decrease by a factor of e.

Diagnostic experiments on the mixing region of the supersonic jets in the HF laser have shown [5.58] that, under the design conditions (equal pressures in the mixing jets) and at a sufficiently small initial boundary layer thickness, this region is divided into three successive zones: a laminar mixing zone, a transition zone, and a zone of well-developed turbulence. Bearing this circumstance in mind and assuming that the basic laminar mixing conditions in the cavity region are satisfied, we may use the following expression [5.96] for the flame front surface coordinate:

$$y_f(x) = \begin{cases} h\sqrt{x/x_D}, & x \leqslant x_D \ , \\ h, & x > x_D \ , \end{cases} \tag{5.45}$$

where h is the jet half-height. According to [5.86], to a first approximation, the diffusion length may be taken to be

$$x_D = \frac{h^2 u}{D} \left(\frac{\varrho_1}{2\varrho_2 f_0} \right)^2 , \qquad (5.46)$$

where $D = CT^{3/2}/p$ is the diffusion coefficient (if the pressure p is measured in mm Hg, $C = 0.22$ for the case of helium diluent and $C = 0.12$ for that of nitrogen diluent), $f_0 = 0.45 + 0.6\log(y_{H_2}/y_F)$, y_F and y_{H_2} are the molar-mass concentrations of the F atoms and H_2 molecules in the starting primary and secondary flows, respectively, and ϱ_1 and ϱ_2 are the total mass densities of the same flows.

Formula (5.45) is based on the obvious relation which corresponds to the laminar mixing conditions and relates the mixing depth y to the mixing time Δt: $y \sim (D\Delta t)^{1/2}$. Since $\Delta t \sim x/u$, the width of the layer in which the two flows undergo laminar mixing grows in proportion to the square root of the distance. The factor defined by the second bracketed term in (5.46) takes account of the effect of the differences in density and the ratio of the fluorine atom and hydrogen molecule concentrations to the mixing length between the jets.

In making estimates of the length scales by formulas (5.44) and (5.46), we will study the case of helium diluent. This case is most interesting because, as we have already seen in Sect. 5.3, dilution with helium (compared to dilution with nitrogen) provides for better energy characteristics of the laser. Obviously this circumstance is primarily associated with the low molecular weight of helium. Moreover, the use of helium provides for faster flow rates of the active medium through the cavity region (because $u \sim (T_0/W_1)^{1/2}$, where T_0 is the stagnation temperature and W_1 the molecular weight of the primary gas flow) and hence wider lasing zones. The latter circumstance is very important from the standpoint of improving the laser beam divergence (the diffraction-limited beam divergence is of the order of $\lambda/\Delta x_l$, where λ is the laser wavelength and Δx_l the lasing zone width).

The initial concentration of helium diluent controls the stagnation temperature, molar composition, and relative concentrations of the end products in the gas generator. Under typical conditions, the stagnation temperature $T_0 \simeq 1500$–2000 K, and the ratio between the active center and diluent concentrations, $\alpha = [F]_1 : [He]_1 \simeq 1 : 5$. In that case, after expanding the primary gas flow through supersonic nozzles, the flow velocity at the exit from the supersonic channel reaches a value of $u_1 \simeq 3.5$ km/s, and the gas temperature $T_1 \simeq 300$–400 K.

With the molar flow rates of the primary flow components being specified, the choice of the molar flow rate of hydrogen (secondary flow) is largely governed by the conditions in the mixing zone. Under the design conditions ($p_1 = p_2$) the particle number densities and temperatures of the primary and secondary flows are related by the relation

$$N_1/N_2 = T_2/T_1 ,$$

where $N_1 = [F]_1 + [He]$ and $N_2 = [H_2]_2$. Hence it follows that

$$\frac{[F]_1}{[H_2]_2} = \frac{\alpha}{1+\alpha} \frac{T_2}{T_1} .$$

The ratio between the molar flow rates of F atoms and H_2 molecules is given by

$$\frac{G_F}{G_{H_2}} = \frac{u_1 S_1 [F]_1}{u_2 S_2 [H_2]_2} \simeq \frac{S_1 [F]_1}{S_2 [H_2]_2} ,$$

where S_1 and S_2 are the cross-sectional areas of the primary and secondary flows, respectively, and for simplicity, u_1 is taken to be approximately equal to u_2. Under the design conditions, we have

$$\frac{G_F}{G_{H_2}} = \frac{S_1}{S_2} \frac{T_2}{T_1} \frac{\alpha}{1+\alpha} .$$

Putting $S_1/S_2 \simeq 4$, $T_2/T_1 \simeq 1/4$ and $\alpha \simeq 1/5$, which is usually the case in HF CWCL's, we have for the fluorine-hydrogen molar flow ratio $G_F/G_{H_2} = 1/6$. As seen, the molar flow rate of H_2 (relative to that of F) greatly exceeds the stoichiometric value.

Then, taking from [5.78] the rate constants k_{chem} and k_{VT} (HF–HF) to be equal to $3 \times 10^{-10} \exp(-805/T)$ and $1.66 \times 10^{-8} T^{-1.43}$ cm^3/s, respectively, we get from (5.44) and (5.45) the following characteristic scale parameter estimates for typical conditions ($p \simeq 5$ mm Hg, $T \simeq 300$ K, $u \simeq 3.5 \times 10^5$ cm/s, $h \simeq 0.2$ cm, and $[F]:[He]:[H_2] = 1:5:6$):

$$x_{chem} \simeq 0.2 \text{ cm}, \quad x_{VT} \simeq 5 \text{ cm}, \text{ and } x_D \simeq 15\text{–}20 \text{ cm} .$$

As can be seen, the reaction zone length is much smaller than the relaxation and diffusion lengths, the relaxation length being in turn smaller than the diffusion length: $x_{chem} \ll x_{VT} < x_D$.

In the general case, diffusive transfer processes are more involved than simple laminar mixing. The character of mixing depends in a great measure on the initial gasdynamic flow parameters and mixing conditions. A number of factors can be indicated, appreciably complicating the picture of mixing: differences in pressure and velocity between the flows undergoing mixing, the presence of jet boundary layers and clearances between the jets, and intensive heat release as a result of chemical processes, the presence of shocks, and so on.

The experimental studies carried out by *Varwig* [5.54] and *Shackleford* et al. [5.58] have demonstrated that the supersonic HF laser may operate under conditions giving rise to a turbulent transfer mechanism. In that case, we have the following empirical relation [5.75, 96] for the mixing front coordinate:

$$y_f(x) = hx/x_{D'}$$

where $h/x_D = C \simeq 10^{-1}$. Depending on conditions, the constant C was found [5.54, 58] to range between 0.07 and 0.14. For these C values and at $h \simeq 0.2$ cm, we get the estimate

$$x_D \simeq 1.5\text{–}3 \text{ cm} .$$

Thus, under turbulent diffusion conditions at a typical pressure of $p \simeq 5$ mm Hg, the condition $x_D \leqslant x_{VT}$ may be satisfied. It should be noted that for each mixing regime, there exists a critical flow pressure p_{cr} for which the length scale of diffusion becomes equal to that of $V \to T$ relaxation ($x_D = x_{VT}$). If $p > p_{cr}$, then $x_D > x_{VT}$, but if $p < p_{cr}$, then $x_D < x_{VT}$. Using the above estimates, one can easily find that $p_{cr} \simeq 2.5$ mm Hg for laminar mixing and $p_{cr} \simeq 10$–15 mm Hg for turbulent mixing.

5.4.3 Mixture-Pressure Dependence of Laser Energy Characteristics

The qualitative behavior features of the energy characteristics of the supersonic-diffusion chemical laser can be understood without specifying the mixing mechanism. For simplicity, the subsequent analysis will be performed within the scope of the isothermal model; also, the gas flow pressure and velocity are taken to be constant. In that case, it is convenient to represent the set of kinetic equations (5.41) in the form

$$\frac{d}{dx}(n_i y_f) = n_{0i} \frac{dy_f}{dx} + y_f \left(\frac{dn}{dx}\right)_i , \tag{5.47}$$

where n_i are expressed in particles/cm^3 and $(dn/dx)_i = (1/u)(dn/dt)_i$.

According to (5.47), the variations of the number densities of the active centers, n_i, and of the reaction products, N, in the HF laser operating on the non-chain reaction $F + H_2 \to HF^* + H$ are described by the equations

$$\frac{d}{dx} N y_f = N_0 \frac{dy_f}{dx} - \frac{d}{dx} n_F y_f = y_f \frac{dN}{dx} ,$$

$$\frac{dN}{dx} = \frac{n_F}{x_{chem}}, \frac{1}{x_{chem}} = k_{chem} n_{H_2} \frac{1}{u} , \tag{5.48}$$

where N_0 is the initial number density of fluorine atoms.

The equations for the number densities of the HF molecules in the various vibrational states v have the form

$$\frac{d}{dx} n_v y_f = y_f [Q_{chem}(v) + Q_{chem}(v) + Q_{VT}(v) + Q_{rad}(v)] ,$$

$$\sum_{n=0}^{m} n_v = N , \tag{5.49}$$

where Q_{VV} allows for the $V \to V$ quantum exchange processes between the HF molecules, and Q_{VT} represents the $V \to T$ relaxation process; for the chemical term $Q_{chem}(V)$ and the radiative term $Q_{rad}(v)$, we may write

$$Q_{chem}(v) = \alpha_v \frac{dN}{dx}, \tag{5.50}$$

$$Q_{rad}(v) = \frac{1}{u}(g_{v+1,v} I_{v+1,v} - g_{v,v-1} I_{v,v-1}) ,$$

where α_v is the probability of population of the vth level in an elementary reaction event, and the intensities $I_{v+1,v}$ are measured in quanta/cm^2 s. In the case of the P-branch transitions, the expressions for the gains $g_{v+1,v}$ have the form

$$g_{v+1,v} = \sigma_{v+1,v}(n_{v+1} - \beta n_v) ,$$

$$\beta = \exp(-2J\theta_{\text{rot}}/T) , \tag{5.51}$$

where $\sigma_{v+1,v} = \sigma_{v,J+1}^{v+1}$ is the cross section of the stimulated transition $v+1$, $J \to v$, $J+1$, J is the rotational quantum number, and θ_{rot} the characteristic rotational temperature of the HF molecule. It is assumed that the rotational levels are in equilibrium with the translational degrees of freedom and that the line broadening is of the Doppler character.

As in Sect. 5.2.2, let us introduce into our analysis the quantities $E = \Sigma_{v=0}^{m} v n_v$ for the density of vibrational quanta stored in the mode v of the molecule and $\varepsilon_{\text{chem}} = \Sigma_{v=0}^{m} v \alpha_v$ for the average number of excitation quanta produced as a result of the chemical pumping reaction (for an elementary $F + H_2$ reaction event, $m = 3$ and $\varepsilon_{\text{chem}} \simeq 2.1$). If the relationship between the $V \to T$ relaxation probabilities is specified by the harmonic oscillator law $P_{v,v-1} = v P_{10}$ (this is not of major importance in our analysis), it is not very difficult to get from (5.49) the energy balance equation in the form

$$\frac{d}{dx} y_f E = y_f \left[\varepsilon_{\text{chem}} \frac{dN}{dx} - \frac{E - E_0}{x_{VT}} - \frac{1}{u} \Sigma_{v=1}^{m} g_{v,v-1} I_{v,v-1} \right] , \tag{5.52}$$

where $E_0 = N\varepsilon_0$ is the equilibrium vibrational energy store, ε_0 is the average number of equilibrium quanta per oscillator,

$$x_{VT}^{-1} = u^{-1} \sum_M k_{VT}(M) n_M ,$$

where n_M is the number density of the species M and $k_{VT}(M)$ the respective constant.

To describe the lasing process, it is convenient to use the technique suggested by *Emanuel* and *Whittier* [5.81]. The quasistationary approximation allows for (1) the lasing process involving all the bands $1 \to 0$, $2 \to 1$, ..., $m \to m-1$ (for simplicity, the rotational quantum number J can be considered to be equal for all the bands) and (2) the lasing process starting and finishing simultaneously in all transitions. Thus, under quasistationary lasing conditions, the following m threshold conditions are satisfied simultaneously:

$$g_{v,v-1} = \sigma_{v,v-1}(n_v - \beta n_{v-1}) = g_{\text{thr}}, \quad v = 1, 2, \ldots, m , \tag{5.53}$$

$$g_{\text{thr}} = (1/2ny_f)\ln[1/(r_0 r_L)^{1/2}] , \tag{5.54}$$

where the expression for the threshold gain takes into account the specific features of the supersonic-diffusion laser stemming from the fact that the reaction zones in it fail to fill the cavity volume to the full (along the full vertical extent of the nozzle bank) as would be the case with instantaneous mixing. Note at once that it is precisely these features that govern the behavior of the laser energy characteristics as a function of pressure. In (5.54), r_0 and r_L are the reflectivities of the cavity mirrors.

If one restricts oneself to the calculation of the total radiation intensity $I = \sum_{v=1}^{m} I_{v, v-1}$, it is advisable to reduce the set of threshold conditions (5.53) to a single one, namely, the condition for the density of vibrational quanta stored in the lasing molecules. Multiplying (5.53) by v and then summing, we get

$$E_{thr}(1-\beta) - \beta N + \beta(m+1)n_m = \sum_{v=1}^{m} v(g_{thr}/\sigma_{v, v-1}) . \qquad (5.55)$$

The number density n_m is found by solving the algebraic equations (5.53). Substituting n_m thus found into (5.55), we have

$$E_{thr} = N\varepsilon_J + (g_{thr}/\sigma_{1,0})B_J , \qquad (5.56)$$

where

$$\varepsilon_J = \frac{\beta}{1-\beta}\left[1 - (m+1)\beta^m \frac{1-\beta}{1-\beta^{m+1}}\right], \qquad (5.57)$$

$$B_J = \frac{1}{1-\beta}\left[\sum_{v=1}^{m} v\frac{\sigma_{1,0}}{\sigma_{v, v-1}}\right.$$

$$\left. -\frac{(m+1)\beta^{m+1}}{1-\beta^{m+1}} \sum_{v=1}^{m} \frac{\sigma_{1,0}}{\sigma_{v, v-1}}\left(\frac{1}{\beta^v}-1\right)\right] . \qquad (5.58)$$

Substituting into (5.52) the threshold vibrational quantum density (5.56) and threshold gain (5.54) and taking into consideration (5.48), we write the expression for the intensity I in the form

$$\frac{I}{u} = \frac{1}{g_{thr}}\left[(\varepsilon_{chem} - \varepsilon_J)\frac{dN}{dx} - \frac{E_{thr} - E_0}{x_{VT}}\right] . \qquad (5.59)$$

The power output and chemical efficiency of the laser can be found from (5.59) in the usual fashion, the lasing onset coordinate x_0 being defined by equation (5.52), with the radiative term "switched off", and expression (5.56). The lasing quenching coordinate x_l is found from (5.59) by equating to zero the expression between the square brackets.

Based on expression (5.59), one can establish some general laws governing the behavior of the power characteristics of the supersonic diffusion HF laser.

Note first of all that if the gas flow pressure exceeds the critical value, the position x_m of the radiation intensity maximum and the characteristic lasing region size x_l are obviously controlled by the collisional processes – relaxation and chemical reaction. Since these processes occur by two-body collisions, it follows from dimensional considerations that x_m is proportional to $1/p$ and $x_l \sim 1/p$. For typical conditions, $x_{chem} \ll x_{VT} \ll x_D$, $x_m \sim x_{chem}$, and $x_l \sim x_{VT}$. Taking into account condition (5.56), one can verify that the expression between the square brackets in (5.59) is proportional to p^2, no matter what the mixing mechanism. On the other hand, the quantity g_{thr} in the extreme cross section x_m of the gas flow is proportional to pressure: $g_{thr}(x_m) \sim p$. Thus, the maximum intensity $I_{max} \sim (1/p)p^2 \sim p$. The laser power output $P_l \sim I_{max} x_l$ is independent of pressure. Inasmuch as the chemical energy store is proportional to pressure, the chemical efficiency and specific energy output of the laser are inversely proportional to pressure: $\eta_{chem} \sim \varepsilon_l \sim 1/p$.

It is not very difficult to satisfy oneself that in the case of premixed gases (pulsed regime analog), the functional dependences have the form $I_{max} \sim p^2$, $P_l \sim p$, and $\eta_{chem} \sim \varepsilon_l \sim p^0$, i.e., the behavior of the laser characteristics is in this case qualitatively different. The physical interpretation of this difference is as follows. In the supersonic-diffusion laser where the deactivation rate is high, lasing is quenched before the oxidizer and fuel flows are mixed completely. As pressure is raised, this tendency becomes more and more manifest: the distance it takes for the active medium excited as a result of the chemical pumping reaction to lose its energy becomes shorter and shorter compared to the mixing length, which reduces the utilization of the starting reactants.

The situation considered above is typical of the supersonic-diffusion laser operating on the non-chain reaction $F + H_2 \rightarrow HF^* + H$, in which the mixing process is one of the chief factors limiting the laser energy.

In this connection, it is of interest to inquire into the ultimate capabilities of this type of laser. Taking into consideration expressions (5.48) and (5.56), we write the intensity relation (5.59) in the form

$$\frac{I g^0_{thr}}{u N_0} = (\varepsilon_{chem} - \varepsilon_J) \left[\frac{d(y_f/h)}{dx} - \frac{d}{dx} \frac{n_F}{N_0} \frac{y_f}{h} \right]$$

$$- \frac{1}{x_{VT}} \left[(\varepsilon_J - \varepsilon_0) \frac{N}{N_0} \frac{y_f}{h} + \frac{g^0_{thr} B_J}{\sigma_{1,0} N_0} \right] , \qquad (5.60)$$

where

$$g^0_{thr} = \frac{1}{2nh} \ln \left(\frac{1}{\sqrt{r_0 r_L}} \right) = \frac{1}{L} \ln \left(\frac{1}{\sqrt{r_0 r_L}} \right) .$$

It is obvious that the case of boosted chemical pumping reaction ($x_{chem} \lesssim x_{VT}$) favors the achievement of the maximum laser output. In the limit $x_{chem}/x_{VT} \rightarrow 0$, expression (5.60) assumes the form

$$\frac{I g^0_{thr}}{u N_0} = (\varepsilon_{chem} - \varepsilon_J) \frac{d}{dx} \frac{y_f}{h}$$

$$- \frac{1}{x_{VT}} \left(\frac{y_f}{h} \varepsilon_J + \frac{g^0_{thr} B_J}{\sigma_{1,0} N_0} \right) , \qquad (5.61)$$

where the equilibrium quanta store ($\varepsilon_0 \ll \varepsilon_J$) is disregarded. It should be noted that for the flow region $x > x_D$ where $y_f/h = 1$, the expression on the right-hand side of (5.61) is negative. This circumstance is due to the approximation $\mu = x_{chem}/x_{VT} \rightarrow 0$ used here. Physically this means that when $\mu \rightarrow 0$, the photon contribution from the region $x > x_D$ is small and can be neglected. In that case, the lasing region should be taken to be limited to the diffusion length x_D.

Now we will study a situation where the lasing region width is smaller than the mixing length $(x_l < x_D)$, i.e., where the starting reactants are obviously not utilized completely.

It follows from (5.61) that in conditions of a low lasing threshold $(g_{thr}^0 B_J / \sigma_{1,0} N_0 \to 0)$, the lasing region width x_l is defined by the equation

$$\frac{1}{y_f} \frac{dy_f}{dx} = \frac{1}{x_{VT}} \frac{\varepsilon_J}{\varepsilon_{chem} - \varepsilon_J} . \tag{5.62}$$

Restricting ourselves to the analysis of the laminar mixing regime and taking into consideration (5.45), we obtain from (5.62)

$$x_l = \frac{x_{VT}}{2} \left(\frac{\varepsilon_{chem}}{\varepsilon_J} - 1 \right) . \tag{5.63}$$

Using (5.61), we get the following expression for the laser power per unit nozzle bank exit face area in conditions of a low lasing threshold:

$$P_l \left(\frac{W}{cm^2} \right) = \hbar \omega N_0 u \left[(\varepsilon_{chem} - \varepsilon_J) \frac{y_f(x_l)}{h} - \frac{\varepsilon_J}{x_{VT}} \int_0^{x_l} \frac{y_f}{h} dx \right] , \tag{5.64}$$

where account is taken of the fact that the lasing onset cross section is close to the zero cross section $(x_0 \simeq 0)$, the mirror losses are taken to be $a_0 = a_L = 0$, the power outcoupling is assumed to be effected through a single mirror, and $\hbar \omega$ denotes the laser energy quantum. Considering (5.63) and (5.45), we have from (5.64) for laminar mixing

$$P_l = \tfrac{2}{3} \hbar \omega N_0 u (\varepsilon_{chem} - \varepsilon_J) \sqrt{x_l / x_D} . \tag{5.65}$$

The chemical efficiency of the laser is defined as

$$\eta_{chem} = \frac{P_l}{Q N_0 u} = \frac{2}{3} \frac{\hbar \omega}{Q} (\varepsilon_{chem} - \varepsilon_J) \sqrt{x_l / x_D} , \tag{5.66}$$

where Q is the heat of the pumping reaction. The supersonic-diffusion laser reaches its ultimate efficiency at $x_l = x_D$, which under laminar mixing conditions comes, as follows from (5.66), to 65% of the ultimate efficiency in the instantaneous mixing regime (i.e., when the condition $x_D \ll x_{chem} \ll x_{VT}$ is satisfied).

5.4.4 Factors Limiting the Energy Performance of the "Cold"-Reaction HF Laser

As has been already noted in Sect. 5.3, the overwhelming majority of publications on continuous-wave supersonic HF lasers consider the excitation of the lasing molecules by the non-chain reaction $F + H_2 \to HF(v) + H$. Since the heat of this process, $(-\Delta H)_1 = 32$ kcal/mole, is considerably less than that of the hydrogen fluorination chain reaction, $(-\Delta H)_1 + (-\Delta H)_2 = 130$ kcal/mole, the HF laser is in this case said to be operating on the "cold" reaction.

Hence the natural question: is the lasing regime in the continuous-wave "cold"-reaction HF chemical laser optimal and what are the prospects for the development of the systems of this class? This question is treated in Sect. 5.5. To conclude this section, we will briefly formulate the basic difficulties limiting the energy performance of the non-chain-reaction continuous-wave HF laser and discuss some possible ways to overcome them.

1) As noted earlier, the fast mixing of the reactants is of major importance. In the case of the "cold"-reaction HF laser, this matter acquires particular significance because the rate of deactivation of the vibrationally excited HF* molecules (mainly as a result of HF*–HF, HF*–F, and HF*–H collisions) is fairly high. In conditions of relatively slow mixing, this means that a material proportion of the F atoms produced is not used at all, and the energy characteristics of the laser thus deteriorate. A natural way to improve the mixing process is to use supersonic channels with small transverse dimensions.

2) The second problem stems from the low operating pressures of the laser. Increased pressures reduce the chemical efficiency of the laser, whereas reduced pressures in the laser cavity region decrease the power output per unit nozzle bank exit face area and considerably complicate the problem of exhausting the waste gases. The latter problem is simplified under supersonic flow conditions characterized by large Mach numbers M, which is tried to be realized in practice [5.61–63]. Large Mach numbers require higher nozzle-area ratios, so that the necessary pressure in the cavity region is ensured by raising the pressure in the gas generator.

3) In operational "cold"-reaction continuous-wave HF lasers, a high temperature ($\simeq 1500$–2300 K) is maintained in the gas generator in order to keep the F atom concentration at an elevated level. This results in the overall efficiency of the laser (which takes account of the heat loss in the generator) remaining low (1–1.5%), despite its high chemical efficiency. Hence the problem of improving the overall laser efficiency and, in the final analysis, fuel economy.

4) The optimum fuel composition in the gas generator is essential to the self-contained supersonic HF laser version. From among the wide choice of the possible fuel mixtures, use should be made of those which burn to produce products which only relatively weakly deactivate the lasing molecules, or of those which have better operational properties while providing the same energy performance of the laser as the others.

5) The high deactivation rates of the excited lasing molecules make the lasing region relatively small (typically around 5 cm). On the other hand, the forced chemical excitation rates give rise to high laser radiation intensities. Thus, the optical system of a high-power laser is subjected to substantial radiation loads, which places heavy engineering requirements on its mirrors. This problem can be solved by devising laser oscillator-amplifier systems [5.110, 111, 113, 116], for the continuous-wave amplification regime, being essentially mirrorless, in a sense eliminates the question of optics. What is more, the laser oscillator-amplifier system makes it possible to control the output emission spectrum of the laser and thus

enhance its efficiency. This possibility has been considered in detail in Chap. 2. Here we would like to mention only the method suggested by *Oraevsky* et al. [5.111] for improving the chemical efficiency of the laser.

Let the temperature of the gas mixture in the HF master oscillator be higher than that in the amplifier. In that case, the frequency spectrum "forced upon" the amplifier by the oscillator is shifted towards the region of higher rotational quantum numbers J not coincident with the maximum amplification region, which, as demonstrated in Chap. 2, makes it possible to raise the CW amplifier efficiency under saturation conditions. In this connection, it is advisable to use in the master oscillator either less diluted active mixtures or a chain-reaction excitation mechanism [5.111]. The specific features of the chain-reaction continuous-wave HF laser are considered in Sect. 5.5.

5.5 Prospects for the Development of the DF–CO₂ and HF CWCL's

The development of high-power chemical lasers largely depends on the effective solution of the exhaust system problem. In solving this problem, we are faced with the following alternatives. On the one hand, conditions should be created in the cavity region that would make it possible to raise the exhaust flow pressure up to the atmospheric value by means of a diffuser. This would radically simplify the exhaust system. On the other hand, it is necessary that the laser under such conditions should not lose its chief advantage – high specific energy characteristics.

In the case of DF–CO₂ laser (Sect. 5.2), the problem of pressure restoration up to the atmospheric value is difficult to solve because of the low (subsonic) flow velocities. In the supersonic HF lasers (Sects. 5.3, 4), the difficulties are due to the low operating pressures in the cavity region.[4]

5.5.1 Supersonic DF–CO₂ Laser

The optimum pressure in the subsonic DF–CO₂ laser operating on the $F_2 + D_2 + CO_2 + He$ mixture ranges between 15 and 30 mm Hg. This pressure in a subsonic flow is obviously too low to be restored to the atmospheric value. The problem can be solved in a supersonic laser system.

Supersonic flow velocities make it possible to raise the hydrodynamic pressure of the flow significantly, and thus allow one to be less exacting with regard to the exhaust system. It will be recalled that the hydrodynamic pressure of a gas is the product of the gas density by the square of its flow velocity, i.e., $P_{hd} = \varrho u^2$. The formula relating the Mach number M to the static pressures in the flow, p, and at the exit from the exhaust system, p_{ex}, has the form

$$p(1 + \gamma M^2) = p_{ex} \ ,$$

---

[4]As we have seen, the pressure in the cavity region is limited mainly because of the efficiency of mixing dropping with the rising pressure.

where $\gamma = c_p/c_v$ is the specific heat ratio. This relation complies with the law of conservation of momentum and holds true for a flow of constant cross-sectional area, skin friction effects being disregarded. Putting, for example, $p_{ex} \simeq 760$ mm Hg, $M \simeq 5$, and $\gamma \simeq 1.5$, we have $p \simeq 20$ mm Hg for the minimum static pressure in the flow that can still be restored to the atmospheric value. It follows from the above relation that even at small Mach numbers the hydrodynamic pressure of a flow may be tens of times the static pressure. It should be noted that it is possible to raise the pressure in the cavity region of the $DF-CO_2$ laser because the mixing of the reactants in it can, in principle, be effected on a time scale shorter than the chain pumping reaction time. The relatively slow rates of chain process (5.3) are due to the comparatively low number densities of the active centers, but the latter circumstance does not result in a low power extraction, the laser chain being well-developed.

One of the possible schemes for the self-contained supersonic $DF-CO_2$ laser was proposed by *Cool* [5.13] and *Basov* et al. [5.27]. A specific feature of this scheme is that the burning fuel in the gas generator yields carbon dioxide, i.e., the lasing molecules are produced in the course of the process that ensures the heating of the gas mixture in the generator. The fuel may be carbon monoxide (CO) or coal dust (C). The fuel in the generator burns in the atmosphere of oxygen; the reaction gives rise to the CO_2 molecules, and the mixture is intensely heated in the process. As usual, the gas temperature is controlled by varying the pressure of the helium diluent injected into the gas generator. The hot $CO_2 + He$ mixture is then mixed with fluorine, which causes a partial dissociation of the latter. Farther downstream the mixture is expanded through a supersonic nozzle at the exit from which D_2 is injected into the flow so that the mixture entering the cavity region downstream of the nozzle has the composition $F + F_2 + D_2 + CO_2 + He$. Taking the stagnation parameters to be $p_0 \simeq 15$ atm. and $T_0 \simeq 1400$ K and the nozzle-area ratio, $f \simeq 12-15$, the gas pressure and temperature in the cavity will be $\simeq 0.1$ atm. and $\simeq 300$ K, respectively. Using a diffuser set up downstream of the cavity, the static pressure in the flow can be raised to the atmospheric value.

The numerical calculations made by *Pimenov* and *Shcheglov* [5.29] demonstrated that the pressure in the cavity region of the supersonic $DF-CO_2$ laser could be raised to some 0.1 atm. without any noticeable deterioration of the efficiency and specific energy characteristics of the laser, as compared to those of its subsonic counterpart. With the pressure in the cavity region so raised, the saturation intensity increased 20-fold, reaching $1-3$ kW/cm², and the specific power per unit nozzle exit area, 40-fold, reaching around 0.5 kW/cm². The calculations were performed for the $F:F_2:D_2:CO_2:He = 0.014:1:1:8:14$ mixture, and the cross-sectional area of the flow channel in the cavity region was taken to be constant. With the conditions at the upstream end of the cavity zone being $p = 60$ mm Hg, $T = 300$ K, and $u = 1.5$ km/s, the active region width amounted to 7 cm, and lasing was quenched at a 50% degree of conversion of the reactants. At the downstream end of the active region, the gas temperature was found to rise to 750 K, the gas pressure to grow to some 170 mm Hg, and the Mach number to decrease to around 2.

To find out whether the quenching of lasing resulted from the heating of the mixture, they calculated the effects of operation under exothermic reaction conditions for the laser reported by *Cool* [5.13]. The calculation showed that the elimination of the temperature increase in the laser cavity region could not be expected to cause any perceptible improvement of the laser performance. According to [5.13], lasing is quenched at about a 60% degree of conversion of the reactants, when the deactivation of the [00^{0}1] state in CO$_2$ by collision with DF becomes a dominant factor. The supersonic regime possesses a number of potentialities but is difficult to realize. The heating of the gas flow by the heat of the reaction may give rise to a shock wave in the lasing region and thus substantially impair the optical properties of the active medium. As far as this aspect is concerned, it would be desirable to limit the temperature of the medium.

Another difficulty is associated with the production in the gas flow of COF$_2$ molecules which strongly absorb radiation at a wavelength of 10.6 μm. As reported in [5.13], the coefficient of absorption of the CO$_2$ laser emission in the $P(20)$ rotational-vibrational transition by the COF$_2$ molecule equals 2×10^{-3} cm^{-1} (mm Hg)$^{-1}$. *Henrici* et al. [5.125] have demonstrated that in the presence of F or F$_2$, any excess CO in the flow is rapidly converted to COF$_2$. The formation of COF$_2$ can be prevented not only by maintaining the balance between CO and O$_2$, but also by providing an individual fluorine injector.

To date, the self-contained supersonic DF–CO$_2$ laser version has been tested by a number of investigators.

The small-scale setup reported by *Cool* [5.13] yielded more than 0.5 kW of laser power at a pressure of 25 mm Hg in the cavity, the chemical efficiency coming to some 3%.

Stregack and *Watt* [5.16] reported a laser system similar in design to the ordinary gasdynamic lasers. The reaction between CO and O$_2$ was effected in the presence of methane serving as a catalyst. The conditions in the gas generator were controlled by injecting cold nitrogen and/or carbon dioxide, and the fluorine-helium mixture was injected downstream of the ignition zone. The stagnation pressure varied between 14 and 20 atm., the mixture temperature, between 950 and 1450 K, and the supersonic nozzles ensured flow velocities with Mach numbers up to $M \simeq 4.5$. The exhaust system was open to the atmosphere.

The above authors [5.16] noted the degrading effect of the COF$_2$ molecules produced in the flow on the laser emission in the case of incomplete oxidation of CO, but failed to present the laser energy characteristics themselves.

The self-contained supersonic DF–CO$_2$ laser reported by *Tregay* et al. [5.18] operated at a stagnation pressure of 7–21 atm. The pressure at the upstream end of the cavity region reached 56 mm Hg. The power output ranged between 6 and 10 kW, the corresponding laser power per unit nozzle bank exit face area ranging between 240 and 400 W/cm^2. These energy characteristics agree well with the theoretical predictions [5.29].

Evers et al. [5.17] demonstrated a modified supersonic DF–CO$_2$ laser version in which the initiation source was the reaction of D$_2$ with excess F$_2$. The reaction took place in the gas generator, and cold CO$_2$ was injected into the flow downstream. The laser generated around 3 kW of power at a specific energy output of $\varepsilon_l \simeq 90$ J/g (850 joules per gram of fluorine), the nozzle bank exit face area amounting to 2.54×7.6 cm^2 and the gas pressure at the exit from the bank to 13 mm Hg. What is striking about this laser is its appreciably higher specific energy output compared to the traditional subsonic $DF-CO_2$ counterpart [5.4, 10, 12, 13]. It will be recalled that the specific energy output of the latter is around 50 J/g.

Unfortunately, the data reported in [5.17] are insufficient for one to explain this fact unambiguously. One can only make one's observations in favor of the laser reported in [5.17]. Obviously the higher the chemical efficiency of the laser, the higher its specific energy output: $\varepsilon_l = \varepsilon_0 \eta_{chem}$ (ε_0 is the specific chemical energy store). The efficiency of a continuous-wave chemical laser depends on both gasdynamic factors (mixing processes in particular) and kinetic factors. What is more, the choice of an adequate optical system is important. According to (5.21), an increase in the chemical efficiency of the laser may be attributed to physically obvious causes: a lowering of the oscillation threshold, an increase in the ratio between the rate of transfer of vibrational quanta from DF to CO_2 and that of deactivation of the excited DF molecules, and an increase in the ratio between the rate of chain reaction (5.3) and that of deactivation of the excited CO_2 molecules.

Emanuel et al. [5.15] reported the development of a supersonic $DF-CO_2$ laser operating at increased gas pressures. As distinct from the laser reported by *Tregay* et al. [5.14], this machine had two specific features: (1) to produce a supersonic flow in the cavity region, use was made of a single nozzle instead of a large number of small-sized nozzles and (2) the gas flow temperature in the gas generator was at room value, i.e., no additional heating of the gas was effected to produce atomic fluorine.

In the gas generator, F_2, CO_2, and He were mixed at room temperature. To eliminate large-scale turbulent fluctuations, the flow of the mixed gases was passed through a sufficiently fine wire gauze. The secondary flow was injected in the nozzle throat plane through an arry of tubes, each with a number of holes 0.24 mm in diameter. The nozzle-area ratio of 1.14 ensured a Mach number of $M = 1.5$, the nozzle exit area being 1.488×4.445 cm^2. The cavity was formed by two 10×10 cm^2 water-cooled gold-coated mirrors. The mass flows of the active mixture components were 5.29 g/s F_2, 57.8 g/s CO_2, 15.1 g/s He, 1.61 g/s D_2, and 1.53 g/s NO. The gas pressure was 631 mm Hg in the gas generator and 233 mm Hg in the cavity. The power output was $P \simeq 1.45$ kW at a chemical efficiency of 1.9%. No research on optimization was performed.

5.5.2 HF CWCL with Chain Excitation Mechanism

Progress in the development of pulsed hydrogen halide chemical lasers (see Chap. 4) was mainly due to the creation and improvement of the HF laser with the chain-type excitation mechanism

a) $F + H_2 \overset{k_1}{\rightarrow} HF(v) + H$, $(-\Delta H)_1 = 31.7$ kcal/mole ,

b) $H + F_2 \overset{k_2}{\rightarrow} HF(v) + F$, $(-\Delta H)_2 = 97.9$ kcal/mole .

This mechanism ensures the high specific energy characteristics of pulsed hydrogen-fluoride lasers. Judging by the available publications [5.69–66], continuous-wave chain-reaction HF lasers so far devised are not as effective; nevertheless, it is believed that further progress in the field of continuous-wave HF chemical lasers depends precisely on the development of the chain-reaction devices [5.68, 108, 109, 126–128].

The material below is concerned with some principal features of the continuous-wave chain-reaction HF laser and its energy capabilities, and also the engineering difficulties involved in the realization of the chain excitation mechanism.

The chain pumping process (a–b) is attractive first of all because the amount of potential chemical energy released in its course in the form of the vibrational energy of excited molecules exceeds that in the case of the non-chain ("cold") pumping reaction (a), the "hot" reaction (b) being energetically dominant.

Since the proportion of the vibrational energy of the products of the chain link (a) amounts to 72% of the heat of reaction $(-\Delta H)_1$, and that of the chain link (b), 52% of the heat of reaction $(-\Delta H)_2$, the vibrational energy equals 23 kcal/mole in the former case and 51 kcal/mole in the latter. Thus, the energy localized in the vibrations of the HF molecules in the course of the "hot" reaction is more than twice that of the "cold" reaction. At the same time, the energy deposited in the translational degrees of freedom of the products in the former case is more than five times that in the latter. These factors are responsible for the special requirements placed upon the design of the continuous-wave chain-reaction HF laser. These requirements stem from the need for an adequate control of the gasdynamic and kinetic processes evolving in the laser cavity region under conditions of increased energy release [5.68, 109, 126–128].

It follows directly from the calculations performed by *Krutova* and co-workers [5.109] within the framework of the quasiunidimensional model (see Sect. 5.4) that the spectral composition of the laser emission depends on the initial degree of dissociation of molecular fluorine, α_F. In the case of "cold"-reaction lasers (molecular fluorine is completely dissociated, $\alpha_F = 1$), the contribution to the laser power comes from the following three bands of the HF molecule: $1 \rightarrow 0$, $2 \rightarrow 1$, and $3 \rightarrow 2$, the major part (around 90%) being accounted for by the vibrational-rotational transitions of the two lower bands $1 \rightarrow 0$ and $2 \rightarrow 1$. As the degree of dissociation of F_2 is reduced, the laser emission spectrum becomes more and more rich, the "hot" chain link (b) becoming more and more manifest. At $\alpha_F \simeq 10^{-2}$, all transitions up to $7 \rightarrow 6$ take part in oscillation [5.109].

With the pressure at the upstream end of the cavity region remaining unchanged, the reduction of the degree of dissociation of fluorine increases the lasing zone width. This situation (Fig. 5.17) is apparently due to a decrease in the pumping reaction rate, which causes the characteristic reaction length x_{chem} to increase. In the case of the chain excitation mechanism, the scale parameter $x_{chem} = u\tau_{chem}$ can easily be estimated considering that the rate of the chain reaction is determined by the product of the rate constant of its slow link by the number density of the active centers $(\tau_{chem}^{-1} = k_2[F])$. Putting $k_2 = 2 \times 10^{-10} \exp(-1210/T) \, cm^3/s$ [5.76], $p \simeq 20$ mm Hg, $T \simeq 200$ K, $u \simeq 2 \times 10^5$ cm/s, $[F_2]:[H_2]:[He] = 1:1:10$, and $\alpha_F = 5\%$, we get the estimate $x_{chem} \simeq 50$ cm. This is comparable with the laminar diffusion length and significantly exceeds the turbulent diffusion length (see Sect. 5.4). In the case of the "cold" reaction ($\alpha_F = 1$) at $p = 20$ mm Hg, $T = 200$ K, and a

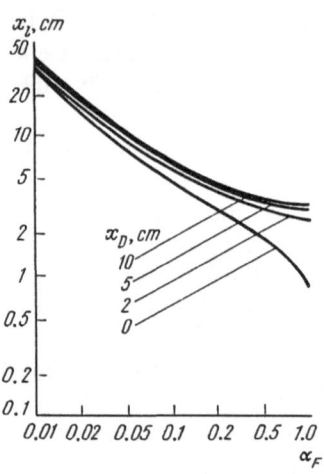

Fig. 5.17. Lasing zone width x_l as a function of the degree of dissociation of fluorine, α_F, for various mixing lengths x_D. (F $+F_2$): $H_2:He = 1.5:1.5:17$; $p = 20$ mm Hg; $T = 200$ K; $g_{thr} = 10^{-3}$ cm^{-1}; x_l corresponds to 95% of the maximum power output

typical molar composition of $[F]:[He]:[H_2] = 1:5:6$, the characteristic pumping reaction length (see (5.44)) is $x_{chem} = 0.15$ cm (the flow velocity u is taken at around 3.5×10^5 cm/s).

It should be noted that in contrast to the "cold"-reaction HF laser in which the active region width is determined by the relaxation length x_{VT} of the excited molecules (Sect. 5.4), the active region width in the case of the chain excitation mechanism depends on the characteristic reaction length.

Two characteristic regions can be singled out in the mixing zone of the continuous-wave chain-reaction HF laser. One of them (pre-reaction region) is adjacent to the exit face of the nozzle bank. This region is characterized by a slow evolution of the reaction, which is associated with low active-center concentrations and gas flow temperatures. The longer the distance it takes for the primary gas flow (fluorine plus diluent) to mix completely with the secondary flow (hydrogen), the greater the pre-reaction region size. Once the mixture has warmed up sufficiently, the reaction proceeds violently. It is precisely in this section of the supersonic flow that the bulk of the active medium is involved in lasing. The process is accompanied by an intensive heat liberation and heating of the mixture. Calculations [5.109] show that with the degree of dilution, $\beta_{He} = [He]/[F_2]$, being around 10, the mixture temperature rises to some 1000–1200 K towards the downstream end of the lasing zone. As can be seen from Fig. 5.17, the lasing zone width in the case of the chain pumping reaction is 10–50 times that in the case of its "cold" counterpart.

Figure 5.18 shows the specific laser energy output as a function of the initial degree of dissociation of fluorine ($\alpha_F = 0.01–1$) for various mixing lengths ($x_D = 0–10$ cm). The curves [5.126] were plotted under the following initial conditions: pressure $p = 20$ mm Hg, temperature $T = 200$ K, threshold gain $g_{thr} = 10^{-3}$ cm^{-1}, and molar mixture composition $(F + F_2):H_2:He = 1.5:1.5:17$. Calculations were performed for a Fabry-Perot cavity with mirrors, the length of which along the flow direction corresponded to 95% of the maximum possible output power level.

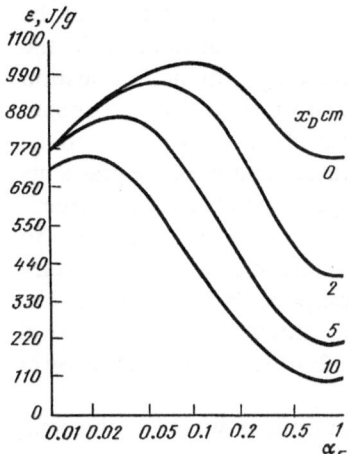

Fig. 5.18. Specific laser energy output ε_l as a function of the degree of dissociation of fluorine, α_F, for various mixing lengths x_D

As one would expect, the calculations showed the factor of mixing to have a substantial effect on the energy characteristics of the laser over a wide range of initial degrees of dissociation of fluorine (except for the region of low α_F at which the mixing length is shorter than the characteristic chemical reaction length). For example, at $\alpha_F = 0.25$, the specific laser energy output ε_l is equal to 880 J/g in the case of instantaneous mixing ($x_D = 0$) and 160 J/g at $x_D = 10$ cm, i.e., the value of ε_l in the latter case is approximately one-fifth that in the former. At lower degrees of dissociation of fluorine, the contrast is not so great: at $\alpha_F = 0.05$, $\varepsilon_l \simeq 1000$ J/g in the case of instantaneous mixing and $\varepsilon_l \simeq 600$ J/g at $x_D = 10$ cm.

It follows from the calculations that there exists an optimum degree of dissociation, α_F, depending on x_D, at which the maximum specific parameters of the laser are realized. For example, at $x_D = 10$ cm, the optimum $\alpha_F \simeq 0.02$. In that case, the specific laser energy output $\varepsilon_l \simeq 760$ J/g, which is about 7 times that in the case of "cold" reaction ($\alpha_F \simeq 1$).

A simple explanation can be provided for the existence of an optimum α_F. In the case of complete fluorine dissociation, the excitation energy is provided by the "cold" reaction alone. As the degree of dissociation of fluorine decreases, the "hot" link of the chain pumping process gains in importance, leading to an increase in the specific laser energy output. On the other hand, when α_F is too low, the chain reaction rate is appreciably reduced, and so is the specific energy output of the laser.

Comparison between the non-chain- and chain-reaction HF CWCL's shows that the latter type possesses some potential advantages. First, the greater store of chemical energy in the chain-reaction laser can be expected to provide a substantial increase in the specific laser energy output. Second, the chain-reaction regime permits higher static pressures in the cavity region, the reasons being the same as in the case of the DF–CO₂ laser. And third, the realization of chain-reaction excitation (i.e, low degrees of dissociation of fluorine) entails a reduction in the expenditure of energy in the gas generator, hence an increase in the overall efficiency of the laser.

The development of an efficient chain-reaction continuous-wave HF laser involves certain difficulties stemming from the kinetic and gasdynamic factors considered above. These difficulties are due to (1) a fairly strong dependence of the specific laser parameters on the initial degree of dissociation of fluorine in the neighborhood of its optimum value (or on the temperature T_{cham} in the combustion chamber) and (2) the sensitivity to T_{cham} of the cavity axis location and mirror size.

What is more, one should bear in mind the substantial chain-reaction heat in the cavity region, which may give rise to increased temperature and pressure gradients leading to the chocking (thermal blockage) of the supersonic channel and quenching of lasing [5.109]. Special measures need to be taken to avoid this effect. In particular, the oxidizer nozzles should be spaced well apart to create the necessary conditions for the supersonic flow of the reacting components to expand in the cavity region. The use of diluent jets in the secondary flows also helps to avoid thermal blockage. And finally, there is a method [5.99, 129–132] involving the use of a cylindrical nozzle bank.

Figure 5.19 is a schematic diagram of an annular continuous-wave HF chemical laser using a cylindrical nozzle bank. The latter comprises a large array of elementary coaxial annular nozzles from which alternate jets of a He-diluted oxidizer (F, F_2) and fuel (H_2) issue radially. Laser radiation propagates along the z-axis of the nozzle bank cylinder in a cavity formed by annular mirrors, the radiation structure being annular as well.

The lasing process in the annular continuous-wave HF chemical laser was analyzed in detail by *Stepanov* and *Shcheglov* [5.129–132] on the basis of the solution of a complete set of the Navier–Stokes equations supplemented with the equations of chemical, vibrational, and radiative kinetics. Both the self-contained and externally excited laser versions were analyzed.

Let us present, without going into details, the main results obtained by the above authors: (1) a certain annular laser geometry makes it possible to solve

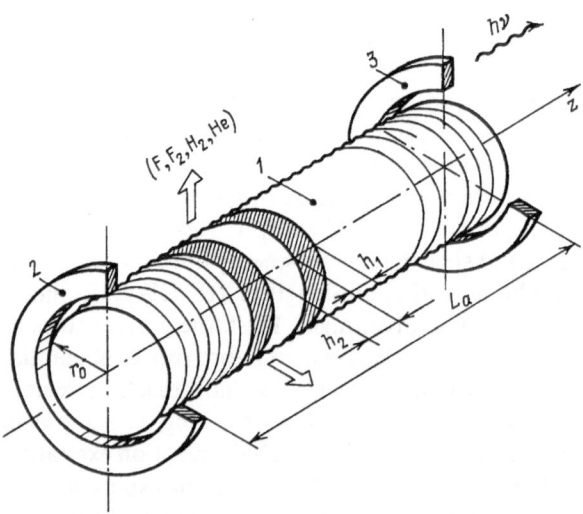

Fig. 5.19. Schematic diagram of an annular continuous–wave HF chemical laser: *1* – cylindrical nozzle bank with alternate oxidizer (h_1) and fuel (h_2) disc nozzles; *2 and 3* – cavity mirrors

the thermal blockage problem effectively, (2) certain operating conditions of the annular laser provide for fairly high chemical efficiencies ($\simeq 10\%$), specific energy outputs ($\simeq 1$ J/g), and power outputs per unit nozzle bank exit face area ($\simeq 1$ kW/cm^2), (3) the use of the chain-reaction excitation mechanism in the annular laser enables one to operate the laser under higher static pressure (up to 15–20 mm Hg) in the cavity compared to those in the non-chain-reaction laser, and (4) the chain-reaction annular continuous-wave HF chemical laser has an appreciably higher overall efficiency (up to 5–8%) compared to its ordinary counterparts. It is thus obvious that the annular type of laser holds much promise.

Note that in the case of low degrees of dissociation of fluorine, special consideration should be given to the question of the optimum mixing method. In particular, the mixing of the chemically active jets can be effected in the throat area of the oxidizer nozzle (provided the pressure is high enough) or somewhat farther downstream. It should be noted that mixing in the throat region of the nozzle was first employed by *Kroshko* et al. [5.133] in a thermally driven gasdynamic N$_2$–CO$_2$ laser, which later on made it possible to develop thermally driven gasdynamic devices [5.134–136] with fairly high specific energy outputs.

In the case under consideration, one can vary the nozzle length or the gas pressure to create conditions in which the flow components at the exit from the nozzle will be mixed well enough, yet remain largely unconverted. This is possible because at sufficiently low degrees of dissociation of fluorine the chain pumping process in the pre-reaction region proceeds at a relatively slow pace, while the high (supersonic) velocities of the gas mixture flow enable it to carry the reactants out of the nozzle region rapidly.

The first successful experiments on the realization of the chain excitation mechanism in continuous-wave HF (DF) lasers were reported by *Cummings* and *Dube* [5.64], *Meinzer* and *Steele* [5.65], and *Sadowski* et al. [5.66]. The most comprehensive information is contained in [5.66]. In this work, the degree of dissociation of fluorine in the gas generator was controlled by an external source, which at the same time eliminated the presence of by-products inhibiting the lasing process. Measurements demonstrated that the laser emission spectrum contained vibrational-rotational lines of the $5 \rightarrow 4$ and $4 \rightarrow 3$ bands, which unambiguously pointed to the participation of the "hot" chain link in the excitation of the HF molecules, hence the realization of the chain pumping regime. The gas pressure at the upstream end of the cavity region was 20 mm Hg and the lasing zone was 23 cm long. The laser power obtained in the experiment amounted to 3.9 kW at a specific energy output of 120 J/g which was equivalent to 330 joules per gram of fluorine (or 4000 joules per gram of the atomic fluorine available).

The laser energy characteristics achieved in [5.66] are worse than predicted by the theory. Nevertheless, bearing in mind the technical difficulties involved in the realization of the optimum chain-reaction pumping regime, the first results obtained on the chain-reaction continuous-wave lasers should be considered reassuring.

5.6 Other Types of CWCL's

The continuous-wave chemical lasers considered in Sects. 5.2–5 have been developed most: the majority of the publications available on CWCL's is devoted to these types of lasers. This is explained first of all by the fact that the power outputs and efficiencies achieved with these lasers are on the scale of immediate practical interest.[5] But in the literature, there are also publications concerning other types of CWCL's using various methods of production of chemically active centers and various reactants.

5.6.1 Hydrogen-Halide Lasers

Airey and *McKay* [5.43] were among the first to report on a supersonic-diffusion HF laser pumped by the reaction

$$F + HCl \rightarrow HF^* + Cl \ ,$$

which used a shock tube for initiation (thermal diffusion of F_2). The shock-tube method of obtaining chemically active centers has not gained wide recognition. The shock tube, however, is convenient for testing new laser mixtures [5.138, 139], for it allows for a simple enough simulation [5.14] and investigation [5.67] of conditions typical of supersonic CWCL's employing conventional initiation techniques.

A number of investigators [5.1, 41, 144–157] reported on subsonic CWCL's operating on hydrogen or deuterium halides, HX or DX (X = F, Cl, Br). In these works, the initial concentration of the active centers was produced by dissociation of various reactants which were flowed through the region of an electrical discharge (glow, microwave or some other type). Initiation in the case of the HF (DF) lasers was usually effected in a flow of F_2 or SF_6 molecules so that F atoms were produced and excitation was provided by the reactions (5.38) proceeding against the background of mixing of subsonic flows (Fig. 5.20).

The continuous-wave chemical lasers of this class can be used for both practical and research purposes; accordingly, their development goes in two directions.

First, work is under way on increasing the power output of subsonic systems [5.141, 143, 149–152]. *Rosen* et al. [5.151] and *Glaze* and *Linford* [5.152] reported power outputs between 5 and 10 W from small-size HF lasers. The exhaust systems used in these lasers allowed the spent gases to be removed at a rate of the order of 200–300 l/s. The subsonic HF laser reported by *Proch* and co-workers [5.150] generated 40 W of power at an engineering efficiency of some 1.5%. The gas-dynamic duct of this laser measured 40×1.5 cm^2. The pumping rate was 1250 l/s at a gas pressure of 3 mm Hg in the cavity region, the average flow velocity being around 2×10^4 cm/s. It should be noted that, though the subsonic laser design avoids some engineering complications associated with supersonic laser systems,

---

[5] *Mayer* et al. [5.137] reported on the use of a supersonic HF laser for hydrogen isotope separation.

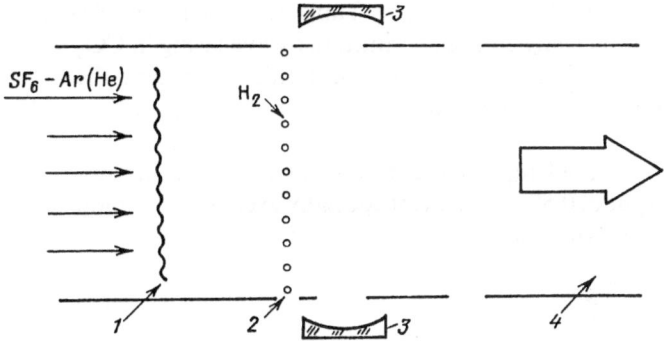

Fig. 5.20. Schematic diagram of an externally initiated subsonic HF laser: *1* – discharge region; *2* – H$_2$ injectors; *3* – cavity mirrors; *4* – exhaust system

the scaling of the subsonic lasers is to a large measure limited by the capabilities of the exhaust system.

Second, new subsonic lasers are being developed which are suitable for laboratory research purposes – gain measurements, determination of kinetic constants, laser tuning, etc. Obviously the power output of the laser is in this case not so important as those characteristics which make it a convenient and reliable research tool: small size, low gas flow rates, frequency selection and tuning capabilities, and so on.

At present, subsonic laser models have already been created satisfying the above requirements. In particular, *Hinchen* [5.149] reported operating an HF laser system having the following specifications: flow cross-sectional area 10×0.3 cm^2, average flow velocity 5×10^3 cm/s, molar flow rates (mole/s) of components $(SF_6 + He):H_2:He = (2.4 + 5.8):0.8:0.4$, gas pressure in the cavity region 10–15 mm Hg, pumping rate 14 l/s. The cavity was formed by a reflection grating and a semitransparent mirror in combination with a piezoelectric crystal. The author managed to isolate the desired vibrational-rotational line from the initial spectrum of the $1 \rightarrow 0$ and $2 \rightarrow 1$ bands of HF at a power level of 0.2–0.3 W by varying the electrical discharge power and active mixture composition, as well as the optical system parameters, and select a single longitudinal mode by varying the cavity length. The laser is distinguished by a high amplitude stability ($\pm 1.5\%$) at a frequency scanning rate of 25 MHz/s. The frequency tuning interval (within the spectral line width) is of the order of 300 MHz.

Note that the above laser was successfully used for measuring gain [5.149, 154] and vibrational and rotational relaxation rate constants of molecules [5.154] (see also [5.155, 156] for information on similar measurements with the use of subsonic HF and DF lasers).

The papers by *Arnold* et al. [5.157–160] are among the most interesting works on subsonic hydrogen-halide chemical lasers.

Arnold et al. [5.157] created a purely chemical HCl laser with a power output of 13 W and a chemical efficiency of 8%. The laser depends for its operation on the reaction of ClO$_2$ with NO to yield atomic chlorine which then reacts with HI to produce vibrationally excited HCl* molecules (see Table 5.8, discussed below for the kinetic scheme). The authors also achieved continuous-wave lasing in CO_2 with a power output of 5 W and a chemical efficiency of 4%. In this experiment, the CO_2 molecules were excited as a result of vibrational energy transfer from the HCl* molecules.

Arnold and co-workers [5.158] reported a purely chemical continuous-wave HBr laser with a power output of 0.58 W. The vibrationally excited HBr* molecules were produced in accordance with the scheme

$$2NO + ClO_2 \rightarrow 2NO_2 + Cl \ ,$$

$$Cl + Br_2 \rightarrow BrCl + Br \ ,$$

$$Br + HI \rightarrow HBr^* + I \ .$$

The experiments in [5.157–160] were conducted at low gas pressures (2–4 mm Hg in the hydrogen-halide lasers and around 10 mm Hg in the HCl–CO_2 laser). *Arnold* et al. [5.161] were able to achieve the supersonic regime of the laser.

5.6.2 Continuous-Wave Chemical CO Lasers

As in the case of the pulsed CO laser, pumping here takes place in the course of burning of a $CS_2 + O_2$ mixture. Of great interest are the works where gain or lasing was achieved in a freely burning flame, i.e., where the purely chemical laser version was realized [5.162–167]. Figure 5.21 shows a schematic diagram of a free flame combustion-driven laser. *Foster* and *Kimbell* [5.165] detected population inversion in vibrational-rotational transitions of CO while burning carbon disulfide in air in a low-temperature flame at low pressures. Lasing in CO was reported for the first time by *Djeu* et al. [5.162, 163], who observed a ladder of vibrational-rotational transitions in the $v \rightarrow v - 1$ ($v = 8, \ldots, 11$) bands of CO.

Additional research on the combustion-driven chemical CO laser was carried out by *Linevsky* and *Carabetta* [5.166] and *Foster* et al. [5.167]. The use of a

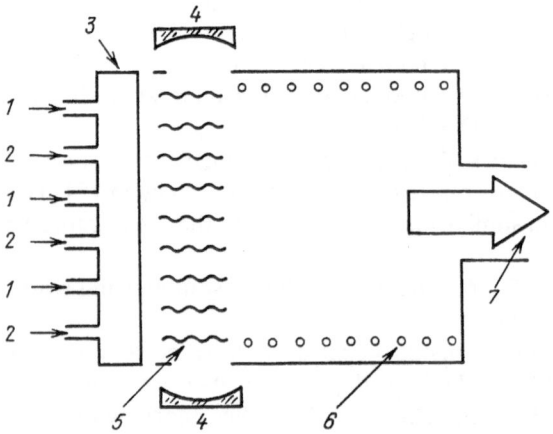

Fig. 5.21. Schematic diagram of a free–flame combustion–driven CO laser: *1*–CS_2 flow; *2*–O_2 flow; *3*–honeycomb mixing system; *4*–cavity mirrors; *5*–$CS_2 + O_2$ mixture ignition region; *6*–cooling system tubes; *7*–to exhaust

honeycomb injector system allowed the authors of [5.166] to achieve a power output of 25 W at a chemical efficiency of some 2.5% and a specific energy output of around 6 J/g. The experiment was conducted at a pressure of 30 mm Hg in a gas flow $60 \times 1.25 \text{ cm}^2$ in cross-sectional area. In addition to CS_2 and O_2, the mixture contained N_2O in a relatively high concentration.

In a great number of works, the continuous-wave chemical CO laser was initiated by means of a microwave or ordinary electrical discharge [5.168–170, 172–177, 179–181], arc discharge [5.171], resistance heater [5.178], and shock tube [5.140]. Under typical experimental conditions, the active centers (oxygen atoms) were preliminarily produced in an $O_2 + He$ flow, and the active components were then mixed in a subsonic flow. The kinetic mechanism responsible for the formation of the active medium in the CO laser is as yet not clearly understood [5.182]. *Hancock* and co-workers [5.183, 184] have established that one of the key processes giving rise to excited CO* molecules is the reaction between O atoms and CS radicals. The most likely scheme for the principal processes is as follows:

$$1) \quad O + CS_2 \xrightarrow{k_1} CS + SO \;,$$

$$2) \quad O + CS \xrightarrow{k_2} CO^* + S \quad (-\varDelta H = 75 \text{ kcal/mole}) \;,$$

$$3) \quad S + O \xrightarrow{k_3} SO + O \;.$$

It is obvious that the more effective the use of the oxygen atoms in the overall chain process (1–3), the higher the efficiency of conversion of the potential chemical energy of the CS_2–O_2 mixture into the laser energy. As can be seen from the above scheme, the oxygen atoms not only perform the usual functions of the active centers in the evolution of the chain process, but also play another part, namely, participate in the production of an intermediate "fuel" component, the CS radicals, for reaction (2). This circumstance, along with chain termination processes, leads to a decrease in the atomic oxygen concentration, hence in the number of the excited CO* molecules per oxygen atom produced by the external initiation source. Thus, generally speaking, for the laser-chemical process to be effective here, the atomic oxygen concentration should be maintained at a sufficiently high level. This object can, in principle, be achieved without increasing the initiation source power by injecting into the gas flow some easy-to-dissociate oxygen-bearing additives.

In the laser reported by *Jeffers* and *Wiswall* [5.175], increased atomic oxygen concentrations were produced in an $O_2 + He$ mixture by a microwave discharge controlled by a magnetron (power 1 kW, frequency 2450 MHz). This made it possible to obtain around 30 W of laser power at a chemical efficiency of some 20% and specific energy output of about 65 J/g. The experimental studies carried out by these authors showed that the laser power output depended on the presence in the active mixture of a number of chemically neutral components. In particular, N_2O was found to be capable of improving the power performance of the laser considerably, whereas NO or COS impaired it.

Despite the fact that good specific performance characteristics were achieved in the above laser [5.175], the effectiveness of atomic oxygen utilization in it was low because the kinetic chain length v under the given experimental conditions did not exceed unity. Hence the question: How could each oxygen atom be made repeatedly to take part in the pumping reaction (2)? The answer was found by *Jeffers* and *Ageno* [5.178] who solved the problem by introducing into the mixture increased concentrations of CS radicals. The radicals were formed by dissociation of carbon disulfide in a flow passed through a carbon resistor oven $(T = 2400°C)$. The schematic diagram of this laser is presented in Fig. 5.22.

If the concentration of CS appreciably exceeds that of CS_2, the role of reaction (1), which in essence has formerly controlled the entire chemical cycle, becomes unimportant so that the "center of gravity" of the cycle is transferred to the chain reaction (2–3). The authors of [5.178] demonstrated experimentally that at $[CS]/[CS_2] = 2$ the chain length increased to 8.5. This method made it possible to raise the specific energy output of the laser to 150 J/g, its power output amounting to some 85 W. Table 5.6 lists some additional data on this laser. For comparison, the table also presents information about the laser reported in [5.175].

The results obtained in [5.178] are the record for systems of this class, the specific laser energy output achieved being of the same order of magnitude as in the case of supersonic HF (DF) lasers. The development of self-contained CO machines in which the necessary active mixture components are produced in the course of supplementary chemical reactions would be a step of principal importance in the progress of these systems.

The problems associated with the realization of a supersonic flow in the chemical CO laser and its operation at increased pressures are also very important, and efforts are now being undertaken to solve them.

Bashkin et al. [5.140] and *Boedeker* et al. [5.171] reported on the creation of supersonic continuous-wave chemical CO_2 lasers. The laser reported in [5.140] used shock-wave initiation, while that reported in [5.171] was based on an ordinary plasma-generator scheme (Sect. 5.3) using an arc heater and Ar diluent. The power output of the latter device was 35 W.

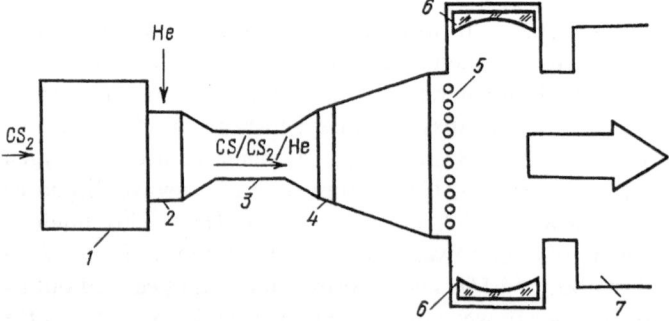

Fig. 5.22. Schematic diagram of the continuous–wave chain–reaction CO chemical laser [5.178]: *1*–carbon resistor oven; *2*–ejector nozzle bank; *3*–supersonic diffuser; *4*–annular injector; *5*–atomic oxygen injectors; *6*–cavity mirrors; *7*–exhaust system

5.6.3 CWCL's with Energy Transfer to Polyatomic Molecules

This group includes first of all the systems based on hydrogen halides and carbon dioxide, i.e., HX (DX)–CO_2 lasers (X = F, Cl, Br), [5.7, 8, 142]. The DF–CO_2 laser considered in Sects. 5.2, 5 stands out among these systems for a number of its kinetic and energy characteristics. Pulsed hydrogen halide-carbon dioxide lasers have been considered in Chaps. 3 and 4.

Questions related to the extension of the set of chemical reactions that can be used to pump suitable lasing molecules were already being discussed in the first works on chemical lasers. In particular, *Dzhidzhoyev* et al. [5.185] considered the possibility of using reactions yielding vibrationally excited CO* molecules with a view to subsequently transferring vibrational quanta from CO* to CO_2, N_2O, COS, HCN, and C_2N_2 which possess vibrational levels close to those of CO. Unfortunately, in most of the potentially suitable reactions, the CO* molecules are produced simultaneously with H_2O, an inevitable product of all oxidation reactions involving hydrogen-bearing substances, and the H_2O molecules rapidly quench vibrations in practically any compound.

The above authors also predicted the possibility of using the carbon disulfide oxidation reaction as a source of excited CO* whose excitation energy could be transferred to CO_2, N_2O, and the like. This possibility was realized by *Petersen* and *Wittig* [5.186, 187] who reported subsonic CO–CO_2 and CO–N_2O lasers depending for their operation on the reaction of CS_2 with O_2.

Bashkin and co-workers [5.138] experimentally tested the OD–CO_2 system. The experiments were conducted with the use of a shock tube to effect thermal dissociation of D_2. The flow carrying the partially dissociated deuterium was expanded through a supersonic channel at the exit of which ozone, O_3, was injected into the flow. The process of excitation of CO_2 molecules is as follows:

$$D + O_2 \rightarrow OD^* + O_2 \ ,$$

$$OD^* + CO_2 \rightarrow OD + CO_2^* \ .$$

These authors registered a gain of around 2×10^{-2} cm^{-1} (unsaturated regime) in CO_2. The power output was a mere 1.5 W.

Later on [5.188], to improve the mixing of the gas flows and hence the laser characteristics, they used an ejector similar to the one employed earlier by *Kroshko* et al. [5.133–135]. Compared to the earlier design, the ejector had an increased ejecting-to-ejected flow area ratio, the injection point of the ejecting gas was moved closer to the nozzle throat, and a single ejector was replaced by an ejector matrix. All these measures increased the mixing rate and improved the homogeneity of the flow. The gain measured under various conditions varied over the range (0.9 to 2.1) $\times 10^{-2}$ cm^{-1}. The maximum gain was registered under the following conditions: the initial pressure of the D_2 : Ar = 1 : 15 mixture in the low-pressure channel of the shock tube was equal to 0.095 atm., the initial pressure of the O_3 : CO_2 : He = 1 : 3 : 30 mixture in the valve was 3.25 atm., and the incident shock wave velocity was 1.43 km/s. The maximum power output was 36 W.

The operation of a laser using the $D(H) + O_3 + CO_2$ mixture has been quite recently studied in detail by *Bashkin* and co-workers [5.189]. The main results obtained by these authors are listed in Table 5.7. These results allow one to draw a number of interesting conclusions. First, the supersonic laser system used provides a sufficiently high specific energy output at a relatively high pressure. It is hoped that this will make it possible to exhaust the flow directly into the atmosphere using only diffusers and no ejectors. Second, despite the comparatively high gas pressure, the lasing region size along the flow direction is fairly large, which, as has been already noted, is essential for narrowing the laser beam divergence and reducing the radiation load on the cavity mirrors. And finally, as demonstrated by experiment, the replacement of deuterium by hydrogen causes no appreciable decrease in the power output of the laser. This is important because deuterium is rather costly compared to hydrogen.

In [5.189], as in [5.138, 188], *Bashkin* and co-workers used a shock tube to obtain atomic deuterium (or hydrogen). To create a purely chemical laser of practical significance, it is necessary to devise an effective method for producing atomic hydrogen. The purely thermal technique used to obtain atomic fluorine is in this case faced with difficulties: perceptible equilibrium concentrations of atomic hydrogen can only be attained at temperatures exceeding 3000 K. But the branching chain reaction of hydrogen with oxygen yields hyperequilibrium concentrations of radicals, hydrogen in the first place [5.190]. Based on the kinetic scheme suggested by *Kondrat'yev* and *Nikitin* [5.191] for the process of burning of hydrogen in oxygen, they analyzed the optimum conditions for the production of atomic hydrogen. Calculations showed this way to hold promise for the solution of the atomic hydrogen problem. Thus, it is felt that in the future it will prove possible to create a purely chemical OH–CO_2 laser with nontoxic combustion products.

Benard et al. [5.192] and *Benson* et al. [5.193] were the first to demonstrate the capabilities of the Na-vapor catalyzed reaction between CO and N_2O. Based on this reaction, they developed a purely chemical N_2O laser. The reaction is a chain process yielding excited products N_2^* and CO_2^*:

$$Na + N_2O \rightarrow NaO + N_2^* \qquad (-\Delta H = 21 \text{ kcal/mole}) ,$$

$$NaO + CO \rightarrow CO_2^* + Na \qquad (-\Delta H = 66.2 \text{ kcal/mole}) .$$

At room temperature, the process proceeds fairly rapidly and is attended by an intense chemiluminescence of the D-line of Na. *Benson* and co-workers [5.193] also successfully tested other alkali metals – K, Cs, and Rb – as a catalyst. Li and Fr can apparently play this part as well. Under typical experimental conditions, the chain length amounted to some 10–30, i.e., this number of CO_2 molecules were produced per sodium atom.

Laser emission was observed only on the $[00^01] \rightarrow [10^00]$ transition in N_2O ($\lambda = 10.8$ μm), the concentration of which was in excess relative to CO. Inasmuch as the vibrational levels of N_2O, CO, N_2, and CO_2 are close to one another, the mechanism of excitation of N_2O may have included the following processes:

$$\left.\begin{array}{l} N_2^* \\ CO_2^* \end{array}\right\} + N_2O \rightarrow \left.\begin{array}{l} N_2 \\ CO_2 \end{array}\right\} + N_2O^* \ ,$$

$$\left.\begin{array}{l} N_2^* \\ CO_2^* \end{array}\right\} + CO + N_2O \rightarrow \left.\begin{array}{l} N_2 \\ CO_2 \end{array}\right\} + CO^* + N_2O \rightarrow \left.\begin{array}{l} N_2 \\ CO_2 \end{array}\right\} + CO + N_2O^* \ .$$

Based on the kinetic scheme suggested by *Derwent* and *Thrush* [5.195] for the excitation of atomic iodine by energy transfer from the singlet oxygen molecule, $O_2(^1\Delta)$, *McDermott* et al. [5.194] demonstrated for the first time the possibility of developing a purely chemical laser operating on the hyperfine-structure transitions in iodine ($\lambda = 1.315$ μm). Later on, when the power output of this laser was brought up to 100 W [5.196], it became clear that a new member of the chemical laser family had made its appearance, with capabilities comparable to those of the HF and DF–CO$_2$ lasers. For this reason, we have devoted an individual chapter (Chap. 6) to the description of the oxygen-iodine laser.

To conclude this section, we present a summary list of CWCL's (Table 5.8).

5.7 Other Possible CWCL's Versions

Today we can state that quite a considerable progress has been made in the field of CWCL's, thanks to the development of highly efficient self-contained systems with high power outputs, primarily the HF (DF), CO, DF–CO$_2$, and O$_2$–I lasers (for detailed information about the O$_2$–I laser, see Chap. 6 and also [5.206, 207]).

The trends in the development of CWCL's are as follows: (1) increasing the efficiency of the existing laser models, (2) extending the spectral range covered by laser emission (in the middle and near infrared and the beginning of the visible region), and (3) improving the optical homogeneity of the active medium and the quality of the output emission. In this connection, we will consider some other possible CWCL versions.

5.7.1 Standing-Detonation-Wave HF Chemical Laser

Speaking about the advantages of the self-contained supersonic HF chemical lasers, one should bear in mind also their shortcomings, due primarily to the gasdynamic effects a consequence of the mixing of a large number of chemically active jets. The interaction of the jets in the laser cavity region gives rise to optical inhomogeneities (boundary layers, weak shocks, dilatation waves) which naturally affect the laser beam divergence. What is more, the finiteness of the jet mixing rate limits the degree of utilization of the oxidizer (atomic fluorine) flow: lasing is quenched before the jets are mixed completely. Obviously to eliminate these shortcomings, one should aim at a CWCL scheme in which the lasing region is spatially separated from the jet mixing region, the system remaining self-contained.

The annular supersonic chain-reaction HF laser considered in Sect. 5.5.2 is close to this scheme. Another approach is based on the use of the initiating reactant technique in conjunction with the "ignition" of a chain reaction on the front of a standing shock wave [5.208, 209]. This method was realized experimentally for the first time by *Moran* and co-workers [5.210, 211]. Schematically, the experiment is as follows. The primary gas flow (F_2–He mixture) issues at room temperature from the chamber of a gas generator and expands in a supersonic channel where it cools in the process. In a particular section of the channel, the secondary gas flow (H_2–NO mixture) is injected into the primary one. Some distance downstream, the cold primary and secondary flows become completely intermixed, but remain unreacted because at low temperatures the "priming" reaction

$$F_2 + NO \xrightarrow{k_0} F + NOF$$

proceeds at an extremely slow rate ($k_0 \simeq 10^{-17}$ cm^3/s at $T \simeq 100$ K). If the supersonic nozzle operates under overexpansion conditions, there can occur in the flow a standing normal shock wave which "ignites" the supplementary reaction, initiating the chain process of excitation of the HF molecules:

$$F + H_2 \rightarrow HF(v) + H \ ,$$

$$H + F_2 \rightarrow HF(v) + F \ .$$

Note that the cavity mirrors are in this case located behind the shock wave front and provision is made for the flow to expand a little in a direction normal to the cavity axis.

L'vov and co-workers [5.208, 209] have analyzed and numerically modelled the amplification, spectral, and energy characteristics of the standing-detonation-wave HF laser. Calculations have shown that to avoid the thermal blockage (choking) of a low-area-ratio supersonic channel due to the large amount of heat liberated in the flow, the mixture should be strongly diluted with an inert gas (helium), which naturally reduces the total chemical energy store of the system.

The calculations have been made for the following initial conditions: the temperature in the gas generator chamber, $T_0 = 300$ K, pressure $p_0 = 50.25$ mm Hg; gas temperature before the shock wave front, $T_1 = 100$ K; mixture composition – variable. The degree of flow expansion in the cavity region has been set in the simplest way: $y(x) = 1 + (x - x_0) \tan \theta$, where 2θ is the flare angle of the supersonic channel in a plane normal to the cavity axis. Under these conditions, the gas pressure at the exit from the nozzle, $p_1 = 2.25$ mm Hg, Mach number $M_1 \simeq 3$; the temperature behind the shock wave front, $T_2 \simeq 270$ K, gas pressure $p_2 \simeq 18.75$ mm Hg, and Mach number $M_2 \simeq 0.45$ (flow velocity $u_2 \simeq 250$ m/s). When calculating the energy parameters of the laser, the threshold gain has been taken to be $g_{thr} = 10^{-3}$ cm^{-1}.

The analysis of the conditions providing for the maximum energy extraction has shown that the optimum concentration of F_2 in the flow is approximately 3–4 times that of H_2, but the effect of the fluorine content of the primary flow on the specific laser energy output is generally weak. The effect of the mixture dilution is much stronger. Thus, reducing the concentration of He in the flow increases the specific laser energy output, but when the dilution $\beta_{He} = $ [He]/[NO] is reduced to a value less than about 30, thermal blockage occurs. For a degree of flow expansion of $\tan \theta = 0.4$ and a molar mixture composition of $F_2 : H_2 : NO : He = 4 : 1 : 1 : 12$, the calculated specific laser energy output is some 300 J/g and chemical efficiency, around 12%. These values are apparently the ultimate for this system.

The analysis of the laser emission spectrum has demonstrated that although lasing is theoretically possible in the transitions of six bands, $1 \rightarrow 0$, $2 \rightarrow 1, \ldots, 6 \rightarrow 5$, the main contribution to the total emission power comes from the transitions of the three lower bands. The unsaturated gain in the $1 \rightarrow 0$ and $2 \rightarrow 1$ bands may reach around 1 cm^{-1} (the most intense lines in the spectrum correspond to the vibrational-rotational transitions with rotational quantum numbers $J = 6$–10). So high a gain (compared to that of the supersonic-diffusion HF laser) is explained by the fact that the gas flow entering the cavity is premixed and the gas pressure in the cavity is several times that in the diffusion laser.

The detonation HF laser considered above is quite attractive, for it requires no high-temperature combustion chamber, and possesses good energy characteristics and a high optical homogeneity of the active medium, thanks to the spatial separation of its mixing and lasing regions. The good amplification properties of the active medium in this laser make it possible to use unstable and "twisted mode" cavities.

Note that when operating a similar laser for 0.75 s under conditions of a slightly underexpanded free jet, *Moran* and co-workers [5.210, 211] obtained an average power output of 55 W, a peak intensity of some 100 W/cm^2, and a chemical efficiency of around 1% at an active zone length of 3.4 cm.

5.7.2 16-μm Flowing-Gas DF–CO$_2$ Chemical Laser

At present, much consideration is being given to gas lasers featuring multiple (cascade) lasing on several successive (cascade) vibrational-rotational transitions in diatomic molecules. Such systems include, for example, electrically, thermally, or chemically driven CO lasers and hydrogen-halide chemical lasers. The systems of this class rely on a partial inversion. In the case of hydrogen-halide lasers, the cascade lasing mechanism results directly from the multiple-level character of excitation, while in the CO lasers, there is also an additional factor associated with the quasiresonant exchange of vibrational quanta.

In recent years, great interest is being shown in the development of high-power CO$_2$ lasers operating on cascade transitions. In these devices, lasing occurs in two successive transitions: either $00^01 \rightarrow 10^00 \rightarrow 01^10$ ($\lambda_1 = 10.4$ μm, $\lambda_2 = 13.9$ μm) or $00^01 \rightarrow 02^00 \rightarrow 01^10$ ($\lambda_1 = 9.4$ μm, $\lambda_2 = 16.2$ μm). In contrast to the lasers operating on diatomic molecules, lasing in this case apparently takes place in conditions of total population inversion. Note that the current interest in a sufficiently powerful laser emission at $\lambda \simeq 16$ μm is stimulated by the need to solve a number of scientific and applied problems (laser spectroscopy, uranium isotope separation, etc.; see, for example, [5.212–215]).

The possibility of realization of a two-frequency cascade lasing in the CO$_2$ laser was demonstrated theoretically for the first time by *Karlov* et al. [5.216] and experimentally by *Manuccia* et al. [5.217] and *Wexler* and *Waynant* [5.218]. A number of questions related to the cascade CO_2 laser were considered by *Kosner* and *Plaasance* [5.219], *Zaroslov* et al. [5.220], *Suzuki* and *Saito* et al. [5.221, 222], *Velikanov* et al. [5.223], and *Biryukov* et al. [5.224–230]. The theoretical works by the latter authors [5.224–230] have been devoted to gas lasers operating on cascade transitions in linear triatomic molecules, consideration being given to the principal features of such lasers, their operation under arbitrary excitation conditions, and methods for calculating their energy characteristics. The authors have calculated the cascade CO_2, N_2O, and CS_2 lasers and suggested concrete ways to improve their efficiency.

It should be emphasized that in contrast to the standard CO_2 lasers (single-frequency emission at $\lambda = 10.4$ μm in the $00^01 \rightarrow 10^00$ transition) which can operate in both pulsed and continuous-wave mode, the cascade lasing mechanism is only operative under pulsed conditions. Indeed, in the former case, continuous-wave lasing is sustained on account of the fast depopulation of the lower vibrational state 10^00 due to the fast redistribution of the molecules among the strongly collisionally bound levels of the symmetric stretching and bending modes of CO_2 (10^00, 02^00, 02^20, 01^10). It is precisely this circumstance that prevents the continuous-wave two-frequency cascade operation of the CO_2 laser, for the molecular redistribution populates the lower level of the low-frequency transition of the cascade, so that the molecular cascade laser is in many respects similar to atomic lasers operating on self-terminating transitions.

The cascade lasing mechanism has the advantages of wider spectral bandwidth and higher quantum yield.

Biryukov and co-workers [5.226] studied the possibility of creating a 16-μm chemically pumped CO_2 laser. Such a laser can rely on exchange processes between vibrationally excited molecules produced in the course of one or other exothermic chemical reaction and CO_2 molecules. These reactions include the processes of formation of hydrogen halides (HF, DF, HCl, HBr), which are typical of chemical lasers, the reaction $CS + O \rightarrow CO^* + S$, and also $D\,(H) + O_3 \rightarrow OD^*(OH^*) + O_2$, $Na + N_2O \rightarrow NaO + N_2^*$ and the like.

The initial object of study was a flowing-gas DF–CO_2 laser operating on a D_2–F_2–CO_2–He mixture. The laser was pumped as usual by the chain hydrogen fluorination reaction, and the vibrational energy of the excited DF^* molecules was then transferred to the asymmetric stretching mode of CO_2. In the course of the chain pumping process, a population inversion is produced in the $00^01 \rightarrow 10^00$ (02^00) transition in CO_2. If the cavity Q-factor is switched on at the moment $t = 0$, the ensuing stimulated transitions may cause overpopulation of the 10^00 (02^00) level with respect to the 01^10 level, and lasing will then be possible on a second, 10^00 (02^00) $\rightarrow 01^10$, transition as well. To obtain 16-μm emission from the $02^00 \rightarrow 01^10$ transition, it is obviously necessary to "block" the $00^01 \rightarrow 10^00$ transition, e.g., by introducing a BCl_3-vapor-filled selective element into the cavity.

It is not very difficult to show that the condition for pulsed two-frequency lasing has the form

$$T_3 > F(T_2) = \theta_3 T_2/(\theta_2 - T_2 \ln 2) \ , \tag{5.67}$$

where T_2 and T_3 are the excitation temperatures of the bending and asymmetric stretching modes of CO_2, respectively, and θ_2 and θ_3 the respective characteristic vibrational temperatures. Note that this condition is more stringent than that for the ordinary single-frequency lasing ($T_3 > \theta_3 T_2/2\theta_2$). In particular, it follows from (5.67) that for $T_2 > \theta_2/\ln 2$, cascade lasing is in principle impossible.

If condition (5.67) is satisfied, the maximum energy densities of the radiation pulses at $\lambda_1 = 16.2$ μm and $\lambda_2 = 9.4$ μm can be estimated by the relations

$$E_1 = N_{CO_2} h\nu_1 \left[\exp(-\theta_3/T_3) - 2\exp(-\theta_2/T_2)\right]/2S ,$$

$$E_2 = N_{CO_2} h\nu_2 \left[3\exp(-\theta_3/T_3) - 2\exp(-\theta_2/T_2)\right]/4S ,$$

where N_{CO_2} and S are the number density and statistical sum of the CO_2 molecules, respectively, and $h\nu_1$ and $h\nu_2$ the energy quanta of the respective transitions.

Actual estimates were made for a self-contained supersonic $DF-CO_2$ laser, the nozzle bank of which consisted as usual of an array of small-size slit nozzles forming alternate primary (F, F_2, CO_2, He) and secondary (D_2, CO_2, He) jets.

The energy parameters of the cascade $DF-CO_2$ laser were estimated by numerically solving a closed set of gasdynamic and chemical and vibrational kinetic equations. For simplicity, an instantaneous mixing model was adopted allowing for a unidimensional approximation. The starting $F_2-D_2-CO_2-He$ mixture composition and degree of dissociation of fluorine, α_F, were selected so as to satisfy the two-frequency lasing condition. It is not very difficult to see that in the situation considered, there exists some threshold value of α_F below which two-frequency cascade lasing is impossible. Obviously this value exceeds the threshold value of α_F for the standard single-frequency lasing. The calculations have shown that a 16 μm pulsed $DF-CO_2$ laser operating on a mixture of typical molar composition ($F_2 : D_2 : CO_2 : He = 1 : 1 : 2 : (30$ to $60)$) at a typical initiation level of $\alpha_F \simeq 1-5\%$ and an initial pressure of $p_0 \simeq 50$ mm Hg and temperature of $T_0 \simeq 250$ K can have the following average characteristics: specific energy $\varepsilon \simeq 0.5-2$ J/g, energy density $E \simeq 10-40$ μJ/cm³, and specific power extraction $P \simeq 0.5-2.5$ W/cm³ (the Q-switching frequency is governed by the time it takes to change the active mixture in the cavity).

5.7.3 Continuous-Wave Optical Resonance Transfer HF Laser

The optical resonance transfer (ORT) HF laser belongs to the optical coherent radiation converters. The necessary population inversion in its active medium is produced in the vibrational-rotational transitions of HF molecules, the rotational quantum numbers of which are greater than those at which the pumping radiation is absorbed. The mechanism responsible for the population inversion is a fast rotational relaxation of the lasing molecules. Note that lasing in the ORT HF laser occurs at frequencies shifted toward the long-wavelength side relative to the resonant pumping radiation.

The expreimental scheme is usually as follows [5.231, 232]. A slow (subsonic) undisturbed flow of hydrogen fluoride strongly diluted with helium is irradiated by a pumping source, as a rule, a pulsed or CW HF chemical laser. For more effective absorption, the pumping radiation is directed into a multiple-passage cavity. To extract the output laser emission, use is made of a second cavity the axis of which can be arranged at an angle to the gas flow direction.

The interest in such systems is due mainly to the fact that they can generate powerful and highly directional radiation, thanks to the high optical homogeneity of their active (re-emitting) medium.

The ORT HF laser has been studied experimentally by *Pummer* et al. [5.231] and *Wang* et al. [5.232] and theoretically by *Yamaguchi* et al. [5.233], *Kwok* and *Wilkins* [5.234], *L'vov* et al. [5.235], *Bashkin* et al. [5.236], and *Bel'dyugin* et al. [5.237]. These investigations have demonstrated that the energy conversion efficiency of the ORT HF laser depends on a number of factors, such as the intensity and spectral composition of the pumping radiation and the pressure, temperature, and HF percentage of the gas flow.

In the case of a continuous-wave ORT HF laser [5.237], if the pumping radiation spectrum contains a single vibrational-rotational line in each of its bands, $1 \rightarrow 0$ and $2 \rightarrow 1$, a higher conversion efficiency is achieved on that line which corresponds to the smaller rotational quantum number J_{pump}. This is explained by the fact that population inversion in the ORT HF laser is formed in vibrational-rotational transitions with greater rotational quantum numbers compared to the pumping radiation, while the rotational relaxation rate of HF decreases with increasing J. For this reason, the high-lying rotational states of the HF molecules are populated comparatively slowly so that the proportion of energy lost by vibrational-translational relaxation grows appreciably higher.

It follows from calculations [5.237] that rather high gains ($\simeq 0.05 \text{ cm}^{-1}$) are attained in several $P(6)$–$P(9)$ vibrational-rotational transitions simultaneously. The gain in transitions with greater J numbers is rather low. This contradicts the results reported in [5.234] where high gain values were observed even in the $P(10)$–$P(11)$ transitions. Note that in [5.234] the rate constants for the rotational relaxation of the excited HF* molecules by collision with unexcited HF molecules and He molecules are taken to be equal, which in the case of strongly diluted mixtures has quite a material effect on the rotational relaxation kinetics, especially for large J numbers.

The calculations in [5.237] were made for typical initial experimental conditions: pressure $p_0 \simeq 50$ mm Hg, temperature $T_0 \simeq 300$ K, flow velocity $u_0 \simeq 50$ m/s, HF concentration $\alpha_{\text{HF}} \simeq 0.3$–1%, and pumping radiation intensity $W_{\text{pump}} = 1 \text{ kW/cm}^2$ (per spectral line in the $1 \rightarrow 0$ and $2 \rightarrow 1$ bands), pumping being effected on the $P(4)$ line. It follows from the calculations that under the above conditions the energy conversion efficiency amounts to some 40%. *L'vov* et al. [5.235] estimated the ultimate conversion efficiency (the case of instantaneous rotational relaxation, $\tau_R/\tau_{VT} \rightarrow 0$) at around 80%. It is qualitatively obvious that the energy conversion coefficient directly depends on the rate of rotational relaxation of HF*, which is the kinetic mechanism responsible for the population of the upper lasing levels. For this reason, to enhance the efficiency of the converter, it is advisable to add to the starting active mixture some buffer gas (e.g., SF_6) increasing the rotational relaxation rate of the HF* molecules.

As the pumping radiation intensity is raised, the energy conversion efficiency increases, the increase being substantial only so long as $W_{pump} \leqslant 0.5 \, \text{kW/cm}^2$. At $W_{pump} \geqslant 1 \, \text{kW/cm}^2$, the conversion efficiency shows little variation.

The active medium of the ORT HF laser requires a strong dilution in order to prevent its intense heating. In the case of strong dilution (less than 1% of HF in the mixture), the active medium temperature varies comparatively little along the gas flow direction, a slight increase of the temperature being even favorable to the improvement of the conversion efficiency. Increasing the HF concentration merely to 3% appreciably raises the temperature and reduces the efficiency.

If the active mixture is strongly diluted with He ($\alpha_{HF} < 1\%$), the rise of the initial gas temperature T_0 to 500 K leads to an increase in the conversion efficiency, the increase being substantial only in the case of single-frequency pumping on spectral lines corresponding to $J_{pump} \geqslant 5$. For weakly diluted mixtures, the influence of the initial gas temperature on the conversion efficiency is insignificant.

Table 5.1. Main parameters of a purely chemical DF-CO_2 laser [5.4, 10, 13]

Power output	160 W
Main lasing transition in CO_2	$P(20)$ $00^00 1 \rightarrow 10^0 0$
Chemical efficiency in terms of the	
$\quad F_2 + D_2 \rightarrow 2DF$ reaction	4.6%
Saturation intensity	140 W/cm^2
Optical system	5-passage cavity
Maximum gain	3.3×10^{-2} cm^{-1}
Average gain (over 5 passages)	2×10^{-2} cm^{-1}
Maximum population density of the $00^0 1$	
$\quad$ level in CO_2	1.3×10^{16} cm^{-3}
Optimum output mirror transmittance	33%
Average flow velocity	200 m/s
Rotational temperature of CO_2 in the reaction region	400 ± 25 K
Static gas pressure in the cavity	15 mm Hg
Optimum reactant flow rates [mmole/s]:	
He	112
CO_2	57
F_2	7.3
D_2	6.5
NO	1.2

Table 5.2. Experimental results obtained with the IRIS-II setup [5.14]

Power [kW]	Pressure [mm Hg]		Molar ratios		
	F atom production channel	Cavity	D_2/F_2	NO/F_2	CO_2/F_2
6	77.2	22.6	1.1	0.45	8.68
5.2	86.5	30.7	0.84	0.39	6.66
7.7	74.7	14.1	1.4	0.45	11.03

Table 5.3. Ratios between gas parameters in supersonic jet and gas generator [5.50]

Heat-carrying agent (diluent)	N_2	He
Nozzle-area ratio A/A^*	15.3	15.3
Mach number M of supersonic jet	4.4	5.9
Ratio between static pressures in jet and gas generator, p/p_0	3.9×10^{-3}	1.8×10^{-3}
Ratio between jet temperature and stagnation temperature, T/T_0	0.2	0.079
Ratio between gas flow velocity and stagnation sonic velocity, u/a_0	1.99	4.8
Ratio between flow velocities of mixtures diluted with helium and nitrogen	2.4	

Table 5.4. Specific energy parameters of HF laser [5.50]

Parameters	Diluent	
	He	N_2
Partial flow densities[1] [g/cm^2 s]		
He or N_2	0.091	0.376
H_2	0.055	0.044
O_2	0.033	0.018
SF_6	0.111	0.08
Total flow density [g/cm^2 s]	0.29	0.518
Specific power [W s/g]	240	95
Chemical efficiency [%]	12	12
Engineering efficiency [%]	3.5	3

[1] Partial flow densities are optimized to obtain maximum specific power and engineering efficiency.

Table 5.5. Nozzle bank parameters [5.61]

Parameters	Oxidizer nozzles	Deuterium nozzles
Number of nozzles, n	36	37
Depth of throat, h^* [mm]	0.07	0.3
Area ratio A/A^*	35	12
Mach number M	7	4
Total exit face area of nozzle bank [cm^2]	2.54×22.86	

Table 5.6. Comparative data on continuous-wave CO chemical lasers reported in [5.175] and [5.178]

Molar flow rates $CS:CS_2:He:O:O_2:N_2O$ [mmole/s]	$O-O_2-CS_2$ laser [5.178] 0.82:0.54:65:0.27:7.2:0	$O-O_2-CS_2$ laser [5.175] 0:0.58:19.9:0.93:5.59:3.11
Cavity pressure [mm Hg]	10.6	5.2
Kinetic chain length v	$\simeq 5$	0.6
Power output P [W]	84	28.8
Specific energy output [J/g]	150	65.6
Chemical efficiency [%]	27	21
Atomic oxygen utilization efficiency $P/\dot{m}_O$ [W s/mmole]	311	31

Table 5.7. Main data on the $D(H)+O_3+CO_2$ laser reported in [5.189]

Parameters	$OD-CO_2$ laser	$OH-CO_2$ laser
Mixture composition	$(O_3+O_2):CO_2:He=1:14:35$	
Lasing region pressure [mm Hg]	15	
Flow velocity [cm/s]	$\simeq 2 \times 10^5$	
Lasing region size along flow direction [cm]	>20	
Power output [W]	960	
Gain [cm^{-1}]	2.65×10^{-2}	2.45×10^{-2}
Specific power output [W s/g]	60	50

Table 5.8. Main types of continuous-wave chemical lasers

Pumping reaction	Operating conditions	Initiation technique	Reference
$H+Cl_2 \rightarrow HCl^*+Cl$ $H+Br_2 \rightarrow HBr^*+Br$ $Cl+HI \rightarrow HCl^*+I$	Mixing of H with Cl_2 or Br_2 and Cl with HI in a flow of very low pressure	Electrical discharge in molecular hydrogen or chlorine flow	[5.1]
$F+H_2 \rightarrow HF^*+H$ $H+F_2 \rightarrow HF^*+F$ $F+HI \rightarrow HF^*+I$ $F+D_2 \rightarrow DF^*+D$ $F+DI \rightarrow DF^*+I$ $H+Cl_2 \rightarrow HCl^*+Cl$ $Cl+HI \rightarrow HCl^*+I$	Mixing of atomic and molecular components in a subsonic flow. Cavity axis aligned with flow	High-frequency discharge in H_2, Cl_2, D_2, F_2, or SF_6 flow	[5.41]
$F+H_2 \rightarrow HF^*+H$ $F+D_2 \rightarrow DF^*+D$	Mixing of atomic components with H_2 or D_2 in a subsonic flow	Electrical discharge in SF_6 flow	[5.144, 143, 149, 151, 153]
$F+H_2 \rightarrow HF^*+H$ $F+D_2 \rightarrow DF^*+D$	Premixing of He, SF_6, and H_2 (D_2)	Magnetic-field stabilized discharge in premixed flow	[5.10, 41]
$F_2+NO \rightarrow NOF+F$ $F+H_2 \rightarrow HF^*+H$ $H+F_2 \rightarrow HF^*+F$	Mixing of NO, F_2, and H_2 (D_2) in a subsonic flow. Cavity axis normal to flow	Purely chemical initiation (priming reactant technique)	[5.10, 41]
$F+H_2 \rightarrow HF^*+H$ $F+D_2 \rightarrow DF^*+D$	Mixing of a subsonic flow of partially dissociated SF_6 and He with a subsonic flow of H_2 (D_2)	Electrical discharge in SF_6 flow	[5.141]
$F+H_2 \rightarrow HF^*+H$ $F+D_2 \rightarrow DF^*+D$ $Cl+HI \rightarrow HCl^*+I$	Mixing of a sonic flow of atomic halogen with secondary flow	Electrical discharge in molecular halogen flow	[5.146, 147]
$Cl+HI \rightarrow HCl^*+I$	Mixing of Cl_2 with HI in a subsonic flow	Microwave discharge in Cl_2+HI flow	[5.148]
$F+HCl \rightarrow HF^*+Cl$	Mixing of atomic fluorine with HCl in a supersonic flow	Shock-wave heating of F_2	[5.43]

Table 5.8 (continued)

Pumping reaction	Operating conditions	Initiation technique	Reference
$F + H_2 \rightarrow HF^* + H$ $H + F_2 \rightarrow HF^* + F$	Mixing of partially dissociated fluorine with hydrogen in a subsonic flow	Electrical discharge in fluorine flow	[5.151, 152]
$F + H_2 \ (D_2)$ $\rightarrow HF^* \ (DF^*) + H \ (D)$	Mixing of F with $H_2 \ (D_2)$ in a supersonic flow. Cavity axis normal to flow	Thermal dissociation of SF_6 by arc discharge to produce F atoms	[5.42, 45, 62]
$F + HBr \rightarrow HF^* + Br$	Mixing of F with HBr in a supersonic flow	Thermal dissociation of SF_6 by arc discharge to produce F atoms	[5.59]
$F + H_2 \ (D_2) \rightarrow$ $\rightarrow HF^* \ (DF^*) + H \ (D)$	Mixing of F with $H_2 \ (D_2)$ in a supersonic flow	Initiation by heat of reaction $F + H_2 \ (D_2) \rightarrow$ 2HF (DF) occurring in combustion chamber	[5.17, 44, 60–63, 198]
$F + H_2 \rightarrow HF^* + H$ $H + F_2 \rightarrow HF^* + F$	Mixing of partially dissociated F_2 with H_2 in a supersonic flow	Initiation by external source	[5.64–66]
$CS_2 + O \rightarrow CS + SO$ $CS + O \rightarrow CO^* + S$	Slow flow of low-pressure $CS_2 + O_2$ mixture	Free burning flame	[5.162–167]
$CS_2 + O \rightarrow CS + SO$ $CS + O \rightarrow CO^* + S$	Mixing of partially dissocidated O_2 with CS_2 in a subsonic flow	Electrical discharge in O_2 flow	[5.168–170, 172–176, 179, 180, 199, 200]
$CS_2 + O \rightarrow CS + SO$ $CS + O \rightarrow CO^* + S$	Mixing of partially dissociated O_2 with CS_2 in a supersonic flow	Arc discharge in Ar followed by mixing with O_2	[5.171]
$CS_2 + O \rightarrow CS + SO$ $CS + O \rightarrow CO^* + S$	Subsonic flow of $CS_2 + O_2$ mixture	Electrical discharge in $CS_2 + O_2$ mixture in cavity region	[5.180, 181]
$CS_2 + O \rightarrow CS + SO$ $CS + O \rightarrow CO^* + S$ $S + O_2 \rightarrow SO + O$	Mixing of $CS + CS_2$ mixture with $O + O_2$ mixture in a subsonic flow	Electrical discharge in CS_2 flow	[5.177]
$CS_2 + O \rightarrow CS + SO$ $CS + O \rightarrow CO^* + S$ $S + O_2 \rightarrow SO + O$	Mixing of $CS + CS_2$ mixture with $O + O_2$ mixture in a subsonic flow	Dissociation of CS_2 in resistance oven	[5.178]
Production of CO^* by reaction of O_2 with a carbon-bearing compound	Slow flow of mixture: $C_3O_2 + O_2 + He$ $He + CH_4 + air$ $He + C_3H_8 + air$ $He + C_2N_2 + air$	Electrical discharge in cavity region in: $C_3O_2 + O_2 + He$ mixture $He + CH_4 + air$ mixture $He + C_3H_8 + air$ mixture $He + C_2N_2 + air$ mixture	[5.201, 202–204] [5.202] [5.205]
$O + CS_2 \rightarrow CS + SO$ $CS + O \rightarrow CO^* + S$	Mixing of partially dissociated O_2 with CS_2 in a supersonic flow	Dissociation of O_2 in a shock tube	[5.140]

Table 5.8 (continued)

Pumping reaction	Operating conditions	Initiation technique	Reference
$N+CS_2 \rightarrow CS+NS$ $NS+N \rightarrow N_2+S$ $S+O_2 \rightarrow SO+O$ $O+CS_2 \rightarrow CS+SO$ $CS+O \rightarrow CO^*+S$	Mixing of N_2 with $(O_2 + CS_2)$ in a slow flow	Electrical discharge in N_2 flow	[5.199]
$Cl+HI \rightarrow HCl^*+I$ $HCl^*+CO_2 \rightarrow HCl+CO_2^*$ $H+Cl \rightarrow HCl^*+Cl$ $HCl^*+CO_2 \rightarrow HCl+CO_2^*$	Subsonic flow mixing of atomic and molecular components	Electrical discharge in Cl_2 or H_2 flow	[5.7]
$H+F_2 \rightarrow HF^*+F$ $HF^*+CO_2 \rightarrow HF+CO_2^*$ $F+D_2 \rightarrow DF^*+D$ $DF^*+CO_2 \rightarrow DF+CO_2^*$	Subsonic flow mixing of atomic and molecular components	Electrical discharge in H_2 or F_2 flow	[5.7, 8]
$H+Br_2 \rightarrow HBr^*+Br$ $HBr^*+CO_2 \rightarrow HBr+CO_2^*$	Subsonic flow mixing of atomic and molecular components	Electrical discharge in H_2 flow	[5.142]
$NO+F_2 \rightarrow NOF+F$ $F+H_2\,(D_2)$ $\quad \rightarrow HF^*\,(DF^*)+H\,(D)$ $H\,(D)+F_2$ $\quad \rightarrow HF^*\,(DF^*)+F$ $HF^*\,(DF^*)+CO_2$ $\quad \rightarrow HF\,(DF)+CO_2^*$	Mixing of NO, F_2, H_2 (D_2), CO_2, and He in a subsonic flow. Cavity axis aligned with or normal to flow	Purely chemical initiation (priming reactant technique)	[5.2–4, 9, 10–14]
$NO+F_2 \rightarrow NOF+F$ $F+D_2+DF^*+D$ $D+F_2 \rightarrow DF^*+F$ $DF^*+CO_2 \rightarrow DF+CO_2^*$	Mixing in a supersonic flow. Heating of primary flow by heat of reaction $NO+F_2 \rightarrow NOF+F$. Helium diluent.	Purely chemical initiation (priming reactant technique)	[5.14]
$NO+F_2 \rightarrow NOF+F$ $F+D_2 \rightarrow DF^*+D$ $D+F_2 \rightarrow DF^*+F$ $DF^*+CO_2 \rightarrow DF+CO_2^*$	No heating in gas generator	Purely chemical initiation (priming reactant technique)	[5.15]
$F+D_2\,(H_2)$ $\quad \rightarrow DF^*(HF^*)+\,D\,(H)$ $D\,(H)+F_2$ $\quad \rightarrow DF^*\,(HF^*)+F$ $DF^*\,(HF^*)+CO_2$ $\quad \rightarrow DF\,(HF)+CO_2^*$	Mixing in a supersonic flow	Initiation by heat of reaction $F_2+D_2\,(H_2) \rightarrow$ 2DF (HF) in combustion chamber	[5.17]
$F+D_2 \rightarrow DF^*+D$ $D+F_2 \rightarrow DF^*+F$ $DF^*+CO_2 \rightarrow DF+CO_2^*$	Supersonic flow mixing. Primary flow: O_2–CO–N_2–CO_2. Methane used as a catalyst in combustion chamber.	Initiation by heat of reaction $CO+(1/2)O_2$ $\rightarrow CO_2$ in combustion chamber	[5.16]

Table 5.8 (continued)

Pumping reaction	Operating conditions	Initiation technique	Reference
$F+D_2 \rightarrow DF^* + D$ $D + F_2 \rightarrow DF^* + F$ $DF^* + CO_2 \rightarrow DF + CO_2^*$	Supersonic flow mixing	Initiation by chemical energy of fuel compositions: $CO-O_2-F_2$, D_2-F_2, $C_6F_6-F_2$, $C_6F_6-NH_3$	[5.18]
$CS_2 + O \rightarrow CS + SO$ $CS + O \rightarrow CO^* + S$ $CO^* + CO_2 \rightarrow CO + CO_2^*$	Mixing of partially dissociated O_2 with (CS_2 $+N_2O$) in a subsonic flow	Electrical discharge in O_2 flow	[5.186, 187]
$Na + N_2O \rightarrow NaO + N_2^*$ $NaO + CO \rightarrow CO_2^* + Na$ $(N_2^*, CO_2^*) + N_2O$ $\rightarrow (N_2, CO_2) + N_2O^*$	Slow flow of $Na-(N_2O,$ CO) mixture	Na-catalyzed combustion (K, Rb, and Cs can be used instead of Na)	[5.192, 193]
$D + O_3 \rightarrow OD^* + O_2$ $OD^* + CO_2 \rightarrow OD + CO_2^*$	Mixing of D atoms with O_3 and CO_2 in a supersonic flow	Heating of $D_2 + Ar$ mixture in a shock tube	[5.138, 188, 189]
$H + O_3 \rightarrow OH^* + O_2$ $OH^* + CO_2 \rightarrow OH + CO_2^*$	Mixing of H atoms with O_3 and CO_2 in a supersonic flow	Heating of $H_2 + Ar$ mixture in a shock tube	[5.189]
$2NO + ClO_2 \rightarrow 2NO_2 + Cl$ $Cl + HI \rightarrow HCl^* + I$	Subsonic flow mixing of NO, ClO_2, and HI	Purely chemical initiation (priming reactant technique)	[5.157, 159, 160]
$2NO + ClO_2 \rightarrow 2NO_2 + Cl$ $Cl + HI \rightarrow HCl^* + I$	Supersonic flow mixing of NO, ClO_2, HI	Purely chemical initiation (priming reactant technique)	[5.161]
$2NO + ClO_2 \rightarrow 2NO_2 + Cl$ $Cl + HI \rightarrow HCl^* + I$ $HCl^* + CO_2 \rightarrow HCl + CO_2^*$	Subsonic flow mixing of NO, ClO_2, HI, and CO_2	Purely chemical initiation	[5.157, 159]
$2NO + ClO_2 \rightarrow 2NO_2 + Cl$ $Cl + Br_2 \rightarrow BrCl + Br$ $Br + HI \rightarrow HBr^* + I$	Subsonic flow mixing of NO, ClO_2, Br_2, and HI	Purely chemical initiation	[5.151]
$O_2\,(^1\Delta) + O_2\,(^1\Delta)$ $\rightarrow O_2\,(^1\Sigma) + O_2\,(^3\Sigma)$ $O_2\,(^1\Sigma) + I_2$ $\rightarrow 2I\,(^2P_{3/2}) + O_2\,(^3\Sigma)$ $O_2\,(^1\Delta) + I\,(^2P_{3/2})$ $\leftrightarrows O_2\,(^3\Sigma) + I\,(^2P_{1/2})$ $I\,(^2P_{1/2}) + O_2\,(^1\Delta)$ $\rightarrow O_2\,(^1\Sigma) + I\,(^2P_{3/2})$	Subsonic flow mixing of $O_2\,(^1\Delta)$ and I_2. Oxygen in $^1\Delta$ state is produced by passing Cl_2 through a concentrated H_2O_2 solution	Purely chemical initiation	[5.194, 196, 197]

6. Oxygen-Iodine Chemical Laser – a New Candidate for Engineering Applications

As far as laser machining and treatment (welding, cutting, surface hardening, drilling, coating, etc.) applications are concerned, the most important laser characteristics are the output power, radiation wavelength, beam divergence and coherence, and the cost of the machine. At present, it is mostly CO_2 and Nd:YAG lasers that are being used for these purposes. The main disadvantages of these lasers are the high operation cost of the Nd:YAG machines with a power output in excess of a few hundred watts and the low efficiency of interaction between metals or alloys and radiation at the CO_2-laser wavelength. The oxygen-iodine chemical laser (OICL) emitting at a wavelength of 1.315 μm may prove to be free from these disadvantages.

The creation of this chemically pumped iodine laser, schematically drawn in Fig. 6.1, is one of the most remarkable recent achievements. The source of energy in the laser is the reaction between the chlorine gas and an alkaline aqueous solution

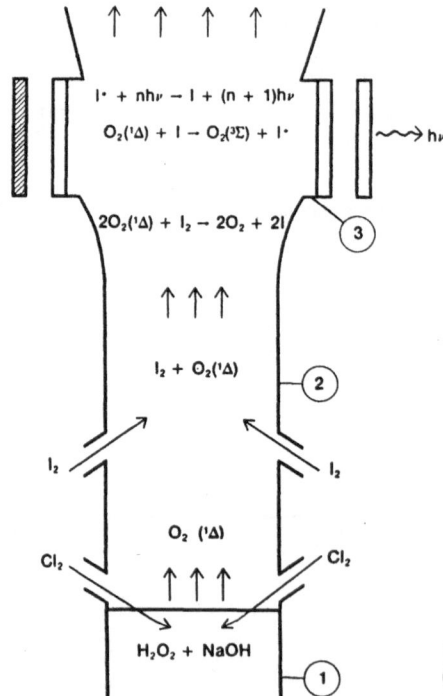

Fig. 6.1. Scheme of the Oxygen Iodine Chemical Laser (OICL): *1*–singlet oxygen generator; *2*–mixing zone; *3*–laser zone

of hydrogen peroxide:

$$H_2O_2 + Cl_2 + 2NaOH \rightarrow O_2(^1\Delta) + 2NaCl + 2H_2O \; . \tag{6.1}$$

The electronically excited singlet oxygen O_2 $(^1\Delta)$ produced in the chemical reactor causes the dissociation of the I_2 molecules and then excites the 1.315 μm lasing transition in the resultant iodine atoms. As the excitation energy here is stored in the singlet oxygen, the laser is usually referred to as the oxygen-iodine chemical laser. The production of population inversion in the OICL owes to the rapid quasiresonant electronic energy exchange between the singlet oxygen and atomic iodine:

$$I(^2P_{3/2}) + O_2(^1\Delta) \leftrightarrow I(^2P_{1/2}) + O_2(^3\Sigma) + \Delta E \; . \tag{6.2}$$

The equilibrium constant of the reaction depends on the gas temperature: $k_e = (3/4) \times \exp(400/T)$. The inversion of the lasing levels in the OICL is possible only if the relative concentration of the singlet oxygen, $\eta_\Delta = [O_2 \, (^1\Delta)]/[O_2]$, exceeds some threshold value, equal to 15% at $T = 300$ K. The main merits of the OICL are the use of inexpensive, readily available starting reactants (hydrogen peroxide, alkali hydroxide, chlorine), the nontoxicity of the reaction products (common salt, oxygen, water), the absence of high-voltage power supplies, and the possibility of using glass instead of salt optics. All this enables one to design less costly and more reliable laser systems. Since the 1.315-μm laser radiation propagates well through light guides, the laser plant can be reliably isolated from the spot where the actual manufacturing operation takes place.

An important part in engineering laser applications is played by the absorption coefficient of the material subject to machining, $1 - R$, where R is the reflectance of the material. At a wavelength of 10.6 μm (CO_2 laser) the value of $1 - R$ is about 0.02 for silver and copper and around 0.05 for steel. In the region of shorter wavelengths, the values of $1 - R$ are appreciably higher than at the CO_2-laser wavelength. In particular, the value of $1 - R$ for steel at $\lambda = 1.3$ μm ranges between 0.3 and 0.35. In practice, this means that welding and heat treatment operations will be much more economical to perform with the oxygen-iodine chemical laser than with the CO_2 laser. The coat of paint or some other material which it is necessary to apply onto the metal surface prior to CO_2 laser machining or treatment does not always prove effective, because of a poor contact between the coating and the metal.

Even the first numerical calculations [6.1, 2] demonstrated that the ultimate specific energy output (per gram of all the reactants consumed) of the oxygen-iodine chemical laser is fairly high (300 J/g). Under actual operating conditions, the maximum specific energy output of the laser reaches around 100 J/g. The laser can operate in continuous-wave, pulsed, and high-repetition-rate modes alike. Because the electronically excited oxygen is metastable, subsonic flow velocities are high enough to change the active medium in the laser cavity. The high-quality optical properties of the active medium ensure narrow beam divergences, which is very important to the laser cutting of metals. The consumption of the starting reactants per kW-hour of the generated laser power is as follows: 5 liters of a 50% aqueous solution of H_2O_2, 70 liters of a 10% aqueous solution of NaOH, and 6.7 kilograms of Cl_2. Hence it follows that a 5 kW OICL operating continuously for 8 hours will consume 0.75 tons of the hydrogen peroxide solution, which is quite acceptable for practical purposes.

It is evident from the foregoing that the oxygen-iodine chemical laser can find wide industrial application where use is now being made of power-intensive Nd:YAG and CO_2 laser systems. In the text below, we will consider the current status and outlook for the development of the OICL.

6.1 Experimental Results

Lasing in a continuous-wave OICL was reported for the first time by *McDermott* et al. [6.3]. Up to now, all the experiments found in the available literature have copied in broad outline the experimental scheme reported by *Benard* et al. [6.4] and *Richardson* and *Wiswall* [6.5].

The singlet oxygen gas is produced in a gas-liquid reactor as a result of the reaction between the chlorine gas and an alkaline solution of hydrogen peroxide:

$$Cl_2 + 2MeOH + H_2O_2 \rightarrow 2MeCl + 2H_2O + O_2(^1\Delta) \ , \tag{6.3}$$

where Me = Na, K, Li. The reaction products MeCl and H_2O remain in solution, whereas the singlet oxygen evolves in the gas phase. The experimental realization of the OICL operating on the above reaction has become possible owing to the development of chemical singlet oxygen generators (SOG) producing singlet oxygen in concentrations above the threshold value. The type of SOG depends on the method of effecting the gas-liquid contact. Such apparatus are widely used in chemical engineering for gas absorption purposes [6.6]. But the SOG differs essentially from the ordinary absorbers in that it produces electronically excited oxygen molecules. This fact compels one to reconsider the question of its optimization [6.7]. By analogy with the gas-liquid absorbers, it is natural to divide SOG's into three types. *Bubbling* SOG's: the gas-liquid contact occurs on the surface of chlorine gas bubbles or jets dispersed in a liquid. Bubbling is effected by passing the gas through a layer of the liquid. Most of the experiments [6.3–5, 8–18] used this type of SOG. *Surface-film* SOG's: the gas-liquid contact occurs on the surface of a film of liquid flowing, for example, down a stack of closely packed thin vertical tubes [6.19, 20]. *Aerosol* SOG's: the gas-liquid contact occurs on the surface of fine liquid particles dispersed in the chlorine gas. Such SOG's were suggested by *Richardson* et al. [6.20], *Zagidullin* et al. [6.21], and *McKnight* and *Stanclife* [6.22]. The first experimental studies were conducted by *Yoshida* et al. [6.23].

The bubbler-type reactors generate singlet oxygen in a relative concentration of 95% at a constant pressure of up to 9 mm Hg [6.9]. The long lifetime of the singlet oxygen allows it to be transported by a gas duct from the reactor to the laser cavity. The generated oxygen contains H_2O and H_2O_2 vapors and unreacted chlorine gas: impurities harmful to the active medium of OICL. For this reason, all the experimental setups described in the literature used cooled traps to purify the oxygen. Cooling with liquid nitrogen froze out all the impurities. Cooling down to around 200 K removed only the H_2O and H_2O_2 vapors. In the cavity region, molecular iodine vapor obtained by heating crystalline iodine to some 50°C is mixed with the singlet oxygen flow. The I_2 molecules in the flow undergo stepwise collisional dissociation, the energy of two to five $O_2(^1\Delta)$ molecules being required to dissociate an I_2 molecule. The ratio between the molar flow rates of iodine and oxygen usually does not exceed 0.01, and so the expenditure of energy on dissociating iodine can be disregarded. All the setups used high-Q single-pass stable cavities, the cavity axis being as a rule perpendicular to the gas flow direction.

The experimental results and conditions have been collected in Table 6.1. The notation in the table is as follows: Cl_2 denotes the flow rate of chlorine, p the active medium pressure in the cavity region, $O_2(^1\Delta)/O_2$ the relative singlet oxygen content, u the gas flow velocity, P the laser power output, ε the laser energy output per unit mass or molar flow rate of chlorine, and η the laser efficiency defined as the ratio between the number of the emitted photons and the number of chlorine molecules consumed. The quantity η is fairly convenient for estimating the efficiency of the apparatus, because any apparatus function, such as the generation of $O_2(^1\Delta)$ and its purification and the mixing of $O_2(^1\Delta)$ with I_2, is included in this parameter.

As can be seen from the above information, the power output of the OICL has reached a kilowatt level at a fairly high specific energy output ($\simeq 200$ joules per gram of CI_2) and reactant utilization efficiency ($\simeq 40\%$). Among the shortcomings of the laser are a low unsaturated gain ($10^{-4} - 10^{-3}$ cm^{-1}) and a rather low power output per unit cross-sectional area of the gas flow (a few watts per square centimeter). These shortcomings are due first of all to the low singlet oxygen pressure in the active medium ($\simeq 1$ m Hg).

The possibility of a material improvement of the performance of the OICL and extension of its application field depends on the answers to the following principal questions:

Table 6.1. Summary of performance characteristics and operating conditions of OICL

Cl_2 [mmole/s]	p [mm Hg]	$O_2(^1\Delta)/O_2$ [%]	u [m/s]	P [W]	$\varepsilon \left[\dfrac{J/g}{kJ/mole} \right]$	η [%]	Reference
19	0.9	40	subsonic	180	$\dfrac{136}{9.47}$	10.4	[6.24]
80	0.8	> 50	subsonic	1080	$\dfrac{193}{13.5}$	14	[6.9, 25]
6.68	1.85	⩾ 50	15	105	$\dfrac{218}{15.2}$	16.7	[6.18]
11	0.75	$\simeq 90 \pm 20$	70	174	$\dfrac{224}{15.8}$	17.4	[6.17]
	2		70	320			[6.17]
5	1.08	80	11.7	192	$\dfrac{540}{38.4}$	42	[6.47]

– Is it possible to raise the SOG pressure and output?
– Is it possible to transport singlet oxygen at increased pressures?
– Are there ways of improving the laser gain?
– Is it possible to operate the OICL without a cooled trap?
– Is it possible effectively to operate the OICL under pulsed conditions?

These issues will be discussed in the sections that follow.

6.2 Singlet Oxygen Generators

The above experimental works provide only scanty information on the operating conditions of SOG's, which makes it difficult to judge their capabilities and the outlook for their development.

To improve the SOG requires optimization of its output parameters – the singlet oxygen content, proportion of unreacted chlorine, total oxygen pressure, and partial H_2O and H_2O_2 vapor pressures. It is possible to construct a general model of the singlet oxygen generation process [6.7] because the residence time of molecular chlorine and singlet oxygen in solution is short compared to the characteristic liquid and gas flow times.

The singlet oxygen generator uses $H_2O : NaOH : H_2O_2$ solutions with typical concentrations of 1–3 mole/l [NaOH] and around 10 mole/l [H_2O_2]. The probability that a Cl_2 molecule will be captured on the solution surface is $\gamma = 10^{-3}$ –10^{-2}. The singlet oxygen $O_2(^1\Delta)$ is produced in the solution by the reaction

$$H_2O_2 + Cl_2 + 2NaOH \rightarrow 2H_2O + 2NaCl + O_2(^1\Delta) + 27 \text{ kcal/mole} . \qquad (6.4)$$

The experiments performed by *Fisk* and *Hays* [6.26] showed the yield of $O_2(^1\Delta)$ in this reaction to be $100 \pm 5\%$. The main singlet oxygen production channel is [6.20]

$$HO_2^- + Cl_2 \xrightarrow{k} O_2(^1\Delta) + Cl^- + HCl; \quad k > 10^7 \text{ 1/mole s} . \qquad (6.5)$$

The HO_2^- ions are formed in the solution as a result of the elementary processes

$$NaOH \leftrightarrow Na^+ + OH^- , \qquad (6.6)$$

$$H_2O_2 + OH^- \leftrightarrow HO_2^- + H_2O + 23 \text{ kcal/mole} . \qquad (6.7)$$

The degree of dissociation of NaOH is close to unity. The equilibrium of the latter reaction is shifted greatly to the right. The hydrochloric acid formed is neutralized in the alkaline solution by the reaction

$$HCl + NaOH \rightarrow NaCl + H_2O . \qquad (6.8)$$

It should be emphasized that sodium hydroxide serves only to supply the OH^- ions, the Na^+ cations taking no part in the singlet oxygen generation reactions. Therefore, the SOG can also use such alkalis as LiOH and KOH and apparently a number of other compounds having an alkaline reaction (e.g., Na_3PO_4). This can be used to obtain low-temperature nonfreezing hydrogen peroxide solutions.

The oxygen formed in the solution evolves violently into the gas phase because its concentration considerably exceeds its equilibrium solubility in the liquid. The relative concentration of $O_2(^1\Delta)$ in the oxygen gas evolving from the solution is determined by the relationship between its quenching time in the solution and the characteristic time of oxygen diffusion from the solution into the gas phase.

The solution of the diffusion problem for the relative yield of the singlet oxygen as a function of kinetic parameters and temperature leads to the following relationship [6.7]:

$$\frac{[O_2(^1\Delta)]}{[O_2]} = \left\{ 1 + \left(\frac{D_{Cl_2}}{\tau_\Delta k [HO_2^-] D_{O_2}} \right)^{1/2} \right\}^{-1}, \tag{6.9}$$

where D_{Cl_2} and D_{O_2} are the diffusion coefficients of Cl_2 and O_2 in the solution, respectively, $\tau_\Delta = 2$ μs is the lifetime of $O_2(^1\Delta)$ in the solution, $[HO_2^-] \simeq 1$ mole/l is the concentration of the OH_2^- ions, and $k = 10^{-8}$ (mole s)$^{-1}$ is the reaction rate constant. It follows from the above expression that the reduction of the pH of the solution from 11 to 7, which corresponds to the decrease of the OH_2^- ion concentration from 3.5 to 3.5×10^{-4} mole/l, will lead to the quenching of lasing ($[O_2(^1\Delta)]/[O_2] < 0.25$).

The above relationship defines the local singlet oxygen yield and holds true for any actual operating conditions of the SOG. To find the integral yield from the entire surface of the solution, it is necessary to know the distribution of the singlet oxygen donors and the OH_2^- ion concentration over the entire gas-liquid interface, the donor concentration being proportional to the chlorine gas concentration near the solution surface.

The problem of calculation of the SOG was generally formulated and solved by *Zagidullin* and co-workers [6.7]. The theory elaborated by the authors enables one to estimate such SOG parameters as the singlet oxygen yield, chlorine utilization efficiency, and ultimate oxygen pressure. The physical reason for the limitation on the operating pressure of the SOG is the depletion of the OH_2^- ions in the surface layer of the solution.

According to the analysis performed by the above authors, in tubular surface-film SOG's (packed absorption columns), there will be no depletion of the OH_2^- ions in the surface layer of the solution at Cl_2 pressures up to 37.5 mm Hg, the characteristic packing tube dimensions being taken to be 10–25 cm (length) and 0.3 mm (radius), the velocity of the solution film flowing down the packing 20–50 cm/s, and the gas flow velocity 4×10^3 cm/s.

In bubbling SOG's, the gas-liquid contact surface is extended as a result of the gas being dispersed as fine bubbles or jets in the liquid. Bubbling conditions are usually characterized by the parameter $m = w/h$, where $w = Q/S$ is the reduced gas flow velocity, Q being the volume flow rate of the gas through the reactor, S the cross-sectional area of the reactor, and h the height of the liquid column in the reactor.

At $m \leqslant 0.1$ s^{-1}, bubbling occurs as single bubbles the ascent velocity of which depends on their size. For bubbles less than 1 cm across, the ascent velocity does not exceed 30 cm/s.

At $m = 0.1-1$ s^{-1}, bubbling is observed to take place in the form of clusters (restricted flow conditions).

The region of $m = 1-50$ s^{-1} is characterized by an intense foaming, the height of the foam layer being comparable with that of the liquid column, h. Under such conditions, the gas-liquid contact surface is the greatest in the foam layer.

At $m > 50$ s^{-1}, there occurs the transition to the jet bubbling regime in which the gas forms channels in the liquid.

Bubbling with low m values is characterized by low gas flow rates through the SOG, the bubble ascent velocities being low. By contrast, jet bubbling apparently holds much promise for the SOG.

Bubbling SOG's are used most frequently, thanks to their simple design and reliable operation. The characteristics of this type of SOG are listed in Table 6.2.

Experiments show that the main decay mechanism of O_2 ($^1\Delta$) in the bubbling SOG is the reaction $O_2(^1\Delta) + O_2(^1\Delta) \rightarrow O_2(^3\Sigma) + O_2(^1\Sigma)$ the rate of which is increased in a two-phase gas-liquid medium on account of the hydrostatic pressure of the liquid [6.29]. The yield of O_2 ($^1\Delta$) rises appreciably as the height of the liquid column is reduced. It can be seen from Table 6.2 that the operating pressure in modern SOG's is in the neighborhood of 1–3 mm Hg. It is hoped that by improving the bubbling SOG design (reducing the liquid column height, increasing the reactor cross section, using Raschig rings to prevent the solution from being carried away by the gas flowing with high rates [6.6]) it will be possible to build high-output plants operating at outlet oxygen pressures up to 10 mm Hg and featuring high singlet oxygen percentages.

Singlet oxygen can be obtained by atomizing an alkaline hydrogen peroxide solution in a chlorine atmosphere. The solution in an aerosol form is injected into a flow of a chlorine-bearing gas. The gas-liquid mixture thus obtained is flowed through a reaction chamber where singlet oxygen is generated. The waste solution is then separated from the gas in a separator. The distribution of Cl_2 along the

Table 6.2. Summary of data on bubbling SOG's

Authors	Year of publication	O_2 ($^1\Delta$) [%]	$P_{O_2}^*$ [mm Hg]	$h_0^\dagger$ [cm]
McDermott et al. [6.3]	1978	35–40	0.975	4.5
Benard et al. [6.4]	1979	35	0.975	9.9
Richardson and *Wiswall* [6.5]	1979	20–40	—	—
Bachar and *Rasenwaks* [6.11]	1982	44	0.323	7.9
Watanabe et al. [6.10]	1983	<44	2.475	7
Bonnet et al. [6.14]	1984	<50	0.3	—
Grigor'yev et al. [6.13]	1984	35–45	1.275	—
Vagin et al. [6.12]	1984	40	0.375	7.1
Watanabe et al. [6.27]	1986	51–64	0.75–0.3	—
Yoshimoto et al. [6.28]	1986	>50	1.8	8.5

$^*P_{O_2}$ is the oxygen pressure in the generator.
$^\dagger h_0$ is the height of the solution column without gas.

reaction chamber is described by the relation [6.7]

$$[Cl_2(z)] = [Cl_2]_0 \exp(-z/z_{abs}) , \tag{6.10}$$

where $z_{abs} = u_{mix}/\pi R^2 \bar{v}\gamma n$ is the characteristic gas absorption length, u_{mix} being the flow velocity of the gas-liquid mixture in the reaction chamber, R the radius of the aerosol droplets, $\bar{v}$ the average thermal velocity of molecules, γ the probability of the Cl_2 molecule being captured on the surface of a droplet ($\gamma < 10^{-2}$), and n the droplet concentration ($nR^3 \ll 1$). The quantity $4\pi R^2 n$ is the specific gas-liquid contact surface area. In atomization-type chemical reactors, it reaches 20 cm^{-1}. For droplets with a radius of $R \geqslant 10^{-3}$ cm, a sufficiently high singlet oxygen yield ($\eta_\Delta > 0.5$) can be achieved at Cl_2 pressures below some maximum [6.7]: $[Cl_2]_0 < 15 \times (4\pi R^2 n)^{1/2}$ (mm Hg). For instance, at $R = 10^{-3}$ cm and $n = 10^6 \text{ cm}^{-3}$, we have $[Cl_2]_0 < 52.5$ mm Hg, the chlorine gas consumption time in this case amounting to 5×10^{-3} s. But the actual operating pressure may prove to be substantially lower than that following from the analysis of the SOG itself, because of the homogeneous quenching of the $O_2(^1\Delta)$ molecules. The characteristic heterogeneous quenching time of the $^1\Delta$ state (on the surface of droplets) is $\tau_Q = (\pi nR^2 \bar{v}\gamma_\Delta)^{-1}$, $\gamma_\Delta \simeq 10^{-5}$ being the heterogeneous quenching probability. For $4\pi R^2 n = 2-10 \text{ cm}^{-1}$, we have $\tau_Q = 0.1-0.5$s. This time sets a limit on the time of the liquid separation from the gas flow. Cyclone or electrostatic separators capable of effectively separating droplets with a radius of $R \geqslant 1$ μm can be used to separate the liquid from the gas. The ratio between the mass flow rates of the liquid and gas in the atomization-type SOG can be determined by the formula $\dot{m}_l/\dot{m}_g = 4\pi R^3 n\varrho_l/3\varrho_g$, where ϱ_l and ϱ_g are the densities of the liquid and gas, respectively. For $p_{Cl_2} = 7.5$ mm Hg, $\varrho_l = 1.5 \text{ g/cm}^3$, $4\pi R^2 n = 2 \text{ cm}^{-1}$, and $R = 10^{-3}$ cm, we have $\dot{m}_l/\dot{m}_g = 50$. This relationship makes it possible to estimate the total rate of flow of the liquid from all atomizers proceeding from the necessary chlorine flow rate. Centrifugal and splash-plate spray atomizers [6.6] capable of handling liquid flow rates as high as a liter per second apparently hold promise for the realization of the atomization-type SOG. The first experiments [6.23] have demonstrated that this type SOG outdoes the bubbling SOG for the oxygen excitation efficiency, but is surpassed by it in the chlorine utilization efficiency. To eliminate this shortcoming, it is necessary to improve the atomizer (atomizer bank) design.

The oxygen gas produced by the SOG of any type contains unreacted chlorine molecules and also H_2O_2 and H_2O vapors. The chlorine content of the gas can be reduced to an acceptable level by properly selecting the operating conditions of the SOG. There are experimental data available [6.23] pointing to the possibility of an almost 100% utilization of chlorine to generate O_2. The situation with the H_2O_2 and H_2O vapors is quite another thing. A certain concentration of these vapors in the active region of the OICL, and lasing is quenched. The use of cooled traps to reduce the content of the harmful impurities makes the laser difficult to operate because of the formation of ice crystals. What is more, a sizable proportion of $O_2(^1\Delta)$ undergoes heterogeneous quenching in the trap. *Zagidullin* et al. [6.30] suggested that the trap should be removed from the OICL scheme and the solution in the SOG be deeply cooled instead, thus allowing the saturated vapor pressure of

H_2O_2 and H_2O to be reduced to a level acceptable in the OICL. An aqueous solution of H_2O_2 stays unfrozen at temperatures down to $-55°C$, and that of KOH can be cooled down to $-62°C$. A mixture of LiOH (7 wt%), H_2O_2 (23 wt. %), and H_2O (70 wt.%) remains liquid down to $-21°C$. The analysis of formula (6.9) shows that the yield of $O_2(^1\Delta)$ will remain at a level of some 50% with the SOG medium cooled down to $-40°C$. According to the kinetic calculations [6.30], an OICL having no cooled trap can provide a gain of the order of $10^{-3} - 10^{-2}$ cm^{-1} with the solution in the SOG being cooled to a temperature in the range $-20°C$ to $-40°C$. However, this approach requires additional experimental investigations, for the viscosity of the solution may rise materially upon cooling.

The most general inference from the studies conducted to date is that the surface-film and aerosol SOG's can operate at pressures up to a few tens of millimeters mercury. The bubbling SOG's are apparently capable of operating at pressures up to 10 mm Hg.

6.3 Kinetics of the Processes Occurring in the Active Medium of OICL

Vinogradova et al. [6.1] and *Zagidullin* et al. [6.2] proposed a unidimensional model of the oxygen-iodine chemical laser. The model includes the main processes taking place in the $O_2(^1\Delta) - I_2$ system: the exchange of energy between the singlet oxygen and atomic iodine and the various quenching processes. The iodine dissociation is apparently most adqeuately described by the model suggested by *Heidner* et al. [6.31], but the available kinetic data are as yet insufficient to use it. Molecular iodine is first excited by collison with an excited I* atom to some state I_2^* that is still not exactly identified (most likely this is a vibrationally excited I_2 ($v = 20$–40) molecule [6.321]. A second collision between I_2^* and $O_2(^1\Delta)$ causes the dissociation of the former.

The model [6.1, 2] allows for the heating of the mixture due to the relaxation of the stored energy. The rate constants of some reactions are taken with due regard for their temperature dependence. To calculate the laser power output, use is as usual made of a cavity model with homogeneously distributed losses. The calculations have been made both within the framework of a two-level atomic iodine excitation scheme and with allowance made for the translational and hyperfine relaxation effects [6.8, 33, 34].

The main elementary processes occurring in the active medium of the OICL are listed in Table 6.3. The complex process of dissociation of I_2 is, in this scheme, described by the model suggested by *Derwent* and *Thrush* [6.35], which involves the $O_2(^1\Sigma)$ molecule. The rate constant of reaction *2* is taken so as to suit the dissociation time of I_2.

In the absence of water vapor, the quenching of lasing is due mainly to reaction *4* because the energy stored in the $O_2(^1\Sigma)$ molecules takes no part in the excitation of iodine. When there is water vapor, the relaxation of energy occurs as a result of reaction *6*.

Table 6.3. Elementary processes occurring in the active medium of OICL [6.8]

No.	Reaction	Rate constant [cm^3/s]; *[cm^6/s]
1	$O_2\,(^1\!\Delta) + O_2\,(^1\!\Delta) \to O_2\,(^1\!\Sigma) + O_2\,(^3\!\Sigma)$	2×10^{-17}
2	$O_2\,(^1\!\Sigma) + I_2 \to O_2\,(^3\!\Sigma) + 2I$	2×10^{-10}
3	$O_2\,(^1\!\Delta) + I \leftrightarrow O_2\,(^3\!\Sigma) + I^*$	7.6×10^{-11}
	$k_e = k_3/k_{-3} = (3/4)\exp(400/T)$	
4	$I^* + O_2\,(^1\!\Delta) \to I + O_2\,(^1\!\Sigma)$	1.1×10^{-13}
5	$O_2(^1\!\Sigma) + H_2O\,(H_2O_2) \to$	
	$O_2\,(^1\!\Delta) + H_2O\,(H_2O_2)$	6.7×10^{-12}
6	$I^* + H_2O \to I + H_2O$	2.3×10^{-12}
7	$I^* + H_2O_2 \to I + H_2O_2$	10^{-11}
8	$I^* + I_2 \to I + I_2$	3.6×10^{-11}
9	$I^* + O_2\,(^3\!\Sigma) \to I + O_2\,(^3\!\Sigma)$	3×10^{-14}
10	$I^* + Cl_2 \to I + Cl_2$	1.4×10^{-14}
11	$I^* + Cl \to I + Cl$	1.5×10^{-11}
12	$I^* + Cl_2 \to ICl + Cl$	6×10^{-15}
13	$I^* + ICl \to I_2 + Cl$	2.3×10^{-11}
14	$ICl + Cl \to I + Cl_2$	8×10^{-12}
15	$I_2 + Cl \to ICl + I$	1.2×10^{-10}
16	$I + I + I_2 \to I_2 + I_2$	$*5 \times 10^{-32}$
17	$I + I + O_2 \to I_2 + O_2$	$*3 \times 10^{-30}$

While the mixture containing singlet oxygen flows from the SOG to the OICL and fills up its active volume, the oxygen inevitably undergoes quenching according to the scheme

$$O_2\,(^1\!\Delta) + O_2(^1\!\Delta) \overset{k_1}{\to} O_2(^1\!\Sigma) + O_2(^3\!\Sigma) + 7.5\ \text{kcal/mole}\ , \tag{6.11}$$

$$O_2(^1\!\Sigma) + M \overset{k_2}{\to} O_2(^3\!\Sigma) + M + 15\ \text{kcal/mole}\ . \tag{6.12}$$

These processes determine the maximum possible accumulation of singlet oxygen.

The quenching of singlet oxygen is accompanied by a substantial heating of the gas. This temperature rise is undesirable because it leads to the rarefaction of the active mixture, an increase in the threshold proportion of $O_2(^1\!\Delta)$, and possibly the decomposition of the iodide. This temperature effect can largely be suppressed by diluting the mixture with some buffer gas of a high heat capacity.

Let a mixture containing oxygen gas with a concentration of n_{O_2}, the relative singlet oxygen content of the gas being η_Δ^0, enter the active volume V_a of the laser at a volume flow rate of Q. During the time it takes to fill up the active volume, $\tau_{\text{fill}} = V_a/Q$, the proportion of the singlet oxygen will drop to

$$\eta_\Delta = \frac{\eta_\Delta^0}{1 + k_1 n_{O_2}\eta_\Delta^0 V_a/2Q}\ . \tag{6.13}$$

Consequently, the energy stored in the volume V_a will be

$$E = h\nu_l n_{O_2} V_a \left(\frac{\eta_\Delta^0}{1 + k_1 n_{O_2}\eta_\Delta^0 V_a/2Q} - \frac{1}{2k_e + 1} \right), \tag{6.14}$$

where $h\nu_l$ is the laser photon energy and k_e the equilibrium constant for the exchange of energy between $O_2(^1\Delta)$ and I^*. It can be seen from this formula that, with the flow rate Q being fixed, increasing the oxygen content or the active volume above certain limits causes no increase in the energy store. It is not very difficult to find that the energy store reaches its maximum at

$$n_{O_2} V_a = \frac{2Q}{k_1\eta_\Delta^0} \left[\sqrt{\eta_\Delta^0(2k_e + 1)} - 1 \right], \tag{6.15}$$

the maximum being

$$E_{max} = \frac{2Qh\nu_l}{k_1\eta_\Delta^0} \left[\eta_\Delta^0 - \frac{1}{(2k_e + 1)^{1/2}} \right]^2 \tag{6.16}$$

and the average proportion of the singlet oxygen in the mixture, $\bar{\eta}_\Delta = [\eta_\Delta^0/(2k_e + 1)]^{1/2} \simeq 35\%$. Thus, with the active laser volume being filled at a fixed rate, increasing n_{O_2} or V_a will not help to store energy in excess of E_{max}.

The main characteristics of the active medium in the OICL are the gain and the store of energy that can be converted to laser emission. At pressures up to 20 mm Hg, the cross section of the stimulated $^2P_{1/2}(F=3)\rightarrow{}^2P_{3/2}(F=4)$ transition is determined mainly by the Doppler broadening mechanism and is equal to 1.3×10^{-17} cm^2. Under typical conditions where the rate of energy transfer from the singlet oxygen molecule to the iodine atom by far exceeds the quenching rate of the upper lasing level I ($^2P_{1/2}$), the gain in the given transition is [6.36]

$$G_{34} = \frac{7}{12} \left\{ \frac{\delta_{34}^D(2k_e + 1)n_l}{2[(k_e - 1)\eta_\Delta + 1]} \right\} \left(\eta_\Delta - \frac{1}{2k_e + 1} \right), \tag{6.17}$$

where $k_e = (3/4)\exp(400/T)$ and n_l is the atomic iodine concentration.

Under zero gain conditions, the amount of energy potentially available for conversion to laser emission in a unit of active medium volume (extractable energy store) is

$$E_v = 10^3 n_{O_2} \left(\eta_\Delta - \frac{1}{k_e + 1} \right) h\nu_l \, , \, (\text{J/l}) \, , \tag{6.18}$$

where n_{O_2} is the total oxygen concentration and $h\nu_l = 1.51 \times 10^{-19}$ J.

Most calculations are restricted to the analysis of subsonic active medium flow conditions in a constant-section channel. The mixing of the reactants is either taken to be instantaneous [6.8] or is allowed for in a most simple way [6.3]. The study by *Azyazov* et al. [6.37] involves a more detailed gasdynamic modelling on the basis of boundary-layer equations. Assuming that there is no friction on and heat transfer to the gasdynamic channel walls, the equations for the concentration, temperature, and velocity of the flow have the form

$$u \frac{dn_M}{dx} = \sum_i W_M^{(i)} - \frac{n_M}{C_P T} \sum_i q_i W_i \, , \tag{6.19}$$

$$u \frac{dT}{dx} = \frac{1}{C_P} \sum_i q_i W_i \, , \tag{6.19b}$$

$$u \frac{du}{dx} = \frac{u}{C_P T} \sum_i q_i W_i \, , \tag{6.19c}$$

where u is the flow velocity, n_M the concentration of the Mth component, T the temperature of the flow, $W_M^{(i)}$ the rate of formation of the Mth component in the ith reaction, q_i and W_i the heat and rate of the ith reaction, respectively, and C_P the specific heat at constant pressure.

When calculating the laser power output, the kinetic and heat balance equations are solved simultaneously with the equation for the photon density in the laser cavity, $\varrho(x)$:

$$u \frac{d\varrho}{dx} = \frac{7}{12} C \sigma \varrho \varDelta - \frac{\varrho}{\tau_{ph}} + \frac{7}{12} \frac{n_{J^*} C \sigma}{V} \, ,$$

where $\varDelta = n_{J^*} - (1/2) n_J$ is the inversion density. The coefficient 7/12 stems from the assumption that lasing occurs between the sublevels with $F = 3$ of the $J^* \equiv J \, (^2P_{1/2})$ state and $F = 4$ of the $J \equiv J \, (^2P_{3/2})$ state. The laser power output per unit active region volume, P_l, is determined by the relation $P_l = \hbar \omega_l \, (\varrho / \tau_{ph})$, and the laser intensity $I_l = P_l l$, where l is the active region size along the cavity axis. The total laser power Q_l as a function of x is given by the equation $dQ_l / dx = P_l S$, where S is the cross-sectional area of the flow.

It is seen from the above relationships that the gain G_{34} is proportional to the energy store and equals $(4.6 \text{ to } 6.7) \times 10^{-18} \, n_1 \, (\text{cm}^{-1})$ at $T = 298$ K and $\eta_A = 0.5\text{--}0.8$. To achieve a gain of $10^{-3} - 10^{-2} \, \text{cm}^{-1}$, the concentration of I atoms, n_1, in the active medium should be around $10^{14} - 10^{15} \, \text{cm}^{-3}$, no matter what the oxygen pressure.

Among the most important characteristics of the active medium in the continuous-wave OICL is the extent of the gain region along the flow direction. The extent of the active zone is characterized by the distance $x_{1/2}$ within which the extractable energy store E_v is halved. To compare regimes with various initial gas velocities, it is convenient to use the parameter $\tau_{1/2} = x_{1/2} / u_0$. Numerical calculations [6.8] have shown that the product $G_{34}^{max} \tau_{1/2}$ is a constant independent of the pressure of O_2 and I_2 over a wide range of parameters. The value of this constant is $6 \times 10^{-6} \, \text{cm}^{-1} \text{s}$.

Let us now consider the effect of vapor pressure of H_2O and H_2O_2 on the active medium parameters. This effect is determined first of all by the parameter $q = (k_6 n_{H_2O} + k_7 n_{H_2O_2}) / k_4 n_{O_2}$. Thus, the energy dissipation in the active medium depends on the ratio of the concentration of the H_2O and H_2O_2 impurities to that of O_2 and not on their absolute content. The analysis [6.8] demonstrates that the presence of H_2O and H_2O_2 vapors starts manifesting itself at $q = 0.5$. Hence one can calculate the maximum permissible temperature of the cryogenic trap, T_c. For example, at $p_{O_2} = 1$ mm Hg and $q = 0.5$, the H_2O vapor concentration is $n_{H_2O} = 8 \times 10^{15}$. For the vapor to be fully in equilibrium with ice, the trap temperature T_c should be equal to 273 K. At $p_{O_2} = 4$ mm Hg and $q = 0.5$, $T_c = 230$ K, and at $p_{O_2} = 10$ mm Hg and $q = 0.5$, $T_c = 240$ K.

The calculations [6.8] have revealed that the inversion density and the stored energy lifetime depend but little on the chlorine content of the active mixture. At an oxygen pressure of 1–8 mm Hg, $200 < p_{O_2}/p_{I_2} < 1000$, and $T_c = 210$ K, the laser gain shows no appreciable change with chlorine content variation in the range 20–40%.

The gas issuing from the SOG can carry solution droplets. The mass proportion γ_m of the disperse phase in the gas flow may reach 1–10% for bubbling SOG's. The solution droplets may evaporate in the active medium of the laser and thus increase the proportion of water vapor by $(32/18)\gamma_m$. What is more, the presence in the active medium of dispersing particles reduces the lasing efficiency. All these factors compel one to give special consideration to the aerosol composition of the active medium of the OICL.

Increasing the oxygen pressure above of 1–3 mm Hg was observed experimentally to reduce the laser power output. The reason may be a sharp increase in the water vapor concentration. The water vapor content of the active medium is usually reduced to an acceptable level by freezing the water out in a condensation trap (CT). *Zagidullin* et al. [6.38] have shown that the efficiency of the oxygen gas purification from water vapor in the trap drops drastically as the total gas pressure is increased. To understand the main laws governing the operation of the CT, consider the following simple problem. Let an unrelaxing $O_2(^1\Delta) + O_2 + H_2O$ mixture enter a condenser made in the form of a tube with cooled walls. Assume that the gas mixture flow is laminar and there is no wall friction. In that case, to analyze the efficiency of the mixture purification from the water vapor, use can be made of the well-known solution to the problem of vapor condensation in the presence of a noncondensing gas:

$$\ln\left[\ln\left(\frac{1-x}{1-x_0}\right) \Big/ \ln\left(\frac{1-x}{1-x_1}\right) \right] = \frac{LD\,\text{Nu}}{R^2 v}, \qquad (6.20)$$

where $x_0 = p^0_{H_2O}/p$ and $x_1 = p^1_{H_2O}/p$ are the relative water vapor concentrations at the condenser inlet and outlet, respectively, $x = p^{sat}_{H_2O}/p$ is the ratio of the saturated water vapor pressure at the condenser wall temperature to the total gas pressure $p = p^0_{H_2O} + p^0_{O_2} = p^1_{H_2O} + p^1_{O_2}$, L the condenser tube length, R the tube radius, v the flow velocity at the outlet of the condenser tube, D the interdiffusion coefficient of O_2 and H_2O, which decreases with the rising total gas pressure as D [cm^2/s] $= 200/p$ [mm Hg], and Nu the Nusselt number which characterizes the mass transfer efficiency and is equal to 3.7 at $4R^2v/LD < 20$.

Consider the variation of x_1 consequent upon the total gas pressure rise from 1 to 4 mm Hg at fixed condenser tube length $L = 50$ cm and pump delivery $Q = \pi R^2 v = 25$ l/s. Assume that $T_c = 200$ K ($p^{sat}_{H_2O} = 1$ μm Hg), $\eta_\Delta = 100\%$, and $x_0 = 0.72$. In that case, it is not very difficult to find that at $p = 1, 2, 3, 4$, mm Hg, $x_1 = 10^{-2}, 0.11, 0.32, 0.38$. Thus, it is seen that a slight increase in the gas pressure raises the water vapor content at the condenser outlet by more than an order of magnitude.

Numerical calculations of the active medium kinetics show that increasing the water vapor concentration from 5% upwards causes a proportionate reduction of the laser gain. For this reason, the laser power output in the above example may start dropping at $p_{O_2}^1 \simeq 3$ mm Hg.

Some or other design features of the CT (e.g. gas duct turns, flow turbulization) may lead to a significant intensification of the mass transfer process and a reduction of the water vapor content of the gas at the outlet of the condenser. For example, the Nusselt number for smooth-walled tubes at Re $\gg 10$ may be several times that under laminar flow conditions with Re $\simeq 1$. It is also possible to provide for forced convection (tube corrugation, flow swirling). All these measures actually give rise to a transverse flow velocity component. In that case, however, one should take account of the hydraulic pressure loss in the condenser.

The water vapor content of the gas at the outlet of the condenser can be reduced by increasing the length of the condenser tube or decreasing the pump delivery. But both these measures lengthen the transportation time of the singlet oxygen $O_2\,(^1\Delta)$. The solution of the appropriate optimization problem shows that the product of the trap tube radius R into the outlet oxygen pressure $p_{O_2}^1$ must satisfy the condition $p_{O_2}^1\,R < 12$ (mm Hg) cm in order to ensure that there is less than 10% water vapor and no less than 50% $O_2(^1\Delta)$ in the active region of the laser. For example, at $R = 3$ cm, the maximum permissible outlet oxygen pressure is equal to some 4 mm Hg. Where the oxygen pressure is around 10 mm Hg, the condenser tube radius must be less than 1.2 mm.

Another way of increasing the maximum possible gas pressure is to reduce the water vapor content of the gas at the inlet of the condenser by cooling the solution in the SOG [6.38].

Modern continuous-wave OICL's use mainly transverse flow arrangements. Longitudinal flow arrangements look promising for the engineering applications of the laser, because such arrangements are simple in design, feature minimal transport losses of singlet oxygen and good optical properties of the active medium, and are suitable for modular construction. The characteristics of the longitudinal flow OICL were studied by *Zaikin* [6.39]. The author achieved a total gain of some 10% per pass at a cavity efficiency of 50% and a specific power output of 80 watts per square centimeter of the cross-sectional area of the gas flow in a 32 cm long active region at a water vapor concentration of 5%, an initial singlet oxygen proportion of 75%, an oxygen pressure of 5 mm Hg, and a flow velocity of 100 m/s.

The power output and specific energy characteristics of the transverse flow OICL were numerically analyzed in a series of papers. [6.3, 4, 8, 15, 33]. The calculations have demonstrated that the cavity efficiency $\eta_{\rm cav}$ can be as high as 50–90% over a wide range of operating conditions at oxygen pressures up to several tens of millimeters mercury. The cavity efficiency characterizes the effectiveness of conversion of the singlet oxygen energy into laser emission in accordance with formula

$$W = u p_{O_2}\,(\eta_\Delta - \eta_\Delta^{\rm thr})\,\eta_{\rm cav}\ , \tag{6.21}$$

where W is the laser power output per unit cross-sectional area of the gas flow, $\eta_\Delta^{\rm thr}$ is the threshold relative singlet oxygen concentration, and η_Δ characterizes the singlet oxygen transportation efficiency in the SOG and gas duct. High cavity efficiency values are obtained where the unsaturated gain substantially exceeds the threshold value and the loss due to relaxation is low.

An important matter in modelling the OICL is consideration of the translational and hyperfine relaxation of the iodine atoms and the effect of these processes on the saturation characteristics. This effect could be responsible for the low values of η_{cav} in a number of experiments. The method of calculating the laser power output with due regard for translational and hyperfine relaxation was developed by *Zagidullin* et al. [6.33]. These processes have a material effect on the extent of the lasing region along the flow direction. Calculations yield exceptionally high values of the saturation intensity in the OICL ($I_{sat} = 10 \text{ kW/cm}^2$ at $p \simeq 2$ mm Hg [6.8]), which is explained by a high rate of electronic energy exchange between the singlet oxygen and atomic iodine.

6.4 Pulsed OICL's

The oxygen-iodine active medium possesses a high energy potential, but the lifetime of the energy store in the $I + O_2 (^1\Delta)$ system is relatively short, and is also inversely proportional to the laser gain. In the absence of water vapor, the main loss channel is the process $I^* + O_2 (^1\Delta) \overset{k}{\to} I + O_2 (^1\Sigma)$ ($k = 10^{-13} \text{ cm}^3/\text{s}$), whereas in the presence of the vapor, it is the process $I^* + H_2O \overset{k}{\to} I + H_2O$ ($k = 2.3 \times 10^{-12} \text{ cm}^3/\text{s}$).

This circumstance substantially limits the active medium volume and power output of the pulsed OICL. The problem of developing highly efficient chemically excited iodine lasers can be solved in principle by using a mixture of singlet oxygen and bound iodine in the active volume [6.40–44]. The laser mixture must be initially free from iodine atoms or molecules in the gas phase. Iodine must be present in the mixture in the form of either molecules of the type RI ($R = H$, CH_3, CF_3, etc.) or small crystals [6.44]. In that case, the most effective stored energy loss channels are excluded while the active volume of the laser is being filled. The active medium is produced by releasing iodine in a pulsed manner, e.g., by flash photolysis.

Zagidullin et al. [6.42] have been the first to give consideration to the merits of pulsed operation of the OICL. It has turned out that the negative effect of water vapor on lasing is in this case eliminated completely, and so the laser can do without any cooled trap or external cooling of the SOG.

Let us consider the following pulsed OICL scheme. The singlet oxygen gas is delivered directly from the SOG into the laser cavity. The water vapor content of the gas depends on the thermal regime of the generator, as described elsewhere in the text. Iodine atoms are produced instantaneously throughout the entire volume of the cavity. This initiation gives rise to a laser pulse.

The process in the SOG is as follows:

$$Cl_2 + 2NaOH + H_2O_2 \to 2H_2O + 2NaCl + O_2(^1\Delta) + 27.5 \text{ kcal/mole} \ . \quad (6.22)$$

If there is no quenching of $O_2(^1\Delta)$ in the generator, 27.5 kilocalories of heat will evolve per mole of the singlet oxygen produced. With the mass yield of oxygen being $\dot{m}_{O_2}$, the heat rate will be Q_T (kW) $= 3.6\, \dot{m}_{O_2}$ (g/s). Under steady-state conditions, the temperature of the solution in the SOG remains constant, the liberation of heat being counterbalanced by its absorption in the process of evaporation of water. The following equality must therefore hold for a heat-insulated SOG: $Q_T = 3.6\, \dot{m}_{O_2}$ $= q\dot{m}_{H_2O}$, where q is the evaporation heat of water (kJ/g) and $\dot{m}_{H_2O}$ the mass yield of water vapour (g/s). Hence we have the following formula for the partial water vapor pressure: $p_{H_2O} = 2.7\, p_{O_2}$.

The numerical calculations [6.42] of lasing have been made for oxygen pressures in the range 2.5–25 mm Hg. The initial atomic iodine concentration has been taken at 10^{15} cm^{-3} and the mixture has been assumed to be diluted with argon at a pressure of $p_{Ar} = p_{O_2}$. The laser energy output per unit active region volume was found to be linear in the oxygen pressure : ε_l (J/1) $= 2.82\, p_{O_2}$ [mm Hg], the energy output per unit oxygen mass being equal to 1.75 kJ/g. The laser pulse duration at half-maxium was the same (9 μs) in all cases.

As can be seen from the numerical calculation results obtained, the loss of energy due to the quenching of excitation by collision with H_2O is not very high (the ultimate energy output in the absence of water vapor amounts to 2.4 kJ/g), despite the very high water vapor pressures.

For engineering laser applications, it is convenient to have a high-repetition-rate machine with the pulse duration τ_l variable over the range from 1 μs to more than 1 ms, the laser intensity remaining high enough. Such a laser would enable one to perform a great variety of manufacturing operations, such as laser-plasma treatment ($\tau_l \simeq 10^{-6}$ s), metal cutting and welding ($\tau_l \simeq 10^{-5}$ s), and surface harden-ing ($\tau_l \simeq 10^{-3}$ s). Both the Nd:YAG and CO_2 lasers fail to meet these requirements. The calculations [6.43, 44] of the dynamics of lasing and energy store relaxation in the oxygen-iodine active medium with bound iodine indicate that the OICL holds promise for the solution of the above problem.

The kinetic model of the pulsed OICL includes the following main processes:

$$I(^2P_{1/2}) + H_2O \xrightarrow{k_1} I(^2P_{3/2}) + H_2O \ , \tag{6.23}$$

$$k_1 = 2.3 \times 10^{-12} \text{ cm}^3/\text{s} \ ,$$

$$I(^2P_{1/2}) + O_2(^1\Delta) \xrightarrow{k_2} I(^2P_{3/2}) + O_2(^1\Sigma) \ , \tag{6.24}$$

$$k_2 = 10^{-13} \text{ cm}^3/\text{s} \ ,$$

$$O_2(^1\Sigma) + H_2O \xrightarrow{k_3} O_2(^1\Delta) + H_2O \ , \tag{6.25}$$

$$k_3 = 6.7 \times 10^{-12} \text{ cm}^3/\text{s} \ ,$$

$$O_2(^1\Delta) + I(^2P_{3/2}) \underset{k_{-4}}{\overset{k_4}{\longleftrightarrow}} O_2(^3\Sigma) + I(^2P_{1/2}) \ , \tag{6.26}$$

$$k_4 = 7.6 \times 10^{-11} \text{ cm}^3/\text{s}; \quad k_{-4} = k_4/k_e; \quad k_e = (3/4)\exp(400/T) \ ,$$

$$RI + h\nu \xrightarrow{k_5} R + I \ . \tag{6.27}$$

Table 6.4. Calculated performance characteristics of a pulsed OICL

n_1 [10^{15} cm^{-3}]	0.0316	0.1	0.316	1	3.16	10
ε_l [J/l]	142	163	168	170	170	170
$\tau_{1/2}$ [µs]	375	115	37.5	12	4	1.7
η_{phys} [%]	45000	16000	5000	1700	500	170

Notes: $p_{O_2} = 100$ mm Hg; $\eta_A = 0.5$; $p_{H_2O} = 10$ mm Hg.

Table 6.4 lists the specific energy output, physical (photolysis) efficiency, and laser pulse duration values calculated for a wide range of I atom concentrations produced by the photoinitiation source. The physical efficiency was calculated by the formula

$$\eta_{phys} = \varepsilon_l / h\nu_{ph} n_1 \eta_i \, , \tag{6.28}$$

where $h\nu_{ph}$ is the photolysis energy quantum and η_i, the photoinitiation efficiency, taken to be 8%. It can be seen from these data that, for high active medium pressures ($\simeq 100$ mm Hg), the laser pulse duration can be varied over the necessary wide range at a high specific energy output ($\simeq 100$ J/l) practically independent of n_1 and a high physical efficiency (10^2–10^4%) increasing as $1/n_1$. This is the advantage of the pulsed OICL over its hydrogen-fluoride counterpart whose pulse duration and specific energy output are rigidly interrelated (Chap. 4). The merits of the pulsed OICL manifest themselves most vividly when the active volume is combined with the zone of the gas-liquid reaction giving rise to the excited O_2^* molecules [6.45]. In that case, the oxygen pressure is not restricted by the active volume filling time and can be raised to 50–100 mm Hg.

Thus, the OICL possesses considerable, as yet unrealized experimentally, potentialities. The laser can be further improved, with the gain increased from 10^{-4} -10^{-3} cm^{-1} to 10^{-2} cm^{-1}, power output per unit cross-sectional area of the flow, from 1–3 W/cm^2 to $(1$ to $3) \times 10$ W/cm^2, specific energy output, from 100 J/g to 300 J/g, and the cavity efficiency, from 15–30% to 50–80%.

It is believed that it will prove possible to realize fairly effective pulsed operating conditions under which the specific energy output will rise from 2–3 J/l [6.46] to some 100 J/l and the pulse duration will be variable over the range 10^{-6}–10^{-3} s. The pulse repetition rate is limited by the active volume filling and purging rate, and it is hoped that it will be possible to bring it to around 100 Hz.

With the theoretical level reached in the understanding of this system, lasers can be designed holding much promise for practical applications.

7. Photon Branching in Chain Reactions and IR-Radiation Initiated Chemical Lasers

At present, it has only proved possible to create a purely chemical laser operating under continuous-wave conditions. Obtaining high pulsed-laser energies from the existing chemical lasers involves the expenditure of an energy of at least 10% of the output laser energy to produce active centers. Thus, with the existing laser systems, to increase the energy of a pulsed chemical laser requires a proportionate increase in the initiation energy, which sets a limit on the energy rise in the pulsed systems. It would be possible to develop purely chemical pulsed lasers based on branching chain reactions, but the search for suitable reactions has so far been unavailing.

An entirely new laser-chemical process initiation scheme has recently been suggested [7.1] allowing purely chemical amplifiers to be built to amplify pulsed IR laser radiation. The laser photons in such amplifiers directly take part in the chemical process by initiating the production of active centers. The present chapter considers this approach.

If a chain reaction leading to stimulated emission of photons can be initiated by the same photons and the emission of the photons in the course of the reaction counterbalances or exceeds their expenditure to initiate the reaction (i.e. to produce active centers), the process on the whole can become self-sustained. The idea that such a process is in principle realizable was advanced for the first time by *Basov* and co-workers [7.1], who discovered the photochemical action of IR radiation. A self-accelerating chemical reaction proceeding in this way can be considered a reaction with a photon branching mechanism, as opposed to the well-known material and energetic branching mechanisms [7.2]. Based on photon-branching reactions, it is possible to build multiple-stage laser amplifiers requiring no supply of external energy in the form of UV radiation, electrical discharge, or electron beam. In such a multiple-stage system, the emission of a preceding stage initiates the reaction in the next stage of a greater size. A continuous, rather than stepwise, amplification is also possible. Here the flow of photons emitted in the preceding region of the amplifier initiates a reaction in the subsequent region and is simultaneously amplified due to the energy released in the reaction. In that case, a sort of photochemical wave propagates through the chemically active medium, with a continuous increase in the number of coherent photons in the wave. The possibility of amplification is due to the fact that the energy required to initiate chain-reaction chemical lasers is in principle much lower than the radiation energy generated. The H_2-F_2 and $D_2-F_2-CO_2$ laser systems are especially notable in this respect, their efficiency in terms of the absorbed e-beam initiation energy reaching around 1000% and 4000–5000%, respectively, at a high specific energy output (see Chap. 3).

Davis [7.3] and *Igoshin* and *Oraevsky* [7.4] have demonstrated the possibility of realizing a photon-branching reaction and amplifying IR radiation energy in a $CH_3F–D_2–F_2–CO_2–He$ mixture. In this molecular system, the CH_3F molecules absorb in a resonant fashion radiation at a wavelength of 9.6 µm, and then react with F_2 to yield F atoms. The laser-chemical process is described by the following kinetic scheme.

Excitation of the CH_3F molecules as a result of the *m*-fold absorption of photons with a wavelength of 9.6 µm:

$$(0) \quad CH_3F + mh\nu_l \overset{k_0}{\rightarrow} CH_3F^\nu \ ,$$

where *m* is the number of photons required to activate the CH_3F molecule and the superscript ν denotes vibrational excitation.

Chemical reaction between· the vibrationally excited CH_3F^ν molecules and F_2 molecules leading to the production of active centers:

$$(1) \quad CH_3F^\nu + F_2 \overset{k_1}{\rightarrow} CH_2F + HF + F \ .$$

Secondary active center nucleation reaction:

$$(2) \quad CH_2F + F_2 \overset{k_2}{\rightarrow} CH_2F_2 + F \ .$$

Quenching of the excited CH_3F^ν molecules:

$$(3) \quad CH_3F^\nu + M \overset{k_3^M}{\rightarrow} CH_3F + M(M^\nu) \ .$$

Chain reaction leading to the production of vibrationally excited DF molecules:

$$(4) \quad F + D_2 \overset{k_4}{\rightarrow} DF(v) + D \ ,$$

$$(5) \quad D + F_2 \overset{k_5}{\rightarrow} DF(v) + F \ .$$

Energy transfer between DF and CO_2:

$$(6) \quad DF(v) + CO_2(00^00) \overset{k_6}{\leftrightarrow} DF(v-1) + CO_2(00^01) \ .$$

Relaxation of the excited $DF(v)$ and $CO_2(00^01)$ molecules:

$$(7) \quad DF(v) + M \overset{k_7^M}{\rightarrow} DF(v-1) + M \ ,$$

$$(8) \quad CO_2(00^01) + M \overset{k_8^M}{\rightarrow} CO_2(nm^l0) + M \ .$$

Emission of coherent radiation by vibrationally excited CO_2 molecules:

(9) $CO_2(00^01) + h\nu_l \leftrightarrow CO_2(02^00) + 2h\nu_l$.

Stages (0) and (3) determine the active center nucleation under the effect of IR radiation, and stages (4–9) are the main stages involved in the operation of the $DF–CO_2$ chemical laser. In this scheme, the laser photons are an equal participant in the chemical process. The process described by the above kinetic scheme is a branching-chain reaction, provided that the photon emission and active center nucleation stages overlap in time and the condition

$$f \geqslant \frac{m}{\phi l}$$

is satisfied, where f is the quantum yield of coherent radiation, i.e. the number of laser photons emitted per active center in the course of the chain reaction, ϕ is the coefficient allowing for the loss of vibrational quanta as a result of relaxation of CH_3F^v (considered in detail later in the text), and l is the number of active centers resulting from the nucleation reactions and in our case equal to 2 [reactions (1) and (2)].

A broader view can be taken of scheme (0–9). In particular, the IR initiation and emission stages can be separated in time (short initiation in comparison with the chain-reaction time). What is more, IR initiation and emission can be effected at different frequencies corresponding to different vibrational-rotational transitions, so that the laser output frequency is at variance with the effective absorption frequency of the CH_3F molecule (e.g., initiation at $\lambda = 9.6$ μm and emission at $\lambda = 10.6$ μm). This version does not allow for multiple-stage amplification. It may, however, prove convenient for the last stage of a multiple-stage amplifier, because it is free from limitations due to absorption by the active medium at the lasing (amplification) frequency. The version in which the initiation and output photons are emitted by different molecular species is also of interest. For instance, the reaction could be initiated by the laser emission from CO_2 molecules at 9.6 or 10.6 μm with a view to obtaining laser emission at 3.8 (2.7) μm from DF (HF) molecules, and vice versa. It is advisable to consider all these versions not as photon-branching-reaction lasers but rather as IR radiation-driven chemical lasers (master oscillators and amplifiers), the former being a particular case of the latter.

A number of experiments have been performed to date on obtaining population inversion and lasing on vibrational-rotational transitions of HF (DF) molecules in the non-chain systems $SF_6 + H_2$ (D_2) [7.5–7], $SF_6 + C_3H_8$ [7.7], and $N_2F_4 + H_2$ (D_2) [7.8, 9] initiated by CO_2-laser radiation. In all the cases studied, the energy conversion coefficient is very small: $10^{-4}–10^{-3}$, the conversion efficiency decreasing with the pressure of the mixture rising over a few mm Hg. Adding molecular fluorine to the active mixture with a view to bringing about a chain reaction has practically no effect on the conversion efficiency [7.5]. The results obtained point to the difficulty of initiating lasing reactions in chemical lasers by making the SF_6 and N_2F_4 molecules dissociate under the effect of resonant irradiation with IR quanta.

It is believed here that active centers are produced as a result of dissociation of the molecules subjected to irradiation, there being some experimental evidence [7.10] to this effect. The low dissociation efficiency of the absorbing molecules in the active medium conditions is apparently explained by the fast relaxation of excitation preventing the accumulation in the molecules of the large enough number (a few tens) of quanta necessary for dissociation. Perhaps it will prove possible to overcome relaxation by increasing the rate of energy deposition in the system [7.6]. However, the use of this approach in multiple-stage amplification may be faced with difficulties due to a mismatch between the initiation and laser pulse durations in the course of the chain reaction.

Anyhow, it is in principle of interest to search for more effective chemical laser initiation mechanisms using IR radiation. The laser excitation of molecules stimulating them to enter into an exothermic chemical reaction accompanied by the production of active centers looks promising for this purpose. Thermoneutral or slightly endothermic reactions should not be ignored either. If the activation energy of the reaction is not very high, the necessary vibrational excitation level of the molecules under irradiation can be substantially reduced and a saving in the initiation energy thus effected (in part at the expense of the exothermicity of the active center production reaction as well), compared with the case where the active centers are produced via the dissociation channel.

The reaction considered in [7.4] between the vibrationally excited methyl-fluoride and molecular fluorine to yield F atoms is attractive for the following reasons.

1) The methyl-fluoride molecules are capable of resonant absorption of the CO_2-laser radiation, the absorption cross section (evaluated at 10^{-18}–10^{-19} cm^2 in [7.10, 11]) allowing the light to penetrate deep enough at the CH_3F concentrations (ranging from 10^{16} to 10^{17} cm^{-3}) necessary to produce the required amount of atomic fluorine. In this case, the distributed loss remains a only small fraction of the unsaturated gain of the DF–CO_2 laser.

2) The $V \rightarrow T$ relaxation of CH_3F is a rather slow process requiring some 15 000 gaskinetic collisions to complete [7.12].

3) As estimated later, the active center nucleation reaction (1) proceeds at a moderate vibrational excitation level of CH_3F and fast enough to ensure the necessary F atom production rate and to be competitive with the relaxation of excitation via channel (3).

4) The CH_2F radical resulting from the reaction of the vibrationally excited methyl-fluoride with fluorine also reacts rapidly with F_2 to yield another F atom. This cuts in half the expenditure of energy to generate an active center.

Let us estimate the maximum attainable energy in the IR radiation driven DF–CO_2 laser. We assume that under the high-pressure conditions considered below, the CH_3F molecules are activated through consecutive absorption of m resonant quanta. The upper level of the resonant absorption transition is rapidly depopulated by transition to other molecular states, while the lower level gets populated as rapidly by molecular collision, so that the activation process is limited by the absorption stage over a wide range of IR radiation intensities.

Accumulation of energy in the molecule can simultaneously proceed by another mechanism: it is a well-known fact that polyatomics can effectively acquire energy in an IR field in the absence of collisions [7.13, 14]. But this fact changes nothing in the essence of the inferences drawn about the energetics of the system.

Let us analyze a simpler energy conversion scheme in which the effect of photons emitted in the course of the reaction on the initiation process can be disregarded, which corresponds to the case where the initiating and emitted photons differ in frequency. With this scheme, all the energy necessary for initiation is supplied by the input IR-radiation pulse. If energy amplification is possible in this scheme, it is even more possible in conditions of a photon-branching reaction, where some of the emitted photons take part in the initiation of the reaction and thus perceptibly reduce, although their number is small, the energy requirements for the external initiation pulse. The photon-branching reaction dynamics is more complex and requires numerical analysis, but such a reaction can be expected to provide for a two- to three-fold increase in the energy gain, thanks to its "milder" excitation conditions.

Within the framework of the adopted assumptions, the kinetic equations describing the chain reaction initiation process have the form

$$\frac{dn_{CH_3F^v}}{dt} = \frac{\sigma I}{m} n_{CH_3F} - n_{CH_3F^v}\left(k_1 n_{F_2} + \sum_M k_3^M n_M\right) , \tag{7.1}$$

$$\frac{dn_a}{dt} = 2k_1 n_{CH_3F^v} n_{F_2} , \tag{7.2}$$

where σ is the absorption cross section of CH_3F at the initiating radiation wavelength, I is the initiating radiation intensity, $n_{CH_3F^v}$ is the concentration of the CH_3F molecules having absorbed m quanta, n_M is the concentration of the M species, k_1 and k_3^M are the rate constants of processes (1) and (3), respectively, and n_a is the total concentration of the D and F atoms.

Following [7.4], we will analyze the reaction initiation process and estimate the energy gain proceeding from the active medium conditions in the DF–CO_2 laser that correspond to the laser regime providing for high quantum yield and specific energy output values (see Chap. 3). These conditions (the partial pressures of the active mixture components, the initial mixture temperature, the concentration of the active centers) and the energy and temporal characteristics of the laser emission are as follows: $D_2 : F_2 : CO_2 : He = 1 : 3.7 : 0.4 : 3.7$, $p = 1$ atm, $T_0 = 300$ K, $n_a = 3.7 \times 10^{15}$ cm^{-3}, $t_l = 3.3$ μs, $\varepsilon_l = 333$ J/l, and $f = 4750$.

We assume that the duration of the initiating IR radiation pulse is somewhat shorter than that of the laser output pulse and amounts to 2×10^{-6} s.

The coefficient 2 on the right-hand side of equation (7.2) allows for the production of a second F atom as a result of reaction (2) following, as one can estimate, almost immediately after reaction (1). Indeed, based on the experimental studies [7.15] of the reactions of fluorination of methane, CH_4, and methyl-fluoride, CH_3F, it may be taken that the rate constant of reaction (2) is close to that of the reaction $CH_3 + F_2 \rightarrow CH_3F + F$, which equals 1.2×10^{-13}cm^3/s, see [7.15]. In that

Note that when analyzing the kinetic scheme, we disregarded the burn-out of CH_3F in the course of the chain reaction

$$F + CH_3F \rightarrow HF + CH_2F$$

$$CH_2F + F_2 \rightarrow CH_2F_2 + F \; ,$$

which goes concurrently with chain process (4), (5). Under the conditions considered, the burn-out time of CH_3F in this reaction is over 30 μs, which is an order of magnitude longer than the duration of the laser-chemical process. An additional contribution to the branching of the chain can be from the decay of the vibrationally excited $CH_2F_2^v$ molecules resulting from reaction (2) the heat of which equals 80 kcal/mole. Under low-pressure (1–10 mm Hg) conditions, the proportion of the $CH_2F_2^v$ molecules undergoing decay ranges, according to [7.15], is between 2×10^{-2} and 5×10^{-2}. Taking this process into account will only increase the estimated maximum attainable energy gain value, but the decay under high-pressure conditions has not been studied.

The question of the stability of the methyl-fluoride + fluorine mixture is of great importance. Experimental studies show that at low reactant pressures, the self-ignition of such mixtures can be suppressed by diluting the reactants with argon. The introduction of CO_2 and O_2 into the mixture is even more effective in this respect [7.15]. This in principle gives reason to hope that it will prove possible to make a high-pressure $CH_3F + CO_2 + D_2 + F_2 + He + O_2$ active mixture stable enough to last for a time of around 10 min, sufficient for a laser experiment. But even if there is failure to achieve the necessary long-term stability of the mixture at high pressures, it will still be possible to realize the above laser-chemical process in a reactor with a fast ($\simeq 10^{-3}$ s) turbulent mixing of the reactants admitted through a nozzle matrix.

The theory of photon-branching chain-reaction lasers has been extensively elaborated during the last few years [7.17–23]. *Igoshin* and *Pichugin* [7.17] suggested and studied a detailed numerical model of the process to produce atomic fluorine by way of resonant vibrational excitation of CH_3F discussed above, and showed it to be possible in principle.

Since no photon-branching reactions have been realized experimentally so far, examination of other approaches seems appropriate. *Igoshin* and *Pichugin* [7.18, 19] and *Basov* et al. [7.22] advanced a well-argued and fundamentally new way to bring about photon-branching chain reactions, based on nonresonant interaction between IR radiation and dispersed chemically active media. It consists in introducing finely dispersed particles (e.g., passivated metal particles) capable of absorbing IR radiation into the active medium of the HF– or DF–CO_2 laser. Under the effect of IR radiation of a certain intensity, these particles are heated to high temperatures within a time of around 1 μs, the temperature of the medium remaining practically unchanged. This heating causes either the evaporation or thermal decomposition of the particles. The metal vapor thus formed reacts with fluorine to yield free F atoms by the reaction $M + F_2 \rightarrow MF + F$. Each metal particle in the IR radiation field will thus be surrounded by active centers tending to diffuse into the laser medium. To make the free atoms spread sufficiently uniformly process (1) measured in [7.15] and also the experimental data [7.12, 16] on the rate constants for the relaxation of CH_3F on various species: $k_1 = 10^{-14}$–10^{-13} cm^3/s, $k_3^M = 2.2 \times 10^{-14}$ cm^3/s ($M = CH_3F$, He; the effectiveness of F_2 and D_2 is taken to be the same), $k_3^{CO_2} = 10^{-12}$ cm^3/s. The main contribution to the relaxation of CH_3F^v is from the quenching of excitation on CO_2, which proceeds by way of a quasiresonant energy transfer. For this reason, to achieve as high as possible values of ϕ, use should preferably be made of mixtures with high n_{F_2}/n_{CO_2} ratios. These requirements are satisfied by the same mixtures that are characterized by high quantum yields of laser emission. Under the conditions indicated above, the coefficient ϕ ranges between 0.073 and 0.43. Assume for the sake of definiteness that the absorption cross section of CH_3F agrees with the above estimate and equals 3 $\times 10^{-19}$ cm^2 and also that the concentration of CH_3F is 10^{17} cm^{-3}. In that case, to produce the necessary active center concentration within 2 μs, the initiating IR radiation intensity must be $(0.2$ to $1.2) \times 10^5$ W/cm^2. It should be noted that in the conditions under consideration, the quasistationary concentration of the CH_3F^v molecules having the necessary vibrational excitation energy is much lower than the total concentration of CH_3F, and comes to around 1%, in accordance with the formula

$$n_{CH_3F^v} = \frac{\sigma I}{m}\left(k_1 n_{F_2} + \sum_M k_3^M n_M\right)^{-1} n_{CH_3F} .$$

A simple calculation shows that the necessary concentration of the CH_3F^v molecules with a vibrational quanta store of $m \geqslant 13$ is achieved with the average number of vibrational quanta per molecule equal approximately to 3. That such a vibrational excitation level of methyl-fluoride can actually be reached in a high-pressure medium was experimentally proved by *McNair* et al. [7.11]. The above analysis makes it possible to estimate the sought-for energy gain, defined as the ratio of the energy emitted by the laser to the initiation energy. The expenditure of photons to generate an active center is equal to $m/2\phi$, and the number of photons emitted by the laser per active center is f. In that case, for the energy gain K, we have

$$K = 2f\phi/m . \tag{7.4}$$

Substituting into (7.4) the values of all parameters, we find that in the scheme under consideration, the energy gain K ranges between 50 and 300. Even if we take the quantum yield f to be lower than the maximum possible value by a factor of 2–3, we nevertheless come to the conclusion that it is possible to build on the basis of this scheme an IR radiation initiated laser with an energy gain of not lower than 10.

We can thus conclude that in conditions of a photon-branching reaction where both the absorbed and emitted photons are of the same frequency, the gain K of a single laser amplifier stage will be in the range 10–10^3. With a three-stage amplification, the energy of the "starting" pulse will be a mere 10^{-3}–10^{-9} of the output energy, which means that the energy requirements for the master oscillator will be very low. It is exactly this fact that allows one to consider pulsed chemical lasers operating on photon-branching reactions as being self-contained.

Note that when analyzing the kinetic scheme, we disregarded the burn-out of CH_3F in the course of the chain reaction

$$F + CH_3F \rightarrow HF + CH_2F$$

$$CH_2F + F_2 \rightarrow CH_2F_2 + F \ ,$$

which goes concurrently with chain process (4), (5). Under the conditions considered, the burn-out time of CH_3F in this reaction is over 30 μs, which is an order of
magnitude longer than the duration of the laser-chemical process. An additional
contribution to the branching of the chain can be from the decay of the vibrationally
excited $CH_2F_2^v$ molecules resulting from reaction (2) the heat of which equals
80 kcal/mole. Under low-pressure (1–10 mm Hg) conditions, the proportion of the
$CH_2F_2^v$ molecules undergoing decay ranges, according to [7.15], is between
2×10^{-2} and 5×10^{-2}. Taking this process into account will only increase the
estimated maximum attainable energy gain value, but the decay under high-
pressure conditions has not been studied.

The question of the stability of the methyl-fluoride + fluorine mixture is of great
importance. Experimental studies show that at low reactant pressures, the self-
ignition of such mixtures can be suppressed by diluting the reactants with argon.
The introduction of CO_2 and O_2 into the mixture is even more effective in this
respect [7.15]. This in principle gives reason to hope that it will prove possible to
make a high-pressure $CH_3F + CO_2 + D_2 + F_2 + He + O_2$ active mixture stable
enough to last for a time of around 10 min, sufficient for a laser experiment. But
even if there is failure to achieve the necessary long-term stability of the mixture at
high pressures, it will still be possible to realize the above laser-chemical process in a
reactor with a fast ($\simeq 10^{-3}$ s) turbulent mixing of the reactants admitted through a
nozzle matrix.

The theory of photon-branching chain-reaction lasers has been extensively
elaborated during the last few years [7.17–23]. *Igoshin* and *Pichugin* [7.17]
suggested and studied a detailed numerical model of the process to produce atomic
fluorine by way of resonant vibrational excitation of CH_3F discussed above, and
showed it to be possible in principle.

Since no photon-branching reactions have been realized experimentally so far,
examination of other approaches seems appropriate. *Igoshin* and *Pichugin* [7.18,
19] and *Basov* et al. [7.22] advanced a well-argued and fundamentally new way to
bring about photon-branching chain reactions, based on nonresonant interaction
between IR radiation and dispersed chemically active media. It consists in
introducing finely dispersed particles (e.g., passivated metal particles) capable of
absorbing IR radiation into the active medium of the HF– or DF–CO_2 laser.
Under the effect of IR radiation of a certain intensity, these particles are heated to
high temperatures within a time of around 1 μs, the temperature of the medium
remaining practically unchanged. This heating causes either the evaporation or
thermal decomposition of the particles. The metal vapor thus formed reacts with
fluorine to yield free F atoms by the reaction $M + F_2 \rightarrow MF + F$. Each metal particle
in the IR radiation field will thus be surrounded by active centers tending to diffuse
into the laser medium. To make the free atoms spread sufficiently uniformly

throughout the active volume during the time τ limited by the duration of the laser-chemical process ($\leqslant 10$ μs), the condition $R_{av} \leqslant 2R_{diff}$ must be satisfied, where R_{av} is the average distance between the particles and $R_{diff} \simeq (D\tau)^{1/2}$, D being the diffusion coefficient of the active centers in the laser medium. This puts a lower bound on the necessary concentration ($\simeq 1/R_{av}^3$) of the dispersed particles. With the concentration of the particles being specified, their maximum size is determined subject to the condition that the coefficient of the laser emission attenuation due to dispersion and absorption by the disperse phase is not in excess of the laser gain. For one to be able to fill the active medium uniformly with dispersed particles without running the risk of the mixture igniting spontaneously, the particles must either be chemically inert at room temperature (NaN_3) or have a strong surface film of an inert oxide (Al_2O_3 in the case of Al particles). The finely dispersed metal particles must be passivated with oxygen in an inert gas or rarefied air in order to prevent them from burning out completely. To introduce the particles into the laser gas mixture when it is being admitted into the reactor, use can be made of a feeder comprising a cylindrical cavity the bottom of which can be vibrated by means of an electrical vibrator [7.24]. The particles dusted as a result of vibration are carried away by the gas flow. To prevent the particles from coalescing, an electric charge can be induced in them.

Igoshin and *Pichugin* [7.20, 21, 23] carried out numerical calculations of photon-branching chain-reaction lasers. The detailed kinetic model calculations [7.20] of the DF–CO_2 laser initiated by IR radiation through resonant vibrational excitation of CH_3F molecules have shown that a sufficiently high energy gain of $K = E_{out}/E_{in} \simeq 26$ can be achieved at a mixture pressure of 7 atm. (mixture $D_2 : F_2 : CO_2 : He = 1$: $3.7 : 0.4 : 3.7$, $p_{CH_3F} = 1$ mm Hg, initiating pulse intensity $I_0 = 6$ mW/cm^2, initiating pulse duration $\tau_i = 2$ μs). In their work reported in [7.21], the authors calculated the lasing characteristics of an IR radiation driven DF–CO_2 laser with a disperse active medium. They performed simultaneous numerical calculation of chemical kinetics equations, relaxation equations for the average numbers of vibrational quanta stored in vibrational modes of CO_2, DF, and D_2, heat balance equation, oscillator rate equation, and also equations for the temperature and radius of the dispersed aluminum particles subject to evaporation in the initiating IR radiation field. For the same mixture composition ($D_2 : F_2 : CO_2 : He = 1 : 3.7 : 0.4 : 3.7$, $p = 7$ atm.), a gain of $K = 26$ is achieved at $I_0 = 20$ mW/cm^2, $\tau_i = 2$ μs, an initial aluminum particle radius of 0.25 μm, and particle concentration of $n = 10^7$ cm^{-3}. The authors also carried out similar calculations [7.23] for an H_2–F_2 laser initiated by IR radiation evaporating fine aluminum particles dispersed in the laser medium. At a laser medium pressure of 1 atm., the energy gain of the H_2–F_2 laser reaches 5–20 and its specific energy output, 200–100 J/l, respectively, when using particles with a radius of 0.1–0.05 μm in a concentration of 10^9 cm^{-3}.

Thus, it is theoretically possible to develop pulsed chemical IR laser amplifiers possessing a sufficiently high gain.

It is of great interest to try to realize in the future a photon-branching chain chemical process involving concurrent and interrelated photon- and energetic- (or material-) branching mechanisms. The quantitative analysis performed in this chapter and also in Chap. 3 gives reasons for positive answers to the question of the feasibility of such a combined process, although the matter requires further research. The implementation of such, as yet hypothetical, process would make it possible to achieve the ultimate in the energy performance characteristics of chemical lasers, because the result of each primary photon-branching process would in this case be multiplied, because of energetic branching and vice versa. It should be noted that there is a difference in principle between the photon-branching chain chemical process considered here and the branching-chain chemical-wave process discussed as far back as the 1960's [7.25]. The chemical-wave process [7.25] involves a chemical chain reaction and a branching-chain "quantum reaction" (stimulated emission of radiation) which proceed concurrently but have no mutual relation.

References

Chapter 1

1.1 J.C. Polanyi: J. Chem. Phys. **34**, 347 (1961)
1.2 N.G. Basov, A.N. Oraevsky: ZhETF **44**, 1742 (1963)
1.3 A.N. Oraevsky: ZhETF **45**, 177 (1963)
1.4 V.L. Tal'roze: Kinet. Katal. **5**, 11 (1964)
1.5 Appl. Opt. Suppl. 2 (Chemical Lasers) (1965)
1.6 J.V. Kasper, G.C. Pimentel: Phys. Rev. Lett. **10**, 352 (1965)
1.7 T.F. Deutsch: Appl. Phys. Lett. **10**, 234 (1967)
1.8 K.J. Kompa, G.C. Pimentel: J. Chem. Phys. **47**, 857 (1967)
1.9 K.G. Anlauf, P.J. Knutz, D.H. Maylott et al.: Discuss. Farad. Soc. **4**, 183 (1967)
1.10 N.G. Basov, A.N. Oraevsky, V.A. Shcheglov: ZhETF **37**, 339 (1967)
1.11 R.W.F. Gross: J. Chem. Phys. **50**, 1887 (1969)
1.12 H.L. Chen, J.C. Stephenson, C.B. Moore: Chem. Phys. Lett. **2**, 593 (1968)
1.13 T.A. Cool, F.J. Falk, R.R. Stephens: Appl. Phys. Lett. **15**, 318 (1969)
1.14 N.G. Basov, E.P. Markin, A.I. Nikitin, A.N. Oraevsky: IEEE J. **QE-6**, 183 (1970)
1.15 D.J. Spencer, H. Mirels, T.A. Jacobs: Appl. Phys. Lett. **16**, 235 (1970)
1.16 J.R. Airey, S.F. McKay: Appl. Phys. Lett. **15**, 401 (1969)
1.17 A.N. Oraevsky: ZhETF **55**, 1423 (1968)
1.18 O.M. Batovskii, G.K. Vasil'yev, E.F. Markov, V.L. Tal'roze: Pis'ma ZhETF **9**, 341 (1969)
1.19 N.G. Basov, L.V. Kulakov, E.P. Markin, A.I. Nikitin, A.N. Oraevsky: Pis'ma ZhETF **11**, 613 (1969)
1.20 G.G. Dolgov-Savel'yev, V.F. Zharov, Yu. S. Neganov, G.M. Chumak: ZhETF **61**, 64 (1971)
1.21 N.G. Basov, S.I. Zavorotnyi, E.P. Markin, A.I. Nikitin, A.N. Oraevsky: Pis'ma ZhETF **15**, 135 (1972)
1.22 N.G. Basov, A.N. Oraevsky: *Author's Certificate No 436413, April 24, 1967.* Bull. Otkr. Izobr. **7** (1974)
1.23 T.A. Cool: IEEE J. **QE 9**, 72 (1973)
1.24 R.A. Meinzer: Int. J. Chem. Kinet. **2**, 335 (1970)
1.25 V.F. Zharov, V.K. Malinovskii, Yu. S. Neganov, G.M. Chumak: Preprint IYaF, Siberian Dept. Akad. Nauk SSSR **45-71** (1974)
1.26 W.E. McDermott, N.R. Pchelkin, D.J. Benard, R.R. Bonsek: Appl. Phys. Lett. **32**, 469 (1978)
1.27 N.G. Basov, A.N. Oraevsky, in: *Nauka i chelovechestvo* (Science and Mankind) (Znaniye, Moscow, 1983), pp. 259–273
1.28 N.G. Basov, V.V. Gromov, E.P. Markin, A.N. Oraevsky, A.K. Piskunov, D.S. Shapovalov: Kvantovaya Elektron. (Moscow) **3**, 1154 (1976)
1.29 B.L. Borovich, V.S. Zuev, V.A. Katulin, L.D. Mikheev, F.A. Nikolayev, O. Yu. Nosach, V.B. Rozanov: *Sil'notochnye izluchayushchie razryady i gazovye lazery s opticheskoi nakachkoi* (High-Current Emitting Discharges and Optically Pumped Gas Lasers), in: *Itogi Nauki i Tekhniki*, Ser. Radiotekh. **15** (VINITI, Moscow, 1978)
1.30 V.A. Danilychev, O.M. Kerimov, I.B. Kovsh: *Molekulyarnye gazovye lazery vysokogo davleniya* (High-Pressure Molecular Gas Lasers), in: *Itogi Nauki i Tekhniki, Ser. Radiotekh.* **12** (VINITI, Moscow, 1978)
1.31 C.B. Moore, Ed.: *Chemical and Biochemical Applications of Lasers* 1 (1974)

1.32 N.G. Basov, V.A. Danilychev, V.I. Dolinina, A.N. Lobanov, A.N. Oraevsky, V.I. Panteleev, A.F. Suchkov, B.M. Urin, F.S. Faizullov, Yu. N. Shebeko, E.V. Gorozhankin, V.V. Kurenkov, V.N. Men'shov: Trudy FIAN **116**, 146 (1980)

Chapter 2

2.1   J.C. Polanyi: J. Chem. Phys. **34**, 347 (1961)
2.2   J.V.V. Kasper, G.C. Pimentel: Phys. Rev. Lett. **14**, 352 (1965)
2.3   A.N. Oraevsky: ZhETF **45**, 177 (1963)
2.4   V.L. Tal'roze: Kinet. Katal. **5**, 11 (1964)
2.5   V.I. Igoshin, A.N. Oraevsky: ZhETF **59**, 1240 (1970)
2.6   A.N. Oraevsky: ZhETF **55**, 1423 (1968)
2.7   J.C. Polanyi: Appl. Opt. Suppl. **2** (Chemical Lasers), 109 (1965)
2.8   N.G. Basov, V.I. Igoshin, E.P. Markin, A.N. Oraevsky: *Laser und ihre Anwendung*, Teil 2, Int. Tag., Dresden, 10.6–17.6.1970, S.91
2.9   N.G. Basov, V.I. Igoshin, E.P. Markin, A.N. Oraevsky, in: *Kvantovaya elektronika* (Quantum Electronics), Iss. 2 (Sovetskoe radio, Moscow, 1971), p. 3
2.10   M.S. Dzhidzhoev, V.T. Platonenko, R.V. Khokhlov: Uspekhi Fiz. Nauk **100**, 641 (1970)
2.11   G.C. Pimentel: Pure Appl. Chem. **18**, 275 (1969)
2.12   K.E. Shuler, F. Carrington, J.C. Light: Appl. Opt. Suppl. **2** (Chemical Lasers), 81 (1965)
2.13   L.T. My: Chim. Mod. **10**, 259 (1965)
2.14   R.A. Young: J. Chem. Phys. **40**, 1848 (1964)
2.15   A.N. Oraevsky: ZhETF **48**, 1150 (1965)
2.16   V.L. Tal'roze, G.K. Vasil'yev, O,M. Batovskii: Kinet. Katal. **11**, 277 (1970)
2.17   M.S. Dzhidzhoev, M.I. Pimenov, V.T. Platonenko, Yu. V. Filippov, R.V. Khokhlov: ZhETF **57**, 411 (1969)
2.18   A.N. Oraevsky: Khim. Vysokikh Energ. **8**, 3 (1974)
2.19   A.N. Bashkin, V.I. Igoshin, A.I. Nikitin, A.N. Oraevsky: *Khimicheskie lazery* (Chemical Lasers), *Itogi Nauk i tekhniki*, Ser. Radiothekh. **8** (VINITI, Moscow, 1975)
2.20   V.N. Kondrat'yev: *Kinetika khimicheskikh gazovykh reaktsii* (Kinetics of Chemical Gas Reactions) (Izd. Akad. Nauk SSSR, Moscow, 1958)
2.21   B.S. Rabinovich, M.S. Flauers: *Khimicheskaya aktivatsiya* (Chemical Activation), in: *Khimicheskaya kinetika i tsepnye reaktsii* (Nauka, Moscow, 1966), p. 66
2.22   T. Carrington, D. Garvin, in: *Comprehensive Chemical Kinetics*, Vol. 3, ed. by C.H. Bamford and C.F.H. Tipper (Elsevier Publ. Co., Amsterdam-London-New York, 1969)
2.23   V.N. Kondrat'yev, E.E. Nikitin: *Kinetika i Mekhanizm gazofaznykh reaktsii* (Kinetics and Mechanism of Gas-Phase Reactions) (Nauka, Moscow, 1974)
2.24   M.J. Berry, G.C. Pimentel: J. Chem. Phys. **53**, 3453 (1970)
2.25   F.M. Tablas, G.C. Pimentel: IEEE J. **QE-6**, 176 (1970)
2.26   T.P. Schaffer, P.E. Siska, J.M. Parson, E.P. Tully, J.C. Wong, J.T. Lee: J. Chem. Phys. **53**, 3385 (1970)
2.27   E.E. Nikitin: *Teoriya elementarnykh atomno-molekulyarnykh protsessov v gazakh* (Theory of Elementary Atomic-Molecular Processes in Gases) (Khimiya, Moscow, 1970)
2.28   J.C. Polanyi, S.D. Rosner: J. Chem. Phys. **38**, 1028 (1963)
2.29   S. Solimeno: Phys. Bull., 1974, November, p. 517
2.30   R.L. Wilkins: J. Chem. Phys. **57**, 912 (1972)
2.31   R.L. Wilkins: J. Chem. Phys. **58**, 2326 (1973)
2.32   J.C. Light, C.C. Rankin: J. Chem. Phys. **51**, 1702 (1969)
2.33   G.C. Schatz, J. M. Bowman, A. Kupperman: J. Chem. Phys. **63**, 674 (1975)
2.34   J.N.L. Connor, W. Jakubetz, J. Manz: Chem. Phys. Lett. **39**, 75 (1976)
2.35   D.L. Miller, R.E. Wyatt: J. Chem. Phys. **67**, 1302 (1977)
2.36   T.F. Deutsch: Appl. Phys. Lett. **11**, 18 (1967)
2.37   E. Guellar, J.H. Parker, G. Pimentel: J. Chem. Phys. **61**, 422 (1974)
2.38   G.D. Downey, D.W. Robinson, J.H. Smith: J. Chem. Phys. **66**, 1685 (1977)

2.39   R.D. Levine, R.B. Bernstein: Acc. Chem. Res. **7**, 393 (1974)
2.40   C.E. Shannon, W. Weaver: *The Mathematical Theory of Communication* (University of Illinois Press, Urbana, 1949)
2.41   J.N.L. Connor, W. Jakubetz, J. Manz: J. Chem. Phys. **17**, 451 (1976)
2.42   M. Rubinson, J.I. Steinfeld: Chem. Phys. **4**, 467 (1974)
2.43   I. Procaccia, R.D. Levine: J. Chem. Phys. **63**, 4261 (1975)
2.44   C.C. Jensen, J.I. Steinfeld, R.D. Levine: J. Chem. Phys. **69**, 1432 (1978)
2.45   G. Herzberg: *Molecular Spectra and Molecular Structure I. Diatomic Molecules* (New York, 1939)
2.46   J.F. Keilkopf: J. Opt. Soc. Am. **63**, 987 (1973)
2.47   S.S. Penner: *Quantitative Molecular Spectroscopy and Gas Emissivities* (Addison-Wesley, Reading, London 1959)
2.48   A.N. Chester, L.D. Hess: IEEE J. **QE-8**, 1 (1972)
2.49   J.R. Airey, S.F. Fried: Chem. Phys. Lett. **8**, 23 (1971)
2.50   R.N. Schwartz, Z.I. Slawsky, K.F. Herzfeld: J. Chem. Phys. **19**, 1591 (1951)
2.51   S. Ormonde: Rev. Mod. Phys. **47**, 193 (1975)
2.52   F.J. Tanczos: J. Chem. Phys. **25**, 439 (1956)
2.53   M.V. Vol'kenshtein, L.A. Gribov, M.A. Yel'yashevich, B.I. Stepanov: *Kolebaniya molecul* (Molecular Vibrations) (Nauka, Moscow, 1972)
2.54   J.D. Lambert, R. Slater: Proc. R. Soc. (London) **253**, 277 (1959)
2.55   T.L. Cottrell, A.J. Matheson: Trans. Faraday Soc. **58**, 2336 (1962)
2.56   C.B. Moore: J. Chem. Phys. **43**, 2979 (1965)
2.57   H.K. Shin: J. Phys. Chem. **75**, 1079 (1971)
2.58   G.A. Kapralova, E.E. Nikitin, A.M. Chaikin: Kinet. Katal. **10**, 974 (1969)
2.59   L.L. Poulsen, P.L. Houston, J.I. Steinfeld: J. Chem. Phys. **58**, 3381 (1973)
2.60   H.K. Shin: Chem. Phys. Lett. **10**, 81 (1971)
2.61   H.K. Shin: Chem. Phys. Lett. **11**, 628 (1971)
2.62   H.K. Shin: J. Chem. Phys. **57**, 3484 (1972)
2.63   H.K. Shin: J. Chem. Phys. **59**, 879 (1973)
2.64   H.K. Shin: J. Phys. Chem. **77**, 1666 (1973)
2.65   H.K. Shin: J. Chem. Phys. **41**, 2864 (1964)
2.66   H.K. Shin: J. Chem. Phys. **60**, 2305 (1974)
2.67   H.K. Shin: J. Chem. Phys. **60**, 2167 (1974)
2.68   W.G. Tam: Can. J. Phys. **52**, 854 (1974)
2.69   H.K. Shin: Chem. Phys. Lett. **26**, 450 (1974)
2.70   H.K. Shin: Chem. Phys. Lett. **50**, 377 (1977)
2.71   G.C. Berend, R.L. Thommarson: J. Chem. Phys. **58**, 3203 (1973)
2.72   J.K. Hancock, A.W. Saunders, Jr.: J. Chem. Phys. **65**, 1275 (1976)
2.73   H.K. Shin, A.W. Young: J. Chem. Phys. **60**, 193 (1974)
2.74   G.C. Berend, R.L. Thommarson: J. Chem. Phys. **58**, 3454 (1973)
2.75   I.W.M. Smith, P.M. Wood: Mol. Phys. **25**, 441 (1873)
2.76   D.L. Thompson: J. Chem. Phys. **57**, 4170 (1972)
2.77   R.L. Wilkins: J. Chem. Phys. **59**, 698 (1973)
2.78   R.L. Wilkins: J. Chem. Phys. **58**, 3038 (1973)
2.79   D.L. Thompson: J. Chem. Phys. **57**, 4164 (1972)
2.80   R.L. Wilkins: Mol. Phys. **29**, 555 (1975)
2.81   R.L. Wilkins: J. Chem. Phys. **63**, 534 (1975)
2.82   R.L. Thommarson, G.C. Berend: Int. J. Chem. Kinet. **6**, 597 (1974)
2.83   R.F. Heidner III, J.F. Bott: J. Chem. Phys. **63**, 1810 (1975)
2.84   J.F. Bott, R.F. Heidner III: J. Chem. Phys. **65**, 1076 (1976)
2.85   J.F. Bott, R.F. Heidner III: J. Chem. Phys. **60**, 2878 (1976)
2.86   H.K. Shin: Chem. Phys. Lett. **14**, 64 (1972)
2.87   H.K. Shin: Chem. Phys. Lett. **32**, 63 (1975)
2.88   H.K. Shin: J. Chem. Phys. **62**, 4230 (1975)
2.89   H.K. Shin: J. Chem. Phys. **68**, 5265 (1978)
2.90   C.B. Moore: Adv. Chem. Phys. **23**, 41 (1973)

2.91   D. Rapp, P.E. Golden: J. Chem. Phys. **40**, 573 (1964)
2.92   B.H. Mahan: J. Chem. Phys. **46**, 98 (1967)
2.93   R.D. Sharma, C.A. Brau: J. Chem. Phys. **50**, 924 (1969)
2.94   R.D. Sharma: Phys. Rev. **177**, 102 (1969)
2.95   W.O. Jeffers, J.D. Kelley: J. Chem. Phys. **55**, 4433 (1971)
2.96   R.D. Sharma, H. Schlossberg: Chem. Phys. Lett, **20**, 5 (1973)
2.97   R.D. Sharma, R.H. Picard: J. Chem. Phys. **62**, 3340 (1975)
2.98   E.E. Nikitin: Nauchn. Tr. Inst. Mech. Mosk. Gos. Univ. **23**, 44 (1973)
2.99   C. Wittig, I.W.M. Smith: Chem. Phys. Lett. **16**, 212 (1972)
2.100 R.D. Sharma: Chem. Phys. Lett. **30**, 261 (1975)
2.101 M. Lev-On, W.E. Ralke, R.C. Millikan: Chem. Phys. Lett. **24**, 59 (1974)
2.102 R.D. Sharma, R.B. Malt, R.R. Hart, R.H. Picard: Chem. Phys. Lett. **35**, 286 (1975)
2.103 T.A. Dillon, J.C. Stephenson: J. Chem. Phys. **58**, 2056 (1973)
2.104 T.A. Dillon, J.C. Stephenson: J. Chem. Phys. **60**, 4286 (1974)
2.105 P.D. Cait: Chem. Phys. Lett. **35**, 72 (1975)
2.106 L.H. Sentman: Chem. Phys. Lett. **18**, 493 (1973)
2.107 R.D. Sharma, H. L. Chen, A. Szöke: J. Chem. Phys. **58**, 3519 (1973)
2.108 H.K. Shin: J. Chem. Phys. **63**, 2901 (1975)
2.109 H.K. Shin, Y.H. Kim: J. Chem. Phys. **64**, 3634 (1976)
2.110 R.L. Wilkins: J. Chem. Phys. **67**, 5838 (1977)
2.111. D.B. Rensch: Appl. Opt. **13**, 2546 (1974)
2.112. D. Rapp, T. Kassal: Chem. Rev. **69**, 61 (1969)
2.113 K.F. Herzfeld, T.A. Litovitz: *Absorption and Dispersion of Ultrasonic Waves* (Academic Press, London, New York, 1959)
2.114 A.B. Callear, J.D. Lambert, in: *Comprehensive Chemical Kinetics*, Vol. 3, ed. by C.H. Bamford & C.F.H. Tippet (Elsevier Publ. Co., Amsterdam-London-New York, 1969)
2.115 E.V. Stupochenko, S.A. Losev, A.I. Osipov: *Relaksatsionnye protsessy v udarnykh volnakh* (Relaxation Processes in Shock Waves) (Nauka, Moscow, 1965)
2.116 J.C. Polanyi, K.B. Woodall: J. Chem. Phys. **56**, 1563 (1972)
2.117 N.C. Lang, J.C. Polanyi, J. Wanner: Chem. Phys. **24**, 219 (1977)
2.118 J.J. Hinchen, R.H. Hobbs: J. Chem. Phys. **65**, 2732 (1976)
2.119 E.B. Turner, W.D. Adams, G. Emanuel: J. Comput. Phys. **11**, 15 (1973)
2.120 V.I. Igoshin, V.S. Masterov: Preprint FIAN (Moscow) **87** (1975)
2.121 W.S. King, H. Mirels: AIAA J. **10**, 1647 (1972)
2.122 R. Tripodi, L.J. Coulter, B.R. Bronfin, L.S. Cohen: AIAA J. **13**, 776 (1975)
2.123 A.P. Kothari, J.D. Anderson Jr., E. Jones: AIAA J. **15**, 92 (1977)
2.124 A.A. Stepanov, V.A. Shcheglov: Preprint FIAN (Moscow) N **182** (1976)
2.125 V.I. Igoshin, A.N. Oraevsky: Preprint FIAN (Moscow) N **162** (1969)
2.126 J.R. Airey: J. Chem. Phys. **52**, 156 (1970)
2.127 G. Emanuel: J. Quant. Spectrosc. Radiat. Transfer **11**, 1481 (1972)
2.128 V.I. Igoshin, V.S. Masterov: Kvantovaya Elektron. (Moscow) **2**, 1638 (1975)
2.129 N.G. Basov, V.T. Galochkin, V.I. Igoshin, L.V. Kulakov, E.P. Markin, A.I. Nikitin, A.N. Oraevsky: Appl. Opt. **10**, 1814 (1971)
2.130 V.I. Igoshin, L.V. Kulakov, A.I. Nikitin, in: *Kvantovaya elektronika* (Quantum Electronics), Iss. 4 (Sovetskoe radio, Moscow, 1973), p. 50
2.131 M.J. Berry: J. Chem. Phys. **59**, 6229 (1973)
2.132 P.S. Perry, J.C. Polanyi: Chem. Phys. **12**, 419 (1976)
2.133 H.J. Parker, G.C. Pimentel: J. Chem. Phys. **51**, 91 (1969)
2.134 J.C. Polanyi, K.B. Woodall: J. Chem. Phys. **57**, 1574 (1972)
2.135 J.C. Polanyi, D.C. Tardy: J. Chem. Phys. **51**, 5717 (1969)
2.136 H.W. Chang, D.W. Setser, H.J. Perona: J. Phys. Chem. **75**, 2070 (1971)
2.137 N. Jonathan, C.W. Melliar-Smith, D.H. Slater: Mol. Phys. **20**, 93 (1971)
2.138 N. Jonathan, C.M. Meliar-Smith, S. Okuda, D.H. Slater, D. Timlin: Mol. Phys. **22**, 561 (1971)
2.139 G.K. Vasil'yev, V.B. Ivanov, E.F. Makarov, A.G. Ryabenko, V.L. Tal'roze: Dokl. Akad. Nauk SSSR **215**, 120 (1974)

2.140 D.J. Douglas, J.C. Polanyi: Chem. Phys. **16** 1 (1976)

2.141 R.D. Coombe, G.C. Pimentel: J. Chem. Phys. **59**, 1535 (1973)

2.142 R.D. Coombe, G.C. Pimentel: J. Chem. Phys. **59**, 251 (1973)

2.143 O.D. Krogh, K. Stone, G.C. Pimentel: J. Chem. Phys. **66**, 368 (1977)

2.144 H.W. Chang, D.W. Setser, M.J. Perona, P.J. Johnson: Chem. Phys. Lett. **9**, 587 (1971)

2.145 W. Duewer, D.W. Setser: J. Chem. Phys. **58**, 2310 (1973)

2.146 H.W. Chang, D.W. Setser: J. Chem. Phys. **58**, 2298 (1973)

2.147 J.T. Muckerman: J. Chem. Phys. **54**, 1155 (1971)

2.148 L.J. Kirsh, J.C. Polanyi: J. Chem. Phys. **57**, 4498 (1972)

2.149 N. Jonathan, C. M. Melliar-Smith, D. Timlin, D.H. Slater: Appl. Opt. **10**, 1821 (1971)

2.150 K.C. Kim, D.W. Setser, C.M. Bogan: J. Chem. Phys. **60**, 1837 (1974)

2.151 N. Jonathan, S. Okuda, D. Timlin: Mol. Phys. **24**, 1143 (1972)

2.152 J. C. Polanyi, J. J. Sloan: J. Chem. Phys. **57**, 4988 (1972)

2.153 J. M. Herbelin, G. Emanuel: J. Chem. Phys. **60**, 689 (1974)

2.154 K.G. Anlauf, P.J. Kuntz, D.HG. Maylotte, P.D. Pacey, J.C. Polanyi: Discuss. Faraday Soc. **44**, 183 (1967)

2.155 H. Hendtmann, J.C. Polanyi: Appl. Opt. **10**, 1738 (1972)

2.156 J.C. Polanyi, W.J. Skrlac: Reprint (University of Toronto, 1977)

2.157 K.G. Anlauf, J.C. Polanyi, W.H. Wong, K.B. Woodall: J. Chem. Phys. **49**, 5189 (1968)

2.158 J.B. Anderson, R.T. Kung: J. Chem. Phys. **58**, 2477 (1973)

2.159 G. Hancock, C. Morley, I.W.M. Smith: Chem. Phys. Lett. **12**, 193 (1971)

2.160 J.L. Ahl, T.A. Cool: J. Chem. Phys. **58**, 5540 (1973)

2.161 M.J. Binda, C.R. Jones: Appl. Phys. Lett. **22**, 44 (1973)

2.162 L.S. Blair, W.D. Breshears, G.L. Schott: J. Chem. Phys. **59**, 1582 (1973)

2.163 J.F. Bott: J. Chem. Phys. **57**, 96 (1972)

2.164 J.F. Bott, N. Cohen: J. Chem. Phys. **58**, 4539 (1973)

2.165 S.S. Fried, J. Wilson, R.L. Taylor: IEEE J. **QE-9**, 59 (1973)

2.166 J.K. Hancock, W.H. Green: J. Chem. Phys. **57**, 4515 (1972)

2.167 J.J. Hinchen: J. Chem. Phys. **59**, 233 (1973)

2.168 R.A. Lucht, T.A. Cool: J. Chem. Phys. **60**, 2554 (1974)

2.169 R.A. Lucht, T.A. Cool: J. Chem. Phys. **60**, 1026 (1974)

2.170 J.J. Hinchen: J. Chem. Phys. **59**, 2224 (1973)

2.171 J.F. Bott: J. Chem. Phys. **61**, 3414 (1974)

2.172 R.M. Osgood, P.B. Sackett, A. Javan: J. Chem. Phys. **60**, 1464 (1974)

2.173 R.R. Stephens, T.A. Cool: J. Chem. Phys. **56**, 5863 (1972)

2.174 M.A. Kwok, R.L. Wilkins: J. Chem. Phys. **63**, 2453 (1975)

2.175 J.F. Bott, N. Cohen: J. Chem. Phys. **55**, 3698 (1971)

2.176 G.K. Vasil'yev, E.F. Makarov, V.G. Papin, V.L. Tal'roze: ZhETF **64**, 2046 (1973)

2.177 V.C. Solomon, J.A. Blauer, F.C. Jaye, J.G. Hnat: Int. J. Chem. Kinet. **3**, 215 (1971)

2.178 N. Cohen, J.F. Bott: Appl. Opt. **15**, 28 (1976)

2.179. J.R. Airey, I.W.M. Smith: J. Chem. Phys. **57**, 1669 (1972)

2.180 P.R. Poole, I.W.M. Smith: J. Chem. Soc. Faraday Trans. II **73**, 1434 (1977)

2.181 D.J. Douglas, C.B. Moore: Chem. Phys. Lett. **57**, 485 (1978)

2.182 E.N. Rityn', A.P. Burtsev, P.V. Slobodskaya: Zh. Prikl. Spektroskop. **24**, 347 (1976)

2.183 G.P. Quigley, G.J. Wolga: Chem. Phys. Lett. **27**, 276 (1974)

2.184 M.A. Kwok, R.L. Wilkins: J. Chem. Phys. **60**, 2189 (1974)

2.185 R.F. Heidner III, J.F. Bott: Ber. Bunsenges. Phys. Chem. **8**, 128 (1977)

2.186 J.F. Bott, R.F. Heidner III: J. Chem. Phys. **68**, 1708 (1978)

2.187 G.P. Quigley, G.J. Wolga: Chem. Phys. **63**, 5263 (1975)

2.188 G.K. Vasil'yev, E.F. Makarov, V.G. Papin: ZhETF **45**, 435 (1975)

2.189 J.A. Blauer, W.C. Solomon, K.H. Sentman, T. W. Owens: J. Chem. Phys. **57**, 3277 (1972)

2.190 J.F. Bott, N. Cohen: J. Chem. Phys. **61**, 681 (1974)

2.191 J.F. Bott: J. Chem. Phys. **65**, 4239 (1976)

2.192 J.F. Bott: J. Chem. Phys. **61**, 2530 (1974)

2.193 J.A. Blauer, W.C. Solomon, T.W. Owens: Int. J. Chem. Kinet. **4**, 293 (1972)

352   References

2.194 W.H. Green, J.K. Hancock: IEEE J. **QE-9**, 50 (1973)
2.195 J.K. Hancock, W.H. Green: J. Chem. Phys. **56**, 2474 (1972)
2.196 G.K. Vasil'yev, E.F. Makarov, V.G. Papin, V.L. Tal'roze: ZhETF **61**, 97 (1971)
2.197 M.A. Kwok, N. Cohen: J. Chem. Phys. **61**, 5221 (1974)
2.198 J.K. Hancock, W.H. Green: J. Chem. Phys. **59**, 6350 (1973)
2.199 K.G. Anlauf, P.H. Dawson, J.A. Herman: J. Chem. Phys. **58**, 5354 (1973)
2.200 S.J. Arnold, G.H. Kimbell: Can. J. Chem. **56**, 387 (1978)
2.201 J.F. Bott, N. Cohen: J. Chem. Phys. **58**, 934 (1973)
2.202 K. Ernst, R.M. Osgood, A. Javan, P.B. Sackett: Chem. Phys. Lett. **23**, 553 (1973)
2.203 J.F. Bott: Chem. Phys. Lett. **23**, 335 (1973)
2.204 J.F. Bott, N. Cohen: J. Chem. Phys. **59**, 447 (1973)
2.205 J.F. Bott: J. Chem. Phys. **63**, 2263 (1975)
2.206 J.F. Bott: J. Chem. Phys. **60**, 427 (1974)
2.207 N.G. Basov, V.T. Galochkin, V.I. Igoshin, L.V. Kulakov, E.P. Markin, A.I. Nikitin, A.N. Oraevsky: Appl. Opt. **10**, 1814 (1971)
2.208 J.J. Hinchen, R.H. Hobbs: J. Chem. Phys. **63**, 353 (1975)
2.209 H.-L. Chen, C.B. Moore: J. Chem. Phys. **54**, 4072 (1071)
2.210 P.F. Zittel, C.B. Moore: J. Chem. Phys. **59**, 6636 (1973)
2.211 C.T. Bowman, D.J. Seery: J. Chem. Phys. **50**, 1904 (1969)
2.212 W.D. Breshears, P.F. Bird: J. Chem. Phys. **50**, 333 (1969)
2.213 V.I. Gorshkov, V.V. Gromov, V.I. Igoshin, E.L. Koshelèv, E.P. Markin, A.N. Oraevsky: Appl. Opt. **10**, 1781 (1971)
2.214 S.R. Leone, C.B. Moore: Chem. Phys. Lett. **19**, 340 (1973)
2.215 Y. Noter, I. Burak, A. Szöke: J. Chem. Phys. **59**, 970 (1973)
2.216 I. Burak, Y. Noter, A.M. Ronn, A. Szöke: Chem. Phys. Lett. **17**, 345 (1972)
2.217 D. Arnoldi, J. Wolfrum: Chem. Phys. Lett. **24**, 234 (1974)
2.218 N.C. Craig, C.B. Moore: J. Phys. Chem. **75**, 1622 (1971)
2.219 B.A. Ridley, I.W.M. Smith: Chem. Phys. Lett. **9**, 457 (1971)
2.220 R.B. Steele, Jr., C.B. Moore: J. Chem. Phys. **60**, 2794 (1974)
2.221 D.J. Seery: J. Chem. Phys. **58**, 1796 (1973)
2.222 J.F. Bott, N. Cohen: J. Chem. Phys. **63**, 1518 (1975)
2.223 H.-L. Chen: J. Chem. Phys. **55**, 5551 (1971)
2.224 H.-L. Chen, C.B. Moore: J. Chem. Phys. **54**, 4080 (1971)
2.225 B.M. Hopkins, H.-L. Chen: J. Chem. Phys. **57**, 3161 (1972)
2.226 B.M. Hopkins, H.-L. Chen, R.D. Sharma: J. Chem. Phys. **59**, 5758 (1973)
2.227 H.C. Stephenson, J. Finzi, C.B. Moore: J. Chem. Phys. **56**, 5214 (1972)
2.228 R.D. Coombe, A.T. Pritt, D. Pilipovich: Chem. Phys. Lett. **35**, 349 (1975)
2.229 P.F. Zittel, C.B. Moore: J. Chem. Phys. **58**, 2004 (1973)
2.230 P.F. Zittel, C.B. Moore: J. Chem. Phys. **58**, 2922 (1973)
2.231 M.Y.-D. Chen. H.-L. Chen: J. Chem. Phys. **56**, 3315 (1972)
2.232 R.J. Donovan, D. Husain, C.D. Stevenson: Trans. Faraday Soc. **66**, 2148 (1970)
2.233 B.M. Hopkins, H.-L. Chen: J. Chem. Phys. **59**, 1495 (1973)
2.234 H.-L. Chen, J.C. Stephenson, C.B. Moore: Chem. Phys. Lett. **2**, 593 (1968)
2.235 I. Burak, Y. Noter, A.M. Ronn, A. Szöke: Chem. Phys. Lett. **16**, 306 (1972)
2.236 A. Ben-Shaul, G.L. Hofacker, K.L. Kompa: J. Chem. Phys. **59**, 4664 (1973)
2.237 D. J. Douglas, C.B. Moore: J. Chem. Phys. **70**, 1769 (1979)
2.238 J.F. Bott: Int. J. Chem. Kinet. **9**, 123 (1977)
2.239 R.G. Macdonald, C.B. Moore, I.W.M. Smith, F.J. Wodarczyk: J. Chem. Phys. **62**, 2934 (1975)
2.240 V.N. Kondrat'yev, E.E. Nikitin, A.I. Reznikov, S. Ya. Umanskii: *Termicheskiye biomolekulyarnye reaktsii v gazakh* (Thermal Biomolecular Reactions in Gases) (Nauka, Moscow, 1976)
2.241 G.M. Jurgich, F.F. Crim: J. Chem. Phys. **74**, 4455 (1981)
2.242 T.J. Foster, F.F. Crim: J. Chem. Phys. **75**, 3871 (1981)
2.243 A.S. Bashkin, V.I. Igoshin, S.P. Sannikov, In: *Tezisy dokladov IV Vsesoyuznogo simpoziuma po lazernoi khimii* (Abstracts of Papers Read at the All-Union Symposium on Laser Chemistry) (Zvenigorod, 1985), p. 114., FIAN, Moscow, 1985

Chapter 3

3.1   V.I. Igoshin, A.N. Oraevsky: Preprint FIAN **162** (1969); Khim. Vysokikh Energ. **5**, 397 (1971)
3.2.   N. Cohen, T. A. Jacobs, G. Emanuel, R.L. Wilkins: Int. J. Chem. Kinet. **1**, 551 (1969)
3.3   O.M. Batovskii, G.K. Vasil'yev, E.F. Makarov, V.L. Tal'roze: Pis'ma ZhETF **9**, 341 (1969)
3.4   N.G. Basov, L.V. Kulakov, E.P. Markin, A.I. Nikitin, A.N. Oraevsky: Pis'ma ZhETF **9**, 613 (1969)
3.5   N.G. Basov, V.T. Galochkin, L.V. Kulakov, E.P. Markin, A.I. Nikitin, A.N. Oraevsky: Kratk. Soobshch. Fiz. FIAN **8**, 10 (1970)
3.6   V.I. Igoshin: Program of Jubilee Sci. Conf. Moscow Physical-Technical Institute, 26–27 Nov., 1971
3.7   V.I. Igoshin: Tr. FIAN **76**, 117 (1974)
3.8   R.L. Kerber, G. Emanuel, J.S. Whittier: Appl. Opt. **11**, 1112 (1972)
3.9   R.L. Kerber, N. Kohen, G. Emanuel: IEEE J. **QE-9**, 94 (1973)
3.10   T.O. Poehler, J.C. Pirkle, Jr., R.E. Walker: IEEE J. **QE-9**, 83 (1973)
3.11   N.G. Basov, A.S. Bashkin, V.I. Igoshin, V. Yu. Nikitin, A.N. Oraevsky: Preprint FIAN **171** (1975)
3.12   V.I. Igoshin, V. Yu. Nikitin, A.N. Oraevsky: Kvantovaya Elektron. (Moscow) **4**, 1282 (1977)
3.13   V.I. Igoshin, V. Yu. Nikitin, A.N. Oraevsky: Kvantovaya Elektron. (Moscow) **3**, 2072 (1976)
3.14   V. Ya. Agroskin, G.K. Vasil'yev, V.I. Kir'yanov, V.L. Tal'roze: Kvantovaya Elektron. **3**, 1932 (1976)
3.15   R.L. Kerber, J.S. Whittier: Appl. Opt. **15**, 2358 (1976)
3.16   V.I. Igoshin, V. Yu. Nikitin, A.N. Oraevsky: Kratk. Soobshch. Fiz. **6**, 20 (1978)
3.17   A.S. Bashkin, V.I. Igoshin, V. Yu. Nikitin. A.N. Oraevsky: Kvantovaya Elektron. (Moscow) **5**, 907 (1978)
3.18   A.S. Bashkin, P.G. Grigor'yev, V.I. Igoshin, V. Yu. Nikitin, A.N. Oraevsky: Kvantovaya Elektron. (Moscow) **4**, 1004 (1977)
3.19   L.V. Kulakov, A.I. Nikitin, A.N. Oraevsky: Kvantovaya Elektron. (Moscow) **3**, 1677 (1976)
3.20   S.N. Suchard, R.L. Kerber, G. Emanuel. J.S. Whittier: J. Chem. Phys. **57**, 5065 (1972)
3.21   H.-L. Chen, R.L. Taylor, J. Wilson, P. Lewis, W. Fyfe: J. Chem. Phys. **61**, 306 (1974)
3.22   J.J.T. Hough: Appl. Opt. **16**, 2297 (1977)
3.23   V.I. Igoshin, A.N. Oraevsky: Kratk. Soobshch. Fiz. FIAN **7**, 27 (1976)
3.24   G.K. Vasil'yev, E.F. Makarov, A.G. Ryabenko, V.L. Tal'roze: ZhETF **71**, 1320 (1976)
3.25   J.B. Moreno: AIAA Pap. **36**, 1 (1975)
3.26   G. Emanuel, J.S. Whittier: Appl. Opt. **11**, 2047 (1972)
3.27   R.L. Kerber: Appl. Opt. **12**, 1157 (1973)
3.28   N.G. Basov, A.S. Bashkin, V.I. Igoshin, A.N. Oraevsky: Kvantovaya Elektron. (Moscow) **3**, 1967 (1976)
3.29   V.I. Igoshin: Kvantovaya Elektron. (Moscow) **6**, 528 (1979)
3.30   G.K. Vasil'yev, E.F. Makarov, Yu. A. Chernyshev: Dokl. Acad. Nauk SSSR **233**, 1118 (1977)
3.31   N.N. Semenov, A.E. Shilov: Kinet. Katal. **6**, 3 (1965)
3.32   G.A. Kapralova, E.M. Trofimova, A.E. Shilov: Kinet. Katal. **6**, 977 (1965)
3.33   G.A. Kapralova, E.M. Margolina, A.M. Chaikin: Kinet. Katal **10**, 32 (1969)
3.34   V.G. Fedotov, A.M. Chaikin: Dokl. Akad. Nauk SSSR **203**, 406 (1972)
3.35   G.A. Kapralova, E.M. Margolina, A.M. Chaikin: Dokl. Akad. Nauk SSSR **196**, 624 (1971)
3.36   V.I. Vedeneyev, V.V. Nosova, V.I. Prppoi, O.M. Sarkisov: Zh. Fiz. Khim. **43**, 1288 (1969)
3.37   V.P. Bulatov, V.I. Vedeneyev, et al.: Izv. Akad. Nauk SSSR **197**, 624 (1971)
3.38   V.L. Tal'roze: Preprint Inst. Khim. Fiz. Akad. Nauk SSSR (1969)
3.39   V. Ch. Bokun, A.M. Chaikin: Dokl. Akad. Nauk SSSR **223**, 1070 (1975)
3.40   V. Ch. Bokun, Yu. F. Pugachev, A. M. Chaikin: Kinet. Katal. **18**, 502 (1977)
3.41   A.F. Dodonov, G.N. Lavrovskaya, I.I. Morozov, V. L. Tal'roze: Dokl. Akad. Nauk SSSR **198**, 6222 (1971)
3.42   K.H. Homann, W.C. Solomon, J. Warnatz, H.C. Wagner, C. Letsch: Ber. Bunsenges. Phys. Chem. **74**, 585 (1970)
3.43   V.I. Igoshin, L.V. Kulakov, A.I. Nikitin, in: *Kvantovaya elektronika* (Quantum Electronics), Iss. 4 (Sovetskoe Radio, Moscow, 1973), p. 50

354   References

3.44 V.I. Igoshin, L.V. Kulakov, A.I. Nikitin: Kratk. Soobshch. Fiz. FIAN **1**, 3 (1973)
3.45 M.A.A. Clyne, D.J. McKenney, R.F. Walker: Can. J. Chem. **51**, 3596 (1973)
3.46 V.P. Bulatov, V.P. Balakhnin, O.M. Sarkisov: Izv. Akad. Nauk SSSR, Ser. Khim. 1734 (1977)
3.47 A. Persky: J. Chem. Phys. **59**, 3612 (1973)
3.48 E.R. Grant, J.W. Root: J. Chem. Phys. **63**, 2970 (1975)
3.49 G. A. Kapralova, A.K. Margolin, A.M. Chaikin: Kinet. Katal. **11**, 811 (1970)
3.50 R.L. Williams, F.S. Rowland: J. Phys. Chem. **75**, 2709 (1971)
3.51 N.F. Chebotarev, G.V. Pukhal'skaya, S. Ya. Pshezhetskii: Kvantovaya Elektron. (Moscow) **4**, 872 (1977)
3.52 W.E. Jones, E.G. Skolik: Chem. Rev. **76**, 563 (1976)
3.53 M.J. Berry, J. Chem. Phys. **59**, 6229 (1973)
3.54 J.B. Levy, B.K.W. Copeland: J. Phys. Chem. **72**, 3168 (1968)
3.55 R.G. Albright, A.F. Dodonov, G.K. Lavrovskaya, I.I. Morozov, V.L. Talroze: J. Chem. Phys. **50**, 3632 (1969)
3.56 S.W. Rabideau, H.G. Hecht, W.B. Lewis: J. Magn. Reson. **6**, 384 (1972)
3.57 I.B. Goldberg, G.R. Schneider: J. Chem. Phys. **65**, 147 (1976)
3.58 W.E. Jones, S.D. Macknight, L. Teng: Chem. Rev. **73**, 407 (1973)
3.59 G.K. Vasil'yev, E.F. Makarov, Yu. A. Chernyshev: Kinet. Katal. **16**, 320 (1975)
3.60 M.J. Kurylo: J. Phys. Chem. **56**, 3518 (1972)
3.61 T. Hikida, J.A. Eyre, L.M. Dorfman: J. Chem. Phys. **54**, 3422 (1971)
3.62 D. Gutman, E. A. Hardwidge, F.A. Dougherty, R.W. Lutz: J. Chem. Phys. **47**, 4400 (1967)
3.63 G.K. Vasil'yev, E.F. Makarov, Yu. A. Chernyshev: Preprint Inst. Khim. Fiz. (Chernogolovka) (1978)
3.64 G.A. Kapralova, E.M. Margolina, A.M. Chaikin: Kinet. Katal. **17**, 292 (1976)
3.65 V. Ch. Bokun, A.M. Chaikin: React. Kinet. Catal. Lett. **3**, 277 (1975)
3.66 A.S. Bashkin, A.F. Konoshenko, A.N. Oraevsky, V.N. Tomashov, N.N. Yuryshev: Preprint No. 274, FIAN, Moscow, 1978.
3.67 V. Ch. Bokun, Yu. F. Pugachev. A.M. Chaikin: Kinet. Katal. **18**, 502 (1977)
3.68 G.K. Vasil'yev, V.V. Vizhin, E.F. Makarov, Yu. A. Chernyshev, V.L. Tal'roze: Khim. Vysokhikh Energ. **9**, 154 (1975)
3.69 V.S. Arutyunov, L.S. Papov. A.M. Chaikin: Kinet. Katal. **17**, 286 (1976)
3.70 V.S. Arutyunov: Abstr. Candidate's Thesis (MFTI, Moscow, 1977)
3.71 P.P. Chegodayev, V.I. Tupikov, E.G. Strukov, S.Ya. Pshezhetskii: Khim Vysokikh Energ. **12**, 116 (1978)
3.72 H.-L. Chen, D.W. Traimor, R.E. Center, W.I. Fyfe: J. Chem. Phys. **66**, 5513 (1977)
3.73 V.N. Kondrat'yev: *Konstanty skorosti gazofaznykh reaktsii* (Rate Constants of Gas-Phase Reactions) (Nauka, Moscow, 1970)
3.74 V.S. Arutyunov, A.M. Chaikin: Kinet. Katal. **18**, 316 (1977)
3.75 V.S. Arutyunov, A.M. Chaikin: Kinet. Katal. **18**, 321 (1977)
3.76 I.P. Guzov, S.B. Kromer, L.V. L'vov, V.T. Punin, M.V. Sinitsyn, E.A. Stankeyev, V.D. Urlin: Kvantovaya Elektron. **3**, 2043 (1976)
3.77 P.S. Canguli, M. Kaufman: Chem. Phys. Lett. **25**, 221 (1974)
3.78 J. H. Sullivan, R.C. Feber, J.W. Starner: J. Chem. Phys. **62**, 1714 (1975)
3.79 A.F. Dodonov, G.N. Lavrovskaya, V.L. Tal'roze: Kinet. Katal. **10**, 701 (1969)
3.80 A.C. Lloyd: Int. J. Chem. Kinet. **6**, 169 (1974)
3.81 V. Ch. Bokun, A.M. Chaikin: Fiz. Goren. Vzryv. **3**, 21 (1978)
3.82 H.-L. Chen, J.D. Daugherty, W. Fyfe: IEEE J. **QE-11**, 648 (1975)
3.83 F.K. Truby: Appl. Phys. Lett. **29**, 247 (1976)
3.84 R.W. Getzinger: Los Alamos Sci. Lab. Rep., 1974
3.85 R. Hoffland, M.L. Lundquist, A. Ching, J.S. Whittier: AIAA Pap. **645** (1973)
3.86 J. Wilson, H.-L. Chen, W. Fyfe, R. Taylor, R. Littl, R. Lowel: J. Appl. Phys. **44**, 5447 (1973)
3.87 J.S. Whittier, M.L. Lundquist, A. Ching, G.E. Thornton, R. Hoffland, Jr.: J. Appl. Phys. **47**, 3542 (1975)
3.88 H.-L. Chen, R.E. Center, D.W. Trainor, W.I. Fyfe: J. Appl. Phys. **48**, 2297 (1977)
3.89 V.I. Igoshin, A.N. Orayevskii: Pis'ma ZhETF **21**, 325 (1975)

3.90   V.I. Igoshin, M.S. Kurdoglyan, A.N. Orayevskii: *Digests, II Symp. Laser Chem.* (Zvenigorod, 15–18 Dec., 1980)

3.91   J. Mangano, R.L. Limpaecher, J.D. Daugherty, R. Russel: Appl. Phys. Lett. **27**, 293 (1975)

3.92   A.S. Denholm, B.S. Quintal: Laser Focus **10**, 41 (1974)

3.93   R.A. Gerber, E.L. Patterson, L.S. Blair, N.R. Greiner: Appl. Phys. Lett. **25**, 281 (1974)

3.94   N.G. Basov, A.S. Bashkin, L.E. Golubev, Yu, I. Kozlov, A.N. Oraevsky, A.K. Piskunov, V.N. Tomashov, V.N. Torgashin, N.N. Yuryshev: Kvantovaya Elektron. (Moscow) **5**, 910 (1978)

3.95   A.S. Bashkin, A.F. Konoshenko, A.N. Oraevsky, V.N. Tomashov, N.N. Yuryshev: Kvantovaya Elektron. (Moscow) **5**, 1608 (1978)

3.96   L.H. Sentman: J. Chem. Phys. **62**, 1608 (1975)

3.97   J.J.T. Hough, R.L. Kerber: Appl. Opt. **14**, 2960 (1975)

3.98   J.R. Kreighton: Ph. D. Thesis, Lowrence Livermor Lab., Univ. Calif. (1975)

3.99   J.G. Skifstad, C.M. Chao: Appl. Opt. **14**, 1713 (1975)

3.100   R.J. Hall: IEEE J. **QE-12**, 453 (1976)

3.101   A. Ben-Shaul, K.L. Kompa, U. Schmailzl: J. Chem. Phys. **65**, 1711 (1976)

3.102   E. Keren, R.B. Gerber. A. Ben-Shaul: Chem. Phys. **21**, 1 (1976)

3.103   A. Ben-Shaul: Chem. Phys. **18**, 13 (1976)

3.104   G.K. Vasil'yev, V.I. Gur'yev, V.L. Tal'roze: ZhETF **72**, 943 (1977)

3.105   L.H. Sentman: Appl. Opt. **17**, 2244 (1978)

3.106   M. Obara, N. Nishida, M. Morimoto, T. Fujioka: Opt. Comm. **26**, 240 (1978)

3.107   R.L. Kerber, J.J.T. Hough: Appl. Opt. **17**, 2369 (1978)

3.108   K.L. Kompa, J.H. Parker, G.C. Pimentel: J. Chem. Phys. **49**, 4257 (1968)

3.109   N.G. Basov, S.I. Zavorotnyi, E.P. Markin, A.I. Nikitin, A.N. Oraevsky: Pis'ma ZhETF **15**, 135 (1972)

3.110   N.G. Basov, A.S. Bashkin, A.S. Grigir'yev, A.N. Oraevsky, O.E. Porodinkov: Kvantovaya Elektron. **3**, 2067 (1976)

3.111   V. Ya. Agroskin, G.K. Vasil'yev, V.I. Kir'yanov, V.L. Tal'roze: Kvantovaya Elektron. **5**, 2436 (1978)

3.112   N.A. Cohen: *A Review of Rate Coefficients in the D_2–F_2 Chemical Laser System*, Techn. Rep. TR-77-152 (Aerospace Corp., El Segundo, Ca., 1977)

3.113   S.W. Zelazny, J.A. Blauer, L. Wood, L.H. Sentman, W.C. Solomon: Appl. Opt. **15**, 1164 (1976)

3.114   E.E. Stark, Jr.: Appl. Opt. **15**, 1164 (1976)

3.115   R.R. Jacobs, K.J. Pettipiece, S.J. Thomas: Phys. Rev. **A11**, 54 (1975)

3.116   R.L. Kerber: Appl. Opt. **12**, 1157 (1973)

3.117   V. Ya. Agroskin, G.K. Vasil'yev, V.I. Kir'yanov: Khim. Vysokikh Energ. **8**, 283 (1974)

3.118   R.L. Taylor, S. Bitterman: Rev. Mod. Phys. **41**, 26 (1969)

3.119   J.A. Blauer, G.R. Nickerson: AIAA Pap. **74**, 536 (1973)

3.120   S.A. Losev: Fiz. Goren. Vzryv. **12**, 163 (1976)

3.121   N.G. Basov, V.G. Mikhailov, A.N. Orayevskii, V.A. Shcheglov: ZhTF **38**, 2031 (1968)

3.122   A.S. Biryukov, B.F. Gordiets: PMTF **6**, 29 (1972)

3.123   S.A. Losev: *Gazodinamicheskie lazery* (Gasdynamic Lasers) (Nauka, Moscow, 1977)

3.124   A. Ben-Shaul, S. Feliks, O. Kraft: Chem. Phys. **36**, 291 (1979)

3.125   Z.B. Alfassi, M. Baer: IEEE J. **QE-15**, 240 (1979)

3.126   V.I. Igoshin, V. Yu, Nikitin, A.N. Orayevskii: Kvantovaya Elektron. **7**, 1438 (1980)

3.127   V.I. Igoshin, M.S. Kurdoglyan, A.N. Orayevskii: Kvantovaya Elektron. (Moscow) **8**, 941 (1981)

3.128   A.S. Bashkin, V.I. Igoshin, S.P. Sannikov, in: *Digests, IV Symp. Laser Chem.* (Zvenigorod, 1985), p. 114.

3.129   P.E. Sojka, R.L. Kerber: Appl. Opt. **25**, 76 (1986)

3.130   A.S. Bashkin, A.N. Orayevskii, V.N. Tomashov, N.N. Yuryshev: Kvantovaya Elektron. (Moscow) **9**, 630 (1982)

3.131   T. Fujioka, F. Kannari, T. Szuki, M. Obara: J. Appl. Phys. **58**, 3975 (1985)

3.132   A.S. Bashkin: Kvantovaya Elektron. (Moscow) **14**, 1563 (1987)

3.133   E.U. Baikov, A.S. Bashkin, N.M. Gamzatov, A.N. Oraevsky, O.E. Porodnikov: Kvantovaya Elektron. (Moscow) **11**, 2336 (1984)

Chapter 4

4.1   J.V.V. Kasper, G.C. Pimentel: Phys. Rev. Lett. **14**, 352 (1965)
4.2   J.V. Parker, L.D. Hess: *IEEE International Electron Devices Meeting*, Washington, 1973 (IEEE, New York, 1973)
4.3   D.B. Nickols, R.B. Hall, J.D. McClure: J. Appl. Phys. **47**, 4026 (1976)
4.4   N. Kh. Petrov, N.F. Chebotarev, S.Ya. Pshezhetskii: Kvantovaya Elektron. **4**, 2248 (1977)
4.5   J.J. De Corpo, R.P. Steiger, J.L. Franklin, J.L. Margrave: J. Chem. Phys. **53**, 936 (1970)
4.6   G.D. Sides, T.O, Tiernan, R.J. Hanrahan: J. Chem. Phys. **65**, 1966 (1976)
4.7   H.-L. Chen, R.E. Center, D.W. Trainor, W.I. Fyfe: J. Appl. Phys. **48**, 2297 (1977)
4.8   W.M. Hickam, R.E. Fox: J. Chem. Phys. **25**, 642 (1956)
4.9   S.J. Corrigan: J. Chem. Phys. **43**, 4381 (1965)
4.10 R. E. Center, A. Mandl: J. Chem. Phys. **57**, 4104 (1972)
4.11 R. Hoffland, M. L. Lundquist, A. Ching, J.S. Whittier: J. Appl. Phys. **45**, 2207 (1974)
4.12 L.E. Christophorou: *Atomic and Molecular Radiation Physics* (Wiley, New York, 1971)
4.13 R.J. Jensen, W.W. Rice: Chem. Phys. Lett. **8**, 214 (1971)
4.14 R.W. F. Gross, R.R. Giedt, T.A. Jacobs: J. Chem. Phys. **51**, 1250 (1969)
4.15 V. M. Akulintsev. A.S. Bashkin, N.M. Gorshunov, Yu. P. Neshchimenko, A.N. Orayevskii, V.I. Trushkin, N.N. Yuryshev: Fiz. Goreniya Vzryv. **12**, 739 (1976)
4.16 J.P. Markiewicz, J.L, Emmett: IEEE J. **QE-2**, 707 (1966)
4.17 I.S. Marshak, A.S. Doinikov, V.P. Zhil'tsov, V.P. Kirsanov, P.E. Rovinskii, L.I. Shchukin, M.G. Feigenbaum: *Impul'snyye istochniki sveta* (Pulsed Light Sources) (Energiya, Moscow, 1978)
4.18 V.N. Volkov, I.G. Zubarev, V.N. Sorokin: Zh. Prikl. Spektrosk. **4**, 735 (1972)
4.19 H.W. Furumoto, H.L. Ceccon: Appl. Opt. **8**, 1613 (1969)
4.20 H.-L. Chen, R.L. Taylor, J. Wilson, P. Lewis, W. Fyfe: J. Chem. Phys. **61**, 306 (1974)
4.21 V.N. Volkov, S.A. Tarasov: Preprint FIAN **3** (1974)
4.22 B.L. Borovich, P.G. Grigor'yev, V.S. Zuyev, V.B. Rozanov, A.V. Startsev, A.P. Shirokikh: Tr. FIAN **76**, 3 (1974)
4.23 A.F. Aleksandrov, A.A. Rukhadze: Usp. Fiz. Nauk **112**, 193 (1974)
4.24 A.S. Bashkin, P.G. Grogor'yev, A.N. Oraevsky, A.B. Skvortsov: Kvantovaya Elektron. (Moscow) **3**, 1824 (1976)
4.25 A.S. Bashkin, A.N. Oraevsky, V.S. Pazyuk, O.E. Porodinkov, N.N. Yuryshev: Kvantovaya Elektron. (Moscow) **6**, 2166 (1979)
4.26 B.R. Belostotskii, Yu. V. Lyubavskii, V.M. Ovchinnikov: *Osnoby lazernoi tekhniki* (Fundamentals of Laser Technology) (Sovetskoye Radio, Moscow, 1972)
4.27 V.A. Danilychev, O.M. Kerimov, I.B. Kovsh: *Molekulyarnye gazovye lazery vysokogo davleniya* (High-Pressure Molecular Gas Lasers) (VINITI, Itogi nauki i tekhniki, Ser. Radiotekhnika, Vol. 12, Moscow, 1977)
4.28 H. Pummer, W. Breitfeld, H. Wedler, G. Klement, K.L. Kompa: Appl. Phys. Lett. **22**, 319 (1973)
4.29 R.G. Wenzel, G.P. Arnold: IEEE J. **QE-8**, 26 (1972)
4.30 P.R. Pearson, H.M. Lamberton: IEEE J. **QE-8**, 145 (1972)
4.31 F. Voignier, M. Gastaud: Appl. Phys. Lett. **25**, 649 (1974)
4.32 J.V. Parker, R.R. Stephens: Appl. Phys. Lett. **22**, 450 (1973)
4.33 V.N. Bagratashvili, I.N. Knyazev, Yu. A. Kudryavtsev, V.S. Letokhov: Pis'ma ZhETF **18**, 110 (1973)
4.34 R.F. Paulson: J. Appl. Phys. **44**, 5633 (1973)
4.35 N.G. Basov, E.M. Belenov, V.A. Danilychev, A.F. Suchkov: Usp. Fiz. Nauk **114**, 213 (1974)
4.36 O.R. Wood, II: Proc. IEEE **62**, 355 (1974)
4.37 R. Hofland, A. Ching, M.L. Lundquist, J.S. Whittier: J. Appl. Phys. **47**, 4543 (1976)
4.38 J.P. Reilly: J. Appl. Phys. **43**, 3411 (1972)
4.39 Y.E. Hill: Appl. Phys. Lett. **22**, 570 (1973)
4.40 G.A. Mesyats: *Generirovaniye moshchnykh nanosekundnykh impul'sov* (Generation of High-Power Nanosecond Pulses) (Sovetskoye Radio, Moscow, 1974)
4.41 R. Aprahamian, J.H.S. Wang, J.A. Betts, R.W. Barth: Appl. Phys. Lett. **24**, 239 (1974)
4.42 G.A. Mesyats, Yu. I. Bychkov, V.V. Kremnev: Usp. Fiz. Nauk **107**, 201 (1972)

4.43 P.A. Miller, J.B. Gerardo: J. Appl. Phys. **43**, 3008 (1972)
4.44 L.I. Rudakov, V.P. Smirnov, A.M. Spektor: Pis'ma ZhETF **15**, 540 (1972)
4.45 A.E. Grün: Zeits. für Naturforschungen **12a**, **2**, 89 (1957)
4.46 T.O. Poehler, R.E. Walker: Appl. Phys. Lett. **22**, 282 (1973)
4.47 V.N. Bagratashvili, I.N. Knyazev, V.S. Letokhov: Opt. Comm. **4**, 154 (1971)
4.48 B.M. Smirnov (ed.): *Khimiya plazmy* (Plasma Chemistry) (Atomizdat, Moscow, 1975)
4.49 N.N. Semenov: *O nekotorykh problemakh khimicheskoi kinetiki i reaktsionnoi sposobnosti* (On Some Problems of Chemical Kinetics and Reactivity) (Izd-vo Akad. Nauk SSSR, Moscow, 1958)
4.50 G.A. Kapralova, E.M. Margolina, A.M. Chaikin: Dokl. Akad. Nauk SSSR, **197**, 624 (1971)
4.51 G.A. Kapralova, E.M. Margolina, A.M. Chaikin: Dokl. Akad. Nauk SSSR **198**, 634 (1971)
4.52 V.L. Tal'roze: Preprint Inst. Khim. Fiz. (Moscow, 1969)
4.53 V. Ch. Bokün, A.M. Chaikin: Fiz. Goreniya Vzryv. **3**, 21 (1978)
4.54 H.-L. Chen, J.D. Daugherty, W. Fyfe: IEEE J. **QE-11**, 648 (1975)
4.55 F.K. Truby: Appl. Phys. Lett. **29**, 247 (1976)
4.56 R.A. Gerber, E.L. Patterson, L.S. Blair, N.R. Greiner: Appl. Phys. Lett. **25**, 281 (1974)
4.57 R.A. Gerber, E.L. Patterson: J. Appl. Phys. **47**, 3524 (1976)
4.58 J.H. Sullivan: J. Phys. Chem. **79**, 1045 (1975)
4.59 J. Wilson, J.S. Stephenson: Appl. Phys. Lett. **20**, 64 (1972)
4.60 J.A. Mangano, R.L. Limpaecher, J.D. Daugherty, F. Russel: Appl. Phys. Lett. **27**, 293 (1975)
4.61 R.N. Greiner: IEEE J. **QE-8**, 872 (1972)
4.62 G.G. Dolgov-Savel'yev, V.F. Zharov, Yu. S. Neganov, G.M. Chumak: ZhETF **61**, 64 (1971)
4.63 V. Ya. Agroskin, G. K. Vasil'yev, V.I. Kir'yanov, E.F. Makarov, Yu. A. Chernyshev: Dokl. Akad. Nauk SSSR **225**, 830 (1975)
4.64 L.D. Hess: J. Chem. Phys. **55**, 2466 (1971)
4.65 G.K. Vasil'yev. E.F. Makarov, Yu. A. Chernyshev: Fiz. Goreniya Vzryv. **16**, 30 (1980)
4.66 O.D. Krogh, G.C. Pimentel: J. Chem. Phys. **56**, 969 (1972)
4.67 J.R. Airey: IEEE J. **QE-3**, 208 (1967)
4.68 K.L. Kompa, G.C. Pimentel: J. Chem. Phys. **47**, 857 (1967)
4.69 K.L. Kompa, J.H. Parker, G.C. Pimentel: J. Chem. Phys. **49**, 4257 (1968)
4.70 J.H. Parker, G.C. Pimentel: J. Chem. Phys. **48**, 5273 (1968)
4.71 J. H. Parker, G.C. Pimentel: J. Chem. Phys. **51**, 91 (1969)
4.72 K.L. Kompa, P. Gensel, J. Wanner: Chem. Phys. Lett. **3**, 210 (1969)
4.73 J. H. Parker, G.C. Pimentel: J. Chem. Phys. **55**, 857 (1971)
4.74 P. Gensel, K.L. Kompa, J. Wanner: Chem. Phys. Lett. **7**, 583 (1970)
4.75 G.G. Dolgov-Savel'yev, I.A. Polyakov, G.M. Chumak: ZhETF **58**, 1197 (1970)
4.76 K.L. Kompa, J. Wanner: Chem. Phys. Lett. **12**, 560 (1972)
4.77 A.N. Chester, L.D. Hess: IEEE J. **QE-8**, 1 (1972)
4.78 R.W.F. Gross, N. Cohen, T.A. Jacobs: J. Chem. Phys. **48**, 3821 (1968)
4.79 L.E. Brus, M.C. Lin: J. Chem. Phys. **75**, 2546 (1971)
4.80 T.D. Padrick, G.C. Pimentel: J. Chem. Phys. **54**, 720 (1971)
4.81 S.N. Suchard, G.C. Pimentel: Appl. Phys. Lett. **18**, 530 (1971)
4.82 R.D. Coombe, G.C. Pimentel, M.J. Berry: *3rd Conf. Chem. Mol. Lasers* (St. Louis, Miss., May 1972)
4.83 M.J. Berry: *3rd Conf. Chem. Mol. Lasers* (St. Louis, Miss., May 1972)
4.84 M.J. Berry, G.C. Pimentel: J. Chem. Phys. **49**, 5190 (1968)
4.85 J.L. Roebber, G.C. Pimentel: *3rd Conf. Chem. Mol. Lasers* (St. Louis, Miss., May 1972)
4.86 M.C. Lin: J. Phys. Chem. **75**, 3642 (1971)
4.87 H. Pummer, K.L. Kompa: Appl. Phys. Lett. **20**, 356 (1972)
4.88 G.P. Arnold, R.G. Wenzel: IEEE J. **QE-9**, 492 (1973)
4.89 D.W. Gregg, B. Krawetz, R.K. Pearson, B.R. Schleicher, S.J. Thomas, E.B. Huss, K.J. Pettipiece, J.R. Creighton, R.E. Niver, Y.-L. Pan: Chem. Phys. Lett. **8**, 609 (1971)
4.90 C.P. Robinson, R.J. Jensen, A. Kolb: IEEE J. **QE-9**, 963 (1973)
4.91 R.W.F. Gross, F. Wesner: Appl. Phys. Lett. **23**, 559 (1973)
4.92 E.L. Patterson, R.A. Gerber, L.S. Blair: J. Appl. Phys. **45**, 1822 (1974)
4.93 R.A. Gerber, E.L. Patterson: IEEE J. **QE-10**, 333 (1974)

4.94   E.L. Patterson, R.A. Gerber: IEEE J. **QE-11**, 642 (1975)
4.95   A.S. Bashkin, A.N. Oraevsky, V.N. Tomashov: Kvantovaya Elektron. **4**, 169 (1977)
4.96   H.W. Chang, D.W. Setser: J. Chem. Phys. **16**, 1 (1976)
4.97   J.C. Polanyi, K.B. Woodall: J. Chem. Phys. **57**, 1574 (1972)
4.98   W. Duewer, D.W. Setser: J. Chem. Phys. **58**, 2298 (1973)
4.99   L.J. Kirsh, J.C. Polanyi: J. Chem. Phys. **57**, 4498 (1972)
4.100 N. Jonathan, C.M. Melliar-Smith, D. Timlin, D.H. Slater: Appl. Opt. **10**, 1821 (1971)
4.101 R.J. Jensen, W.W. Rice: Chem. Phys. Lett. **7**, 627 (1970)
4.102 N.R. Greiner: IEEE J. **QE-11**, 844 (1975)
4.103 S.W. Mayer, D. Taylor, M.A. Kwok: Appl. Phys. Lett. **23**, 434 (1973)
4.104 N.G. Basov, L.V. Kulakov, E.P. Markin, A.I. Nikitin, A.N. Orayevskii: Pis'ma ZhETF **9**, 613 (1969)
4.105 O.M. Batovskii, G.K. Vasil'yev, E.F. Makarov, V.L. Tal'roze: Pis'ma ZhETF **9**, 341 (1969)
4.106 S.N. Suchard, A. Ching, J.S. Whittier: Appl. Phys. Lett. **21**, 274 (1972)
4.107 T.O. Poehler, M. Shandor, R.E. Walker: Appl. Phys. Lett. **20**, 497 (1972)
4.108 N.G. Basov, S.I. Zavorotnyi, E.P. Markin, A.I. Nikitin, A.N. Oraevsky: Pis'ma ZhETF **15**, 135 (1972)
4.109 N.G. Basov, S.I. Zavorotnyi, E.P. Markin, A.I. Nikitin, A.N. Oraevsky, B.L. Borovich, P.G. Grigor'yev, V.S. Zuyev: Kvantovaya Elektron. (Moscow) **1**, 560 (1974)
4.110 N.G. Basov, A.S. Bashkin, P.G. Grigor'yev, A.N. Oraevsky, O.E. Porodinkov: Kvantovaya Elektron. (Moscow) **3**, 2067 (1976)
4.111 L.V. Kulakov, A.I. Nikitin, A.N. Oraevsky: Kvantovaya Elektron. (Moscow) **3**, 1677 (1976)
4.112 V. Ya. Agroskin, G.K. Vasil'yev, V.I. Kir'yanov, V.L. Tal'roze: Kvantovaya Elektron. (Moscow) **5**, 2436 (1978)
4.113 L.D. Hess: J. Appl. Phys. **43**, 1157 (1972)
4.114 O.M. Batovskii, V.I. Gur'yev: Kvantovaya Elektron. (Moscow) **1**, 1446 (1974)
4.115 V. Ya. Agroskin, G.K. Vasil'yev, V.I. Kir'yanov, V.L. Tal'roze: Kvantovaya Elektron. (Moscow) **3**, 1932 (1976)
4.116 N.R. Greiner: IEEEE J. **QE-9**, 1123 (1973)
4.117 A.S. Bashkin, N.P. Vagin, O.R. Nazyrov, A.N. Oraevsky, V.S. Pazyuk, O.E. Porodinkov, N.N. Yuryshev: Kvantovaya Elektron. (Moscow) **7**, 1821 (1980)
4.118 J.S. Whittier, R.L. Kerber: IEEEE J. **QE-10**, 844 (1974)
4.119 R. Turner, T.O. Poehler: J. Appl. Phys. **47**, 3038 (1976)
4.120 V.F. Zharov, V.K. Malinovskii, Yu. S. Neganov, G.M. Chumak: Pis'ma ZhETF **16**, 219 (1972)
4.121 N.R. Greiner, L.S. Blair, P.E. Bird: IEEEE J. **QE-10**, 646 (1974)
4.122 R.A. Gerber, E.L. Patterson, L.S. Blair, N.R. Greiner: Appl. Phys. Lett. **25**, 281 (1974)
4.123 A.S. Bashkin, A.F. Konoshenko, A.N. Oraevsky, V.N. Tomashov, N.N. Yuryshev: Preprint FIAN **274** (1978); Kvantovaya Elektron. (Moscow) **6**, 2277 (1979)
4.124 A.S. Bashkin, A.N. Oraevsky, V.N. Tomashov, N.N. Yuryshev: Kvantovaya Elektron. (Moscow) **7**, 1357 (1980)
4.125 K.G. Anlauf, P.E. Charters, D.S. Home, R.G. Mac Donald, D.H. Maylotle, J.C. Polanyi, W.J. Skiloc, D.C. Tardy, K.B. Woodall: J. Chem. Phys. **53**, 4091 (1971)
4.126 J.C. Polanyi, Sato Hiroshi: J. Chem. Phys. **57**, 4988 (1972)
4.127 N.G. Basov, V.T. Galochkin, V.I. Igoshin, L.V. Kulakov, E.P. Markin, A.I. Nikitin, A.N. Oraevsky: Appl. Opt. **10**, 1814 (1971)
4.128 S.N. Suchard: Appl. Phys. Lett. **23**, 68 (1973)
4.129 O.M. Batovskii, V.I. Gur'yev: Kvantovaya Elektron. (Moscow) **1**, 676 (1974)
4.130 G.K. Vasil'yev, E.F. Makarov, A.G. Ryabenko, V.L. Tal'roze: ZhETF **71**, 1320 (1976)
4.131 V.I. Igoshin, A.N. Oraevsky: Kratk. Soobshch. Fiz, FIAN **7**, 27 (1976)
4.132 J.J. Hinchen: Appl. Phys. Lett. **27**, 672 (1975)
4.133 G.K. Vasil'yev, V.I. Gur'yev, V.L. Tal'roze: ZHETF **72**, 943 (1977)
4.134 J. Wilson, H.-L. Chen, W. Fyfe, R.L. Taylor, R. Little, R. Lowell: J. Appl. Phys. **44**, 5447 (1973)
4.135 J.S. Whittier, M.L, Lundquist, A. Ching, G.E. Thornton, R. Hofland: J. Appl. Phys. **47**, 3542 (1976)
4.136 N.G. Basov, V.T. Galochkin, L.V. Kulakov, E.P. Markin, A.I. Nikitin, A.N. Oraevsky: Kratk. Soobshch. Fiz. FIAN **8**, 10 (1970)

4.137 T.O. Poehler, J.C. Pirkle, R.E. Walker: IEEEE J. **QE-9**, 1, Pt. **2**, 83 (1973)
4.138 V. Ya. Agroskin, B.G. Bravyi, G.K. Vasil'yev, V.I. Kir'yanov: Kvantovaya Elektron. (Moscow) 7, 229 (1980)
4.139 J. Munch, L.W. Casperson, E.C. Rea: Appl. Opt. **18**, 869 (1979)
4.140 A.V. Belotserkovets, L.I. Zykov, G.A. Kirillov, S.B. Kormer, Yu. V. Kuratov, Yu. V. Savin, S.A. Sukharev: Kvantovaya Elektron. (Moscow) 3, 1102 (1976)
4.141 L.I. Zykov, G.A. Kirillov, S.B. Kormer, V.D. Nikolayev, S.A. Sukharev: Kvantovaya Elektron. (Moscow) **4**, 1336 (1977)
4.142 S.S. Penner: *Quantitative Molecular Spectroscopy and Gas Emissivities* (Addison-Wesley, Reading-London, 1959)
4.143 D.E. Mann, B.A. Thrush, D.R. Lide, J.J. Ball, N. Acquista: J. Chem. Phys. **34**, 420 (1961)
4.144 J.F. Hon, J.R. Novak: IEEEE J. **QE-11**, 698 (1975)
4.145 A.S. Bashkin, V.I. Igoshin, Yu. S. Leonov, A.N. Oraevsky, O.E. Porodinkov: Kvantovaya Elektron. (Moscow) **4**, 1112 (1977)
4.146 N.G. Basov, A.S. Bashkin, V.I. Igoshin, A.N. Oraevsky: Kvantovaya Elektron. (Moscow) **3**, 1967 (1976)
4.147 R.W. Getzinger, N.R. Greiner, K.D. Ware, J.P. Carpenter, R.G. Wenzel: IEEE J. **QE-12**, 556 (1976)
4.148 G.C. Tisone, J.M. Hoffman: J. Appl. Phys. **47**, 3530 (1976)
4.149 N.G. Basov, A.S. Bashkin, L.E. Golubev, Yu. I. Kozlov, A.N. Oraevsky, A.K. Piskunov, V.N. Tomashov, V.N. Troshagin, N.N. Yuryshev: Kvantovaya Elektron. (Moscow) **5**, 910 (1978)
4.150 R. Aprahamian, J.A. Betts, J.H. Wang, R.H. Herret, R.W. Barth: Appl. Phys. Lett. **24**, 384 (1974)
4.151 T.V. Jacobson, G.H. Kimbell, D.R. Snelling: IEEE J. **QE-9**, 496 (1973)
4.152 *Laser Focus* **11**, No 1, 28 (1975)
4.153 J.A. Woodroffe, R. Limpaecher: Appl. Phys. Lett. **30**, 195 (1977)
4.154 P.H. Corneil. G.C. Pimentel: J. Chem. Phys. **49**, 1379 (1968)
4.155 P. Gensel K.L. Kompa, J. Wanner: Chem. Phys. Lett. **5**, 179 (1970)
4.156 Y.-L. Pan, C.E. Turner, Jr., K.J. Pettipiece: Chem. Phys, Lett. **10**, 577 (1971)
4.157 S.N. Suchard: J. Chem. Phys. **58**, 1269 (1973)
4.158 G.C. Pimentel: J. Chem. Phys. **58**, 1270 (1973)
4.159 G.G. Dolgov-Savel'yev, G.M. Chumak, in: *Kvantovaya Elektronika* (Quantum Electronics), No. 4 (10) (Sovetskoe radio, Moscow, 1972), p. 108
4.160 N.F. Chebotarev, S.Ya. Pshezhetskii, E.P. Poltolyarnyi: Kvantovaya Elektron. (Moscow) **1**, 1551 (1974)
4.161 N.F. Chebotarev, S.Ya. Pshezhetskii: Kvantovaya Elektron. (Moscow) **3**, 2232 (1976)
4.162 O.D. Krogh, G.C. Pimentel: IEEE J. **QE-11**, Pt. **2**, 706 (1975)
4.163 J.K. Hancock, W.H. Green: J. Chem. Phys. **59**, 6350 (1974)
4.164 J.F. Bott, N. Cohen: J. Chem. Phys. **61**, 681 (1974)
4.165 N.F. Chebotarev, L.I. Trakhtenberg, S. Ya. Pshezhetskii: Kvantovaya Elektron. (Moscow) **3**, 1331 (1976)
4.166 V.I. Igoshin: Kvantovaya Elektron. (Moscow) **6**, 528 (1979)
4.167 A.S. Bashkin. A.F. Konoshenko, A.N. Oraevsky, S. Ya. Pshezhetskii, V.N. Tomashov, N.F. Chebotarev, N.N. Yuryshev: Kvantovaya Elektron. (Moscow) **5**, 2657 (1978)
4.168 V.G. Voronkov, N.N. Semenov: Zh. Fiz. Khim. **13**, 1695 (1939);
    G.V. Tompson: Zh. Fiz. Khim. **2**, 447 (1931);
    F.D. Leicester: J. Soc. Chem. Ind. **52**, 341 (1933)
4.169 M.A. Pollack: Appl. Phys. Lett. **8**, 237 (1966)
4.170 D.W. Howgate, T.A. Barr, Jr.: J. Chem. Phys. **59**, 2815 (1973)
4.171 V.N. Kondrat'yev: Kinet. Katal. **13**, 1367 (1972)
4.172 D.W. Gregg, S.J. Thomas: J. Appl. Phys. **39**, 4399 (1968)
4.173 R.D. Suart, S.J. Arnold. C.H. Kimbell: Chem. Phys. Lett. **7**, 337 (1970)
4.174 H.S. Pilloff, S.K. Searles, N. Djeu: Appl. Phys. Lett. **19**, 9 (1971)
4.175 G. Hancock, C. Morby, I.W.M. Smith: Chem. Phys. Lett. **12**, 193 (1971)
4.176 K.D. Foster: J. Chem. Phys. **57**, 2451 (1972)
4.177 W.Q. Jeffers, C.E. Wiswall: Appl. Phys. Lett. **23**, 626 (1973)

4.178 W.E. Woodmansee, J.C. Decuis: J. Chem. Phys. **36**, 1831 (1962)
4.179 M. Margottin-Maclou, L. Doyennette: Appl. Opt. **10**, 1768 (1971)
4.180 M.A. Kovacs: IEEE J. **QE-9**, 189 (1973)
4.181 E.B. Gordon, V.S. Pavlenko, Yu. L. Moskvin, I.S. Drozdov, P.S. Vinogradov, V.L. Tal'roze: ZhETF **63**, 1159 (1972)
4.182 A.S. Bashkin, A.N. Oraevsky, V.N. Tomashov, N.N. Yuryshev: Kvantovaya Elektron. (Moscow) **3**, 362 (1976)
4.183 J.J. Tiee, C.R. Quick, Jr., C.D. Harper, A.B. Petersen, C. Wittig: J. Appl. Phys. **46**, 5191 (1975)
4.184 S.J. Arnold, G.H. Kimbell: Appl. Phys. Lett. **15**, 351 (1969)
4.185 A.S. Bashkin, N.N. Yuryshev, in: *Kvantovaya Elektronika* (Quantum Electronics) No. 5 (11) (Sovetskoe radio, Moscow, 1972), p. 129
4.186 J. Stricker, S.H. Bauer: Chem. Phys. Lett. **28**, 98 (1974)
4.187 T. Jacobson, G.H. Kimbell: J. Appl. Phys. **41**, 5210 (1970)
4.188 J.L. Ahl, W.R. Binns: IEEE J. **QE-12**, 20 (1976)
4.189 G. Hancock, D.F. Starr, W.H. Green: J. Chem. Phys. **61**, 3017 (1974)
4.190 J.W. Rabalais, J.M. McDonald, V. Scherr, S.P. McGlynn: Chem. Rev. **71**, No. 1 (1971)
4.191 K.H. Homann, G. Krome, H. Gg. Wagner: Ber. Bunsenges. Phys. Chem. **74**, 654 (1970)
4.192 S. Rosenwaks, S. Yatsiv: Phys. Lett. **9**, 226 (1971)
4.193 S. Rosenwaks, I.W.M. Smith: J. Chem. Soc. Faraday Trans. **69**, 1416 (1973)
4.194 Yu. A. Gerasimov. G.N. Kashnikov, V.K. Orlov, A.K. Piskunov, Yu. V. Romanenko, A.V. Taran: Kvantovaya Elektron. (Moscow) **2**, 140 (1975)
4.195 W.Q. Jeffers, H.Y. Ageno, C.E. Wiswall: J. Appl. Phys. **47**, 2509 (1976)
4.196 C. Wittig, I.W.M. Smith: Appl. Phys. Lett. **21**, 536 (1972)
4.197 M.C. Lin, L.E. Brus: J. Chem. Phys. **54**, 5423 (1971); Int. J. Chem. Kinet. **5**, 173 (1973); Int. J. Chem. Kinet. **6**, 1 (1974)
4.198 N.G. Basov, A.S. Bashkin, V.I. Igoshin, A.N. Oraevsky, N.N. Yuryshev: Pis'ma ZhETF **16**, 551 (1972)
4.199 A.S. Bashkin, A.N. Oraevsky, O.E. Porodinkov, N.N. Yuryshev: Khim. Vysokikh Energ. **8**, 513 (1974)
4.200 A.S. Bashkin, A.N. Oraevsky, V.N. Tomashov, N.N. Yuryshev: Kvantovaya Elektron. (Moscow) **2**, 2092 (1975)
4.201 S.K. Searles, J.R. Airey: Appl. Phys. Lett. **22**, 513 (1973)
4.202 A.S. Bashkin, A.N. Oraevsky, V.N. Tomashov, N.N. Yuryshev: Kvantovaya Elektron. (Moscow) **9**, 625 (1982)
4.203 F. Kannari, T. Suzuki, M. Obara, T. Fujioka: Jap. J. Appl. Phys. **22**, L736 (1983)
4.204 V.M. Borisov, E.B. Gordon, V.I. Matyushenko, V.D. Sizov, O.B. Khristoforov: Kvantovaya Elektron. (Moscow) **9**, 434 (1982)
4.205 D.B. Nichols, K.H. Wrolstad, J.D. McClure: J. Appl. Phys. **45**, 5360 (1974)
4.206 A.S. Bashkin, N.P. Vagin, V.A. Zolotarev, A.L. Kiselevskii, M.P. Frolov: Kvantovaya Elektron. (Moscow) **10**, 1693 (1983)
4.207 A.S. Bashkin, V.A. Zolotarev, L.V. Kulakov, M.P. Frolov: Kvantovaya Elektron. (Moscow) **13**, 1065 (1986)
4.208 P.P. Chegodayev: *Candidate's Thesis* (Fil. NIFKhI, Obninsk, 1974)
4.209 O.E. Porodinkov: *Candidate's Thesis* (FIAN, Moscow, 1984)
4.210 A.S. Bashkin, A.L. Kiselevskii, A.N. Oraevsky, V.N. Tomashov, M.P. Frolov. N.N. Yuryshev: Kvantovaya Elektron. (Moscow) **10**, 2126 (1983)
4.211 A.S. Bashkin, A.N. Oraevsky, V.N. Tomashov, N.N. Yuryshev: Kvantovaya Elektron. (Moscow) **9**, 630 (1982)
4.212 A.S. Bashkin, A.N. Oraevsky, V.N. Tomashov, N.N. Yuryshev: Kvantovaya Elektron. (Moscow) **9**, 628 (1982)
4.213 H. Mirels, R. Hofland, J.S. Whittier: J. Appl. Phys. **50**, 6660 (1979)
4.214 G.W. Cooper: IEEE J. **QE-17**, 1840 (1981)
4.215 E.U. Baikov, A.S. Bashkin, A.N. Oraevsky: Preprint FIAN 225 (1984)
4.216 A.S. Bashkin, V.A. Zolotarev, V.N. Tomashov, M.P. Frolov: Kvantovaya Elektron. (Moscow) **14**, 1563 (1987)

4.217 S.D. Velikanov, S.B. Kormer, M.V. Sinitsyn, V.D. Urlin, G.V. Tachayev, V.V. Shchurov: Pis'ma ZhETF **9**, 134 (1983)
4.218 S.T. Amimoto, J.S. Whittier, M.L. Lundquist, F.G. Ronkowski, R. Hofland, P.J. Ortwerth: Appl. Phys. Lett. **40**, 20 (1982)
4.219 T. Fujioka, F. Kannari, T. Suzuki, M. Obara: J. Appl. Phys. **58**, 3975 (1985)
4.220 A.S. Baskhin, V. Yu, Nikitin, A.N. Oraevsky, O.E. Porodinkov, V.N. Tomashov, M.P. Frolov, N.N. Yuryshev: Izv. Akad. Nauk SSSR, ser. fiz. **46**, 1528 (1982)
4.221 E.B. Gordon, V.I. Matyushenko, V.D. Sizov: Kvantovaya Elektron. (Moscow) **9**, 2186 (1982)
4.222 S.D. Velikanov, A.F. Zapol'skii, M.V. Sinitsyn, Yu. N. Sheremet'yev: Kvantovaya Elektron. (Moscow) **11**, 2381 (1984)
4.223 V.S. Bosamykin, E.B. Gordon, V.V. Gorokhov, V.I. Karelin, V.I. Matyushenko, P.B. Repin, V.D. Sizov: Kvantovaya Elektron. (Moscow) **9**, 1489 (1982)
4.224 E.L. Patterson, G.N. Hays, F.K. Truby, R.A. Gerber: J. Appl. Phys. **50**, 2643 (1979)
4.225 V.P. Borisov, S.D. Velikanov, V.D. Kvachev, S.B. Kormer, M.V. Sinitsyn, G.V. Tachayev, Yu. N. Frolov: Kvantovaya Elektron. (Moscow) **8**, 1208 (1981)
4.226 A.S. Bashkin, O.E. Porodinkov, E.V. Khoroshilov: Kvantovaya Elektron. (Moscow) **13**, 2344 (1986)

Chapter 5

5.1   K.G. Anlauf, D.H. Maylotte, P.D. Pacey, J.C. Polanyi: Phys. Lett. **24A**, 208 (1967)
5.2   T.A. Cool, R.R. Stephens: J. Chem. Phys. **51**, 5175 (1969)
5.3   T.A. Cool, R.R. Stephens: Appl. Phys. Lett. **16**, 55 (1970)
5.4   T.A. Cool, J.A. Shirley, R.R. Stephens: Appl. Phys. Lett. **17**, 278 (1970)
5.5   N.G. Basov, A.N. Oraevsky, V.A. Shcheglov: ZhTF **37**, 339 (1967)
5.6   N.G. Basov, A.N. Oraevsky, *Author's Certificate NO. 436413 (USSR), April 24, 1967.* Bull. Otkr. Izobr. **7** (1974)
5.7   T.A. Cool, R.R. Stephens, T.J. Falk: Int. J. Chem. Kinet. **1**, 495 (1969)
5.8   T.A. Cool, T.J. Falk, R.R. Stephens: Appl. Phys. Lett. **15**, 318 (1969)
5.9   N.G. Basov, V.V. Gromov, E.L. Koshelev, E.P. Markin, A.N. Oraevsky, D.S. Shapovalov, V.A. Shcheglov: Pis'ma ZhETF **13**, 496 (1971)
5.10 J.A. Shirley, R.N. Sileo, R.R. Stephens, T.A. Cool: AIAA Pap. No. 71–27 (1971); *AIAA 9th Aerospace Sciences Meeting* (New York, 1971)
5.11 H. Brunet, M. Mabru: C.R. Acad. Sci. Paris Ser. **B 272**, 232 (1971)
5.12 T.J. Falk: Rep. No. AFWL-TR-71-96 (Cornell Aerospace Lab., Buffalo, New York 1971); Rep. No. AFWL-TR-72-169 (Cornell Aerospace Lab., Buffalo, New York 1972)
5.13 T.A. Cool: IEEE J. **QE-9**, 72 (1973)
5.14 G.W. Tregay, M.G. Drexhage, I.M. Wood, S.J. Andrysiak: IEEE J. **QE-11**, 672 (1975)
5.15 G. Emanuel, W.G. Gaskill, R.J. Reiner, C. Shelton, W. Watkins: IEEE J. **QE-12**, 739 (1976)
5.16 J.A. Stregack, W.S. Watt: IEEE J. **QE-11**, 711 (1975)
5.17 W.H. Evers, L.S. Forman, J.J. Vieceli: IEEE J. **QE-11**, 711 (1975)
5.18 G.W. Tregay, T.E. Furner, R.J. Driscoll, W.C. Solomon: *5th Conf. Chem. Mol. Lasers* (St. Louis, Missouri 1977)
5.19 T.O. Poehler, J.C. Pirkle, Jr., R.E. Walker: IEEE J. **QE-9**, 833 (1973)
5.20 R.L. Kerber: Appl Opt. **12**, 1157 (1973)
5.21 V.N. Kondrat'yev: *Konstanty skorosti gasofaznykh reaktsii* (Rate Constants of Gas-Phase Reactions) (Nauka, Moscow 1970)
5.22 G.J. Wolga: references cited in [5.13]
5.23 W. Baumann, J.A. Blauer, S.W. Zelazny, W.S. Solomon: Appl. Opt. **13**, 2823 (1974)
5.24 A.S. Bashkin, V.I. Igoshin, A.I. Nikitin, A.N. Oraevsky: *Khimicheskie lazery* (Chemical Lasers), in: *Itogi nauki i tekhniki*, Ser. Radiotekh. 8 (VINITI, Moscow 1975)
5.25 A.L. Mikaelyan, M.L. Ter-Mikaelyan, Yu. G. Turkov: *Opticheskiye kvantovye generatory na tverdom tele* (Solid State Lasers) (Sovetskoe radio, Moscow 1967)
5.26 V.P. Pimenov, V.A. Shcheglov: Rep. No. 149, f. 2 (FIAN, Moscow 1975)
5.27 N.G. Basov, A.N. Oraevsky, V.A. Shcheglov: Rep. No. 152, f. 2 (FIAN, Moscow 1971)

5.28   S.W. Zelazny, J.A. Blauer, L. Wood, L.H. Sentman, W.S. Solomon: Appl. Opt. **15**, 1164 (1976)
5.29   V.P. Pimenov, V.A. Shcheglov: Rep. No. 150, f. 2 (FIAN, Moscow 1975)
5.30   J.J. Vieceli, W.H. Evers, A.G. Negro: IEEE J. **QE-11**, 672 (1975)
5.31   R.L. Kerber, N. Cohen, G. Emanuel: IEEE J. **QE-9**, 94 (1973)
5.32   N.G. Basov, V.G. Mikhailov, A.N. Oraevsky, V.A. Shcheglov: ZhTF **38**, 2031 (1968)
5.33   J.D. Anderson: *Gasdynamic Lasers: An Introduction* (Academic, London, New York 1976)
5.34   S.A. Losev: *Gazodinamicheskiye lazery* (Gasdynamic Lasers), (Nauka, Moscow 1977)
5.35   A.S. Biryukov: *Candidate's Thesis* (FIAN, Moscow 1973)
5.36   R.I. Soloukhin, V.P. Chebotayev, eds.: *Gazovye lazery* (Gas Lasers), (Nauka, Novosibirsk 1977)
5.37   V.I. Igoshin: Tr. FIAN **76**, 114 (1974)
5.38   J. Thoens, A.W. Ratliff: AIAA Pap. No. 72-147 (1972); AIAA Pap. No. 73-644 (1973)
5.39   L.D. Landau, E.M. Lifshits: *Mekhanika sploshnykh sred* (Mechanics of Continuous Media), (Gostekhizdat, Moscow 1953)
5.40   A.A. Townsend: *The Structure of Turbulent Shear Flow* (Cambridge University Press, Cambridge 1956)
5.41   T.A. Cool, R.R. Stephens, H.A. Shirley: J. Appl. Phys. **41**, 4038 (1970)
5.42   D.J. Spencer, T.A. Jacobs, H. Mirels, R.W.F. Gross: Int. J. Chem. Kinet. **1**, 493 (1969)
5.43   J.R. Airey, S.F. McKay: Appl. Phys. Lett. **15**, 401 (1969)
5.44   R.A. Meinzer: Int. J. Chem. Kinet. **2**, 335 (1970)
5.45   D.J. Spencer, T.A. Jacobs, H. Mirels, R.W.F. Gross: Appl. Phys. Lett. **16**, 235 (1970)
5.46   D.J. Spencer, H. Mirels, T.A. Jacobs: Appl. Phys. Lett. **16**, 384 (1970)
5.47   M.A. Kwok, R.R. Giedt, R.W.F. Gross: Appl. Phys. Lett. **16**, 386 (1970)
5.48   D.J. Spencer, H. Mirels, T.A. Jacobs: Opto-Electronics **2**, 155 (1970)
5.49   H. Mirels, D.J. Spencer: IEEE J. **QE-7**, 501 (1971)
5.50   D.J. Spencer, H. Mirels, D.A. Durran: J. Appl. Phys. **43**, 1151 (1972)
5.51   D.J. Spencer, H.A. Bixer, D.A. Durran: Appl. Phys. Lett. **20**, 164 (1972)
5.52   D.J. Spencer, R.L. Varwig: AIAA J. **11**, 1000 (1973)
5.53   R.L. Varwig, M.A. Kwok: AIAA J. **12**, 208 (1974)
5.54   R.L. Varwig: AIAA J. **12**, 1448 (1974)
5.55   M.A. Kwok, D.J. Spencer, R.W.F. Gross: J. Appl. Phys. **45**, 3500 (1974)
5.56   R.L. Varwig: Rep. No TR-0073 (3250-10)-8 (Aerospace Corp., El Segundo, California 1973)
5.57   R.A. Chodzko, H. Mirels, F.S. Roehrs, R.J. Pedersen: IEEE J. **QE-9**, 523 (1973)
5.58   W.L. Shackleford, A.B. Witte, J.E. Broadwell, J.E. Trost, T.A. Jacobs: AIAA J. **12**, 1009 (1974)
5.59   J.C. Cummings, C.M. Dube, A.B. Witte: Appl. Phys. Lett. **25**, 89 (1974)
5.60   E.R. Schulman, W.G. Burwell, R.A. Meinzer: AIAA Pap. No. 74-546 (1974)
5.61   C.K. Nagai, L.W. Carlson, R.R. Giedt, R.D. Klopotek: AIAA Pap. No. 74-684 (1974)
5.62   R.R. Giedt: IEEE J. **QE-11**, 718 (1975)
5.63   J.C. Hyde, R. Knight: IEEE J. **QE-11**, 718 (1975)
5.64   J.C. Cummings, C.M. Dube: IEEE J. **QE-11**, 712 (1975)
5.65   R.A. Meinzer, R.V. Steele: IEEE J. **QE-11**, 712 (1975)
5.66   T.J. Sadowski, C.E. Kepler, B.R. Bronfin, M.D. Kroskey, R. Roback: IEEE J. **QE-11**, 705 (1975)
5.67   J.O. Bunting, P.C. Steel: AIAA Pap. No. 75-722 (1975)
5.68   D. Finklemann, R.A. Greenberg: AIAA Pap. No. 74-297 (1974)
5.69   R.J. Cavalleri, H.O. Laeger: AIAA Pap. No. 74-548 (1974)
5.70   R. Roback, L. Lynds: AIAA Pap. No. 74-1142 (1974)
5.71   A.W. Ratliff, A.J. McDanal, Sh. C. Krizius: AIAA Pap. No. 75-721 (1975)
5.72   V.A. Shcheglov: Rep. No. 151, f. 2 (FIAN, Moscow, 1971)
5.73   B. Lewis, R.N. Pease, H.S. Taylor, eds.: *High Speed Aerodynamics and Jet Propulsion*, Vol. 2: *Combustion Processes* (Princeton University Press, Princeton 1956)
5.74   A.S. Predvodityelyev, ed.: *Issledovaniya po fizicheskoi gazodinamike* (Studies on Physical Gas Dynamics) (Nauka, Moscow 1966)
5.75   G.N. Abramovich: *Teoriya turbulentnykh strui* (Theory of Turbulent Jets) (Fizmatgiz, Moscow, 1960)
5.76   P.F. Pohyl, V.M. Maltsev, V.M. Zaitsev: *Metody issledovaniya protsessov goreniya i detonatsii* (Methods for Studying Combustion and Detonation Processes) (Nauka, Moscow 1969)
5.77   G. Emanuel: JQSRT **11**, 1481 (1971)

5.78   G. Emanuel, N. Cohen, T.A. Jacobs: JQSRT **13**, 1365 (1973)
5.79   E.B. Turner, W.D. Adams, G. Emanuel: J. Comp. Phys. **11**, 15 (1973)
5.80   G. Emanuel, W.D. Adams, E.B. Turner: Rep. No NTR-0172 (2776)-1 (Aerospace Corp., El Segundo, California 1972)
5.81   G. Emanuel, J.S. Whittier: Appl. Opt. **11**, 2047 (1972)
5.82   J.G. Skifstad: Combust. Sci. Technol. **6**, 287 (1973)
5.83   L.H. Sentman: J. Chem. Phys. **62**, 3523 (1975); Appl. Opt. **15**, 744 (1976)
5.84   J.G. Skifstad, C.M. Chao: Appl. Opt. **14**, 1713 (1975)
5.85   G. Emanuel: JQSRT **12**, 913 (1972)
5.86   R. Hofland, H. Mirels: AIAA J. **10**, 420 (1972)
5.87   R. Hofland, H Mirels: AIAA J. **10**, 1271 (1972)
5.88   W.S. King, H. Mirels: AIAA J. **10**, 1647 (1972)
5.89   R.R. Mikatarian, J.W. Benefield: AIAA Pap. No. 74-148 (1974)
5.90   R. Tripodi, L.J. Coulter, B.R. Bronfin, N. Cohen: AIAA J. **13**, 776 (1975); AIAA Pap. No. 74-224 (1974)
5.91   A.W. Ratliff, J. Thoenes: AIAA Pap. No 74–225 (1974)
5.92   R.R. Mikatarian, A.J. McDanal: AIAA Pap. No 75-39 (1975)
5.93   R.R. Mikatarian: AIAA Pap. No 74-547 (1974)
5.94   V.I. Golovichev, N.G. Preobrazhenskii: Fiz. Goreniya Vzryv. **3**, 366 (1977)
5.95   V.K. Bayev, V.I. Golovichev, V.A. Yasakov: *Dvumernye turbulentnye techeniya reagiruyushchikh gazov* (Two-Dimensional Turbulent Flows of Reacting Gases) (Nauka, Novosibirsk, 1976)
5.96   H. Schlichting: *Grenzschicht-Theorie* (Braun, Karslruhe, 1965)
5.97   G.N. Abramovich, S. Yu. Krashennikov, A.N. Sekundov, I.P. Smirnova: *Turbulentnoye smesheniye gazovykh strui* (Turbulent Mixing of Gas Jets) (Nauka, Moscow, 1974)
5.98   I.P. Ginzburg: *Aerogazodinamika* (Aerogasdynamics) (Vysshaya shkola, Moscow, 1966)
5.99   A.A. Stepanov, V.A. Shcheglov: Preprint FIAN **182** (1976)
5.100 A.P. Kothari, J.D. Anderson, E. Tohnes: AIAA J. **15**, 92 (1977)
5.101 H. Mirels, R. Hofland, W.S. King: AIAA J. **11**, 156 (1973); AIAA Pap. No 72–145 (1972)
5.102 J.E. Broadwell: Appl. Opt. **13**, 962 (1974)
5.103 P.M. Chung: AIAA J. **11**, 1040 (1973)
5.104 A.A. Stepanov, V.A. Shcheglov: Kvantovaya Elektron. (Moscow) **2**, 1379 (1975)
5.105 H. Mirels: AIAA J. **13**, 785 (1975)
5.106 A.A. Stepanov, V.A. Shcheglov: ZhTF **46**, 563 (1976)
5.107 H. Mirels: AIAA J. **14**, 930 (1976)
5.108 A.N. Oraevsky, V.P. Pimenov, A.A. Stepanov, V.A. Shcheglov: Kvantovaya Elektron. (Moscow) **3**, 136 (1976)
5.109 V.G. Krutova, A.N. Oraevsky, A.A. Stepanov, V.A. Shcheglov: Kvantovaya Elektron. (Moscow) **3**, 1919 (1976)
5.110 V.P. Pimenov, V.A. Shcheglov: Kvantovaya Elektron. (Moscow) **3**, 1041 (1976)
5.111 A.N. Oraevsky, V.P. Pimenov, A.A. Stepanov, V.A. Shcheglov: Kvantovaya Elektron. (Moscow) **3**, 1896 (1976)
5.112 R.J. Hall: IEEE J. **QE-12**, 453 (1976)
5.113 V.G. Krutova, A.N. Oraevsky, A.A. Stepanov, V.A. Shcheglov: ZhETF **47**, 2383 (1977)
5.114 Ya. Z. Virnik, V.G. Krutova, A.I. Mashchenko, A.N. Oraevsky, A.A. Stepanov, V.A. Shcheglov: Kvantovaya Elektron. (Moscow) **4**, 2234 (1977)
5.115 Ya. Z. Virnik, V.G. Krutova, A.A. Stepanov, V.A. Shcheglov: Kvantovaya Elektron. (Moscow) **4**, 2527 (1977)
5.116 Ya. Z. Virnik, V.G. Krutova, A.I. Mashchenko, A.N. Oraevsky, A.A. Stepanov, V.A. Shcheglov: Kvantovaya Elektron. (Moscow) **5**, 883 (1978)
5.117 Ya. Z. Virnik, A.K. Piskunov, A.A. Stepanov, V.A. Shcheglov: Kvantovaya Elektron. (Moscow) **6**, 236 (1979)
5.118 S.W. Zelazny, R.J. Driscoll, J.W. Raymonda, J.A. Blauer, W.C. Solomon: AIAA J. **16**, 297 (1978)
5.119 F.A. Williams: *Combustion Theory* (Addison-Wesley, Reading, Massachusetts 1965)
5.120 V.A. Shvab, in: *Issledovaniye protsessov goreniya natural'nogo topliva* (Investigation into the Burning of Natural Fuel) (Gosenergoizdat, Moscow-Leningrad, 1948)
5.121 Ya. B. Zel'dovich: ZhETF **19**, 1199 (1949)

5.122 R.J. Driscoll: AIAA J. **14**, 1571 (1976)

5.123 G. Emanuel: AIAA J. **15**, 120 (1977)

5.124 A.A. Stepanov, V.A. Shcheglov: ZhTF **49**, 581 (1979)

5.125 H. Henrici, M.C. Lin, S.H. Bauer: J. Chem. Phys. **52**, 5834 (1970)

5.126 B.R. Bronfin: Rep. No AFQL-TR-73-48 (Air Force Weapons Lab., 1973)

5.127 W.R. Warren: Acta Astronaut. **1**, 813 (1974)

5.128 W.R. Warren: Astronaut. Aeronaut. **13**, 36 (1975)

5.129 A.A. Stepanov, V.A. Shcheglov: Preprint FIAN **269** (1978)

5.130 A.A. Stepanov, V.A. Shcheglov: Kvantovaya Elektron. (Moscow) **6**, 1476 (1979)

5.131 A.A. Stepanov, V.A. Shcheglov: Preprint FIAN **59** (1979)

5.132 A.A. Stepanov, V.A. Shcheglov: Kvantovaya Elektron. (Moscow) **6**, 747 (1979)

5.133 V.N. Croshko, R.I. Soloukhin, P. Wolansky: Opt. Comm. **6**, 275 (1972)

5.134 V.N. Kroshko, R.I. Soloukhin, N.A. Fomin: Fiz. Goreniya Vzryv. **9**, 473 (1973)

5.135 V.N. Kroshko, R.I. Soloukhin: Dokl. Akad. Nauk SSSR **211**, 829 (1973)

5.136 R. Borgi, A.P. Carrega, M. Charpenel, J.P.E. Taran: Appl. Phys. Lett. **22**, 661 (1973)

5.137 S.W. Mayer, M.K. Kwok, R.W.F. Gross, D.J. Spencer: Appl. Phys. Lett. **17**, 516 (1970)

5.138 A.S. Bashkin, N.M. Gorshunov, Yu. A. Kunin, Yu. P. Neshchimenko, A.N. Oraevsky, N.N. Yuryshev: Kvantovaya Elektron. (Moscow) **3**, 1142 (1976)

5.139 P.B. Scott, S. Johnson, G. Watson: IEEE J. **QE-11**, 693 (1975)

5.140 A.S. Bashkin, N.M. Gorshunov, Yu. A. Kunin, Yu. P. Neshchimenko, A.N. Oraevsky, V. N. Tomashov, N.N. Yuryshev: Kvantovaya Elektron. (Moscow) **3**, 462 (1976)

5.141 R.R. Stephens, T.A. Cool: Rev. Sci. Instr. **42**, 1489 (1971)

5.142 T.A. Cool, R.R. Stephens: J. Chem. Phys. **52**, 3304 (1970)

5.143 J.J. Hinchen, C.M. Banas: Appl. Phys. Lett. **17**, 386 (1970)

5.144 J.J. Hinchen, C.J. Ultee: AIAA Pap. No 71–216 (1971)

5.145 C.J. Buczek, R.J. Freiberg, J.J. Hinchen, P.P. Chenausky, R.J. Wayne: Appl. Phys. Lett. **17**, 514 (1970)

5.146 J.A. Glaze: Appl. Phys. Lett. **19**, 135 (1971)

5.147 J.A. Glaze, J. Finzi, W.F. Krupke: Appl. Phys. Lett. **18**, 173 (1971)

5.148 D.W. Naegeli, C.J. Ultee: Chem. Phys. Lett. **6**, 121 (1970)

5.149 J.J. Hinchen: J. Appl. Phys. **45**, 1818 (1974)

5.150 D. Proch, H. Pummer, K.L. Kompa: Rev. Sci. Instr. **46**, 1101 (1975)

5.151 D.I. Rosen, R.N. Sileo, T.A. Cool: IEEE J. **QE-9**, 163 (1973)

5.152 J.A. Glaze, G.J. Linford: Rev. Sci. Instr. **44**, 600 (1973)

5.153 J.M. Gagne, S.Q. Mah, Y. Conturie: Appl. Opt. **13**, 2835 (1974)

5.154 J.J. Hinchen: IEEE J. **QE-11**, 695 (1975)

5.155 J.L. Ahl, T.A. Cool: J. Chem. Phys. **58**, 5540 (1973)

5.156 R.A. Lucht, T.A. Cool: J. Chem. Phys. **60**, 1026 (1974)

5.157 S.J. Arnold, K.D. Foster, D.R. Snelling, R.D. Suart: Appl. Phys. Lett. **30**, 637 (1977)

5.158 S.J. Arnold, K.D. Foster: Appl. Phys. Lett. **33**, 716 (1978)

5.159 S.J. Arnold, K.D. Foster, D.R. Snelling, R.D. Suart: *High-Power Lasers and Applications* (Proc. 4th Coll. Electron Transit. Lasers, Munich, 1977)

5.160 S.J. Arnold, K.D. Foster, D.R. Snelling, R.D. Suart: IEEE J. **QE-14**, 293 (1978)

5.161 S.J. Arnold, K.D. foster, D.R. Snelling: J. Appl. Phys. **50**, 1189 (1979)

5.162 N. Djeu, H.S. Piloff, S.K. Searles: Appl. Phys. Lett. **18**, 538 (1971)

5.163 N. Djeu, H.S. Piloff, S.K. Searles: Appl. Phys. Lett. **19**, 9 (1971)

5.164 S.K. Searles, N. Djeu: Chem. Phys. Lett. **12**, 53 (1971)

5.165 K.D. Foster, G.H. Kimbell: J. Chem. Phys. **53**, 2539 (1970)

5.166 M.J. Linevsky, R.A. Carabetta: Appl. Phys. Lett. **22**, 288 (1973)

5.167 K.D. Foster, G.H. Kimbell, D.R. Snelling: IEEE J. **QE-11**, 700 (1975)

5.168 C. Wittig, J.C. Hassler, P.D. Coleman: J. Chem. Phys. **55**, 5523 (1971); Nature **226**, (1970); Appl. Phys. Lett. **16**, 117 (1970)

5.169 R.D. Suart, S.J. Arnold, G.H. Kimbell: Chem. Phys. Lett. **5**, 519 (1970)

5.170 W.Q. Jeffers, C.E. Wiswall: Appl. Phys. Lett. **17**, 67 (1970)

5.171 L.R. Boedeker, J.A. Shirley, B.R. Bronfin: Appl. Phys. Lett. **21**, 247 (1972)

5.172 R.D. Suart, P.H. Dawson, G.H. Kimbell: J. Appl. Phys. **43**, 1022 (1972)
5.173 K.D. Foster: J. Chem. Phys. **57**, 2451 (1972)
5.174 C.J. Ultee, P.A. Bonczyk: IEEE J. **QE-10**, 105 (1974)
5.175 W.Q. Jeffers, C.E. Wiswall: IEEE J. **QE-10**, 860 (1974)
5.176 W.Q. Jeffers, C.E. Wiswall: Appl. Phys. Lett. **17**, 67 (1973)
5.177 W.Q. Jeffers, C.E. Wiswall, J.D. Kelley, R.J. Richardson: Appl. Phys. Lett. **22**, 587 (1973)
5.178 W.Q. Jeffers, H.Y. Ageno: Appl. Phys. Lett. **27**, 227 (1975)
5.179 Y. Hirose, Y. Nachshon, T.A. De Temple, P.D. Coleman: Appl. Phys. Lett. **23**, 195 (1973)
5.180 A. Leckyer, N. Legay-Sommaire: C.R. Acad. Sci. Paris, Ser. B **271**, 1212 (1970)
5.181 C.J. Ultee: Appl. Phys. Lett. **19**, 535 (1971)
5.182 D.W. Howgate, T.A. Barr, Jr.: J. Chem. Phys. **59**, 2815 (1973)
5.183 G. Hancock, I.W.M. Smith: Chem. Phys. Lett. **3**, 573 (1969)
5.184 G. Hancock, C. Morley, I.W.M. Smith: Chem. Phys. Lett. **12**, 193 (1971)
5.185 M.S. Dzhidzhoyev, V.T. Platonenko, R.V. Khokhlov: Usp. Fiz. Nauk **100**, 641 (1970)
5.186 A.B. Petersen, C. Wittig: Chem. Phys. Lett. **27**, 442 (1974)
5.187 A.B. Petersen, C. Wittig: IEEE J. **QE-11**, 699 (1975)
5.188 A.S. Bashkin, A.G. Velikanov, N.M. Gorshunov, Yu. A. Kunin, Yu. P. Neshchimenko, A.N. Oraevsky, N.N. Yuryshev: Kvantovaya Elektron. (Moscow) **5**, 2656 (1978)
5.189 A.S. Bashkin, N.M. Gorshunov, Yu. A. Kunin, Yu. P. Neshchimenko, A.N. Oraevsky, N.N. Yuryshev: Preprint FIAN **140** (1979)
5.190 V.F. Kochubei, F.B. Monin: Fiz. Goreniya Vzryv. **13**, 476 (1977)
5.191 V.N. Kondrat'yev, E.E. Nikitin: *Kinetika i mekhanizm gazofaznykh reaktsii* (Kinetics and Mechanism of Gas-Phase Reactions) (Nauka, Moscow, 1974)
5.192 D.J. Benard, R.C. Benson, R.E. Walker: Appl. Phys. Lett. **23**, 82 (1973)
5.193 R.C. Benson, C.B. Bargeron, R.E. Walker: IEEE J. **QE-11**, 688 (1975)
5.194 W.E. McDermott, N.R. Pchelkin, D.J. Benard, R.R. Bousek: Appl. Phys. Lett. **32**, 469 (1978)
5.195 R.G. Derwent, B.A. Thrush: Faraday Discuss. Chem. Soc. **53**, 162 (1972)
5.196 D.J. Benard, W.E. McDermott, N.R. Pchelkin, R.R. Bousek: Appl. Phys. Lett. **34**, 40 (1979)
5.197 R.J. Richardson, C.E. Wiswall: Appl. Phys. Lett. **33**, 296 (1978)
5.198 L.E. Wilson, D.L. Hook: AIAA Pap. No 76-344 (1976)
5.199 F.G. Sadie, P.A. Büger, O.G. Malan: J. Appl. Phys. **43**, 2906, 5141 (1972)
5.200 P.S. Vinogradov, E.B. Gordon, M.S. Drozdov, Yu. L. Moskvin, V.S. Pavlenko, V.L. Tal'roze: ZhETF **63**, 1159 (1972)
5.201 M.C. Lin, S.H. Bauer: Chem. Phys. Lett. **7**, 223 (1970)
5.202 J.D. Barry, W.E. Boney, J.E. Brandelik: Appl. Phys. Lett. **18**, 15 (1971)
5.203 J.D. Barry, W.E. Boney, J.E. Brandelik: IEEE J. **QE-7**, 208 (1971)
5.204 J.D. Barry, W.E. Boney, J.E. Brandelik: IEEE J. **QE-7**, 461 (1971)
5.205 J.D. Barry, W.E. Boney, J.E. brandelik: Appl. Phys. Lett. **19**, 141 (1971)
5.206 A.A. Stepanov, V.A. Shcheglov, N.N. Yuryshev: Kvantovaya Elektron. (Moscow) **12**, 1127 (1985)
5.207 V.K Ablekov, Yu. N. Denisov: *Gazoprotochnye khimicheskiye lazery* (Flowing Gas Chemical Lasers), (Energoatomizdat, Moscow, 1987)
5.208 V.I. L'vov, A.A. Stepanov, V.A. Shcheglov: Kvantovaya Elektron. (Moscow) **12**, 1035 (1985)
5.209 N.A. Konoplev, V.I. L'vov, A.A. Steppanov, V.A. Shcheglov: Fiz. Goreniya Vzryv. **22**, 78 (1986)
5.210 J.P. Moran: *5th Conf. Chem. Mol. Lasers* (St. Louis, Missouri, 1977)
5.211 J.P. Moran, A.C. Stanton, R.B. Doan: AIAA J. **18**, 958 (1980)
5.212 A. Kaldar, P. Rabinowitz, D.M. Cox, J. Horsley, R. Brickman: *10th Int. Conf. Quant. Electron.* (Atlanta, Georiga, 1978)
5.213 S.S. Alimpiyev, A.L. Babichev, G.S. Baranov: Kvantovaya Elektron. (Moscow) **6**, 2155 (1979)
5.214 V.G. Averin, A.L. Babichev, G.S. Baranov: Kvantovaya Elektron. (Moscow) **6**, 2637 (1979)
5.215 B.I. Vasil'yev, A.P. Dyad'kin, A.N. Sukhanov: Pis'ma ZhTF **6**, 311 (1980)
5.216 N.V. Karlov, Yu. B. Konev, I.V. Kochetov, V.P. Pevgov, A.M. Prolkhorov: Pis'ma ZhTF **2**, 1062 (1976)
5.217 T.J. Manuccia, J.A. Stregack, N.W. Harris, B.L. Wexler: Appl. Phys. Lett. **29**, 360 (1976)
5.218 B.L. Wexler, R.W. Waynant: Appl. Phys. Lett. **34**, 674 (1979)
5.219 W.H. Kosner, L.D. Plaasance: Appl. Phys. Lett. **31**, 82 (1977)

5.220 D. Yu Zaroslov, N.V. Karlov, I.O. Kovalev, G.R. Kuz'min, A.M. Prokhorov: Pis'ma ZhTF **5**, 759 (1979)
5.221 K. Suzuki, S. Saito, M. Obara, T. Fujioka: J. Appl. Phys. **51**, 4003 (1980)
5.222 S. Saito, M. Obara, T. Fujioka: Appl. Opt. **20**, 2838 (1981)
5.223 A.G. Velikanov, N.M. Gorshunov, Yu. P. Neschimenko, A.V. Shcherbo: Kvantovaya Elektron. (Moscow) **8**, 156 (1981)
5.224 A.S. Biryukov, N.A. Konoplev, V.A. Shcheglov: Pis'ma ZhTF **7**, 482 (1981)
5.225 A.S. Biryukov, V.A. Shcheglov: Kvantovaya Elektron. (Moscow) **8**, 2371 (1981)
5.226 A.S. Biryukov, A.A. Stepanov, V.A. Shcheglov: Kvantovaya Elektron. (Moscow) **8**, 2066 (1981)
5.227 A.S. Biryukov, I.V. Karachanova, N.A. Konoplev, V.A. Shcheglov: Kvantovaya Elektron. (Moscow) **10**, 1667 (1983)
5.228 A.S. Biryukov, I.V. Karachanova, N.A. Konoplev, V.A. Shcheglov: Kvantovaya Elektron. (Moscow) **10**, 2501 (1983)
5.229 A.S. Biryukov, N.A. Konoplev, V.A. Shcheglov: Kratk. Soobshch. Fiz. FIAN **2**, 12 (1984)
5.230 A.S. Biryukov, R.I. Serikov, A.M. Starik, V.A. Shcheglov: Kvantovaya Elektron. (Moscow) **11**, 849 (1984)
5.231 H. Pummer, D. Proch, V. Schmeilzl, K.L. Kompa: Opt. Comm. **19**, 273 (1976)
5.232 J.H.S. Wang, J. Finzi, P.K. Baily, G.H. Holleman, K.K. Hui, F.N. Mastrup: Appl. Phys. Lett. **36**, 24 (1980)
5.233 S. Yamaguchi, K. Kumamoto, F. Kannari, M. Obara, T. Fujioka: J. Appl. Phys. **54**, 1675 (1983)
5.234 M.A. Kwok, R.L. Wilkins: Appl. Opt. **22**, 2721 (1983)
5.235 V.I. L'vov, A.A. Stepanov, V.A. Shcheglov: Preprint FIAN (Moscow) **260** (1984); Opt. Spektroskop. **61**, 159 (1986)
5.236 A.S. Bashkin, M.S. Kurdoglyan, A.N. Orayevskii: Kvantovaya Elektron. (Moscow) **12**, 1758 (1985)
5.237 I.M. Bel'dyugin, Yu. P. Vysotskii, A.A. Stepanov, V.A. Shcheglov: Khim. Fiz. **5**, 1018 (1986)

Chapter 6

6.1 L.G. Vinogradova, M.V. Zagidullin, V.I. Igoshin, V.A. Katulin, N.L. Kupriyanov: Kvantovaya Elektron. (Moscow) **9**, 1193 (1982)
6.2 M.V. Zagidullin, V.I. Igoshin, V.A. Katulin, M.L. Kupriyanov: Kvantovaya Elektron. (Moscow) **9**, 1899 (1982)
6.3 W.E. McDermott, N.R. Pchelkin, D.J. Benard, R.R. Bousek: Appl. Phys. Lett. **32**, 496 (1978)
6.4 D.J. Benard, W.E. McDermott, N.R. Pchelkin, R.R. Bousek: Appl. Phys. Lett. **34**, 40 (1978)
6.5 R.J. Richardson, C.E. Wiswall: Appl. Phys. Lett. **35**, 138 (1979)
6.6 V.M. Ramm: *Adsorbtsiya gazov* (Adsorption of Gases) (Khimiya, Moscow, (1976)
6.7 M.V. Zagidullin, V.I. Igoshin, V.A. Katulin, N.L. Kupriyanov: Preprint FIAN (Moscow) **211** (1982)
6.8 N.G. Basov, M.V. Zagidullin, V.I. Igoshin, V.A. Katulin, N.L. Kupriyanov: Tr. FIAN **171**, 30 (1986)
6.9 D.J. Miller, J.G. Berg, J.C. Brock *et al.*: *Int. Conf. Laser-82, Techn. Dig.*, 1982, Pap. No R2
6.10 K. Watanabe, S. Kashiwabata, K. Sawai, S. Toshima, R. Fujimoto: J. Appl. Phys. **54**, 1228 (1983)
6.11 J. Bachar, S. Rasenwaks: Appl. Phys. Lett. **41**, 16 (1982)
6.12 N.P. Vagin, A.F. Konoshenko, P.G. Kryukov *et al.*: Kvantovaya Elektron. (Moscow) **11**, 1688 (1984)
6.13 F.V. Grigor'yev, L.V. Goryachev, V.A. Yeroshenko *et al.*: Izv. Akad. Nauk SSSR, Ser. Fiz. **48**, 1383 (1984)
6.14 J. Bonnet, D. David, E. Georges, B. Leporcq, D. Pigache, C. Verdier: Appl. Phys. Lett. **45**, 1009 (1984)
6.15 D. David, E. Georges, B. Leporcq *et al.*: *Proceed. 1st Int. Workshop Iodine Laser and Applications* (Bechyne, Czechoslovakia, 1986), p. 236
6.16 P.V. Avizonis "*Chemical Oxygen Iodine Laser Review*" *Paper presented at 7-th Int. Symposium on Gas Flow and Chemical Lasers*", Vienna, 22–26 Aug. 1988. Book of Abstracts, p. 58.
6.17 A.F. Konoshenko, P.G. Kryukov, D.Kh. Nurligareyev *et al.*: *Proceed 1st Int. Workshop Iodine Laser and Applications* (Bechyne, Czechoslovakia, 1986), p. 236

6.18 H. Yoshimoto, H. Yamakoshi, Y. Shibukawa, T. Uchiyama: *Proceed. 1st Int. Workshop Iodine Laser and Applications* (Bechyne, Czechoslovakia, 1986), p. 268

6.19 C.E. Wiswall, R.J. Richardson, K.V. Reddy: IEEE J. **QE-17**, No 12, Pt. 2, 225 (1981)

6.20 R.J. Richardson, J.D. Kelley, C.E. Wiswall: J. Appl. Phys. **52**, 1066 (1981)

6.21 M.V. Zagidullin, V.I. Igoshin, V.A. Katulin, N.L. Kupriyanov: Kvantovaya Elektron. (Moscow) **10**, 797 (1983)

6.22 A.K. McKnight, A.C. Stanclife: US Pat No 4342116 (1982)

6.23 S. Yoshida, M. Iizuka, H. Yamakoshi, K. Kobayashi, H. Saito, T. Fujioka: *Proceed 1st Int. Workshop Iodine Laser and Applications* (Bechyne, Czechoslovakia, 1986), p. 244

6.24 C.E. Wiswall, R.J. Richardson, K.V. Reddy: J. Appl. Phys. **58**, 115 (1985)

6.25 D.J. Miller, C.W. Clendering, W.D. English, J.C. Berg. J.E. Trost: Pap. No FS2, CLEO 82, *Conf. Lasers and Electro-Optics*, Phoenics, Az. (Optical Soc. Amer., Washington, D.C., 1982), p. 170

6.26 G.N. Fisk, G.N. Hays: IEEE J. **QE-17**, 1823 (1981)

6.27 K. Watanabe, S. Kashivabara, R. Fujimoto: J. Appl. Phys. **59**, 42 (1986)

6.28 H. Yoshomoto, H. Yamakoshi, Y. Shibukawa, T. Uchiyama: J. Appl. Phys. **59**, 3965 (1986)

6.29 J. Kodymova, O. Shpalek, A. Hirshl: *Proceed. 1st Int. Workshop Iodine Laser and Applications* (Bechyne, Czechoslovakia, 1986), p. 230

6.30 M.V. Zagidullin, V.I. Igoshin, V.A. Katulin, N.L. Kupriyanov: Kvantovaya Elektron. (Moscow) **10**, 131 (1983)

6.31 R.F. Heidner, C.E. Gardner, G.I. Segal, T.M. El Sayed: J. Phys. Chem. **87**, 2348 (1983)

6.32 D. David, V. Joly, A. Fausse: Reprint ONERA, France (1986)

6.33 M.V. Zagidullin, V.I. Igoshin, N.L. Kupriyanov: Kvantovaya Elektron. (Moscow) **11**, 382 (1984)

6.34 M.V. Zagidullin, V.I. Igoshin, N.L. Kupriyanov: Kvantovaya Elektron. (Moscow) **11**, 1379 (1984)

6.35 R.G. Derwent, D.A. Thrush: Chem. Phys. Lett. **9**, 591 (1971)

6.36 M.V. Zagidullin, V.I. Igoshin, V.A. Katulin, N.L. Kupriyanov: Preprint FIAN (Moscow) **215**, (1984)

6.37 V.N. Azyazov, V.I. Igoshin, N.L. Kupriyanov, T. Yu. Nemkova, V.P. Sirochenko: Preprint FIAN **115**, (1984)

6.38 M.V. Zagidullin, V.I. Igoshin, N.L. Kupriyanov: Kvantovaya Elektron. (Moscow) **14**, 516 (1987)

6.39 A.P. Zaikin: *Program 19th Sci. Conf. KuGU* (Kuibyshev, 1988), p. 29

6.40 M.V. Zagidullin, A.P. Zaikin, N.L. Kupriyanov, V.I. Igoshin, S. Yu. Pichugin: Preprint FIAN (Moscow) **226** (1986)

6.41 N.G. Basov, N.P. Vagin, P.G. Kryukov, D.N. Nurligareyev, V.S. Pazyuk, N.N. Yuryshev: Kvantovaya Elektron. (Moscow) **11**, 1893 (1984)

6.42 M.V. Zagidullin, V.I. Igoshin, V.A. Katulin, N.L. Kupriyanov, N.N. Yuryshev: Kvantovaya Elektron. (Moscow) **11**, 201 (1984)

6.43 M.V. Zagidullin, A.P. Zaikin, V.I. Igoshin, N.L. Kupriyanov: Preprint FIAN (Moscow) **151** (1986)

6.44 M.V. Zagidullin, V.I. Igoshin, S.Yu. Pichugin: Preprint FIAN (Moscow) **226** (1986)

6.45 V.I. Igoshin, V.A. Katulin, N.L. Kupriyanov, M.V. Zagidyullin: *Proc. 1st Int. Laser-Sci. Conf.* (Dallas, 1985), p. 145

6.46 A. Yu. Kurov, V.A. Katulin, V.D. Nikolayev: *Program 19th Sci. Conf. KuGU* (Kuibyshev, 1988), p. 29

6.47 S. Yoshida, M. Endo, T. Sawano: J. Appl. Phys., **65**, 870 (1989)

Chapter 7

7.1 N.G. Basov, E.P. Markin, A.N. Oraevsky, A.V. Pankratov: Dokl. Akad. Nauk SSSR **198**, 1034 (1971)

7.2 N.N. Semenov: Preprint OIKhFAN (Chernogolovka, 1975)

7.3 S.J. Davis: Proc. Soc. Photo-Opt. Instrum. Eng. **540**, 188 (1985)

7.4 V.I. Igoshin, A.N. Oraevsky: Kvantovaya Elektron. (Moscow) **6**, 2517 (1979)

.7.5 N.N. Akinfiyev, N.G. Basov, V.T. Galochkin, S.I. Zavorotnyi, E.P. Markin, A.N. Oraevsky, A.V. Pankratov: Pis'ma ZhETF **19**, 745 (1974)

7.6 V.T. Galochkin, S.I. Zavorotnyi, V.N. Kosinov, A.A. Ovchinnikov, A.N. Oraevsky, N.F. Starodubtsev: Kvantovaya Elektron. (Moscow) **3**, 125 (1976)

7.7 V.I. Balykin, Yu. R. Kolomiiskii, O.A. Tumanov: Kvantovaya Elektron. (Moscow) **2**, 819 (1975)

368   References

7.8   J.L. Lyman, R.J. Jensen: J. Phys. Chem. **77**, 883 (1973)
7.9   A.V. Belotserkovets, G.A. Kirillov, S.B. Kormer, G.G. Kochemasov Yu. V. Kuratov, V.I. Mashendzhiyev, Yu. V. Savin, E.a. Stankeyev, V.D. Urlin: Kvantovaya Elektron. (Moscow) **2**, 2412 (1975)
7.10 T.K. Plant, T.A. De Temple: J. Appl. Phys. **46**, 3042 (1976)
7.11 R.E. McNair, S.F. Fulghum, G.W. Flynn, M.S. Feld, B.J. Feldman: Chem. Phys. Lett. **43**, 241 (1977)
7.12 E. Weitz, G. Flynn: J. Chem. Phys. **58**, 2679 (1973)
7.13 R.V. Ambartsumyan, Yu. A. Gorokhov, V.S. Letokhov, G.N. Makarov: ZhETF **69**, 1956 (1975)
7.14 D.S. Frankel, Jr., T.J. Manuccia: Chem. Phys. Lett. **54**, 451 (1978)
7.15 N.G. Fedotov: *Candidte's Thesis* (MFTI, Moscow, 1978)
7.16 J.C. Stephenson, C.B. Moor: J. Chem. Phys. **52**, 2333 (1970)
7.17 V.I. Igoshin, S. Yu. Pichugin: Kvantovaya Elektron. (Moscow) **10**, 370 (1983)
7.18 V.I. Igoshin, S. Yu. Pichugin: Kvantovaya Elektron. (Moscow) **10**, 458 (1983)
7.19 V.I. Igoshin, S. Yu. Pichugin: Kvantovaya Elektron. (Moscow) **10**, 1922 (1983)
7.20 V.I. Igoshin, S. Yu. Pichugin: Preprint FIAN (Moscow) **75** (1983)
7.21 V.I. Igoshin, S. Yu. Pichugin: Preprint FIAN (Moscow) **79** (1983)
7.22 N.G. Basov, V.I. Igoshin, V.A. Kartulin, A.N. Oraevsky, S.Yu. Pichugin: Preprint FIAN (Moscow) **121** (1984)
7.23 V.I. Igoshin, S. Yu. Pichugin: *Program 18th Sci. Conf. KuGU* (Kuibyshev, 1987)
7.24 P.D. Pokhil, A.F. Belyayev, Yu.V. Frolov, V.S. Logachev, A.I. Korotkov: *Goreniye poroshkoobraznykh metallov v aktivnykh sredakh* (Burning of Powdered Metals in Active Media) (Nauka, Moscow, 1972)
7.25 V.L. Tal'roze: Kinet. Katal. **5**, 11 (1964)

Subject Index

S. M. Rytov, Y. A. Kravtsov, V. I. Tatarskii,
Moscow, USSR

Principles of Statistical Radiophysics 1

Elements of Random Process Theory

1987. X, 253 pp. 28 figs. ISBN 3-540-12562-0

Principles of Statistical Radiophysics 2

Correlation Theory of Random Processes

Translated from the Russian by A. P. Repyev

1988. X, 234 pp. 54 figs. ISBN 3-540-16186-4

Principles of Statistical Radiophysics 3

Elements of Random Fields

1989. X, 239 pp. 36 figs. ISBN 3-540-17829-5

Principles of Statistical Radiophysics 4

Wave Propagation Through Random Media

Springer-Verlag Berlin
Heidelberg New York London
Paris Tokyo Hong Kong

Translated from the Russian by A. P. Repyev

1989. X, 188 pp. 43 figs. ISBN 3-540-17828-7

Springer

V. S. Butylkin, A. E. Kaplan,
Yu. G. Khronopulo, E. I. Yakubovich

Resonant Nonlinear Interactions of Light with Matter

Translated from the Russian by
O. A. Germogenova

1989. XIV, 342 pp. 70 figs.
ISBN 3-540-12109-9

A. Hasegawa, Murray Hill, NJ, USA

Optical Solitons in Fibers

1989. X, 75 pp. 22 figs. (Springer
Tracts in Modern Physics, Volume 116)
ISBN 3-540-50668-3

O. Sild, K. Haller, Tartu, USSR (Eds.)

Zero-Phonon Lines

and Spectral Hole Burning in Spectroscopy and Photochemistry

1988. IX, 183 pp. 64 figs.
ISBN 3-540-19214-X

Springer-Verlag Berlin
Heidelberg New York London
Paris Tokyo Hong Kong

Springer